FOR STUDENTS

- Career Opportunities
- Career Fitness Program
- Becoming an Electronics Technician
- Free On-Line Study Guides (companion web sites)

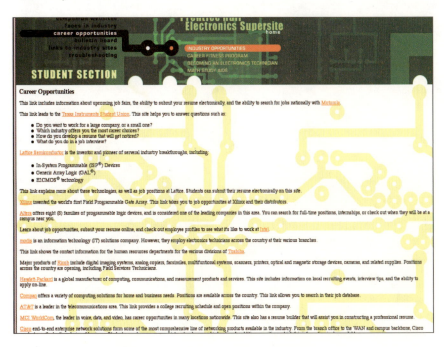

FOR FACULTY*

- Supplements
- On-Line Product Catalog
- Electronics Technology Journal
- Prentice Hall Book Advisor

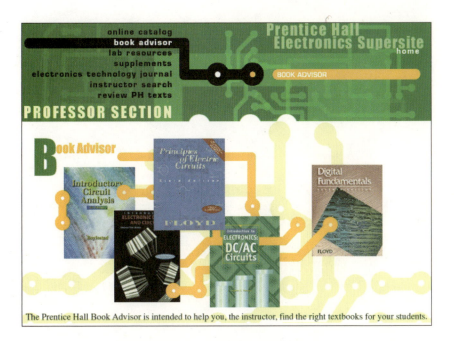

Please contact your Prentice Hall representative for passcode

Practical Electronics

Second Edition

Nigel P. Cook

Prentice Hall

Upper Saddle River, New Jersey
Columbus, Ohio

Library of Congress Cataloging-in-Publication Data
Cook, Nigel P.,
    Practical electronics / Nigel P. Cook.
      p. cm.
    Includes index.
    ISBN 0-13-042082-4
      1. Electronics I. Title.

TK7816 .C654 2002
621.381—dc21   2001036772

Editor in Chief: Stephen Helba
Product Manager: Scott J. Sambucci
Production Editor: Rex Davidson
Design Coordinator: Karrie Converse-Jones
Cover Designer: Tom Mack
Cover Photo: PhotoDisc
Illustrations: Rolin Graphics
Production Manager: Pat Tonneman
Project Management: Holly Henjum, Clarinda Publication Services

This book was set in Times Roman by The Clarinda Company and was printed and bound by Von Hoffmann Press, Inc. The cover was printed by The Lehigh Press, Inc.

Electronics Workbench® and MultiSim® are registered trademarks of Electronics Workbench.

Pearson Education Ltd., *London*
Pearson Education Australia Pty. Limited, *Sydney*
Pearson Education Singapore Pte. Ltd.
Pearson Education North Asia Ltd., *Hong Kong*
Pearson Education Canada, Ltd., *Toronto*
Pearson Educación de Mexico, S.A. de C.V.
Pearson Education—Japan, *Tokyo*
Pearson Education Malaysia Pte. Ltd.
Pearson Education, *Upper Saddle River, New Jersey*

10 9 8 7 6 5 4 3 2 1
ISBN: 0-13-042082-4

To Dawn, Candy, and Jon

Books by Nigel P. Cook

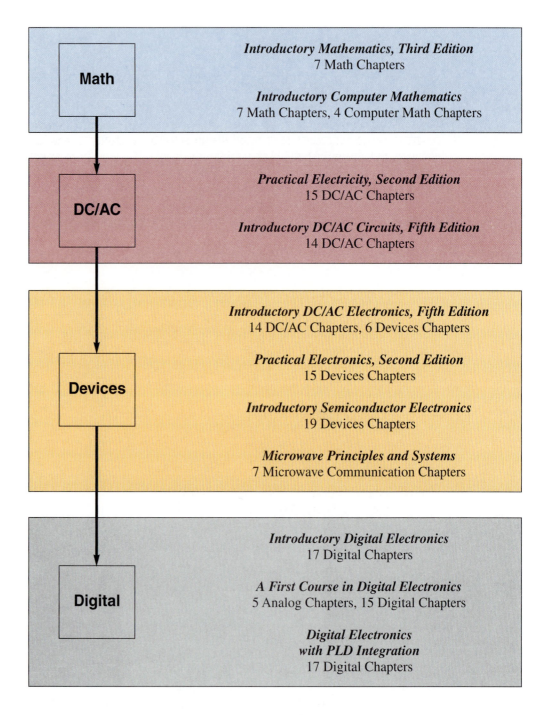

Math

Introductory Mathematics, Third Edition
7 Math Chapters

Introductory Computer Mathematics
7 Math Chapters, 4 Computer Math Chapters

DC/AC

Practical Electricity, Second Edition
15 DC/AC Chapters

Introductory DC/AC Circuits, Fifth Edition
14 DC/AC Chapters

Devices

Introductory DC/AC Electronics, Fifth Edition
14 DC/AC Chapters, 6 Devices Chapters

Practical Electronics, Second Edition
15 Devices Chapters

Introductory Semiconductor Electronics
19 Devices Chapters

Microwave Principles and Systems
7 Microwave Communication Chapters

Digital

Introductory Digital Electronics
17 Digital Chapters

A First Course in Digital Electronics
5 Analog Chapters, 15 Digital Chapters

*Digital Electronics
with PLD Integration*
17 Digital Chapters

For more information on any of the other textbooks by Nigel Cook, see his web page at www.prenhall.com/cook or ask your local Prentice Hall representative.

Preface

TO THE STUDENT

The early pioneers in electronics were intrigued by the mystery and wonder of a newly discovered science, whereas people today are attracted by its ability to lend its hand to any application and accomplish almost anything imaginable. If you analyze exactly how you feel at this stage, you will probably discover that you have mixed emotions about the journey ahead. On the one hand, imagination, curiosity, and excitement are driving you on, while apprehension and reservations may be slowing you down. Your enthusiasm will overcome any indecision you have once you become actively involved in electronics and realize that it is as exciting as you ever expected it to be.

ORGANIZATION OF THE TEXTBOOK

This textbook has been divided into four basic parts. Chapter 1 introduces semiconductor electronics. Chapters 2 through 6 cover diodes and diode circuits. Chapters 7 through 12 cover transistors and transistor circuits, and Chapters 13 and 14 cover integrated circuits, thyristors and transducers.

The material covered in this book has been logically divided and sequenced to provide a gradual progression from the known to the unknown and from the simple to the complex. A great deal of effort has been made to ensure that the style, format, approach, and content of this book are compatible with you, the student.

ANCILLARIES ACCOMPANYING THIS TEXT

The following ancillaries accompanying this text provide extensive opportunity for further study and support:

- *Laboratory Manual.* Co-authored by Nigel Cook and Gary Lancaster, the lab manual offers numerous experiments designed to translate all of the textbook's theory into practical experimentation.

- *Companion Website,* located at http://www.prenhall.com/cook. Numerous interactive study questions are provided on this site to reinforce the concepts covered in the book.

- Many of the text's circuits have been rendered in Multisim from Electronics Workbench and are available on the Companion Website.

To complete the ancillary package, the following supplements are essential elements for any instructor using this text for a course:

- *Instructor's Solutions Manual*

- *Solutions Manual to Accompany Laboratory Manual*

- *PowerPoint*^{TM} *Transparencies.* This CD-ROM includes a full set of **lecture presentations** as well as transparencies for all schematics appearing in the text.

- *Test Item File*

- *PH Test Manager*

DEVELOPMENT, CLASS TESTING, AND REVIEWING

The first phase of development for this manuscript was conducted in the classroom with students and instructors as critics. Each topic was class-tested by videotaping each lesson, and the results were then evaluated and implemented. This invaluable feedback enabled me to fine-tune my presentation of topics and instill understanding and confidence in the students.

The second phase of development was to forward a copy of the revised manuscript to several instructors at schools throughout the country. Their technical and topical critiques helped to mold the text into a more accurate form.

The third and final phase was to class-test the final revised manuscript and then commission the last technical review in the final stages of production.

ACKNOWLEDGMENTS

My appreciation and thanks are extended to the following instructors who have reviewed and contributed greatly to the development of this textbook: Venkata Anadu, Southwest

Texas State University; Don Barrett, Jr., DeVry Institute of Technology; Charles Dale, East-field College; Robert Diffenderfer, DeVry Institute of Technology; Lynnette Garetz, Heald College; Joe Gryniuk, Lake Washington Technical College; Jerry M. Manno, DeVry Institute of Technology; John Njemini, DeVry Institute of Technology; Gerald Schickman, Miami Dade Community College; George Sweiss, ITT Technical Institute; and Bradley J. Thompson, State University of New York College of Technology at Alfred.

Nigel P. Cook

Timeline Photo Credits

Page 4, Sir Joseph Thomson, Library of Congress; p. 6, Dr. John Mauchly and Dr. J. Presper Eckert, Unisys Corporation; p. 23, René Descartes, Library of Congress; p. 140, Karl Friedrich Gauss, Steven S. Nau/Pearson Education/PH College; p. 287, Nikola Tesla, Nikola Tesla Musemum, Belgrade, Yugoslavia; p. 367, Gottfried Wilhelm von Liebniz, Library of Congress; p. 370, John Baird, Hulton Getty/Archive Photos; p. 371, Ted Hoff, Intel Corporation Museum Archives and Collections

Photos in Introduction to Electronics, pp. xii–xxiv, courtesy of Hewlett-Packard Company.

Contents

10

Communication Circuits   260

11

Field Effect Transistors   287

12

Field Effect Transistors MOS   317

PART IV
Integrated Circuits, Thyristors, and Transducers   338

13

Operational Amplifiers   338

14

Timers, Thyristors, and Transducers   367

Leibniz's Language of Logic   367
Introduction   367

Appendixes

Index   416

Introduction to Electronics

From World War II onward, no branch of science has contributed more to the development of the modern world than electronics. It has stimulated dramatic advances in the fields of communication, computing, consumer products, industrial automation, test and measurement, and health care. It has now become the largest single industry in the world, exceeding the automobile and oil industries, with annual sales of electronic systems exceeding $2 trillion.

Your Course in Electronics

Your future in the electronics industry begins with this text. To give you an idea of where you are going and what we will be covering, Figure I-1 acts as a sort of road map, breaking up your study of electronics into four basic steps.

Step 1:   Basics of Electricity
Step 2:   Electronic Components
Step 3:   Electronic Circuits
Step 4:   Electronic Systems

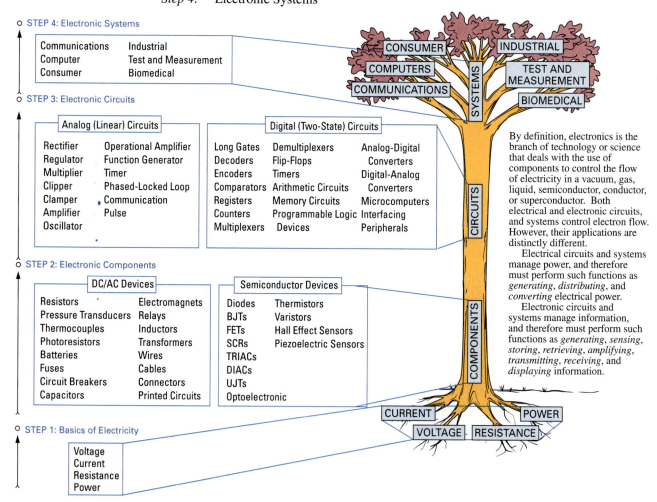

STEP 4: Electronic Systems

Communications	Industrial
Computer	Test and Measurement
Consumer	Biomedical

STEP 3: Electronic Circuits

Analog (Linear) Circuits

Rectifier	Operational Amplifier
Regulator	Function Generator
Multiplier	Timer
Clipper	Phased-Locked Loop
Clamper	Communication
Amplifier	Pulse
Oscillator	

Digital (Two-State) Circuits

Long Gates	Demultiplexers	Analog-Digital
Decoders	Flip-Flops	Converters
Encoders	Timers	Digital-Analog
Comparators	Arithmetic Circuits	Converters
Registers	Memory Circuits	Microcomputers
Counters	Programmable Logic	Interfacing
Multiplexers	Devices	Peripherals

STEP 2: Electronic Components

DC/AC Devices

Resistors	Electromagnets
Pressure Transducers	Relays
Thermocouples	Inductors
Photoresistors	Transformers
Batteries	Wires
Fuses	Cables
Circuit Breakers	Connectors
Capacitors	Printed Circuits

Semiconductor Devices

Diodes	Thermistors
BJTs	Varistors
FETs	Hall Effect Sensors
SCRs	Piezoelectric Sensors
TRIACs	
DIACs	
UJTs	
Optoelectronic	

STEP 1: Basics of Electricity

Voltage
Current
Resistance
Power

CONSUMER   INDUSTRIAL
COMPUTERS   TEST AND MEASUREMENT
COMMUNICATIONS   BIOMEDICAL
SYSTEMS

CIRCUITS

COMPONENTS

CURRENT   POWER
VOLTAGE   RESISTANCE

By definition, electronics is the branch of technology or science that deals with the use of components to control the flow of electricity in a vacuum, gas, liquid, semiconductor, conductor, or superconductor. Both electrical and electronic circuits, and systems control electron flow. However, their applications are distinctly different.

Electrical circuits and systems manage power, and therefore must perform such functions as *generating*, *distributing*, and *converting* electrical power.

Electronic circuits and systems manage information, and therefore must perform such functions as *generating*, *sensing*, *storing*, *retrieving*, *amplifying*, *transmitting*, *receiving*, and *displaying* information.

FIGURE I-1 **The Steps Involved in Studying Electronics.**

The main purpose of this introduction is not only to introduce you to the terms of the industry, but also to show you why the first two chapters in this text begin at the very beginning with "voltage and current," and then "resistance and power." **Components,** which are the basic electronic building blocks, were developed to control these four roots or properties, and when these devices are combined they form **circuits.** Moving up the tree to the six different branches of electronics, you will notice that just as components are the building blocks for circuits, circuits are in turn the building blocks for **systems.** Take a glance at the following pages, which will list many of the different types of components, circuits, and systems in the electronics industry.

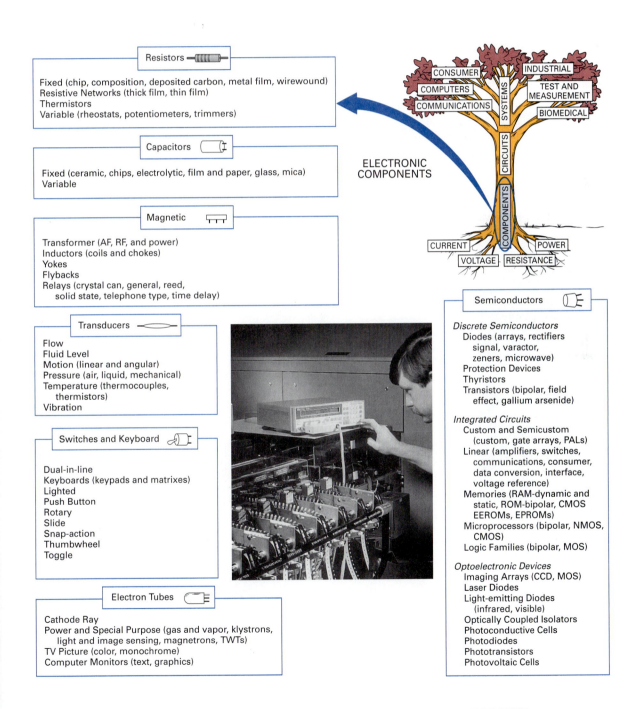

Resistors
Fixed (chip, composition, deposited carbon, metal film, wirewound)
Resistive Networks (thick film, thin film)
Thermistors
Variable (rheostats, potentiometers, trimmers)

Capacitors
Fixed (ceramic, chips, electrolytic, film and paper, glass, mica)
Variable

Magnetic
Transformer (AF, RF, and power)
Inductors (coils and chokes)
Yokes
Flybacks
Relays (crystal can, general, reed,
    solid state, telephone type, time delay)

Transducers
Flow
Fluid Level
Motion (linear and angular)
Pressure (air, liquid, mechanical)
Temperature (thermocouples,
    thermistors)
Vibration

Switches and Keyboard
Dual-in-line
Keyboards (keypads and matrixes)
Lighted
Push Button
Rotary
Slide
Snap-action
Thumbwheel
Toggle

Electron Tubes
Cathode Ray
Power and Special Purpose (gas and vapor, klystrons,
    light and image sensing, magnetrons, TWTs)
TV Picture (color, monochrome)
Computer Monitors (text, graphics)

CONSUMER
COMPUTERS
COMMUNICATIONS
INDUSTRIAL
TEST AND MEASUREMENT
BIOMEDICAL
SYSTEMS
CIRCUITS
COMPONENTS
CURRENT
POWER
VOLTAGE
RESISTANCE

ELECTRONIC COMPONENTS

Semiconductors
Discrete Semiconductors
Diodes (arrays, rectifiers
    signal, varactor,
    zeners, microwave)
Protection Devices
Thyristors
Transistors (bipolar, field
    effect, gallium arsenide)

Integrated Circuits
Custom and Semicustom
    (custom, gate arrays, PALs)
Linear (amplifiers, switches,
    communications, consumer,
    data conversion, interface,
    voltage reference)
Memories (RAM-dynamic and
    static, ROM-bipolar, CMOS
    EEROMs, EPROMs)
Microprocessors (bipolar, NMOS,
    CMOS)
Logic Families (bipolar, MOS)

Optoelectronic Devices
Imaging Arrays (CCD, MOS)
Laser Diodes
Light-emitting Diodes
    (infrared, visible)
Optically Coupled Isolators
Photoconductive Cells
Photodiodes
Phototransistors
Photovoltaic Cells

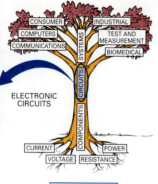

ELECTRONIC
CIRCUITS

Signal Generators and Timers

Sine-wave Generators (oscillators)
Square- and Pulse-wave Generators
Ramp- and Triangular-wave Generators

Power Supplies

Switching
Linear
Uninterruptible

Miscellaneous

Detectors and Mixers
Filters
Phase-locked Loops
Converters
Data Acquisition
Synthesizers

Digital Circuits

Oscillators and Generators   Latches
Gates and Flip-flops   Registers
Display Drivers   Multiplexers
Counters and Dividers   Demultiplexers
Encoders and Decoders   Gate Arrays
Memories
Input/Output
Microprocessors

Amplifiers

Bipolar Transistor
Field-effect Transistor
Operational Amplifier

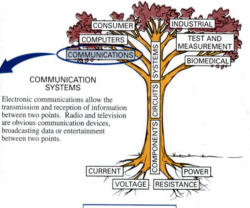

COMMUNICATION
SYSTEMS

Electronic communications allow the
transmission and reception of information
between two points. Radio and television
are obvious communication devices,
broadcasting data or entertainment
between two points.

Telecommunications

Switching Systems
Data and Voice Switching
Voice-only Switching
Cellular Systems
Telephones
Corded
Cordless
Telephone/Video Equipment
Facsimile Terminals
Fiber Optic Communication
   Systems

Television

Broadcast Equipment   Satellite TV Equipment
CATV Equipment   HDTV (High Definition
CCTV Equipment   TV) Equipment

Data Communications

Concentrators
Front-end Communications Processors
Message-switching Systems
Modems
Multiplexers
Network Controllers
Mixed Service (combining voice, data,
   video, imaging)

Radio

Amateur (mobile and base stations)
Aviation Mobile and Ground Support Stations
Broadcast Equipment
Land Mobile (mobile and base stations)
Marine Mobile (ship and shore stations)
Microwave Systems
Satellite Systems
Radar and Sonar Systems

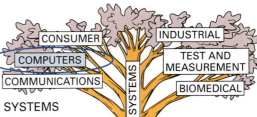

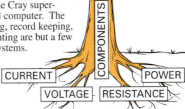

COMPUTER SYSTEMS

The computer is proving to be one of the most useful of all systems. Its ability to process, store, and manipulate large groups of information at an extremely fast rate makes it ideal for almost any and every application. Systems vary in complexity and capability, ranging from the Cray super-computer to the home personal computer. The applications of word processing, record keeping, inventory, analysis, and accounting are but a few examples of data processing systems.

Data Terminals

CRT Terminals
  ASCII Terminals
  Graphics Terminals (color, monochrome)
Remote Batch Job Entry Terminals

Computer Systems

Microcomputers and Supermicrocomputers
Minicomputers (personal computers) and
  Superminicomputers (technical
  workstations, multiuser)
Mainframe Computers
Supercomputers

I/O Peripherals

Computer Microfilm
Digitizers
Graphics Tablets
Light Pens
Trackball and Mice
Optical Scanning Devices
Plotters
Printers
  Impact
  Nonimpact (laser, thermal, electrostatic, inkjet)

Data Storage Devices

Fixed Disk (14, 8, $5\frac{1}{4}$, and $3\frac{1}{2}$ in.)
Flexible Disk (8, $5\frac{1}{4}$, and $3\frac{1}{2}$ in.)
Optical Disk Drives (read-only, write once, erasable)
Cassette
Cartridge Magnetic Tape ($\frac{1}{4}$ in.)
Cartridge Tape Drives ($\frac{1}{2}$ in.)
Reel-type Magnetic Tape Drives

Audio Equipment

Car
Stereo Equipment
   Compact Systems (miniature components)
   Components (speakers, amps, turntables,
      tuners, tape decks)
Phonographs and Radio Phonographs
Radios (table, clock, portable)
Tape Players/Recorders
Compact Disk Players
Digital Tape Players

Video Equipment

TV Receiver (color, monochrome)
Projection TV Receivers
Video Cassette Recorders (VCRs)
Video Disk Players
Camcorders (8 mm, $\frac{1}{2}$ in.)
Home Satellite Receivers

Personal

Calculators, Cameras, Watches
Telephone Answering Equipment
Personal Computers
Microwave Ovens
Musical Equipment and Instruments
Pacemakers and Hearing Aids
Alarms and Smoke Detectors

Automobile Electronics

Dashboard
Engine Monitoring and Analysis
Computer Navigation Systems
Alarms
Telephones

CONSUMER
COMPUTERS
COMMUNICATIONS

INDUSTRIAL
TEST AND MEASUREMENT
BIOMEDICAL

SYSTEMS
CIRCUITS
COMPONENTS

CONSUMER
SYSTEMS

CURRENT
VOLTAGE
RESISTANCE
POWER

From the smart computer-controlled automobiles, which provide navigational information and monitor engine functions and braking, to the compact disc players, video camcorders, satellite TV receivers, and wide-screen stereo TVs, this branch of electronics provides us with entertainment, information, safety, and, in the case of the pacemaker, life.

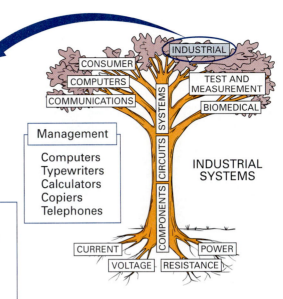

Manufacturing Equipment

Energy Management Equipment

Inspection Systems

Motor Controllers (speed, torque)

Numerical-control Systems

Process Control Equipment (data-acquisition systems, process instrumentation, programmable controllers)

Robot Systems

Vision Systems

Computer-aided Design and Engineering CAD/CAE

Hardware Equipment
Design Work Stations (PC based, 32 microprocessor based platform, host based)
Application Specific Hardware

Design Software
Design Capture (schematic capture, logic fault and timing simulators, model libraries)
IC Design (design rule checkers, logic synthesizers, floor planners–place and route, layout editors)
Printed Circuit Board Design Software
Project Management Software
Test Equipment

Management

Computers
Typewriters
Calculators
Copiers
Telephones

INDUSTRIAL SYSTEMS

Almost any industrial company can be divided into three basic sections, all of which utilize electronic equipment to perform their functions. The manufacturing section will typically use power, motor, and process control equipment, along with automatic insertion, inspection, and vision systems, for the fabrication of a product. The engineering section uses computers and test equipment for the design and testing of a product, while the management section uses electronic equipment such as computers, copiers, telephones, and so on.

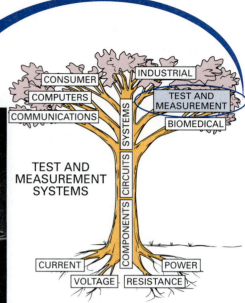

The rising complexity of electronic components, circuits, and equipment is causing a demand for sophisticated automatic test equipment for both the manufacturer and the customer to test their products.

Automated Test Systems

Active and Discrete Component Test Systems
Automated Field Service Testers
IC Testers (benchtop, general purpose, specialized)
Interconnect and Bare Printed Circuit-Board Testers
Loaded Printed Circuit-Board Testers (in-circuit, functional, combined)

General Test and Measurement Equipment

Amplifiers (lab)
Arbitrary Waveform Generators
Analog Voltmeters, Ammeters, and Multimeters
Audio Oscillators
Audio Waveform Analyzers and Distortion Meters
Calibrators and Standards
Dedicated IEEE–488 Bus Controllers
Digital Multimeters
Electronic Counters (RF, Microwave, Universal)
Frequency Synthesizers
Function Generators
Pulse/Timing Generators
Signal Generators (RF, Microwave)

Logic Analyzers
Microprocessor Development Systems
Modulation Analyzers
Noise-measuring Equipment
Oscilloscopes (Analog, Digital)
Panel Meters
Personal Computer (PC) Based Instruments
Recorders and Plotters
RF/Microwave Network Analyzers
RF/Microwave Power-measuring Equipment
Spectrum Analyzers
Stand-alone In-circuit Emulators
Temperature-measuring Instruments

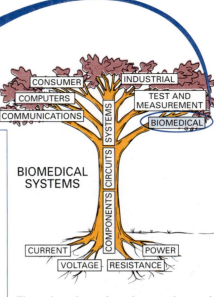

Patient Care

Surgical Equipment
  Endoscope
  Laser Equipment
  Anesthesia Equipment
  Isotripters
  Electrocautery
  Microscopic Equipment

Monitoring Equipment (computer-controlled care units)
  Intensive Care Unit (ICU): Arrhythmia
    Computer Unit (heart wave, blood pressure,
    respiration rate, heart rate, blood gas)
  Cardiac Care Unit (CCU): Heart/Lung Machines and ICU Equipment
  Neonatal Intensive Care Unit (NICU): Temperature monitors (probes,
    incubators, open radiant warmers)

**BIOMEDICAL
SYSTEMS**

Electronic equipment is used more and more within the biological and medical fields, which can be categorized simply as being either patient care or diagnostic equipment. In the operating room, the endoscope, which is an instrument used to examine the interior of a canal or hollow organ, and the laser, which is used to coagulate, cut, or vaporize tissue with extremely intense light, both reduce the use of invasive surgery. A large amount of monitoring equipment is used both in and out of operating rooms, and the equipment consists of generally large computer-controlled systems that can have a variety of modules inserted (based on the application) to monitor, on a continuous basis, body temperature, blood pressure, pulse rate, and so on. In the diagnostic group of equipment, the clinical laboratory test results are used as diagnostic tools. With the advances in automation and computerized information systems, multiple tests can be carried out at increased speeds. Diagnostic imaging, in which a computer constructs an image of a cross-sectional plane of the body, is probably one of the most interesting equipment areas.

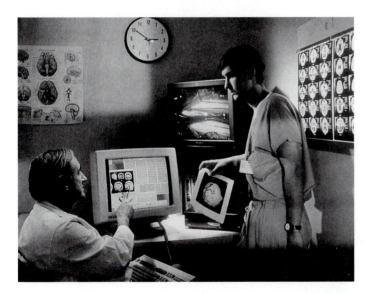

Diagnostic

Diagnostic Equipment
  X-Ray (computed tomography)
  Magnetic Resonance Imaging (MRI)
  Diagnostic Sounder
  Electrocardiograph (EKG)
  Electromyograph
  Electroencephalograph (EEG)
  Coagulograph
  Ultrasound (computed sonography)
  Nuclear Medicine (isotopes, spectroscopy)

Clinical Laboratory
  Automated Clinical Analyzers
  Centrifuge Incubators
  Cell Counters

Development of an Electronic Product

The flow chart below shows the order, from top to bottom, in which an electronic product is developed from conception to shipping.

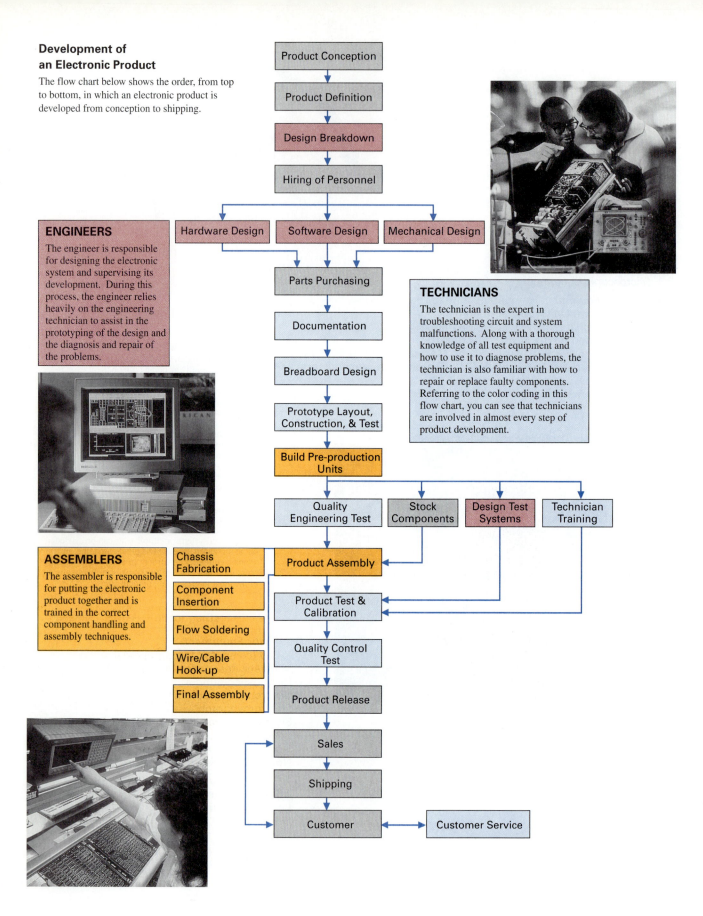

Product Conception → **Product Definition** → **Design Breakdown** → **Hiring of Personnel**

Hardware Design | **Software Design** | **Mechanical Design**

→ **Parts Purchasing** → **Documentation** → **Breadboard Design** → **Prototype Layout, Construction, & Test** → **Build Pre-production Units**

Quality Engineering Test | **Stock Components** | **Design Test Systems** | **Technician Training**

→ **Product Assembly** → **Product Test & Calibration** → **Quality Control Test** → **Product Release** → **Sales** → **Shipping** → **Customer** → **Customer Service**

Chassis Fabrication · **Component Insertion** · **Flow Soldering** · **Wire/Cable Hook-up** · **Final Assembly**

ENGINEERS

The engineer is responsible for designing the electronic system and supervising its development. During this process, the engineer relies heavily on the engineering technician to assist in the prototyping of the design and the diagnosis and repair of the problems.

TECHNICIANS

The technician is the expert in troubleshooting circuit and system malfunctions. Along with a thorough knowledge of all test equipment and how to use it to diagnose problems, the technician is also familiar with how to repair or replace faulty components. Referring to the color coding in this flow chart, you can see that technicians are involved in almost every step of product development.

ASSEMBLERS

The assembler is responsible for putting the electronic product together and is trained in the correct component handling and assembly techniques.

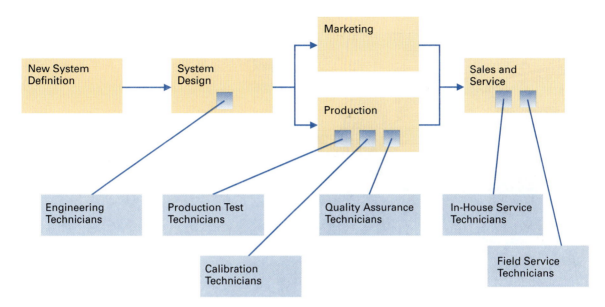

ENGINEERING TECHNICIAN

Here you can see an engineering technician breadboarding the design. From sketches supplied by the engineers, a breadboard model of the design is constructed. The breadboard model is an experimental arrangement of a circuit in which the components are temporarily attached to a flat board. In this arrangement, the components can be tested to prove the feasability of the circuit. A breadboard facilitates making easy changes when they are necessary.

Working under close supervision, the engineering technician performs all work assignments as given by all levels of engineers.

RESPONSIBILITIES:

• Breadboard electronic circuits from schematics.

• Test, evaluate, and document circuits and system performance under the engineer's direction.

• Check out, evaluate, and take data for the engineering proto-types including mechanical assembly of prototype circuits.

• Help to generate and maintain preliminary engineering docu-mentations. Assist engineers and senior Engineering Techni-cians in ERN documentation.

• Maintain the working station equipment and tools in orderly fashion.

• Support the engineers in all aspects of the development of new products.

REQUIREMENTS:

AS/AAS Degree in Electronics or equivalent plus 1–2 years technician experience. Ability to read color code, to solder properly, and to bond wires where the skill is required to com-plete breadboards and prototypes. Ability to use common machinery required to build prototype circuits. Working knowl-edge of common electronic components: TTL logic circuits, op-amps, capacitors, resistors, inductors, semiconductor devices.

Once the system is fully operational, it is calibrated by a calibration technician. The more complex problems are handled by the production test technicians seen in this photograph.

TECHNICIAN I

Working under close supervision, the production test technician performs all work assignments, and exercises limited decision-making.

RESPONSIBILITIES:

• Perform routine, simple operational tests and fault isolation on simple components, circuits, and systems for verification of product performance to well-defined specifications.

• May perform standard assembly operations and simple alignment of electronic components and assemblies.

• May set up simple test equipment to test performance of products to specifications.

REQUIREMENTS:

AS/AAS Degree in Electronic Technology or equivalent work experience.

TECHNICIAN II

Working under moderate supervision, the production test technician exercises general decision-making involving simple cause and effect relationships to identify trends and common problems.

RESPONSIBILITIES:

• Perform moderately complex operational tests and fault isolation on components, circuits, and systems for verification of product performance to well-defined specifications.

• Perform simple mathematical calculations to verify test measurements and product performance to well-defined specifications.

• Set up general test stations, utilizing varied test equipment, including some sophisticated equipment.

• Perform standard assembly operations.

REQUIREMENTS:

AS/AAS Degree in Electronic Technology or equivalent plus 2–4 years directly related experience. Demonstrated experience working mathematical formulas and equations. Working knowledge of counters, scopes, spectrum analyzers, and related industry standard test equipment.

CALIBRATION TECHNICIAN

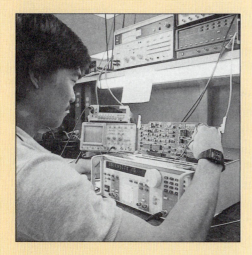

The calibration technician shown here is undertaking a complete evaluation of the newly constructed breadboard's mechanical and electrical form, design, and performance.

Working under general supervision, the calibration technician interfaces with test equipment in a system environment requiring limited decision-making.

RESPONSIBILITIES:

• Work in an interactive mode with test station. Test, align, and calibrate products to defined specifications.

• May set up own test stations and those of other operators.

• Perform multiple alignments to get products to meet specifications.

• May perform other manufacturing-related tasks as required.

REQUIREMENTS:

1–2 years experience with test and measurement equipment, experience with multiple alignment and calibration of assemblies including test station setups. Able to follow written instructions and write clearly.

QUALITY ASSURANCE TECHNICIAN

The quality assurance (QA) technician takes one of the pre-production units through an extensive series of tests to determine whether it meets the standards listed. This technician is evaluating the new product as it is put through an extensive series of tests.

Working under direct supervision, the quality assurance technician performs functional tests on completed instruments.

RESPONSIBILITIES:

• Using established Acceptance Test Procedures, perform operational tests of all completed systems to ensure that all functional and electrical parameters are within specified limits.

• Perform visual inspection of all completed systems for cleanliness and absence of cosmetic defects.

• Reject all systems that do not meet specifications and/or established parameters of function and appearance.

• Make appropriate notations on the system history sheet.

• Maintain the QA Acceptance Log in accordance with current instructions.

• Refer questionable characteristics to supervisor.

REQUIREMENTS:

AS/AAS Degree in Electronics or equivalent including the use of test equipment. Must know color code and be able to distinguish between colors. Must have a working knowledge of related test equipment. Must know how to read and interpret drawings.

IN-HOUSE SERVICE TECHNICIAN

This photograph shows some in-house service technicians troubleshooting problems on returned units. Once the customer has received the electronic equipment, customer service provides assistance in maintenance and repair of the unit through direct in-house service or at service centers throughout the world.

Working under moderate supervision, the in-house service technician performs all work assignments given by lead tech or direct supervisor. Work necessary overtime as assigned by supervisors.

RESPONSIBILITIES:

• Utilizing all appropriate tools, troubleshoot and repair customer systems in a timely, quality manner, to the general component level.

• Working with basic test equipment, perform timely quality calibration of customer systems to specifications.

• Timely repair of QA rejects.

• Solder and desolder components where appropriate, meeting company standards.

• Aid marketing in solving customer problems via the telephone.

• When appropriate, instruct customers in the proper methods of calibration and repair of products.

REQUIREMENTS:

AS/AAS Degree in Electronics or equivalent plus 2–3 years experience troubleshooting analog and/or digital systems, at least 6–12 months of which should be in a service environment. Must be able to read and understand flow charts, block diagrams, schematics, and truth tables. Must be able to operate and utilize test equipment such as oscilloscopes, counters, voltmeters, and analyzers.

Ability to effectively communicate and work with customers.

FIELD SERVICE TECHNICIAN

The field service technician seen here has been requested by the customer to make a service call on a malfunctioning unit that currently is under test.

Working under moderate supervision, the field service technician performs all work assignments given by the lead tech or direct supervisor. Work necessary overtime as assigned by supervisors.

RESPONSIBILITIES:

• Utilizing all appropriate tools, troubleshoot and repair customer systems in a timely, quality manner, to the general component level.

• Working with basic test equipment, perform timely quality calibration of customer systems to specifications.

• When appropriate, instruct others in proper soldering techniques meeting company standards.

• Timely repair of QA rejects.

• Aid marketing in solving customer problems via the telephone or at the customers' facility at marketings' discretion.

• When appropriate, aid marketing with sales applications.

• Help QA, Production, and Engineering in solving field problems.

• Evaluate manuals and other customer documents for errors or omissions.

REQUIREMENTS:

AS/AAS Degree in Electronics or equivalent plus 4–5 years experience troubleshooting analog and/or digital systems, at least 6–12 months of which should be in a service environment. Must be able to read and understand flow charts, block diagrams, schematics, and truth tables. A demonstrated ability to effectively communicate and work with customers, and suggest alternative applications for product utilization is also required.

Semiconductor Principles

1

The Turing Enigma

During the Second World War, the Germans developed a cipher-generating apparatus called "Enigma." This electromechanical teleprinter would scramble messages with several randomly spinning rotors that could be set to a predetermined pattern by the sender. This key and plug pattern was changed three times a day by the Germans and cracking the secrets of Enigma became of the utmost importance to British Intelligence. With this objective in mind, every brilliant professor and eccentric researcher was gathered at a Victorian estate near London called Bletchley Park. They specialized in everything from engineering to literature and were collectively called the Backroom Boys.

By far the strangest and definitely most gifted of the group was an unconventional theoretician from Cambridge University named Alan Turing. He wore rumpled clothes and had a shrill stammer and crowing laugh that aggravated even his closest friends. He had other legendary idiosyncrasies that included setting his watch by sighting on a certain star from a specific spot and then mentally calculating the time of day. He also insisted on wearing his gas mask whenever he was out, not for fear of a gas attack, but simply because it helped his hay fever.

Turing's eccentricities may have been strange but his genius was indisputable. At the age of twenty-six he wrote a paper outlining his "universal machine" that could solve any mathematical or logical problem. The data or, in this case, the intercepted enemy messages could be entered into the machine on paper tape and then compared with known Enigma codes until a match was found.

In 1943 Turing's ideas took shape as the Backroom Boys began developing a machine that used 2,000 vacuum tubes and incorporated five photoelectric readers that could process 25,000 characters per second. It was named "Colossus," and it incorporated the stored program and other ideas from Turing's paper written seven years earlier.

Turing could have gone on to accomplish much more. However, his idiosyncrasies kept getting in his way. He became totally preoccupied with abstract questions concerning machine intelligence. His unconventional personal lifestyle led to his arrest in 1952 and, after a sentence of psychoanalysis, his suicide two years later.

Before joining the Backroom Boys at Bletchley Park, Turing's genius was clearly apparent at Cambridge. How much of a role he played in the development of Colossus is still unknown and remains a secret guarded by the British Official Secrets Act. Turing was never fully recognized for his important role in the development of this innovative machine, except by one of his Bletchley Park colleagues at his funeral who said, "I won't say what Turing did made us win the war, but I daresay we might have lost it without him."

Introduction

Materials can be divided into three main types according to the way they react to current when a voltage is applied across them. **Insulators** (nonconductors), for example, are materials that have a very high resistance and therefore oppose current, whereas **conductors** are materials that have a very low resistance and therefore pass current easily. The third type of material is the **semiconductor,** which, as its name suggests, has properties that lie between the insulator and the conductor. Semiconductor materials are not good conductors or insulators and so the next question is: What characteristic do they possess that makes them so useful in electronics? The answer is that they can be controlled to either increase their resistance and behave more like an insulator or decrease their resistance and behave more like a conductor. *It is this ability of a semiconductor material to vary its resistive properties that makes it so useful in electrical and electronic applications.*

In this chapter we will examine the characteristics of semiconductor materials so that we can better understand the operation and characteristics of semiconductor devices.

1-1 SEMICONDUCTOR DEVICES

Semiconductor materials such as *germanium* and *silicon* are used to construct semiconductor devices like the *diodes, transistors,* and *integrated circuits* (*ICs*) shown in Figure 1-1. These devices are used in electrical and electronic circuits to control current and voltage so as to produce a desired result. For example, a diode could be used as the controlling element in a rectifier circuit that would convert ac to pulsating dc. A transistor, on the other hand, could be made to act like a variable resistance so it could amplify a radio signal. Conversely, an integrated circuit could be used to generate an oscillating signal or be made to perform arithmetic operations.

The most significant development in electronics since World War II has been a small semiconductor device called the *transistor.* It was first introduced in 1948 by its inventors William Schockley, Walter Bratten, and John Bardeen in the Bell Telephone Laboratories and was described as a **solid state device.** This term was used because the transistor contained a solid semiconductor material between its input and output pins, unlike its predecessor the vacuum tube, which had a vacuum between its input and output pins.

Insulators
Materials that have a very high resistance and oppose current.

Conductors
Materials that have a very low resistance and pass current easily.

Semiconductors
Materials that have properties that lie between insulators and conductors.

Solid State Device
Uses a solid semiconductor material, such as silicon, between the input and output, whereas a vacuum tube has vacuum between input and output.

TIME LINE
In the early days of World War II, German scientist Konrad Zuse (1910–1995), who designed and built the first general-purpose computer, proposed constructing a computer that would operate 1,000 times faster than anything else at that time. This proposal was rejected by Hitler, who was not interested in this long-term, two-year project, as he was sure that the war was going to be, for him, a certain, quick victory. Due to Hitler's shortsightedness, this powerful computer, which could have been used to break British communication codes, was never developed. However, unknown to both Hitler and Zuse, the British code-breaking computer project, called Ultra, had highest priority and was moving rapidly toward completion.

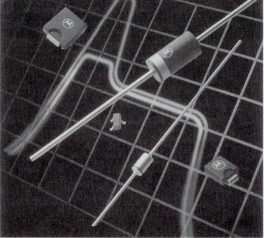

FIGURE 1-1   Semiconductor Devices. (Copyright of Motorola, Inc. Used by permission.)

The first *point-contact transistor* unveiled in 1948 was extremely unreliable, and it took its inventors another 12 years to develop the superior *bipolar junction transistor* (*BJT*) and make it available in commercial quantities.

In 1960, many electronic system manufacturers began to use the bipolar junction transistor instead of the vacuum tube in low-power and low-frequency applications. Research and development into semiconductor or solid state devices mushroomed, and a variety of semiconductor devices began to appear. A different type of transistor emerged, called the *field effect transistor* (*FET*), which had characteristics similar to those of the vacuum tube. Once it was discovered that semiconductor materials could also generate and sense light, a new line of optoelectronic devices became available. Later it was discovered that semiconductor materials could sense magnetism, temperature, and pressure, and as a result, a variety of sensor devices or transducers (energy converters) appeared on the market. Along with all these different types of semiconductor devices, a wide variety of semiconductor diodes emerged that could rectify, regulate, and oscillate at high frequencies. Even to this day it is clear that we have not yet seen all the potential value of semiconductors. Figure 1-2 illustrates many of these semiconductor, or solid state, devices.

Although semiconductor diodes and transistors are still widely used as individual or **discrete components,** in 1959 Robert Noyce discovered that more than one transistor could be constructed on a single piece of semiconductor material. Soon other components such as resistors, capacitors, and diodes were added along with transistors and then interconnected to form a complete circuit on a single chip or piece of semiconductor material. This integrating of various components on a single chip of semiconductor was called an *integrated circuit* (*IC*) or *IC chip*. Today, the IC is used extensively in every branch of electronics, with hundreds of thousands of transistors and other components being placed on a chip of semiconductor no bigger than this ■. Figure 1-3 illustrates some of the different types of integrated circuits.

Like an evolving species, semiconductors have come to dominate the products of which they used to be only a part. For example, there used to be 400 components in a typical cellular telephone. Now there are 40, and soon only 3 or 4 IC chips will make up the entire phone circuitry. Today the semiconductor business—once regarded as a technical sideshow—occupies center stage and is key to the development of new products for all industries.

TIME LINE
In 1904, John A. Fleming, a British scientist, saw the value of an effect that was discovered by Thomas Edison but for which he saw no practical purpose. The "Edison effect" permitted Fleming to develop the "Fleming value," which passes current in only one direction. Its operation made it the first device able to convert alternating current into direct current and to detect radio waves.

Discrete Components
Separate active and passive devices that were manufactured before being used in a circuit.

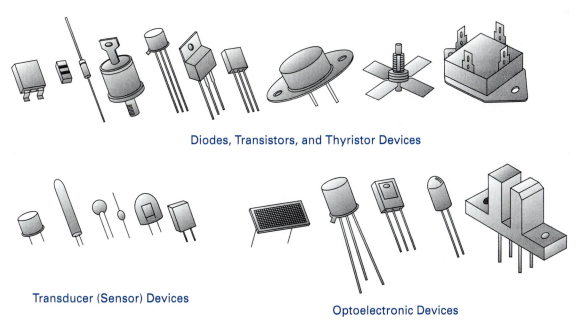

Diodes, Transistors, and Thyristor Devices

Transducer (Sensor) Devices

Optoelectronic Devices

FIGURE 1-2 Discrete Semiconductor (Solid State) Devices.

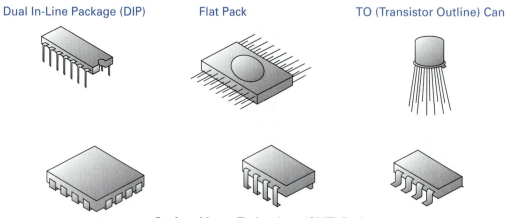

Dual In-Line Package (DIP)   Flat Pack   TO (Transistor Outline) Can

Surface Mount Technology (SMT) Packages

FIGURE 1-3   Semiconductor Integrated Circuits (ICs).

SELF-TEST EVALUATION POINT FOR SECTION 1-1

Use the following questions to test your understanding of Section 1-1.

1. Name the three most frequently used semiconductor devices in electrical and electronic equipment.
2. The main function of a semiconductor device is to control the _____ or _____ in an electrical or electronic circuit.

1-2   SEMICONDUCTOR MATERIALS

A semiconductor material is one that is neither a conductor nor a nonconductor (insulator). This means simply that it will not conduct current as well as a conductor or block current as well as an insulator. Some semiconductor materials are pure or natural elements such as carbon (C), germanium (Ge), and silicon (Si), whereas other semiconductor materials are compounds.

Silicon and germanium are used most frequently in the construction of semiconductor devices for electrical and electronic applications. Germanium is a brittle grayish-white element that may be recovered from the ash of certain types of coals. Silicon, the most popular semiconductor material due to its superior temperature stability, is a white element normally derived from sand. Let us now examine the silicon, germanium, and carbon semiconductor atoms in more detail.

1-2-1   *Semiconductor Atoms*

Figure 1-4 illustrates the silicon, germanium, and carbon atoms. The silicon atom has 14 protons in its nucleus and 14 electrons in three orbital paths distributed as 2, 8, and then 4 electrons in its valence shell. The germanium atom has 32 protons within its nucleus and 32 electrons in four orbital paths distributed as 2, 8, 18, and finally 4 electrons in the valence band or shell. The carbon atom has 6 protons in its nucleus and 6 orbiting electrons in two orbital paths distributed as 2 and 4 electrons in the valence shell. The question is: What do all these atoms have in common? The answer is that all semiconductor atoms have *four valence electrons.*

The valence shell of an atom can contain up to 8 electrons, and it is the number of electrons in this valence shell that determines the conductivity of the atom. For example, an atom with only 1 valence electron would be classed as a good conductor, whereas an atom

TIME LINE

Jean Perrin discovered that cathode rays consisted of negatively charged particles, and these particles, which later became known as electrons, were measured by an English physicist, Joseph Thomson (1846–1914).

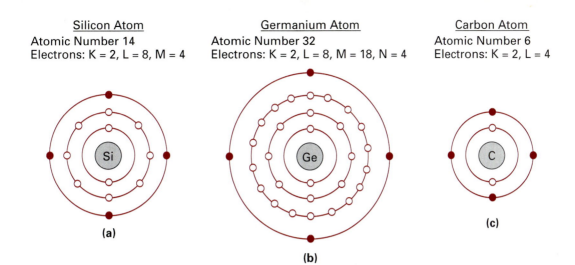

FIGURE 1-4 **Semiconductor Atoms with Their Four Valence Shell Electrons.**

having 8 valence electrons, and therefore a complete valence shell, would be classed as an insulator.

To summarize, Figure 1-4 shows three semiconductor atoms, all of which contain 4 valence electrons. Because the number of valence electrons determines the conductivity of the element, semiconductor atoms are midway between conductors (which have 1 valence electron) and insulators (which have 8 valence electrons). Silicon and germanium are used to manufacture semiconductor devices, whereas carbon is combined with other elements to construct resistors.

1-2-2 *Crystals and Covalent Bonding*

So far we have discussed only isolated atoms. When two or more similar semiconductor atoms are combined to form a solid element, they automatically arrange themselves into an orderly lattice-like structure or pattern known as a **crystal,** as shown in Figure 1-5(a). This pattern is formed because each atom shares its 4 valence electrons with its four neighboring atoms. Since each atom shares 1 electron with a neighboring atom, two atoms will share 2, or a pair, of electrons between the two cores. These two atom cores are pulling the 2 electrons with equal but opposite force, and it is this pulling action that holds the atoms together in this solid crystal lattice structure. The joining together of two semiconductor atoms is called an **electron-pair bond** or **covalent bond.** When many atoms combine, or bond, in this way, the result is a crystal (smooth, glassy, solid) lattice structure. To illustrate this bonding process, each atom in Figure 1-5(a) has been drawn as a square and each valence shell has been drawn as an octagon (eight-sided figure) so that we can easily see which electrons belong to which atom. As you can see, the atom in the center of the diagram has 4 valence electrons (shown at the corners of the Si square), and shares 1 electron from each of its four neighbors.

Figure 1-5(b) shows a larger view of a silicon crystal structure. All of the atoms in this structure are electrically stable because all of their valence shells are complete (they all contain 8 electrons). These completed valence shells cause the pure semiconductor crystal structure to act as an insulator because it will not easily give up or accept electrons. Pure semiconductor materials, which are often called **intrinsic** materials, are therefore very poor conductors. Once this pure material is available, it must then be modified by a *doping* process to give it the qualities necessary to construct semiconductor devices. Silicon is most frequently used to construct solid state or semiconductor devices such as diodes and transistors because germanium has poor temperature stability and carbon crystals (diamonds) are too expensive to use.

Crystal

A solid element with an orderly lattice-like structure.

Electron-Pair Bond or Covalent Bond

A pair of electrons shared by two neighboring atoms.

Intrinsic Semiconductor Materials

Pure semiconductor materials.

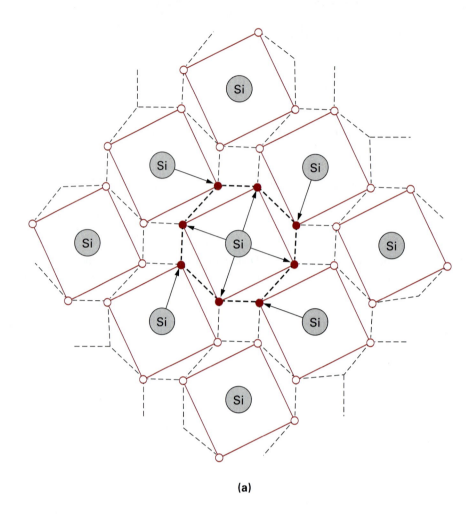

TIME LINE

Although the Fleming valve was an advance, it could not amplify or boost a signal. The "audion," developed by U.S. inventor Lee de Forest (1873–1961), sparked an era known as "vacuum-tube electronics" that brought about transcontinental telephony in 1915, radio broadcasting in 1920, radar in 1936, and television between 1927 and 1946 because of this triode vacuum tube's ability to amplify small signals.

(a)

TIME LINE

In 1947, J. Presper Eckert (right) and John Mauchly (left) unveiled ENIAC, which used over 300,000 vacuum tubes. ENIAC, which is an acronym for "electronic numerical integrator and computer," was the first large-scale electronic digital computer.

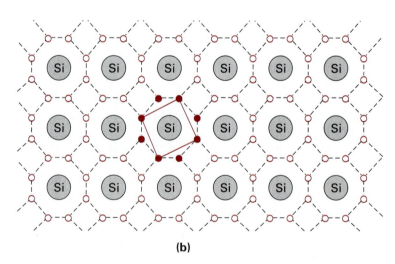

(b)

FIGURE 1-5   Covalent Bonding. (a) Silicon Atoms Sharing Valence Electrons. (b) Silicon Crystal Lattice Structure.

1-2-3   *Energy Gaps and Energy Levels*

Let us now examine the relationships between electrons and orbital shells in a little more detail so that we can better understand charge and conduction within a semiconductor material.

There are seven shells available for electrons around the nucleus. Electrons must travel or orbit in one of these orbital paths because they cannot exist in any of the spaces between orbital shells. Each orbital shell has its own specific energy level. Therefore, electrons traveling in a specific orbital shell will contain the shell's energy level. Figure 1-6 shows an example of an atom's orbital shell energy levels. The energy levels for each shell increase as they move away from the nucleus of the atom. The valence shell and the valence electrons will always have the highest energy level for a given atom. The space between any two orbital shells is called the **energy gap.** Electrons can jump from one shell to another if they absorb enough energy to make up the difference between their initial energy level and the energy level of the shell that they are jumping to. For example, in Figure 1-6 the valence shell has an energy level of 1.0 **electron-volts (eV).** Because this atom has three orbital shells, the valence shell will be energy level 3 (e3). The second energy level or orbital shell (e2) has an energy level of 0.6 eV. Therefore, for an electron to jump from energy level 2 (shell 2) to

Energy Gap

The space between two orbital shells.

Electron-Volt (ev)

A unit of energy equal to the energy acquired by an electron when it passes through a potential difference of 1 V in a vacuum.

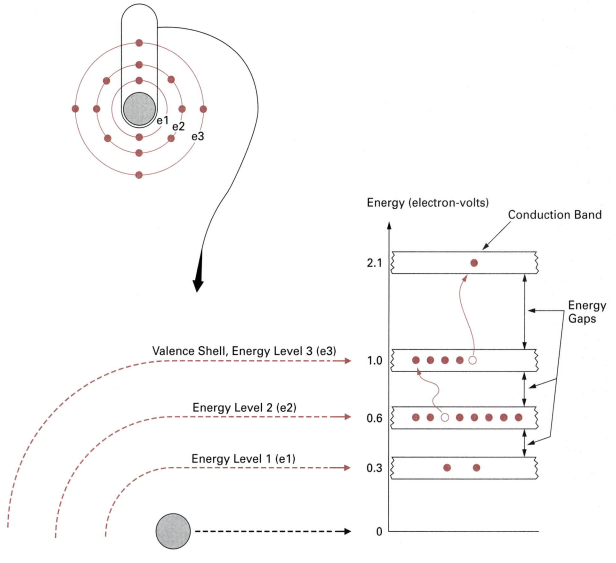

FIGURE 1-6   An Atom's Orbital Shell Energy Levels.

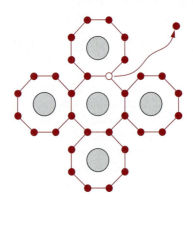

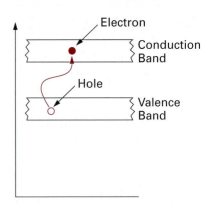

(a)

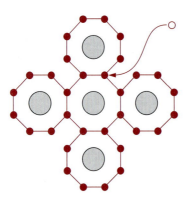

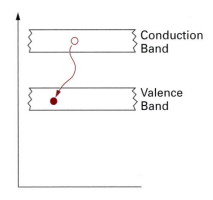

(b)

Conduction Band

An energy band in which electrons can move freely within a solid.

Excited State

An energy level in which an electron may exist if given sufficient energy to reach this state from a lower state.

Hole

The gap in the covalent bond left when an electron jumps from the valence shell or band to the conduction band.

Electron-Hole Pair

When an electron jumps from the valence shell or band to the conduction band, it leaves a gap in the covalent bond called a *hole*. This action creates an electron-hole pair.

FIGURE 1-7   **Valence Band and Conduction Band Actions. (a) Generating an Electron-Hole Pair. (b) Recombination.**

energy level 3 (e3 or valence shell), it will have to absorb a value of energy equal to the difference between e2 and e3. This will equal:

$$1.0 \text{ eV} - 0.6 \text{ eV} = 0.4 \text{ eV}$$

In this example, when either heat, light, or electrical energy was applied, one of the electrons in shell 2 (e2) absorbed 0.4 electron-volts of energy and jumped to valence shell (e3).

If a valence (e3) electron absorbs enough energy, it can jump from the valence shell into the **conduction band.** The conduction band is an energy band in which electrons can move freely or wander within a solid. When an electron jumps from the valence shell into the conduction band, it is released from the atom and no longer travels in one of its orbital paths. The electron is now free to move within the semiconductor material and is said to be in the **excited state.** An excited electron in the conduction band will eventually give up the energy it absorbed in the form of light or heat and return to its original energy level in the atom's valence shell.

When an electron jumps from the valence shell or band to the conduction band, it leaves a gap in the covalent bond called a **hole.** This action is shown in Figure 1-7(a). A hole is created every time an electron enters the conduction band. This action creates an **electron-hole pair.**

It only takes a few microseconds before a free electron in the conduction band will give up its energy and fall into one of the valence shell holes in the covalent bond. This action is called **recombination** and is shown in Figure 1-7(b). The time difference between an electron jumping into the conduction band (becoming a free electron) and then falling back into a hole (recombination) is called the **lifetime** of the electron-hole pair.

1-2-4 *Temperature Effects on Semiconductor Materials*

At extremely low temperatures, the valence electrons are tightly bound to their parent atoms, preventing valence electrons from drifting between atoms. Therefore, pure or intrinsic semiconductor materials function as insulators at temperatures close to absolute zero ($-273.16°C$ or $-459.69°F$).

At room temperature, however, the valence electrons absorb enough heat energy to break free of their covalent bonds, creating electron-hole pairs, as shown in Figure 1-8. Therefore, *the conductivity of a semiconductor material is directly proportional to temperature, in that an increase in temperature will cause an increase in the semiconductor material's conductance.* This means: *an increase in temperature ($T\uparrow$) will cause an increase in a semiconductor's conductivity ($G\uparrow$) and current ($I\uparrow$).* This is why all circuits containing a semiconductor device tend to consume more current once they have warmed up.

Stated another way, *semiconductor materials, and therefore semiconductor devices, have a negative temperature coefficient of resistance, which means as temperature increases ($T\uparrow$), their resistance decreases ($R\downarrow$).*

1-2-5 *Applying a Voltage Across a Semiconductor*

If a voltage was applied across a room-temperature section of intrinsic semiconductor material, **free electrons** in the conduction band would make up a small electrical current, as

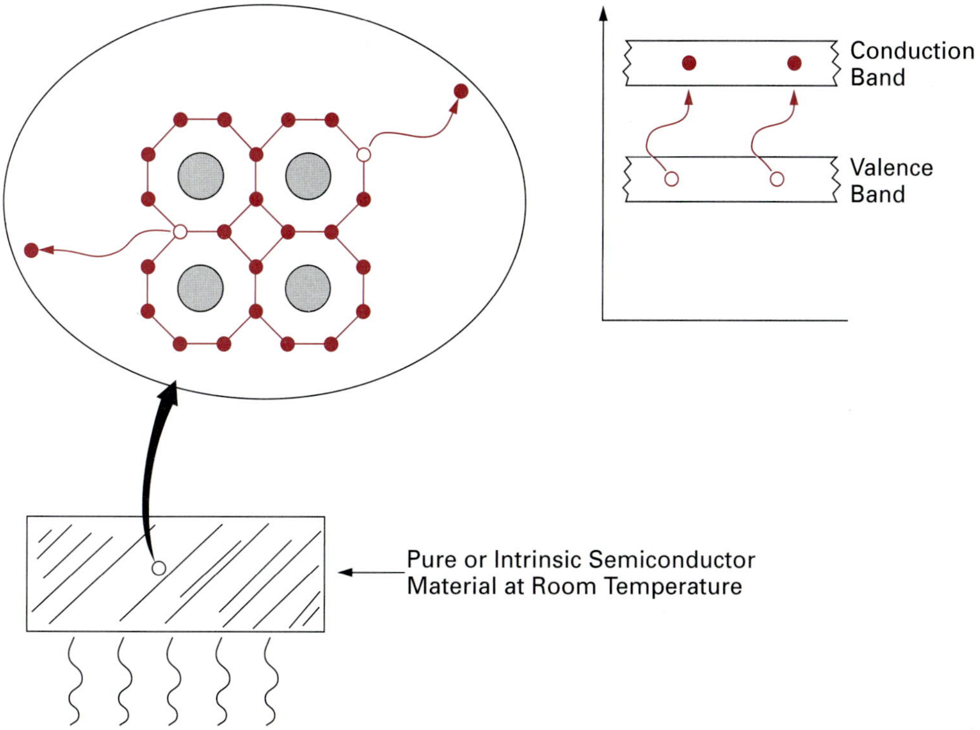

Recombination

An action occurring in microseconds as an electron in the conduction band gives up its energy and falls into one of the valence shell holes in the covalent bond.

Lifetime

The time difference between an electron's jumping into the conduction band and then falling back into a hole.

Free Electron

An electron that is able to move freely when an external force is applied.

FIGURE 1-8 **Temperature Effects on Semiconductor Materials.**

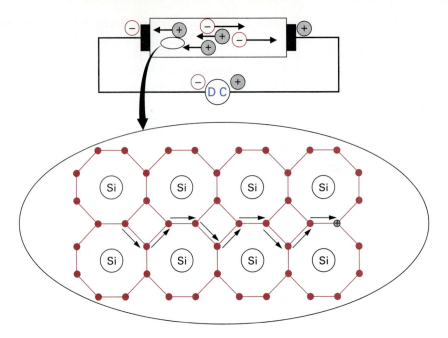

FIGURE 1-9   **Electron Flow and Hole Flow in an Intrinsic Semiconductor.**

shown in Figure 1-9. In this illustration you can see how the negatively charged free electrons are attracted to the positive terminal of the voltage source. For every free electron that leaves the semiconductor material on the right side and travels to the positive terminal of the source, another electron is generated at the negative terminal of the voltage source and is injected into the left side of the semiconductor material. These injected electrons are captured by holes in the semiconductor material (recombination). As you can see from this illustration, current in a semiconductor material is made up of both electrons and holes. The holes act like positively charged particles, while the electrons act like negatively charged particles. As electrons jump between atoms in a migration to the positive terminal of the source voltage, they leave behind them holes, which are then filled by other advancing electrons. These advancing electrons leave behind them other holes, making it appear as though these holes are traveling toward the negative terminal of the source voltage. This **hole flow** is a new phenomenon to us, and it is one of the key differences between a semiconductor and a conductor. With conductors, we were only interested in free-electron flow, but with semiconductors, we must consider the movement of free electrons (negative charge carriers) and the apparent movement of holes (positive charge carriers).

In summary, therefore, *when a potential difference is applied across a semiconductor, the electrons move toward the positive potential and the holes travel toward the negative potential. The total current flow is equal to the sum of the electron flow and the hole flow currents.*

Hole Flow

Conduction in a semiconductor as electrons move into holes when a voltage is applied.

Use the following questions to test your understanding of Section 1-2.

1. What do all semiconductor atoms have in common?

2. The electrical conductivity of an element is determined by the number of electrons in the valence shell. Semiconductor atoms are midway between conductors, which have _____ valence electron(s), and insulators, which have _____ valence electron(s).

3. Each of the semiconductor atoms in a crystal lattice shares its electrons with four neighboring atoms. This joining of atoms is called a _____.

4. Semiconductor materials have a _____ temperature coefficient of resistance, which means that as temperature increases, resistance _____.

5. The number of electron-hole pairs within a semiconductor will increase as temperature _____.

6. When a pure or _____ semiconductor is connected across a voltage, free electrons travel toward the _____ terminal of the applied voltage, whereas holes appear to travel toward the _____ terminal of the applied voltage.

1-3 DOPING SEMICONDUCTOR MATERIALS

At room temperature, pure or intrinsic semiconductors will not permit a large-enough value of current. Therefore, some modification has to be applied in order to increase the semiconductor's current-carrying capability, or conductivity. **Doping** is a process wherein impurities are added to the intrinsic semiconductor material either to increase the number of free electrons (negative doping) or to increase the number of holes (positive doping).

Basically, there are two types of impurities that can be added to semiconductor crystals. One type of impurity is called a *pentavalent material* because its atom has five (*penta*) valence electrons. The second type of impurity is called a *trivalent material* because its atoms have three (*tri*) valence electrons. A doped semiconductor material is referred to as an **extrinsic semiconductor** material because it is no longer pure.

1-3-1 n-*Type Semiconductor*

Figure 1-10(a) shows how a semiconductor material's atoms will appear after pentavalent atom impurities have been added. The pentavalent atoms, which are listed in Figure 1-10(b), can be added to molten silicon to create, when cooled, a crystalline structure that has an extra electron due to the pentavalent (5 valence-electron impurity) atoms. The fifth pentavalent electron is not part of the covalent bonding and requires little energy to break free and enter the conduction band, as shown in Figure 1-10(c). Because millions of pentavalent atoms are added to the pure semiconductor, there will be millions of free electrons available for flow through the material.

Even though the doped semiconductor material has millions of free electrons, the material is still electrically neutral. This is because each arsenic atom has the same number of protons as electrons, as do the silicon atoms. Therefore, the overall numbers of protons and electrons in the semiconductor are still equal and the result is a net charge of zero. However, because we now have more electrons than valence-band holes, the material is called an **n-type semiconductor.** n-Type semiconductors have more conduction-band electrons than valence-band holes. The electrons are therefore called the **majority carriers** and the valence-band holes are called the **minority carriers.** In Figure 1-10(c) you can see the abundance of conduction-band electrons. The holes in the valence band are few and are generated by thermal energy because the semiconductor is at room temperature.

When a voltage is applied across an *n*-type semiconductor, as shown in Figure 1-10(d), the additional free conduction-band electrons travel toward the positive terminal of the dc source. The applied voltage will cause extra electrons to break away from their covalent

Doping

The process wherein impurities are added to the intrinsic semiconductor material either to increase the number of free electrons or to increase the number of holes.

Extrinsic Semiconductor

A semiconductor whose electrical properties are dependent on impurities added to the semiconductor crystal.

n-Type Semiconductor

A material that has more conduction-band electrons than valence-band holes.

Majority Carriers

The type of carrier that makes up more than half the total number of carriers in a semiconductor device.

Minority Carriers

The type of carrier that makes up less than half of the total number of carriers in a semiconductor device.

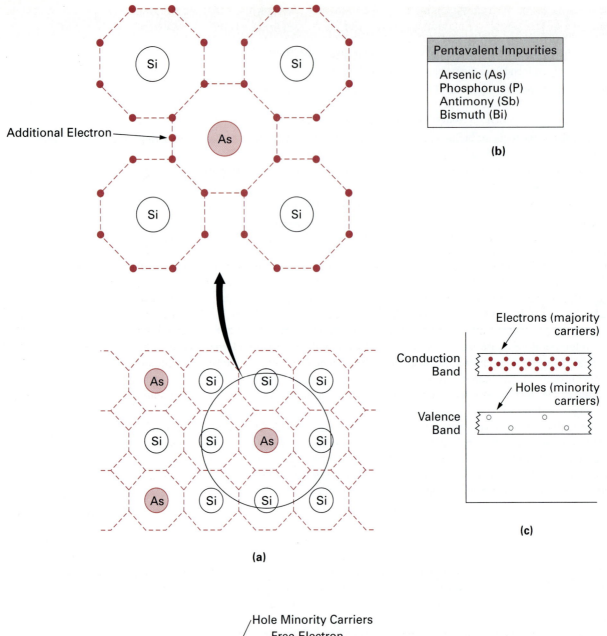

Additional Electron

Pentavalent Impurities

Arsenic (As)
Phosphorus (P)
Antimony (Sb)
Bismuth (Bi)

(b)

Electrons (majority carriers)

Conduction Band

Holes (minority carriers)

Valence Band

(c)

(a)

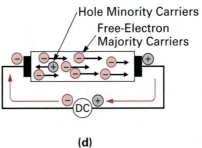

Hole Minority Carriers
Free-Electron Majority Carriers

DC

(d)

FIGURE 1-10   **Adding Pentavalent Impurities to Create an *n*-Type Semiconductor Material.**

bonds to create holes, resulting in an increase in current and conductivity. Although the total current flow in this *n*-type semiconductor is the sum of the electron and hole currents, the conduction-band electrons make up the majority of the flow.

1-3-2 p-*Type Semiconductor*

Figure 1-11(a) on page 14 shows how a semiconductor material's atoms will appear after trivalent atom impurities have been added. The trivalent atoms, which are listed in Figure 1-11(b), can be added to molten silicon to create, when cooled, a crystalline structure that has a hole in the valence band of every trivalent (3 valence-electron impurity) atom. Instead of an excess of electrons, we now have an excess of holes. Because millions of trivalent atoms are added to the pure semiconductor, there will be millions of holes available for flow through the material.

Even though the doped semiconductor material has millions of holes, the material is still electrically neutral. This is because each aluminum atom has the same number of protons as electrons, as do the silicon atoms. Therefore, the overall numbers of protons and electrons in the semiconductor are still equal and the result is a net charge of zero. However, because we now have more valence band holes than electrons, the material is called a ***p*-type semiconductor.** *p*-Type semiconductors have more valence-band holes than conduction-band electrons. The holes are called the *majority carriers* and the electrons are called the *minority carriers*. In Figure 1-11(c) you can see the abundance of valence-band holes. The few electrons in the conduction band are generated by thermal energy because the semiconductor is at room temperature.

p-**Type Semiconductor**
A material that has more valence-band holes than conduction-band electrons.

When a voltage is applied across a *p*-type semiconductor, as illustrated in Figure 1-11(d), the large number of holes within the material will attract electrons from the negative terminal of the dc source into the *p*-type semiconductor. These holes appear to move because each time an electron moves into a hole it creates a hole behind it, and the holes appear to move in the opposite direction to the electrons (toward the negative terminal of the dc source). The applied voltage will cause some electrons to break away from the covalent bond, resulting in an increased current and conductivity. Although the total current flow in this *p*-type semiconductor is the sum of the hole and electron currents, the valence-band holes make up the majority of the flow.

SELF-TEST EVALUATION POINT FOR SECTION 4-1

Use the following questions to test your understanding of Section 1-3.

1. Why are impurities added to pure semiconductor materials?
2. Pentavalent atoms add _____ to semiconductor crystals to create _____-type semiconductors.
3. Trivalent atoms add _____ to semiconductor crystals to create _____-type semiconductors.
4. In an *n*-type semiconductor the majority carriers are _____, whereas in a *p*-type semiconductor the majority carriers are _____.

1-4   THE P-N JUNCTION

On their own, *n*-type semiconductor materials and *p*-type semiconductor materials are of little use. Together, however, these two form a **P-N semiconductor junction.** Semiconductor devices such as diodes and transistors are constructed using these P-N junctions, which give specific current flow characteristics. In this section we will examine the characteristics of the P-N junction in detail.

P-N Junction
The point at which two opposite doped materials come in contact with one another.

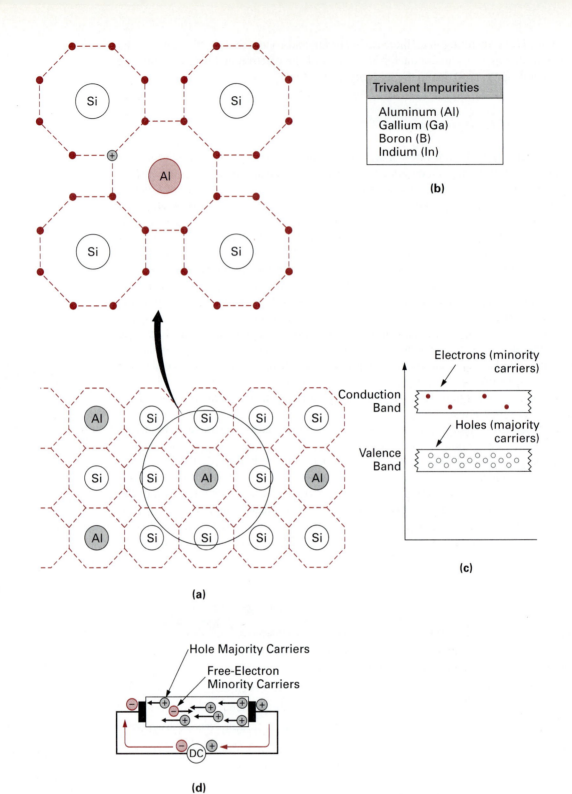

(b)

Trivalent Impurities
Aluminum (Al)
Gallium (Ga)
Boron (B)
Indium (In)

(a)

(c)

(d)

FIGURE 1-11 **Adding Trivalent Impurities to Create a *p*-Type Semiconductor Material.**

1-4-1   *The Depletion Region*

Figure 1-12(a) shows the individual *n*-type and *p*-type materials. The *n*-type material is represented as a block containing an excess of electrons (solid circles), while the *p*-type material is represented as a block containing an excess of holes (open circles). The energy diagrams below the two semiconductor sections show the differences between the two materials. Because different impurity atoms were added to the pure semiconductor material, the atomic make-up of the *n*-type and *p*-type materials is slightly different, which is why the valence bands and conduction bands are at slightly different energy levels.

Figure 1-12(b) shows the two *n*-type and *p*-type semiconductor sections joined together. A manufacturer of semiconductor devices would not join two individual pieces in this way to create a P-N junction. Instead, a single piece of pure semiconductor material would have each of its halves doped to create a *p*-type and *n*-type section.

The point at which the two oppositely doped materials come in contact with one another is called the *junction*. This junction of the two materials now permits the free electrons in the *n*-type material to combine with the holes in the *p*-type material, as shown in Figure 1-12(c). As free electrons in the *n* material cross the junction and combine with holes in the *p* material, they create negative ions (atoms with more electrons than protons) in the *p* material and leave behind positive ions (atoms with fewer electrons than protons) in the *n* material, as shown in Figure 1-12(d). An area or region on either side of the junction becomes emptied or depleted of free electrons and holes. This small layer containing positive and negative ions is called the **depletion region.**

As the ion layer on either side of the junction builds up, it has the effect of diminishing and eventually preventing any further recombination of free electrons and holes across the junction. In other words, the negative ions in the *p* region near the junction repel and prevent free electrons in the *n* region from recombination. This action prevents the depletion region from becoming larger and larger.

These positive ions or charges and negative ions or charges accumulate a certain potential. Because these charges are opposite in polarity, a potential difference, or voltage, called the **barrier potential** or **barrier voltage,** exists across the junction, as shown in Figure 1-12(e). At room temperature, the barrier voltage of a silicon P-N junction is approximately 0.7 V and a germanium P-N junction is approximately 0.3 V.

Depletion Region

A small layer on either side of the junction that becomes empty, or depleted, of free electrons or holes.

Barrier Potential or Barrier Voltage

The potential difference, or voltage, that exists across the junction.

1-4-2   *Biasing a P-N Junction*

Semiconductor devices are constructed using P-N junctions. These P-N junctions need voltages of a certain amplitude and polarity to control their operation. These voltages, which incline or cause the device to operate in a certain manner, are known as **bias voltages.** Bias voltages control the width of the depletion region, which in turn controls the resistance of the P-N junction and, therefore, the amount of current that can pass through the P-N junction or semiconductor device.

To be specific, a small depletion region ($dr\downarrow$) will offer a small P-N junction resistance ($R\downarrow$) and therefore permit a large P-N junction current ($I\uparrow$). In this instance, the P-N semiconductor junction is said to be **forward biased** and acts like a conductor.

On the other hand, a large depletion region ($dr\uparrow$) will offer a large P-N junction resistance ($R\uparrow$) and therefore permit only a small P-N junction current ($I\downarrow$). In this instance, the P-N semiconductor junction is said to be **reverse biased** and acts like an insulator.

Bias Voltages

The dc voltages applied to control a device's operation.

Forward Biased

A small depletion region at the junction will offer a small resistance and permit a large current. Such a junction is forward biased.

Forward Biasing a P-N Junction

Figure 1-13 shows in detail why a forward-biased P-N junction will pass current with almost no opposition (act like a conductor). To begin, Figure 1-13(a) shows a P-N junction with wires attached. A resistor has been included to limit the amount of current passing through the P-N junction to a safe level. Energy diagrams have also been included on the

Reverse Biased

A large depletion region at the junction will offer a large resistance and permit only a small current. Such a junction is reverse biased.

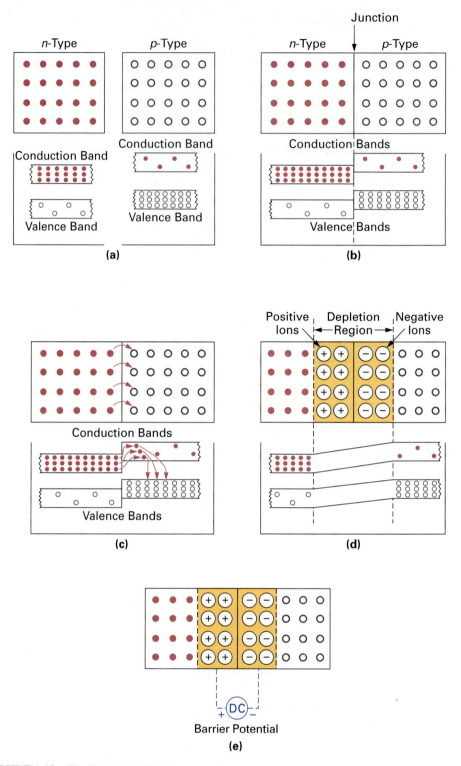

FIGURE 1-12 **The Depletion Region.**

right side of each part of Figure 1-13 to show the relationship between the conduction band and valence band of the *p* and *n* regions.

Let us now connect a dc voltage across the P-N junction to see how it reacts. This is shown in Figure 1-13(b). The negative potential of the dc source has been applied to the *n* region, and the positive potential of the dc source has been applied to the *p* region. Referring to the energy diagram in Figure 1-13(b), you can see that the conduction band electrons in

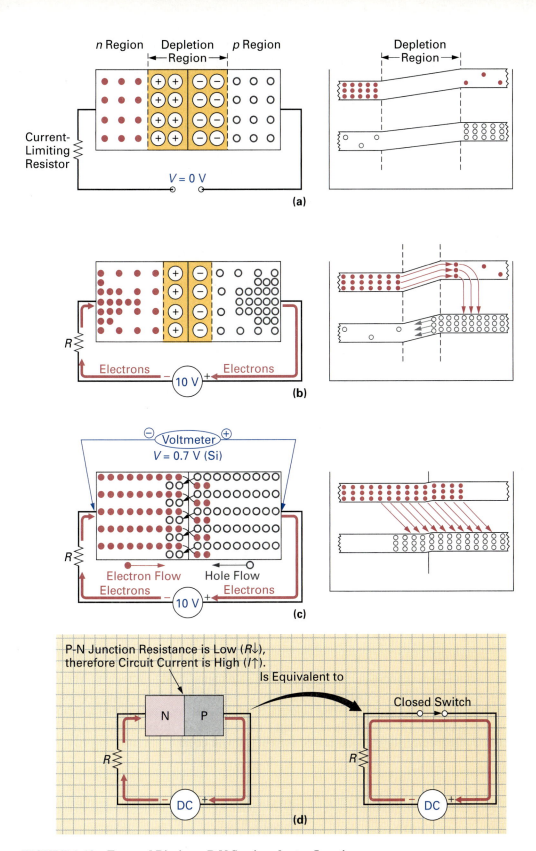

FIGURE 1-13   Forward Biasing a P-N Semiconductor Junction.

the *n* region are repelled by the negative voltage source toward the junction. On the opposite side, valence band holes in the *p* region are repelled by the positive voltage source toward the junction. A forward-conducting current will begin to flow if the external source voltage is large enough to overcome the internal barrier voltage of the P-N junction. In this example we will assume that the dc source voltage is 10 volts and therefore this will be more than enough to overcome the silicon P-N junction's barrier potential of 0.7 volts.

Conduction through the P-N junction is shown in Figure 1-13(c). When forward biased, a P-N junction will act as a conductor and have a low but finite resistance value that will cause a corresponding voltage drop across its terminals. This **forward voltage drop** (**V_F**) is approximately equal to the P-N junction's barrier voltage:

$$\text{Forward Voltage Drop } (V_F) \text{ for Silicon} = 0.7 \text{ V}$$
$$\text{Forward Voltage Drop } (V_F) \text{ for Germanium} = 0.3 \text{ V}$$

Figure 1-13(c) shows how a voltmeter can be used to measure the forward voltage drop of 0.7 V across a silicon P-N junction when it is forward biased.

To summarize, Figure 1-13(d) shows that when a P-N junction is forward biased ($+V \rightarrow p$ region, $-V \rightarrow n$ region), the P-N junction resistance is low ($R\downarrow$) and therefore the circuit current is high ($I\uparrow$). When forward biased, therefore, the P-N junction acts like a conductor and is equivalent to a closed switch.

■ **EXAMPLE:**

Calculate the current for the circuit in Figure 1-14.

FIGURE 1-14   A P-N Junction Circuit.

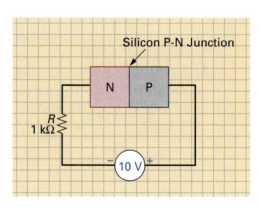

■ *Solution:*

The silicon P-N junction is forward biased ($+V \rightarrow p$ region, $-V \rightarrow n$ region). The applied voltage of 10 V will be more than enough to overcome the silicon P-N junction forward voltage drop of 0.7 volts ($V_F = 0.7$ V for silicon). Since 10 volts are applied, and the P-N junction is dropping 0.7 V, the remaining voltage of 9.3 V is being dropped across the 1 kΩ resistor. Consequently, the forward-biased current (I_F) will equal

$$I_F = \frac{V_S - V_{P\text{-}N}}{R}$$
$$= \frac{10 \text{ V} - 0.7 \text{ V}}{1 \text{ k}\Omega}$$
$$= 9.3 \text{ mA}$$

Reverse Biasing a P-N Junction

Figure 1-15 shows in detail why a reverse-biased P-N junction will reduce current to almost zero (act like an insulator). To begin, Figure 1-15(a) shows a P-N junction with wires

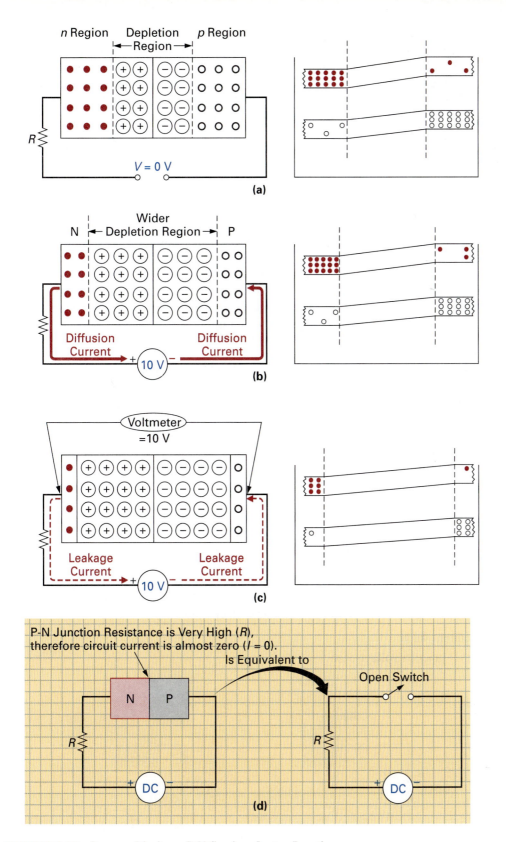

FIGURE 1-15 Reverse Biasing a P-N Semiconductor Junction.

attached and no voltage being applied. Energy diagrams have again been included to show the relationship between the conduction band and valence band of the *p* and *n* regions.

Let us now connect a dc voltage across the P-N junction to see how it reacts. This is shown in Figure 1-15(b). The positive potential of the dc source is now being applied to the *n* region, and the negative potential of the dc source has been applied to the *p* region.

A forward-biased P-N junction is able to conduct current because the external bias voltage forces the majority carriers in the *n* and *p* regions to combine at the junction. In this instance, however, the dc bias voltage polarity has been reversed, causing free electrons in the *n* region to travel to the positive terminal of the voltage source and leaving behind a large number of positive ions at the junction. This increases the width of the depletion region. At the same time, electrons from the negative terminal of the source are attracted to the holes in the *p* region of the P-N junction. These electrons fill the holes in the *p* region near the junction, creating a large number of negative ions. This further increases the width of the depletion region. The current that is present at the time the depletion layer is expanding is called the **diffusion current.** Referring to Figure 1-15(b), you can see that the depletion region is now wider than the unbiased P-N junction shown in Figure 1-15(a).

The ions on either side of the junction build up until the P-N junction's internal-barrier voltage is equal to the external-source voltage, as shown by the voltmeter in Figure 1-15(c). When reverse biased, therefore, the **reverse voltage drop (V_R)** across a P-N junction is equal to the source or applied voltage. At this time the resistance of the junction has been increased to a point that current drops to zero.

Actually, an extremely small current called the **leakage current** or **reverse current (I_R)** will pass through the P-N junction, as shown in Figure 1-15(c). It is present because the minority carriers (holes in the *n* region, electrons in the *p* region) are forced toward the junction, where they combine, producing a constant small current. The current in the P-N junction is still considered to be at zero because the leakage or reverse current is so small (nanoamps in silicon diodes).

To summarize, Figure 1-15(d) shows that when a P-N junction is reverse biased ($+V \rightarrow n$ region, $-V \rightarrow p$ region), the P-N junction resistance is extremely high ($R \uparrow \uparrow$) and the circuit current is effectively zero ($I = 0$ amps). When reverse biased, therefore, the P-N junction acts like an insulator and is equivalent to an open switch.

Diffusion Current

The current that is present when the depletion layer is expanding.

Reverse Voltage Drop (V_R)

The reverse voltage drop is equal to the source voltage (applied voltage).

Leakage Current or Reverse Current (I_R)

The extremely small current present at the junction.

■ **EXAMPLE:**

Referring to Figure 1-16(a) and (b), calculate each circuit's:

a. current value

b. P-N junction voltage drop

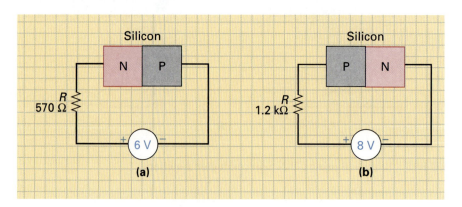

FIGURE 1-16 **P-N Junction Circuit Examples.**

■ *Solution:*

The P-N junction in Figure 1-16(a) is reverse biased ($+V \rightarrow n$ region, $-V \rightarrow p$ region); therefore, the P-N junction resistance is extremely high ($R \uparrow\uparrow$) and the circuit current is effectively zero ($I = 0$). When reverse biased, the P-N junction acts like an insulator and is equivalent to an open switch, and the voltage developed across the open P-N junction will equal the source voltage applied. For Figure 1-16(a):

$$\text{Circuit Current} = 0$$
$$\text{P-N Junction Voltage Drop} = V_S = 6 \text{ V}$$

The P-N junction in Figure 1-16(b) is forward biased ($+V \rightarrow p$ region, $-V \rightarrow n$ region). Therefore, the P-N junction resistance is low ($R\downarrow$) and the circuit current is high ($I\uparrow$). When forward biased, the P-N junction acts like a conductor and is equivalent to a closed switch, and the P-N junction's voltage drop will equal the forward voltage drop (V_F) for a silicon P-N junction. For Figure 1-16(b):

$$\text{Circuit Current} = I_F = \frac{V_S - V_{P\text{-}N}}{R}$$
$$= \frac{8 \text{ V} - 0.7 \text{ V}}{1.2 \text{ k}\Omega}$$
$$= 6.08 \text{ mA}$$
$$\text{P-N Junction Voltage Drop} = V_F = 0.7 \text{ V}$$

SELF-TEST EVALUATION POINT FOR SECTION 1-4

Use the following questions to test your understanding of Section 1-4.

1. When a P-N junction is formed, a _____ region is created on either side of the junction.

2. The barrier voltage within a silicon diode is
   a. 700 mV   **b.** 7.0 V   **c.** 0.3 V   **d.** None of the above

3. True or false: A P-N junction is forward biased when its P terminal is made positive relative to its N terminal.

4. A reverse-biased P-N junction acts like a/an _____ switch, whereas a forward-biased P-N junction acts like a/an _____ switch.

REVIEW QUESTIONS

Multiple-Choice Questions

1. What is the atomic number of silicon?
   a. 14   **b.** 16   **c.** 10   **d.** 32

2. How many valence electrons are normally present in the valence shell of a semiconductor material?
   a. 2   **b.** 4   **c.** 6   **d.** 8

3. Adding trivalent impurities to an intrinsic semiconductor will produce a/an _____ material.
   a. Extrinsic
   b. *n*-type
   c. *p*-type
   d. Both (a) and (c) are true

4. What is the majority carrier in an *n*-type material?
   a. Holes   **b.** Electrons   **c.** Neutrons   **d.** Protons

5. Adding pentavalent impurities to an intrinsic semiconductor will produce a/an _____ material.
   a. Extrinsic
   b. *n*-type
   c. *p*-type
   d. Both (a) and (b) are true

6. What are the majority carriers in a *p*-type semiconductor?
   a. Holes   **b.** Electrons   **c.** Neutrons   **d.** Protons

7. A semiconductor material has a _____ temperature coefficient of resistance, which means that as temperature increases its resistance _____.
   a. Positive, increases   **c.** Negative, increases
   b. Positive, decreases   **d.** Negative, decreases

8. A hole is considered to be _____.
   a. Negative   **c.** Neutral
   b. Positive   **d.** Both (b) and (c) are true

9. Intrinsic semiconductors are doped to increase their _____.

    a. Resistance   **c.** Inductance

    b. Conductance   **d.** Reactance

10. As temperature increases, a semiconductor acts more like a/an _____.

    a. Conductor   **b.** Insulator

11. A negative ion has more:

    a. Protons than electrons   **c.** Neutrons than protons

    b. Electrons than protons   **d.** Neutrons than electrons

12. A positive ion has:

    a. Lost some of its electrons   **c.** Lost neutrons

    b. Gained extra protons   **d.** Gained more electrons

13. The resistance of a semiconductor material is more than the resistance of:

    a. Glass   **c.** Ceramic

    b. Copper   **d.** Both (a) and (c) are true

14. The basic function of a semiconductor device in an electrical or electronic circuit is to:

    a. Control current

    b. Control voltage

    c. Increase the price of the equipment

    d. Both (a) and (b) are true

15. For a silicon P-N junction, $V_F = ?$

    a. The value of the applied voltage   **c.** 0.7 V

    b. 300 mV   **d.** 10 V

Practice Problems

16. Which of the silicon P-N junctions in Figure 1-17 are forward biased, and which are reverse biased?

17. Determine the current for the circuits shown in Figure 1-18.

18. What would be the voltage drop (V_F) across each of the P-N junctions shown in Figure 1-18?

19. What would be the voltage drop across each of the resistors in Figure 1-18?

20. Which of the P-N junctions in Figure 1-18 are equivalent to open switches, and which are equivalent to closed switches?

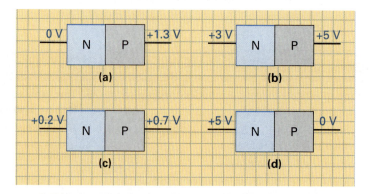

FIGURE 1-17   **Biased P-N Junctions.**

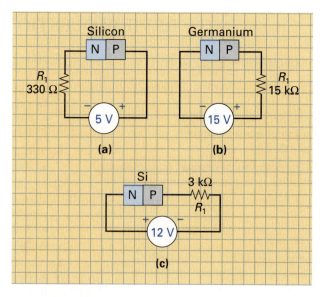

FIGURE 1-18   **P-N Junction Circuit.**

Junction Diodes

2

A Problem with Early Mornings

René Descartes was born in Brittany, France, in 1596. At the age of eight he had surpassed most of his teachers at school and was sent on to the Jesuit College in La Flèche, one of the best in Europe. It was here that his genius in mathematics became apparent; however, due to his extremely delicate health, his professors allowed him to study in bed until midday.

In 1616, he had an urge to see the world and so he joined the army, which made use of Descartes' mathematical genius in military engineering. While traveling, Descartes met Dutch philosopher Isaac Beekman, who convinced him to leave the army and, in his words, "turn his mind back to science and more worthier occupations."

After leaving the army, Descartes traveled, looking for some purpose, and then on November 10, 1619, he found it. Descartes was in Neuberg, Germany, where he had shut himself in a well-heated room for the winter. It was on the eve of St. Martin's that a freezing blizzard forced Descartes to retire early. That night he described having an extremely vivid dream that clarified his purpose and showed him that physics and all sciences could be reduced to geometry and were therefore all interconnected like a chain.

In his time, and to this day, he is heralded as an analytical genius. In fact, Descartes' procedure can still be used as a guide to solving any problem. Descartes' four-step procedure for solving a problem:

1. Never accept anything as true unless it is clear and distinct enough to exclude doubt from your mind.
2. Divide the problem into as many parts as necessary to reach a solution.
3. Start with the simplest things and proceed step-by-step toward the complex.
4. Review the solution so completely and generally that you are sure nothing was omitted.

For me, this four-step procedure has been especially helpful as a troubleshooting guide for system and circuit malfunctions.

Descartes' fame was so renowned that he was asked in 1649 to tutor Queen Christina of Sweden. The queen demanded that her lessons begin at 5 o'clock in the morning, which conflicted with Descartes' lifetime practice of remaining in bed until midday. After several unsuccessful attempts to change her majesty's mind, and with pressure being applied by the French ambassador, Descartes agreed to the early-morning lessons. A short time later on his way to the palace one cold winter morning, Descartes caught a severe chill and died within two weeks.

It was philosopher René Descartes who first stated that to solve any problem you should start with the simple and then proceed to the complex. For Descartes, there were three approximations: *The first approximation was the simplest, the second approximation contained more detail, and the third approximation was the complex.* I have applied this method to the problem of learning any new topic.

In this chapter you will be introduced to your first semiconductor component: the *diode.* To help you gain a clear understanding of this device, we will begin with a *first approximation description of the diode,* in which we will discuss the diode's schematic symbol, physical appearance, basic operation, and basic application. Following this basic complete picture description, the *second approximation description of the diode* will cover the diode in more detail, addressing its characteristics, analog and digital circuit applications, data sheet specifications, and testing procedure.

As a technician, you will not need to examine the *third approximation description of a diode,* because these topics include more detail on specific device specifications and how to implement diodes into new circuit designs. This area of understanding is needed only for engineering students specializing in design. Throughout this text we will be concentrating on semiconductor device operation, characteristics, applications, and testing, along with analog and digital circuit applications and troubleshooting—the necessary knowledge for a good *electronics technician.*

2-1 FIRST APPROXIMATION DESCRIPTION OF A DIODE

The first diode was accidentally created by Edison in 1883 when he was experimenting with his light bulb. At this time he did not place any importance on the device and its effect, as he could not see any practical application for it. The word *diode* is derived from the fact that the device has two (*di*) electrodes (*ode*).

Once the importance of diodes was realized, construction of the device began. The first diodes were vacuum-tube devices having a hot-filament negative cathode, which released free electrons that were collected by a positive plate called the *anode.* Today's diode is made of a P-N semiconductor junction but still operates on the same principle. The *n*-type region (cathode) is used to supply free electrons, which are then collected by the *p*-type region (anode). The operation of both the vacuum tube and semiconductor diode is identical in that the device will pass current in only one direction. That is, it will act as a conductor and pass current easily in one direction when the bias voltage across it is of one polarity, yet it will block current and imitate an insulator when the bias voltage applied is of the opposite polarity.

2-1-1 *Diode Schematics Symbol and Packaging*

The two electrodes or terminals of the diode are called the *anode* and *cathode,* as seen in Figure 2-1(a), which shows the schematic symbol of a diode. To help you remember which terminal is the anode and which is the cathode, and which terminal is positive and which is negative, Figure 2-1(b) shows how a line drawn through the triangle section of the symbol will make the letter "A" and indicate the "anode" terminal. Similarly, if the vertical flat side of the diode symbol is aligned horizontally "—", as in Figure 2-1(b), it becomes the "nega-

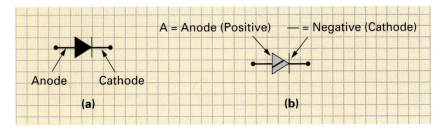

FIGURE 2-1 Schematic Symbol of a Diode.

tive" symbol. This memory system helps us to remember that the anode terminal of the diode is next to the triangle part of the symbol and is positive, while the cathode terminal of a diode is next to the vertical line of the symbol and is negative.

The diode is generally mounted in one of the three basic packages shown in Figure 2-2. These packages are designed to protect the diode from mechanical stresses and the environment. The difference in the size of the packages is due to the different current rating of the diodes. A black band or stripe is generally placed on the package closest to the cathode terminal for identification purposes, as seen in Figure 2-2(a) and (b). Larger diode packages, like the one seen in Figure 2-2(c), usually have the diode symbol stamped on the package to indicate anode/cathode terminals.

2-1-2 *Diode Operation*

As far as operation is concerned, the diode operates like a switch. If you give the diode what it wants—that is, make the anode terminal positive with respect to the cathode terminal, as seen in Figure 2-3(a)—the device is equivalent to a closed switch, as seen in Figure 2-3(b). In this condition, the diode is said to be ON or *forward biased*.

On the other hand, if you do not give the diode what it wants—that is make the anode terminal negative with respect to the cathode, as seen in Figure 2-3(c)—the device is equivalent to an open switch, as seen in Figure 2-3(d). In this condition, the diode is said to be OFF or *reverse biased*.

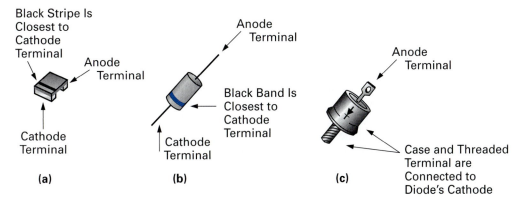

FIGURE 2-2 Diode Packaging. (a) Chip Package—1/4 A. (b) Small Current Package—less than 3 A. (c) Large Current Package—greater than 3 A.

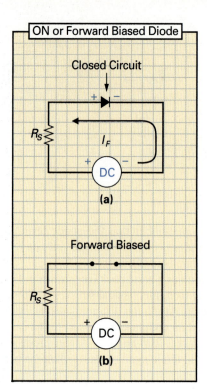
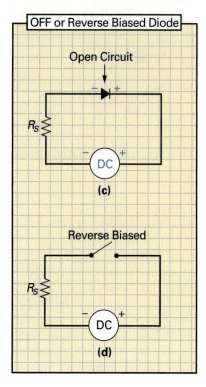

FIGURE 2-3 Diode Operation. (a)(b) Forward Biased (ON) Diode. (c)(d) Reverse Biased (OFF) Diode.

■ **EXAMPLE:**

Determine whether the diodes in Figure 2-4 are ON or OFF.

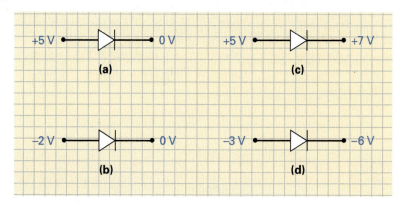

FIGURE 2-4 ON or OFF Diodes?

■ *Solution:*

    a. Diode is ON since anode is positive relative to cathode.
    b. Diode is OFF since anode is negative relative to cathode.
    c. Diode is OFF since anode is less positive than cathode.
    d. Diode is ON since anode is more positive than cathode.

Encoder Circuit

A circuit that produces different output voltage codes depending on the position of a rotary switch.

2-1-3   *Diode Application*

As an application, Figure 2-5 shows how the diode can be used as a switch within an **encoder circuit.** The pull-up resistors R_1, R_2, and R_3 ensure that lines *A, B,* and *C* are normally

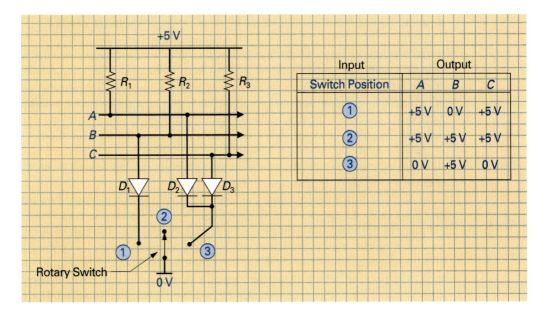

FIGURE 2-5 Diode Application: A Switch Encoder Circuit.

all at +5 V. This is the output voltage on each line when the rotary switch is in position 2, as seen in the table in Figure 2-5.

When the rotary switch is turned to position 1, D_1 is connected in circuit and, because its anode is made positive via R_2 and its cathode is at 0 V, the diode D_1 will turn ON and be equivalent to a closed switch. The 0 V on the cathode of D_1 will be switched through to line *B* (all of the 5 volts will be dropped across R_2), producing an output voltage code of $A = +5$ V, $B = 0$ V, $C = +5$ V, as seen in the table in Figure 2-5.

When the rotary switch is turned to position 3, D_2 and D_3 are connected in circuit, and because both anodes are made positive via R_1 and R_3 and both diode cathodes are at 0 V, D_2 and D_3 will turn ON. These forward-biased diodes will switch 0 V through to lines *A* and *C*, producing an output voltage code of $A = 0$ V, $B = +5$ V, $C = 0$ V, as seen in the table in Figure 2-5.

This *code generator* or *encoder* circuit will produce three different output voltage codes for each of the three positions of the rotary switch. These codes could then be used to initiate one of three different operations based on the operator setting of the rotary control switch.

SELF-TEST EVALUATION POINT FOR SECTION 2-1

Use the following questions to test your understanding of Section 2–1.

1. Name the two terminals of a diode.
2. The diode operates like a _____ .
3. An ON diode is said to be _____ biased, while an OFF diode is said to be _____ biased.
4. Would a diode be ON or OFF if its anode had +6 V applied and its cathode had +9 V applied?

2-2   SECOND APPROXIMATION DESCRIPTION OF A JUNCTION DIODE

Now that we understand the basic operation and application of the diode, let us examine its characteristics in more detail.

2-2-1 The P-N Junction Within a Junction Diode

In the previous chapter, we discussed how a pure or intrinsic semiconductor material could be doped with a pentavalent element or trivalent element to obtain the two basic semiconductor types. The first type is called an *n*-type semiconductor because its majority carriers are electrons, while the second type is called a *p*-type semiconductor because its majority carriers are holes. On their own, the *n*-type semiconductor and *p*-type semiconductor are of little use. Together, however, these two form a P-N semiconductor junction. A manufacturer of semiconductor devices would not join two individual pieces to create a P-N junction. Instead, a single piece of pure semiconductor material would have each of its halves doped to create a *p*-type and *n*-type section or region.

Semiconductor devices such as diodes and transistors are constructed using these P-N junctions. A diode, for example, has only one P-N junction and is created by doping a single piece of pure semiconductor to produce an *n*-type and *p*-type region. A bipolar junction transistor, on the other hand, has two P-N junctions and is created by doping a single piece of pure semiconductor with three alternate regions (NPN or PNP). As mentioned in Chapter 1, the point at which these two opposite-doped materials come in contact with each other is called a *junction,* which is why these devices are called **junction diodes** and *bipolar junction transistors.*

Figure 2-6 illustrates the schematic symbol for the *junction diode,* and the inset shows how it contains one P-N junction. The *n*-type region is called the *cathode,* while the *p*-type region is called the *anode.*

A junction diode is basically a P-N semiconductor junction, so the diode will operate in exactly the same way as the P-N junction described in Chapter 1.

2-2-2 Biasing a Junction Diode

Semiconductor diodes are constructed using P-N junctions. These P-N junctions need voltages of a certain amplitude and polarity to control their operation. These voltages, which incline or cause the diode to operate in a certain manner, are known as **bias voltages.** Bias voltages control the width of the depletion region, which in turn controls the resistance of the junction and, therefore, the amount of current that can pass through the P-N junction diode.

Forward Biasing a Diode

Figure 2-7(a) shows how a junction diode can be forward biased. Like the P-N junction, the junction diode's operation is determined by the polarity of the applied voltage. In this figure

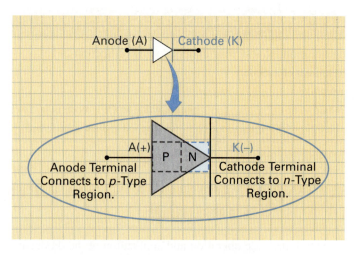

FIGURE 2-6 **The P-N Junction Within a Junction Diode.**

Junction Diode

A semiconductor diode whose ON/OFF characteristics occur at a junction between the *n*-type and *p*-type semiconductor materials.

Bias Voltage

Voltage that inclines or causes the diode to operate in a certain manner.

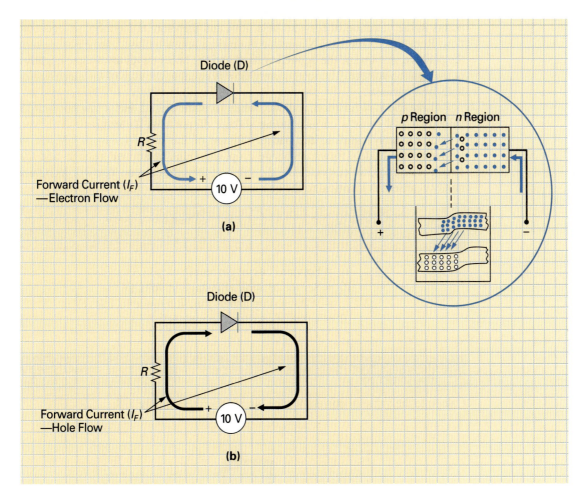

FIGURE 2-7 **A Forward-Biased P-N Junction Diode. (a) Electron Flow. (b) Hole (Conventional) Flow.**

the negative terminal of the applied voltage is connected to the *n* region of the diode, and the positive terminal of the applied voltage is connected to the *p* region of the diode ($+V \rightarrow p$ region, $-V \rightarrow n$ region). Free electrons are repelled from the *n* region by the negative source and attracted to the positive terminal of the voltage source. This forward-conducting electron flow will occur only if the external source voltage is large enough to overcome the internal barrier voltage of the junction diode. For a silicon diode, the external source voltage must be equal to or greater than 0.7 V, whereas for a germanium diode, the applied voltage must be equal to or greater than 0.3 V. The resistor is added in series with the diode to limit the forward current to a safe level; an excessive current will generate more heat than the diode can dissipate, causing the diode to burn out.

A forward-biased diode will conduct current as long as the external bias voltage is of the correct polarity and amplitude. Figure 2-7(a) shows the direction of forward current (I_F). This electron-flow current is from the negative terminal of the applied voltage to the *n* region of the diode, through the diode to the *p* region, and then to the positive terminal of the applied voltage source. This means that forward (electron-flow) current passes through the diode symbol from the bar to the triangle. In other words, forward electron flow is actually traveling against the arrow formed by the diode's symbol. To further explain this, let us compare Figure 2-7(a), which shows forward electron flow, to Figure 2-7(b), which shows forward hole flow. As previously discussed in Chapter 1, apparent hole flow is in the opposite direction to electron flow. That is, electron flow is from negative to positive, while hole flow, or conventional current flow, is from positive to negative. The diode symbol actually points in the direction of forward hole flow, and therefore the symbol reflects conventional flow. Remember, when a P-N junction or diode is forward biased, the electrons move toward the positive terminal of the applied voltage and the holes travel toward the negative terminal of the applied voltage. The total current flow is equal to the sum of the electron-flow and the hole-flow currents.

Like the P-N junction, a diode, when forward biased, has a low but finite resistance value that will cause a corresponding voltage drop across its terminals. This voltage drop is approximately equal to the barrier voltage of the diode (Si = 0.7 V, Ge = 0.3 V). Knowing the diode's forward voltage drop, the value of applied voltage, and the value of circuit resistance, we can calculate the value of forward current. You will recognize this formula because it is identical to the one used in Chapter 1 for the P-N junction.

$$I_F = \frac{V_S - V_{\text{diode}}}{R}$$

I_F = Forward circuit current in amps (A)
V_S = Source or applied voltage in volts (V)
V_{diode} = Voltage drop across junction diode in volts (V)
R = Value of resistor in ohms (Ω)

EXAMPLE:

Calculate the value of current for the circuit shown in Figure 2-8.

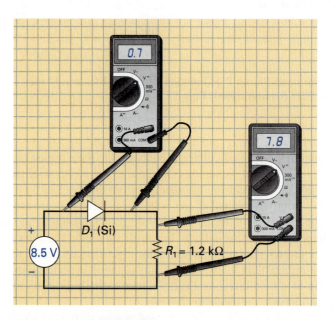

FIGURE 2-8   A P-N Junction Diode Circuit.

Solution:

The diode is forward biased because the applied voltage is connected so that its positive terminal is applied to the *p* region (anode) and the negative terminal is applied to the *n* region (cathode). Because a silicon diode is being used, the forward voltage drop will be 0.7 V. With an applied voltage of 8.5 V and a circuit resistance of 1.2 kΩ, the circuit current will equal

$$I_F = \frac{V_S - V_{\text{diode}}}{R}$$

$$I_F = \frac{8.5\text{V} - 0.7\text{V}}{1.2\Omega}$$

$$I_F = \frac{7.8\text{V}}{1.2\text{k}\Omega}$$

$$I_F = 6.5\text{mA}$$

Reverse Biasing a Diode

Figure 2-9 shows how a junction diode can be reverse biased. Once again, the junction diode's operation is determined by the polarity of the applied voltage. In this figure the positive terminal of the applied voltage is connected to the n region of the diode, and the negative terminal of the applied voltage is connected to the p region of the diode ($+V \rightarrow n$ region, $-V \rightarrow p$ region). This applied voltage polarity will reverse bias the junction diode because free electrons in the n region are attracted and therefore travel to the positive terminal of the applied source voltage. At the same time, holes feel the attraction of the negative terminal of the applied voltage. As a result, the depletion region will increase in width until its internal voltage is equal, but opposite to, the external applied voltage. At this time the diode will stop conducting, cutting off all current flow.

In actual fact, a very small current leaks through the diode when it is reverse biased. This current is extremely small (microamps to nanoamps in value) and in most cases can be ignored. It is caused by minority carriers, which are holes in the n region and electrons in the p region, that are forced toward the junction by the applied source voltage.

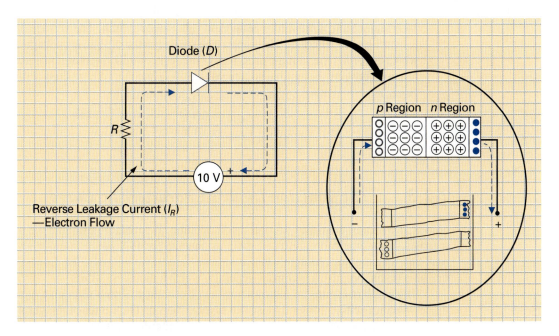

FIGURE 2-9   A Reverse-Biased P-N Junction Diode.

EXAMPLE:

Calculate the voltage drop across the diodes and the resistors in Figure 2-10.

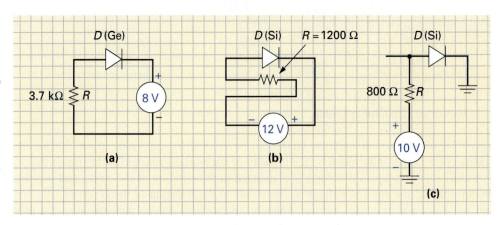

FIGURE 2-10 P-N Junction Diode Biasing Examples.

Solution:

a. The germanium diode in Figure 2-10(a) is reverse biased ($+V \rightarrow n$ region, $-V \rightarrow p$ region) and is therefore equivalent to an open switch. Because all of the applied voltage will always appear across an open in a series circuit,

$$V_{\text{diode}} = V_S = 8 \text{ V}$$
$$V_R = 0 \text{ V}$$

b. The silicon diode in Figure 2-10(b) is also reverse biased ($+V \rightarrow n$ region, $-V \rightarrow p$ region), and is therefore equivalent to an open switch. Because all of the applied voltage will always appear across an open in a series circuit,

$$V_{\text{diode}} = V_S = 12 \text{ V}$$
$$V_R = 0 \text{ V}$$

c. The silicon diode in Figure 2-10(c) is forward biased ($+V \rightarrow p$ region, $-V$ or ground $\rightarrow n$ region) and is therefore equivalent to a closed switch. The voltage drop across a forward-biased silicon diode is approximately equal to the barrier voltage of the diode, which is 0.7 V. The voltage drop across the resistor will therefore be equal to the difference between the applied source voltage (V_S) and the voltage drop across the diode (V_{diode}).

$$V_{\text{diode}} = 0.7 \text{ V}$$
$$V_R = V_S - V_{\text{diode}} = 10 \text{ V} - 0.7 \text{ V} = 9.3 \text{ V}$$

2-2-3 *The Junction Diode's Characteristic Curve*

Now that we know how the junction diode operates, it is time to examine the diode's characteristics in a little more detail. To help us analyze the P-N junction diode's voltage, current, and temperature characteristics at various values, we will plot the diode's characteristics on a graph, because a picture is generally worth a thousand words.

The graph in Figure 2-11 shows how much current will pass through a typical junction diode when it is either forward biased or reverse biased. The center of the graph, where the horizontal axis and vertical axis cross, is called the **graph origin.** This origin is a zero point for both voltage and current. For instance, voltage is plotted on the horizontal axis, with forward bias voltages (V_F) increasing positively to the right of the origin and reverse bias

Graph Origin

Center of the graph where the horizontal axis and vertical axis cross.

voltages (V_R) increasing negatively to the left of the origin or zero voltage point. Conversely, current is plotted on the vertical axis of this graph, with forward current (I_F) increasing positively above the origin and reverse current (I_R) increasing negatively below the origin or zero current point. Manufacturers of diodes create a graph like the one seen in Figure 2-11 by applying various values of forward and reverse voltages. The result is a continuous curve called a *voltage–current* or *V–I characteristic curve*. Let us now examine the forward diode characteristics (upper-right quadrant) and the reverse diode characteristics (lower-left quadrant) in more detail.

Forward Characteristics

The upper-right quadrant of the four sections in Figure 2-11 shows what forward current will pass through the diode when a forward bias voltage is applied. As you can see from the inset, the diode is forward biased by applying a positive potential to its anode and a negative potential to its cathode. To review, when the forward bias voltage exceeds the diode's internal barrier voltage, its resistance drops to almost zero, resulting in a rapid increase in forward current. In this instance, the diode is said to be ON and is equivalent to a closed switch.

These characteristics can be seen in the forward curve in Figure 2-11. Beginning at the graph origin and following the curve into the forward quadrant, you can see that the forward current through a diode is extremely small until the forward bias voltage exceeds the diode's internal barrier voltage, which for silicon is 0.7 V and for germanium is 0.3 V. Once the forward bias voltage exceeds the diode's internal barrier voltage, the forward current through the diode increases rapidly at a linear rate. The point on the forward voltage scale at

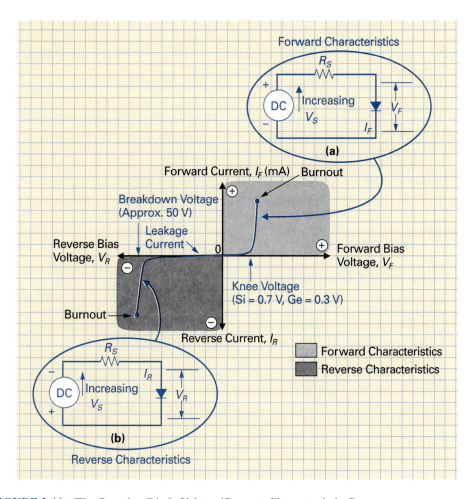

FIGURE 2-11   The Junction Diode Voltage/Current Characteristic Curve.

Knee Voltage

The point in the forward voltage scale of the V–I characteristic curve at which the curve suddenly rises and resembles the shape of a human knee.

which the curve suddenly rises resembles the shape of a human knee, which is why this point is called the **knee voltage.** This knee voltage is just another name for the diode's internal barrier voltage, which for a silicon diode is about 0.7 V.

Referring back to the linearly increasing current portion of the forward curve in Figure 2-11, you will notice that although there is a large change in forward current, the forward voltage drop across the diode remains almost constant between 0.7 V to 0.75 V.

The amount of heat produced in the diode is proportional to the value of current through the diode ($P\uparrow = I^2\uparrow \times R$). For example, an IN4001 diode, which is a commonly used low-power silicon diode, has a manufacturer's maximum forward (I_F max.) rating of 1 A. If this value of current is exceeded, the diode will begin generating more heat than it can dissipate and burn out. A series current limiting resistor (R_S) is generally always included to limit the forward current, as shown in the inset in Figure 2-11. Although the series resistor will limit forward current, it cannot prevent a damaging forward current if enough pressure or forward voltage is applied ($V\uparrow = I\uparrow \times R$).

Reverse Characteristics

The lower-left quadrant of the four sections in Figure 2-11 shows what reverse current will pass through the diode when a reverse bias voltage is applied. As you can see from the inset, a diode is reverse biased by applying a negative potential to its anode and a positive potential to its cathode. To review, when reverse biased, the internal barrier voltage of a diode will increase until it is equal to the external voltage. In this instance, current is effectively reduced to zero and the diode is said to be OFF and equivalent to an open switch.

These characteristics can be seen in the reverse curve in Figure 2-11. Beginning at the graph origin and following the curve into the reverse quadrant, you can see that the reverse current through the diode increases only slightly (approximately 100 μA). Throughout this part of the curve the diode is said to be blocking current because the leakage current is generally so small it is ignored for most practical applications. If the reverse voltage (V_R) is further increased, a point will be reached where the diode will break down, resulting in a sudden increase in current. This excessive current is due to the large external reverse bias voltage that is now strong enough to pull valence electrons away from their parent atoms, resulting in a large increase in minority carriers. The point on the reverse voltage scale at which the diode breaks down and there is a sudden increase in reverse current is called the **breakdown voltage.** Referring to the reverse curve in Figure 2-11, you can see that most silicon diodes break down as the reverse bias voltage approaches 50 V. For example, the IN4001 low-power silicon diode has a reverse breakdown voltage (which is sometimes referred to as the **Peak Inverse Voltage** or **PIV**) of 50 V listed on its manufacturer's data sheet. If this reverse bias voltage is exceeded, an avalanche of continuously rising current will eventually generate more heat than can be dissipated, resulting in the destruction of the diode.

Breakdown Voltage or Peak Inverse Voltage (PIV)

The point on the reverse voltage scale at which the diode breaks down and there is a sudden increase in the reverse current.

Specification Sheet or Data Sheet

Details the characteristics and maximum and minimum values of operation of a device.

Temperature Characteristics

As discussed previously in Chapter 1, semiconductor materials, and therefore diodes, have a negative temperature coefficient of resistance. This means as temperature increases ($T\uparrow$), their resistance decreases ($R\downarrow$). Let us now consider these temperature effects on diodes because they may be critical in some applications.

Figure 2-12 shows how the forward and reverse currents through a diode can be affected by changes in temperature. The forward voltage drop across a conducting diode is inversely proportional to temperature, which means that the voltage drop across a diode will be less at higher temperatures. This change in forward voltage drop, however, is very small and does not greatly affect the diode's operation in most applications.

On the other hand, the diode's reverse current is greatly affected by changes in temperature. To review, the leakage current in a diode is caused by the minority carriers in the n and p regions. At low temperatures, the reverse leakage current is almost zero. At room temperature,

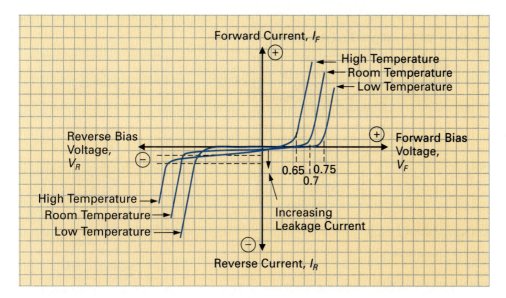

FIGURE 2-12 Temperature Effects on a Junction Diode's *V–I* Characteristic Curve.

the reverse leakage current has increased; however, it is still too small to have any adverse effects on circuit applications. At higher temperatures, however, the reverse leakage current increases to a point that the reverse-biased diode is no longer equivalent to an open switch. This means that if a circuit malfunction generated excessive heat, or if a system's cooling fan failed, the system's diodes might not switch OFF when reverse biased, causing additional system problems. As a general rule, the reverse leakage current tends to double for every temperature increase of 10°C.

2-2-4   *A Junction Diode's Specification Sheet*

A manufacturer's **specification sheet** or **data sheet** details the characteristics and maximum and minimum values of operation for a given device. Generally, engineers will study these details to determine whether a specific device can be incorporated into a circuit design. As a technician, you should be familiar with some of the basic operating limits of a device so that you can isolate component malfunctions within a circuit by determining whether the device is operating to specifications. For example, if a specific diode is placed in a circuit in which its maximum rating is exceeded in some instances, the fault is not with the component but with the circuit design. In this situation, simply replacing the component will not solve the problem. The device will have to be replaced with a diode that has a greater maximum rating.

Figure 2-13 (see page 36) shows the specific specification sheet for the IN4001 through IN4007 series of silicon junction diodes. This data sheet serves as a good example of the amount of detail that is normally supplied to design engineers by device manufacturers. As a technician, you would mainly be interested in the diode's maximum reverse voltage, maximum forward current, and average voltage drop. For the IN4001, these values are

Maximum reverse voltage $(V_R) = 50$ V

Maximum forward current $(I_O) = 1$ A

Average voltage drop $(V_{F(av)}) = 0.8$ V

Specification Sheet or Data Sheet

Details the characteristics and maximum and minimum values of operation of a device.

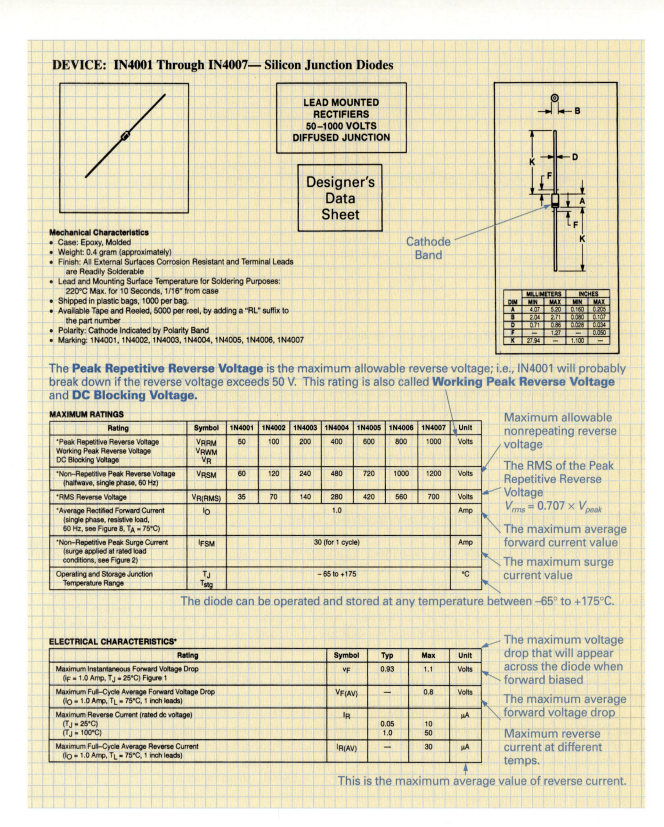

FIGURE 2-13 Specific Specification Sheet for the IN4001 through IN4007 Junction Diodes. (Copyright of Motorola. Used by permission.)

■ EXAMPLE:

Would any of the maximum ratings of the IN4001 diode in Figure 2-14 be exceeded?

FIGURE 2-14 An IN4001 Diode Example Circuit.

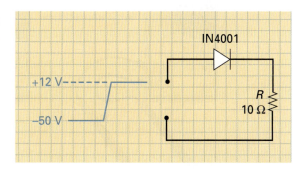

■ Solution:

The maximum reverse voltage for an IN4001 is 50 V, and since this reverse voltage is not being exceeded by the applied voltage, the diode is being operated within this specification.

The maximum forward current will be:

$$I = \frac{V_S - V_{\text{diode}}}{R}$$

$$I = \frac{12V - 0.8V}{10\Omega} = 1.12A$$

Because this is in excess of the 1 A maximum listed in the specification sheet, the diode will more than likely burn out due to excessive current and, therefore, heat.

2-2-5   *Junction Diode Applications*

It would be safe to say that the junction diode is used in almost every electronic and electrical system. Like resistors and capacitors, the list of diode applications is endless. However, certain uses of the diode predominate. As you proceed through this text you will see the junction diode used in a wide variety of analog and digital circuit applications. The junction diode is most used in a basic analog circuit and a basic digital circuit.

2-2-6   *Testing Junction Diodes*

Diodes that are operated beyond their maximum forward and reverse ratings will, more than likely, malfunction. These malfunctions can result in one of two types of failure. The diode may burn out and then act as a permanently open switch (less common) or effectively melt the semiconductor material and act as a permanently closed switch (more common). A defective diode will not be able to be switched ON or OFF. Instead, it will remain either permanently OFF or permanently ON.

Figure 2-15 shows how an ohmmeter can be used to check whether a diode has malfunctioned or is operating correctly. A good diode should display a very low resistance when it is biased ON and a very high resistance when it is biased OFF. Figure 2-15(a) shows how a diode can be forward biased by an ohmmeter's internal battery (+ lead to anode, – lead to cathode) and, if good, should display a low value of resistance (typically less than 10 Ω). Figure 2-15(b) shows how the diode is then flipped over and reverse biased by the ohmmeter's internal battery (+ lead to cathode, – lead to anode). If the diode is good, the ohmmeter should display a very

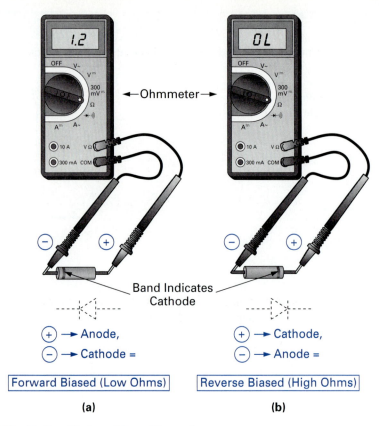

FIGURE 2-15 **Testing Diodes with an Ohmmeter.**

high resistance (typically, greater than 1000 MΩ). Because this value is generally off the ohmmeter's scale, the display will probably show OL or OR. This is what you should expect because it means that the reverse-biased diode's resistance is so high that it is over the range, or off the scale, selected.

A diode that has been damaged and is permanently shorted between its anode and cathode will display a LOW resistance on the ohmmeter when it is both forward and reverse biased by the ohmmeter. Finding the shorted diode in a circuit may not be the end of your troubles, because one of the neighboring components may have initially caused the diode's ratings to be exceeded. Furthermore, the shorted diode may have allowed a damaging current to pass through to another part of the circuit, causing additional damage.

A diode that has been damaged and is permanently opened between its anode and cathode will display a HIGH resistance on the ohmmeter when it is both forward and reverse biased by the ohmmeter. Once again, the destruction of the diode generally occurs when the maximum voltage or current ratings of the diode have been exceeded. Remember that the diode may not have just malfunctioned on its own; the problem may have been caused by one of the neighboring components. In this instance, however, an open diode rarely damages other associated components because current is blocked by the fuse-like action of the diode's destruction.

As you proceed through this text, you will see diodes used in a variety of circuit applications, and you will see what symptoms a malfunctioning diode will display in these circuits.

■ EXAMPLE:

Figure 2-16 shows the results from testing four diodes with an ohmmeter. Which of the diodes tested good, which tested bad, and for what reason?

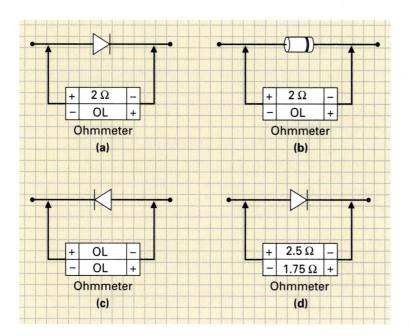

FIGURE 2-16 **Testing Diode Examples.**

■ *Solution:*

a. The diode in Figure 2-16(a) has a low forward resistance and a high reverse resistance and is switching ON and OFF correctly.

b. The diode in Figure 2-16(b) has a low forward resistance and a high reverse resistance and is switching ON and OFF correctly.

c. The diode in Figure 2-16(c) has a high forward resistance and a high reverse resistance and is not switching ON and OFF. It seems to be remaining permanently OFF, indicating an open between its anode and cathode terminals.

d. The diode in Figure 2-16(d) has a low forward resistance and a low reverse resistance and is not switching ON and OFF. It seems to be remaining permanently ON, indicating a short between its anode and cathode terminals.

SELF-TEST EVALUATION POINT FOR SECTION 2-2

Use the following questions to test your understanding of Section 2–2.

1. What is the typical forward voltage drop across a silicon diode?

2. What barrier voltage has to be overcome in order to forward bias a silicon diode?

3. How many P-N junctions are within a junction diode?

4. What is PIV, and what would it be for a typical silicon diode?

5. As the temperature of a diode increases, the forward voltage drop _____. (increases or decreases)

Multiple-Choice Questions

1. What is the barrier voltage for a silicon junction diode?
   a. 0.3 V   **b.** 0.4 V   **c.** 0.7 V   **d.** 2.0 V

2. Which of the following junction diodes are forward biased?
   a. Anode = +7 V, cathode = +10 V
   b. Anode = +5 V, cathode = +3 V
   c. Anode = +0.3 V, cathode = +5 V
   d. Anode = −9.6 V, cathode = −10 V

3. The junction diode _____ current when it is forward biased, and _____ current when it is reverse biased.
   a. Blocks, conducts   **c.** Blocks, prevents
   b. Conducts, passes   **d.** Conducts, blocks

4. The *n*-type region of a junction diode is connected to the _____ terminal and the *p*-type region is connected to the _____.
   a. Cathode, anode   **b.** Anode, cathode

5. When reverse biased, a junction diode has a leakage current passing through it, which is typically measured in:
   a. Amps   **b.** Milliamps   **c.** Microamps   **d.** Kiloamps

6. The black band on a diode's package is always closest to the _____.
   a. Anode   **c.** *p*-type material
   b. Cathode   **d.** Both (a) and (c) are true

7. When current dramatically increases, the voltage point on the diode's forward *V–I* characteristic curve is called the:
   a. Breakdown voltage   **c.** Barrier voltage
   b. Knee voltage   **d.** Both (b) and (c) are true

8. When current dramatically increases, the voltage point on the diode's reverse *V–I* characteristic curve is called the:
   a. Breakdown voltage   **c.** Barrier voltage
   b. Knee voltage   **d.** Both (b) and (c) are true

9. What resistance should a good diode have when it is reverse biased?
   a. Less than 10 Ω   **c.** Between 120 Ω and 1.2 kΩ
   b. More than 1000 MΩ   **d.** Both (b) and (c) are true

10. What resistance should a good diode have when it is forward biased?
    a. Less than 10 Ω
    b. More than 1000 MΩ
    c. Between 120 Ω and 1.2 kΩ
    d. Both (b) and (c) are true

Practice Problems

11. Which of the silicon diodes in Figure 2-17 are forward biased and which are reverse biased?
12. Calculate I_F for the circuits in Figure 2-18.
13. What would be the voltage drop across each of the diodes in Figure 2-18?
14. What would be the voltage drop across each of the resistors in Figure 2-18?

Troubleshooting Questions

15. Referring to the switch encoder circuit in Figure 2-19, first determine what digital codes will be generated at *A*, *B*, and *C* for each of the switch positions 1, 2 and 3. Next, determine what problems would occur for each of the following circuit malfunctions or conditions:
    a. What would happen if D_1 were to open permanently?
    b. Would a 200 mA fuse protect the +5 V supply voltage?

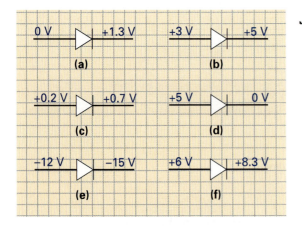

FIGURE 2-17   **Biased Junction Diodes.**

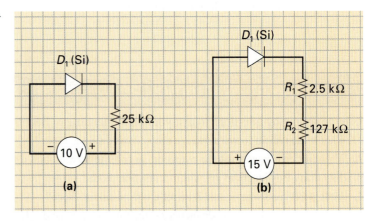

FIGURE 2-18   **Forward Current Examples.**

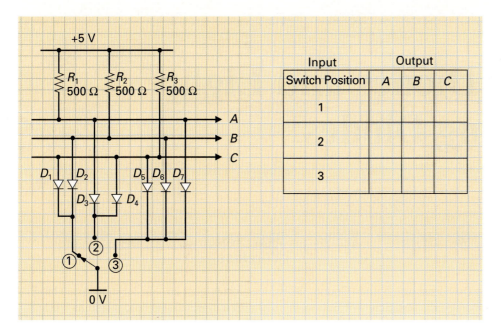

Input	Output		
Switch Position	A	B	C
1			
2			
3			

FIGURE 2-19 **A Switch Encoder Circuit.**

Zener Diodes and Light-Emitting Diodes

3

Wireless

Lee De Forest was born in Council Bluffs, Iowa on August 26, 1873. He obtained his Ph.D. from Yale Sheffield Scientific School in 1899 with a thesis that is recognized as the first paper on radio communications in the United States.

De Forest was an astounding practical inventor; however, he wasn't a good businessman and many of his inventions were stolen and exploited by business partners.

In 1902, he invented an electrolytic radio detector and an alternating-current transmitter while working for the Western Electric Company. He also developed an optical sound track for movies, which was rejected by film producers. Later another system was used that was based on De Forest's principles.

In 1907, De Forest patented his audion detector or triode (three-electrode) vacuum tube. This tube was a more versatile vacuum tube than the then-available diode (two-electrode) vacuum tube, or *Fleming valve*, invented by English electrical engineer, Sir John Fleming. By adding a third electrode, called a *control grid*, De Forest made amplification possible because the tube could now be controlled. The triode's ability to amplify as well as rectify led to the development of radio communications, or wireless as it was called, and later to television. De Forest also discovered that by cascading (connecting end-to-end) amplifiers, a higher gain could be attained. This ability led to long-distance communications because weak long-range input signals could be increased in magnitude to a usable level.

In 1905, he began experimenting with speech and music broadcasts. In 1910, De Forest transmitted the singing voice of Enrico Caruso and, in so doing, became one of the pioneers of radio broadcasting.

Slowly the usefulness of the triode as a generator, amplifier, and detector of radio waves became apparent. However, it was not until World War I that the audion became an invaluable electronic device and was manufactured in large quantities.

As well as working on the technical development of radio, De Forest gave many practical public demonstrations of wireless communications. He became internationally renowned and was often referred to as "the father of radio." Before his fame and fortune, however—just before everyone became aware of the profound effect that his audion tube would have on the world—an incident occurred. While trying to sell stock in his company, De Forest was arrested on charges of fraud and his device, which was to launch a new era in electronics and make possible radio and television, was at this time called "a strange device like an incandescent lamp . . . which has proved worthless."

Introduction

In the previous chapter we studied the *basic P-N junction diode.* In 1960, many electronic system manufacturers began to use the semiconductor P-N junction diode instead of the vacuum tube diode in low-power and low-frequency applications. Like all inventions, once the basic P-N junction diode appeared on the market and was in wide use, it inspired others to investigate all of its possibilities. Research and development into all semiconductor, or solid state, devices mushroomed at this time, resulting in a variety of new semiconductor devices, including several different types of diodes. Today, there are many different types of semiconductor diodes available, all of which have different characteristics and are suited to different applications. After the basic P-N junction diode, the two most widely used diodes are the zener diode and the light-emitting diode.

In this chapter, we will be examining the operation and characteristics of the zener diode and light-emitting diode in preparation for their many different circuit applications. For example, in Chapter 4 we will see how the basic P-N junction, zener, and light-emitting diode can be used to construct dc power supply circuits. In later chapters, we will see how these and other diodes are used in a variety of circuit applications such as rectifying, regulating, detecting, clipping, clamping, multiplying, switching, displaying, and oscillating.

3-1 THE ZENER DIODE

In the previous chapter we discussed the forward and reverse characteristics of a basic P-N junction diode. When forward biased, a basic junction diode will turn ON and be equivalent to a closed switch. When reverse biased, a basic junction diode will turn OFF and be equivalent to an open switch. In all basic P-N junction diode circuit applications, the external bias voltage was always kept below the maximum forward current rating and reverse breakdown voltage rating; otherwise, a damaging value of current would result.

It was American inventor Clarence Zener who first began investigating in detail the reverse breakdown of a diode. He discovered that the basic P-N junction diode would conduct a high reverse current when a sufficiently high reverse bias voltage was applied across its terminals. This large external reverse voltage would cause an increase in the number of minority carriers in the *n* and *p* regions as valence electrons are pulled away from their parent atoms. Clarence Zener was fascinated by the reverse breakdown characteristics of a basic P-N junction diode and, after many years of study, developed a special diode that could handle high values of reverse current and still safely dissipate any heat that was generated. These special diodes—constructed to operate at voltages that are equal to or greater than the reverse breakdown voltage rating—were named **zener diodes** after their inventor.

Let us now examine the zener diode's schematic symbol and package types and then the operation and characteristics of the zener diode.

Zener Diode

Diodes constructed to operate at voltages that are equal to or greater than the reverse breakdown voltage rating.

3-1-1 *Zener Diode Symbol and Packages*

Figure 3-1(a) shows the two schematic symbols used to represent the zener diode. As you can see, the zener diode symbol resembles the basic P-N junction diode symbol in appearance; however, the zener diode symbol has a zig-zag bar instead of the straight bar. This zig-zag bar at the cathode terminal is included as a memory aid because it is "Z"-shaped and will always remind us of zener.

Figure 3-1(b) shows two typical low-power zener diode packages and one high-power zener diode package. The surface-mount low-power zener package has two metal pads for direct mounting to the surface of a circuit board; the axial-lead low-power zener package has

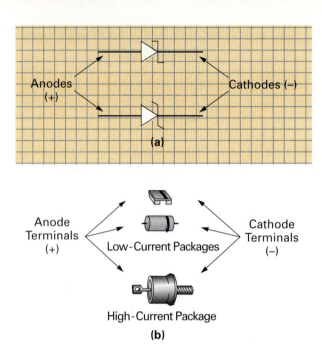

Anodes (+) Cathodes (−)

(a)

Anode Terminals (+) Low-Current Packages Cathode Terminals (−)

High-Current Package

(b)

FIGURE 3-1   The Zener Diode. (a) Zener Diode Schematic Symbols. (b) Packages.

the zener mounted in a glass or epoxy case. The high-power zener package is generally stud mounted and contained in a metal case. These packages are identical to the basic P-N junction diode low-power and high-power packages. Once again, a band or stripe is used to identify the cathode end of the zener diode in the low-power packages, whereas the threaded terminal of a high-power package is generally always the cathode.

3-1-2   *Zener Diode Voltage–Current (V–I) Characteristics*

Figure 3-2 shows the *V–I* (voltage–current) characteristic curve of a typical zener diode. This characteristic curve is almost identical to the basic P-N junction diode's characteristic curve. For example, when forward biased at or beyond 0.7 V, the zener diode will turn ON and be equivalent to a closed switch; whereas, when reverse biased, the zener diode will turn OFF and be equivalent to an open switch. The main difference, however, is that the zener diode has been specifically designed to operate in the reverse breakdown region of the curve. This is achieved, as can be seen in the inset in Figure 3-2, by making sure that the external bias voltage applied to a zener diode will not only reverse bias the zener diode (+ → cathode, − → anode) but also be large enough to drive the zener diode into its reverse breakdown region. Because the zener diode will generally always be operated in the reverse breakdown region, let us now discuss this reverse breakdown region of the *V–I* curve in a little more detail.

Zener Reverse Breakdown Region

Referring to the zener reverse breakdown region of the curve in Figure 3-2, you can see that when the external reverse bias voltage is high enough, the zener diode will break down and conduct a high reverse current. This reverse current through the diode will increase at a rapid rate as the external reverse bias voltage is increased. The rapid increase in current through the diode is caused by the zener diode's sudden drop in resistance because the external reverse-bias voltage is now large enough to tear valence electrons away from their parent atoms.

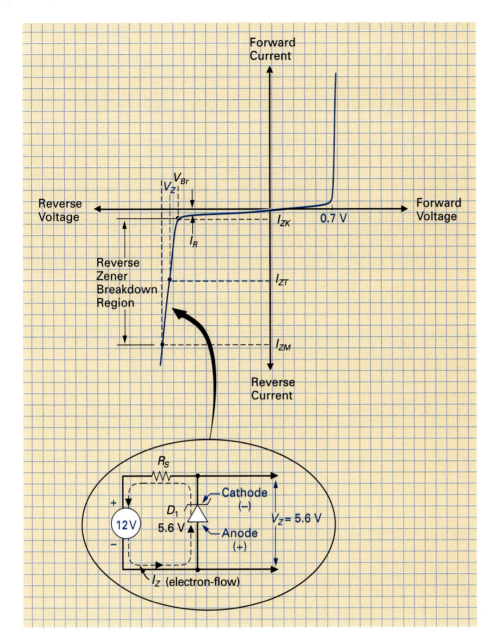

FIGURE 3-2 **The Zener Diode Voltage-Current Characteristic Curve.**

As the reverse voltage across the zener diode is increased from the graph origin (which represents 0 volts), the value of **reverse leakage current (I_R)** begins to increase. The reverse zener breakdown region of the curve begins when the current through the zener suddenly increases. This point on the reverse bias voltage scale is called the **zener breakdown voltage (V_{Br}).** At this knee-shaped point, a minimum value of reverse current, known as the **zener knee current (I_{ZK}),** will pass through the zener. If the applied voltage is increased beyond V_{Br}, the value of **zener current (I_Z)** will increase. The voltage drop developed across the zener when it is being operated in the reverse zener breakdown region is called the **zener voltage (V_Z).** Comparing the voltage developed across the zener (V_Z) to the value of current through the zener (I_Z), you may have noticed that *the voltage drop across a zener diode (V_Z) remains almost constant when it is operated in the reverse zener breakdown region, even though current through the zener (I_Z) can vary considerably. This ability of the zener diode to maintain a relatively constant voltage regardless of variations in zener current is the key characteristic of the zener diode.*

Reverse Leakage Current (I_R)
The undesirable flow of current through a device in the reverse direction.

Zener Breakdown Voltage (V_{Br})
The point on the reverse bias voltage scale where the current through the zener suddenly increases.

Zener Knee Current (I_{ZK})
The knee-shaped point on the V–I curve at which a minimum value of reverse current will pass through the diode.

Zener Current (I_Z)
The current through the zener diode.

Zener Voltage (V_Z)
The voltage drop across the zener when it is being operated in the reverse zener breakdown region.

Zener Voltage Rating

The breakdown voltage of a zener diode is controlled by adjusting the doping to control the width of the zener diode's depletion layer. By controlling the depletion layer's width, a manufacturer can create a zener that will break down at a specific voltage. Generally, manufacturers rate zener diodes based on their zener voltage (V_Z) rather than on their breakdown voltage (V_{Br}). A wide variety of zener diode voltage ratings are available, ranging from 1.8 V to several hundred volts. For example, many of the frequently used low-voltage zener diodes have ratings of 3.3 V, 4.7 V, 5.1 V, 5.6 V, 6.2 V, and 9.1 V. All of these zener voltage ratings are nominal values that represent the reverse voltage that will be developed across the zener diode when a specified value of current, called the **zener test current** (I_{ZT}), is passing through the zener diode. As can be seen in Figure 3-2, the zener test current is typically at a midpoint in the reverse zener breakdown region, midway between the low zener knee current (I_{ZK}) and the *maximum zener current* (I_{ZM}).

Like other components such as resistors and capacitors, a zener diode's voltage rating (V_Z) has a specified value and a tolerance. The following example shows how to calculate the tolerance for a typical zener diode.

> **Zener Test Current (I_{ZT})**
>
> The specified value of current that is passing through the zener diode when the reverse voltage rating is measured.

◼ EXAMPLE:

What will be the zener voltage range for a 5.6 V $\pm$ 10% zener diode?

◼ *Solution:*

$$10\% \text{ of } 5.6 \text{ V} = 0.56 \text{ V}$$
$$5.6 \text{ V} - 0.56 \text{ V} = 5.04 \text{ V}$$
$$5.6 \text{ V} + 0.56 \text{ V} = 6.16 \text{ V}$$
$$5.04 \text{ volts to } 6.16 \text{ volts}$$

Figure 3-3 shows the small *change in zener voltage* (ΔV_Z) and the corresponding larger *change in zener current* (ΔI_Z).

Zener Power Dissipation

> **Maximum Power Dissipation (P_D)**
>
> Maximum power that the zener diode can safely dissipate.

Zener diodes have a **maximum power dissipation (P_D)** rating that indicates the maximum power that the zener diode can safely dissipate. A variety of zener diodes are available with maximum power ratings from several hundred milliwatts up to 100 watts. However, most of the more frequently used zener diodes have maximum power ratings of 500 mW and 1 W. This power rating is generally given for a specific operating temperature of 50°C. Like most semiconductors, zener diodes have a negative temperature coefficient of resistance, which means as temperature increases their resistance will decrease, resulting in an increase in current and therefore heat. This means that the amount of power that a zener diode can safely dissipate will decrease as the operating temperature increases.

Zener Maximum Reverse Current

If the maximum zener current (I_{ZM}) of a zener diode is exceeded, the diode will more than likely be destroyed because this value of current will generate more heat than the zener can safely dissipate (P_D). Manufacturers of zener diodes generally specify this value of maximum zener current (I_{ZM}) in the device data sheet. If a value of maximum zener current (I_{ZM}) is not specified, it can be calculated by using the zener diode's power dissipation rating (P_D), the zener voltage rating (V_R), and the power formula,

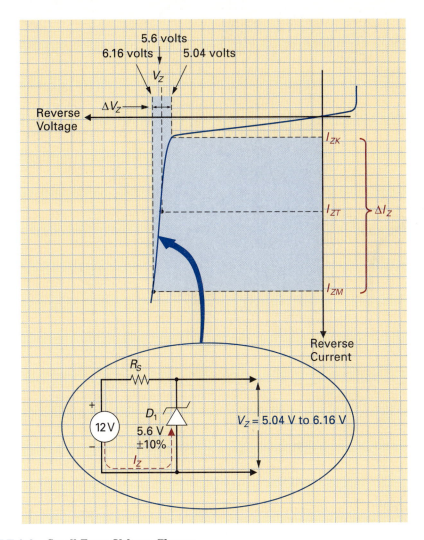

FIGURE 3-3 Small Zener Voltage Change.

$$I_{ZM} = \frac{P_D}{V_Z}$$

■ **EXAMPLE:**

Suppose the zener diode shown in the inset in Figure 3-4 has a power rating of 500 mW, and a zener voltage of 5.6 V with a ±10% tolerance. Calculate the value of maximum zener current.

■ *Solution:*

As previously calculated, the maximum value of zener voltage would be 5.6 V, plus 10% of 5.6 V, which equals 6.16 V. Now that we have the maximum voltage limit and maximum power dissipation value, we can use the power formula to calculate the maximum zener current

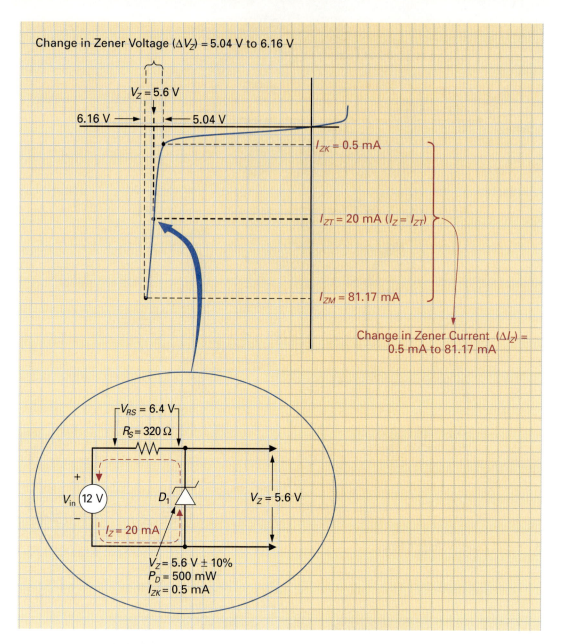

Change in Zener Voltage (ΔV_Z) = 5.04 V to 6.16 V

$V_Z = 5.6$ V

6.16 V → | ← 5.04 V

$I_{ZK} = 0.5$ mA

$I_{ZT} = 20$ mA ($I_Z = I_{ZT}$)

$I_{ZM} = 81.17$ mA

Change in Zener Current (ΔI_Z) = 0.5 mA to 81.17 mA

$V_{RS} = 6.4$ V

$R_S = 320 \, \Omega$

V_{in} (12 V)

D_1

$V_Z = 5.6$ V

$I_Z = 20$ mA

$V_Z = 5.6$ V ± 10%
$P_D = 500$ mW
$I_{ZK} = 0.5$ mA

FIGURE 3-4   **Zener Circuit Voltage and Current Calculations.**

$$I_{ZM} = \frac{P_D}{V_Z}$$
$$I_{ZM} = \frac{500\text{mW}}{6.16\text{V}}$$
$$I_{ZM} = 81.17\text{mA}$$

Referring to the reverse zener breakdown region in Figure 3-4, you can see I_{ZM} shown at the lower end of the reverse characteristic curve.

Zener Circuit Voltage and Current

In the example in the inset in Figure 3-4, a 5.6-V zener diode is connected across a 12-V source. The 12-V input voltage polarity is connected so that it reverse biases the zener diode.

CHAPTER 3 / ZENER DIODES AND LIGHT-EMITTING DIODES

In order for this circuit to operate properly, the input voltage (V_{in}) must be higher than the breakdown voltage of the zener diode (V_Z). In this instance, the input voltage of 12-V is large enough to send the 5.6-V zener into its reverse breakdown region. As far as the voltage drops across each of the components in the inset in Figure 3-4, the voltage across the zener diode will be equal to the zener diode's voltage rating (5.6 V), and the voltage developed across the series resistor (V_{RS}) will be equal to the difference between the diode's zener voltage and the input voltage. Therefore,

$$V_{RS} = V_{in} - V_Z$$

$$= 12\,V - 5.6\,V$$
$$= 6.4\,V$$

As we know, once the series resistor's voltage drop (V_{RS}) and resistance (R_S) are known, we can calculate current.

$$I = \frac{V_{RS}}{R_S}$$
$$I = \frac{6.4V}{320\Omega}$$
$$I = 20\,mA$$

Because the series resistor (R_S) and the zener diode (D_1) are in series with one another, the current through the zener diode (I_Z) will be the same as the current through the series resistor ($I_{RS} = I_Z = 20$ mA).

Zener Impedance

Zener impedance (Z_Z) is the opposition offered by a zener diode to current. A manufacturer calculates the value of zener impedance by varying the zener current above and below the zener test current value and then monitoring the change in zener voltage. Once the change in zener current (ΔI_Z) and corresponding change in zener voltage (ΔV_Z) are known, Ohm's law can be used to calculate zener impedance with the following formula:

$$Z_Z = \frac{\Delta V_Z}{\Delta I_Z}$$

Zener Impedance (Z_Z)
Opposition offered by a zener diode to current.

Applying this formula to the example in the inset in Figure 3-2, we can calculate the zener diode's impedance. Previously, we determined that the change in zener voltage (ΔV_Z) is from 5.04 V to 6.16 V. If the change in zener current (ΔI_Z) is from 0.5 mA (zener knee current, I_{ZK}) to 81.17 mA (maximum zener current, I_{ZM}), the zener diode's impedance will be

$$Z_Z = \frac{\Delta V_Z}{\Delta I_Z}$$
$$Z_Z = \frac{6.16V - 5.04V}{81.17mA - 0.5mA}$$
$$Z_Z = \frac{1.12V}{80.67mA}$$
$$Z_Z = 13.9\Omega$$

3-1-3 *Zener Diode Data Sheets*

Like the basic P-N junction diode's data sheet, a zener diode data sheet details the characteristics and maximum and minimum values of operation for a given device. Figure 3-5 contains the data sheet for a series of silicon zener diodes. By studying this specification sheet

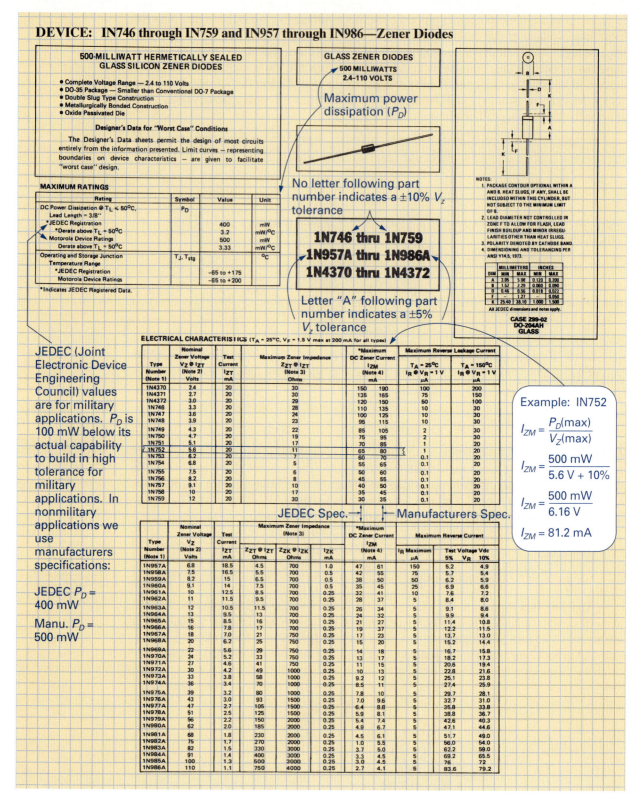

FIGURE 3-5 Device Data Sheet for a Series of Zener Diodes. (Copyright of Motorola. Used by permission.)

along with the associated notes, you will be able to find the characteristics and maximum and minimum ratings previously discussed. Remember that, as a technician, you should be familiar with some of the basic operating limits of a device so that you can isolate component malfunctions within a circuit by determining whether the device is operating to specifications. For example, the 5.6 V IN752 zener diode has the following specifications when operated at a zener test current (I_{ZT}) of 20 mA:

Zener voltage (V_Z) = 5.6 V $\pm$ 10%
Zener impedance (Z_Z) = 11 Ω
Manufacturer's recommended maximum zener current (I_{ZM}) = 80 mA
Manufacturer's recommended power dissipation (P_D) = 500 mW

3-1-4   *Zener Diode Applications*

Like the basic P-N junction diode, the zener diode is used in almost every electronic and electrical system. Although the zener diode can be forward and reverse biased, it is most widely used in applications where it is continually reverse biased and operating in the reverse zener breakdown region. As you proceed through this text, you will see the zener diode used in a wide variety of analog and digital circuit applications. For now, let us see how the junction diode is used in a basic circuit.

A Voltage Regulator

All electronic systems that use the 120-V ac power from the wall outlet as a power source have an internal dc power supply that converts the ac input into the needed dc supply voltages for all of the circuits within the system. The ac from the electric company fluctuates during the day as the demand by consumers changes. For example, in the early morning hours when demand is low, the ac input could be as high as 125 V ac, whereas at midday when the demand is at its peak, the ac is being heavily loaded and the voltage could be pulled down to as low as 105 V ac. The circuits within electronic systems require dc power supplies that have a very tight tolerance. For example, many digital electronic circuits use a 5-V dc power supply that must only vary plus or minus a quarter of a volt (4.75 V to 5.25 V). If the dc supply voltage exceeds this tolerance, the circuit may not operate properly or be damaged.

A circuit is therefore needed that will stabilize or regulate the input voltage to provide an unvarying output voltage. This is achieved with a **voltage regulator circuit,** which maintains the output voltage of a voltage source constant despite variations in the input voltage and the load resistance. Figure 3-6(a) shows how a zener diode can be used to regulate a dc source voltage, which varies between 20 V and 30 V, to produce an unvarying 12-V output. To begin with, the zener diode must be connected so that the polarity of the source voltage reverse biases the diode. Second, the source voltage must exceed the zener diode's reverse breakdown voltage (V_Z) in order to send the zener into its reverse breakdown region. A series resistor (R_S) is included to allow enough current to pass through the diode so that it operates within its reverse zener breakdown region and yet limit current so that it does not exceed the maximum zener current. Because R_S and the zener diode are in series with one another, the current through both devices can be calculated with the formula:

Voltage Regulator Circuit

Maintains the output voltage of a voltage source constant despite variations in the input voltage and load resistance.

$$I_s = \frac{V_{\text{in}} - V_Z}{R_S}$$

I_S = Series Current
V_{in} = Input Source Voltage
V_Z = Zener Voltage
R_S = Series Resistor Value

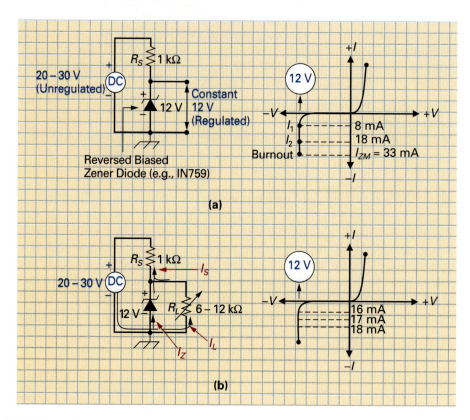

FIGURE 3-6 The Zener Diode in a Voltage Regulator Circuit. (a) Regulating the 12-V Output Despite Variations in the Source Voltage. (b) Regulating the 12-V Output Despite Variations in the Load Resistance.

In the example circuit in Figure 3-6(a), the input source voltage can vary anywhere from 20 V to 30 V. Let us calculate the value of series current for the two extreme low- and high-input voltages:

When $V_{in} = 20$ V:

$$I_S = \frac{V_{in} - V_Z}{R_S}$$

$$I_S = \frac{20V - 12V}{1 \text{ k}\Omega} = 8 \text{ mA}$$

When $V_{in} = 30$ V:

$$I_S = \frac{V_{in} - V_Z}{R_S}$$

$$I_S = \frac{30V - 12V}{1k\Omega} = 18mA$$

As the dc input voltage increases and decreases, it causes an increase and decrease in the dc current through the zener diode and the series resistor. Referring to the V–I curve in Figure 3-6(a), you can see that because the zener is operating within its reverse break-down region, a large swing in the dc input voltage (20 V to 30 V) will only result in a change in zener current (8 mA to 18 mA) and not in zener voltage. The zener voltage, which is applied to the output and is therefore the regulated output voltage, remains constant at 12 V.

The zener diode is able to maintain the voltage drop across its terminals constant by continually changing its resistance or impedance (Z_Z) in response to a change in input voltage. For example, an input voltage increase ($V_{in}\uparrow$) will result in a circuit current increase and zener current increase ($I_Z\uparrow$), which will be countered by an equal but opposite zener diode impedance decrease ($Z_Z\downarrow$). An input voltage decrease ($V_{in}\downarrow$) will result in a circuit current decrease and zener current decrease ($I\downarrow$), which will be countered by a zener diode impedance increase ($Z_Z\uparrow$). Therefore, a change in zener diode current (I_Z) is always accompanied by an equal but opposite change in zener diode impedance (Z_Z). It is this equal but opposite change in I_Z and Z_Z that causes a cancellation, resulting in a constant output zener voltage. Because the zener diode's voltage remains constant due to its changing impedance, all of the input voltage change will appear across the series resistor, which has a fixed or constant resistance. In summary:

When $V_{in}\downarrow$ **to 20 V:**
$I_Z\downarrow$ to 8 mA, causes $Z_Z\uparrow$ to 1,500 Ω ($Z_Z = V_Z/I_Z$ = 12 V/8 mA = 1,500 Ω) and therefore V_Z remains constant.

$$V_Z = I_Z \times Z_Z = 8 \text{ mA} \times 1,500 \text{ Ω} = 12 \text{ V} \qquad (V_Z = I_Z\downarrow \times Z_Z\uparrow)$$
$$V_{RS} = V_{in} - V_Z = 20 \text{ V} - 12 \text{ V} = 8 \text{ V}$$

When $V_{in}\uparrow$ **to 30 V:**
$I_Z\uparrow$ to 18 mA, causes $Z_Z\downarrow$ to 666.7 Ω ($Z_Z = V_Z/I_Z$ = 12 V/18 mA = 666.7 Ω) and therefore V_Z remains constant.

$$V_Z = I_Z \times Z_Z = 18 \text{ mA} \times 666.7 \text{ Ω} = 12 \text{ V} \qquad (V_Z = I_Z\uparrow \times Z_Z\downarrow)$$
$$V_{RS} = V_{in} - V_Z = 30 \text{ V} - 12 \text{ V} = 18 \text{ V}$$

As mentioned in the beginning of this section, *a voltage regulator circuit must maintain the output voltage of a voltage source constant despite variations in the input voltage and the load resistance.* Up to now, we have seen voltage regulation from only one side of the coin. We have seen how the voltage regulator circuit maintains the output voltage constant despite variations in the input voltage. However, as the voltage regulator circuit definition states, it must also maintain the output voltage constant despite changes in load resistance.

In Figure 3-6(b), a load resistance has been connected to the output of the voltage regulator circuit. This load resistance represents the overall resistance of a system, circuit, or device when it is connected to the output of the regulator circuit. The load, which could be a television, for example, is represented in Figure 3-6(b) as a variable resistor. This is because most loads do change their resistance when they are in operation. For example, a television draws more current, and therefore has a lower resistance, when its volume or brightness controls are turned up. Because the voltage drop across a load is determined by the resistance of the load, a continual change in load resistance would normally cause a continual change in voltage drop and therefore supply voltage. If the load was connected directly across the input voltage without the regulator circuit, the voltage supplied would change continuously with every change in load resistance. Our zener regulator circuit now has two variables to regulate—a continuous change in input voltage (between 20 V to 30 V) and a continuous change in output load resistance (between 6 kΩ to 12 kΩ).

To operate properly, the load, which in this example we will think of as a pocket television, must be supplied with a constant 12-V power supply. Before beginning our calculations, let us study the basics of the circuit in Figure 3-6(b). Because the zener diode and load resistance are in parallel with one another, the current through the zener diode (I_Z) will combine with the current through the load (I_L) to form the total series current (I_S), which will all pass through the series resistor (R_S). Therefore,

$$I_Z + I_L = I_S$$

Let us now calculate all the possible combinations of maximum and minimum values of input voltage and load resistance:

When $V_{in} = 20$ V: $V_{RS} = V_{in} - V_Z = 20$ V $- 12$ V $= 8$ V, and therefore

$$I_{RS} = \frac{V_{RS}}{R_S} = \frac{8\text{V}}{1\text{ k}\Omega} = 8\text{ mA}$$

$$R_L = 6\text{ k}\Omega$$

$$I_L = \frac{V_L}{R_L} = \frac{12\text{V}}{6\text{ k}\Omega} = 2\text{ mA}$$

$$I_L = I_S - I_L = 8\text{ mA} - 2\text{ mA} = 6\text{ mA}$$

$$R_L = 12\text{ k}\Omega$$

$$I_L = \frac{V_L}{R_L} = \frac{12\text{V}}{12\text{ k}\Omega} = 1\text{ mA}$$

$$I_Z = I_S - I_L = 8\text{ ma} - 1\text{ mA} = 7\text{ mA}$$

When $V_{in} = 30$ V: $V_{RS} = V_{in} - V_Z = 30$ V $- 12$ V $= 18$ V, and therefore

$$I_{RS} = \frac{V_{RS}}{R_S} = \frac{18\text{V}}{1\text{ k}\Omega} = 18\text{ mA}$$

$$R_L = 6\text{ k}\Omega$$

$$I_L = \frac{V_L}{R_L} = \frac{12\text{V}}{6\text{ k}\Omega} = 2\text{ mA}$$

$$I_Z = I_S - I_L = 18\text{ mA} - 2\text{ mA} = 16\text{ mA}$$

$$R_L = 12\text{ k}\Omega$$

$$I_L = \frac{V_L}{R_L} = \frac{12\text{V}}{12\text{ k}\Omega} = 1\text{ mA}$$

$$I_Z = I_S - I_L = 18\text{ mA} - 1\text{ mA} = 17\text{ mA}$$

As you can see from these calculations, the zener voltage, and therefore the output load voltage (because the zener and load are in parallel), is kept at a constant 12 V despite changes in the input voltage and the output load resistance. Voltage regulator circuits will be discussed in more detail in the following dc power supply circuits chapter and in later circuits chapters.

3-1-5   *Testing Zener Diodes*

Because a zener diode is designed to conduct in both directions, we cannot test it with the ohmmeter as we did the basic P-N junction diode. The best way to test a zener diode is to connect the voltmeter across the zener while it is in circuit and power is applied, as seen in Figure 3-7. If the voltage across the zener is at its specified voltage, then the zener is func-

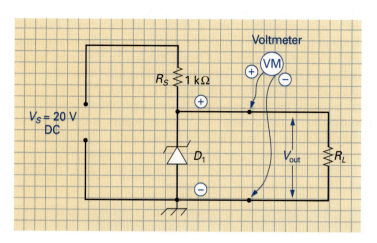

FIGURE 3-7   Zener Diode Testing.

tioning properly. If the voltage across the zener is not at the nominal value, then the following checks should be made:

1. Check the source input voltage. If this voltage (V_{in}) does not exceed the zener voltage (V_Z), the zener diode will not be at fault because the source voltage is not large enough to send the zener into its reverse breakdown region.

2. Check the series resistor (R_S) to determine that it has not opened or shorted. An open series resistor will have all of the input voltage developed across it and there will be no voltage across the zener. A shorted series resistor will not provide any current-limiting capability and the zener could possibly burn out.

3. Check that there is not a short across the load. This would show up as 0 V across the zener and make the zener look faulty. To isolate this problem, disconnect the load and see if the zener functions normally.

   If these three tests check out okay, the zener diode is probably at fault and should be replaced.

SELF-TEST EVALUATION POINT FOR SECTION 3-1

Use the following questions to test your understanding of Section 3–1.

1. The zener diode is designed specifically to operate at voltages exceeding breakdown. True or false?

2. In most applications, a zener diode is _____ biased.

3. What is the difference between a zener diode's schematic symbol and a basic P-N junction's symbol?

4. Will the voltage across the series resistor in a zener voltage regular circuit remain constant as the input voltage changes?

5. A voltage regulator circuit should deliver a constant output voltage despite variations in the _____ and the _____ .

3-2   THE LIGHT-EMITTING DIODE

The **light-emitting diode (LED)** is a semiconductor device that produces light when an electrical current or voltage is applied to its terminals. In other words, it converts electrical energy into light energy. It is classified as an **optoelectronic** device because it combines optics and electronics. The LED is probably the most widely used light source in electronic equipment, having replaced the incandescent lamp due to its longer life expectancy and lower operating power.

The LED is basically a P-N junction diode and, like all semiconductor diodes, it can be either forward biased or reverse biased. When forward biased, it will emit energy in response to a forward current. This emission of energy may be in the form of heat energy, light energy, or both heat and light energy, depending on the type of semiconductor material used. It is the type of material used to construct the LED that determines the color, and therefore frequency, of the light emitted. For example, different compounds are available that will cause the LED to emit red, yellow, green, blue, white, orange, or infrared light when it is forward biased.

Light-Emitting Diode (LED)
A semiconductor device that produces light when an electrical current or voltage is applied to its terminals.

Optoelectronic
Combines optics and electronics.

3-2-1   *LED Symbol and Packages*

Figure 3-8(a) shows the two schematic symbols used most frequently to represent an LED. The two arrows leaving the diode symbol represent light.

A typical LED package is shown in Figure 3-8(b). The package contains the two terminals for connection to the anode and cathode and a semiclear case that contains the light-

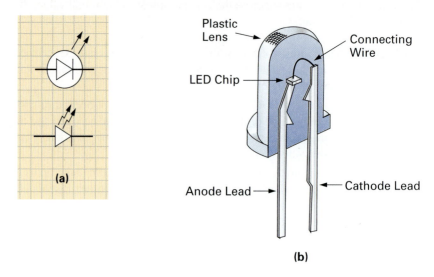

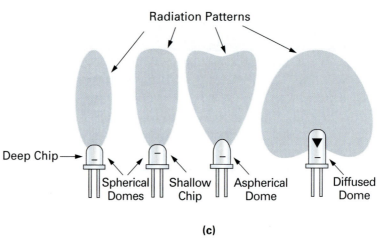

Radiation Patterns

FIGURE 3-8 **The Light-Emitting Diode (LED). (a) Schematic Symbol. (b) Construction. (c) Radiation Patterns. (d) Lead Identification. (e) Package Types.**

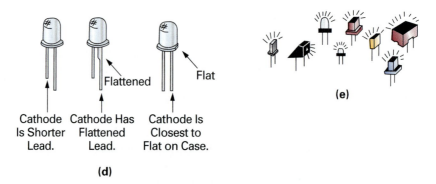

emitting diode and a lens. Looking at this illustration, you can see that the LED chip is directly connected to the cathode lead, while the anode lead is connected to the LED chip by a thin wire. The dome-shaped top of the plastic (epoxy) case serves as the lens and acts as a magnifier to conduct light away from the LED chip. By adjusting the lens material, lens shape, and the distance the LED chip is from the lens, manufacturers can obtain a variety of radiation patterns, some of which are shown in Figure 3-8(c).

There are three methods used to identify the anode and cathode leads of the LED, and these are shown in Figure 3-8(d). In all three cases, the cathode lead is distinguished from the anode lead by being shorter, flattened, or nearest to the flat side of the case.

To give you some idea of the variety of LEDs available, Figure 3-8(e) shows many of the more frequently used package styles.

3-2-2   *LED Characteristics*

Light-emitting diodes have *V–I* characteristic curves that are almost identical to the basic P-N junction diode, as shown in Figure 3-9.

By studying the forward characteristics, you can see that LEDs have a high forward voltage rating (V_F is normally between +1 V to +3 V) and a low maximum forward current rating (I_F is normally between 20 mA to 50 mA). In most circuit applications, the LED will have a forward voltage drop of 2 V and a forward current of 20 mA. The LED will usually always need to be protected from excessive forward current damage by a series current-limiting resistor, which will limit the forward current so that it does not exceed the LED's *maximum forward current* (I_{FM} or I_{FMAX}) rating, as seen in the inset in Figure 3-9. The circuit in the inset shows how to forward bias an LED. In this example circuit, the source voltage of 5 V is being applied to a 130-Ω series resistor (R_S) and an LED with an I_{FM} rating of 50 mA and a V_F rating of 1.8 V. To calculate the value of current in this circuit, we would first deduct the 1.8 V developed across the LED (V_{LED}) from the source voltage (V_S) to determine the voltage drop across the resistor (V_{RS}). Then using Ohm's law, we divide V_{RS} by R_S to determine current.

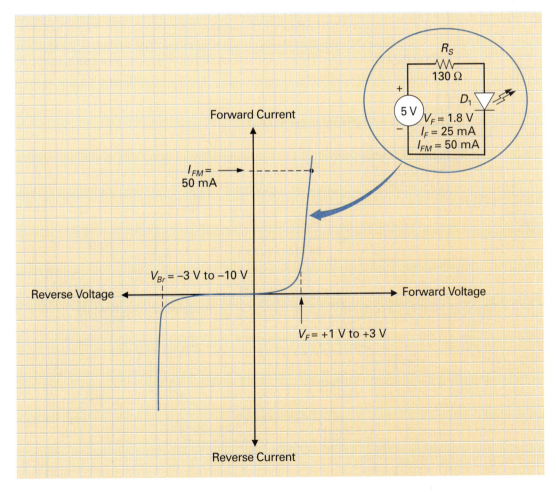

FIGURE 3-9   **The LED Voltage-Current Characteristic Curve.**

$$I_S = \frac{V_S - V_{LED}}{R_S}$$

$$= \frac{5V - 1.8IV}{130 \ \Omega} = \frac{3.2V}{130 \ \Omega} = 24.6 \ mA$$

By studying the reverse characteristics, you can see that LEDs have lower reverse breakdown voltage values than do junction diodes (V_{Br} is typically -3 V to -10 V), which means that even a low reverse voltage will cause the LED to break down and become damaged.

3-2-3   LED Operation

Figure 3-10 shows the basic operation of a light-emitting diode. When forward biased, the negative terminal of the dc source injects electrons into the *n*-type region, or cathode, of the LED. These electrons travel toward the P-N junction. At the same time, electrons are attracted out of the *p*-type region, or anode, and travel toward the positive terminal of the source voltage making it appear as though holes are moving toward the P-N junction. The electrons from the *n*-type region and the holes from the *p*-type region combine at the P-N junction. Because the electrons are at a higher energy level (conduction band), they release energy as they drop into a valence hole. The energy is a small packet of light called a **photon.** Most forward biased LEDs release billions of photons per second as billions of electrons combine with holes at the P-N junction. The semitransparent semiconductor material used to construct LEDs allows the light energy to escape, where it is captured and focused by the dome-shaped lens of the LED package.

The basic P-N junction diode also generates photons when holes and electrons recombine at the P-N junction. However, because silicon is opaque (not transparent), the light simply cannot escape.

Photon
A small packet of light.

3-2-4   LED Data Sheet

Figure 3-11 shows a typical LED data sheet. By studying this specification sheet along with the associated notes, you will be able to find the LED's key characteristics. Remember that, as a technician, you should be familiar with some of the basic operating limits of a device so

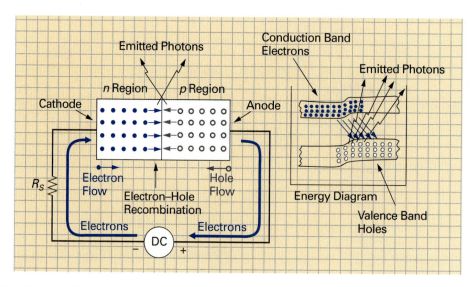

FIGURE 3-10   Basic Operation of an LED.

DEVICE: HLMP- 3000 Series—Red LED Lamps

T-1¾ (5 mm) Red Solid State Lamps

Technical Data

HLMP-3000
HLMP-3001
HLMP-3002
HLMP-3003
HLMP-3050

Features

- **Low Cost, Broad Applications**
- **Long Life, Solid State Reliability**
- **Low Power Requirements:** 20 mA @ 1.6 V
- **High Light Output:** 2.0 mcd Typical for HLMP-3000 4.0 mcd Typical for HLMP-3001
- **Wide and Narrow Viewing Angle Types**
- **Red Diffused and Non-diffused Versions**

Description

The HLMP-3000 series lamps are Gallium Arsenide Phosphide light emitting diodes intended for High Volume/Low Cost applications such as indicators for appliances, smoke detectors, automobile instrument panels, and many other commercial uses.

The HLMP-3000/-3001/-3002/-3003 have red diffused lenses whereas the HLMP-3050 has a red non-diffused lens. These lamps can be panel mounted using mounting clip HLMP-0103. The HLMP-3000/-3001 lamps have 0.025" leads and the HLMP-3002/-3003/-3050 have 0.018" leads.

Package Dimensions

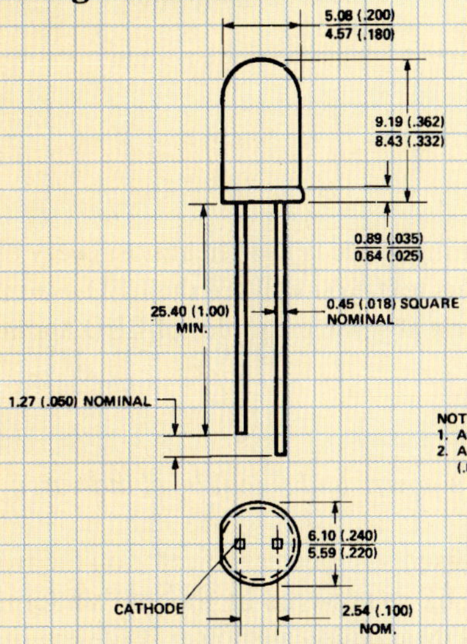

HLMP-3002/-3003/-3050

HLMP-3000/-3001

NOTES:
1. ALL DIMENSIONS ARE IN MILLIMETRES (INCHES).
2. AN EPOXY MENISCUS MAY EXTEND ABOUT 1mm (.040") DOWN THE LEADS.

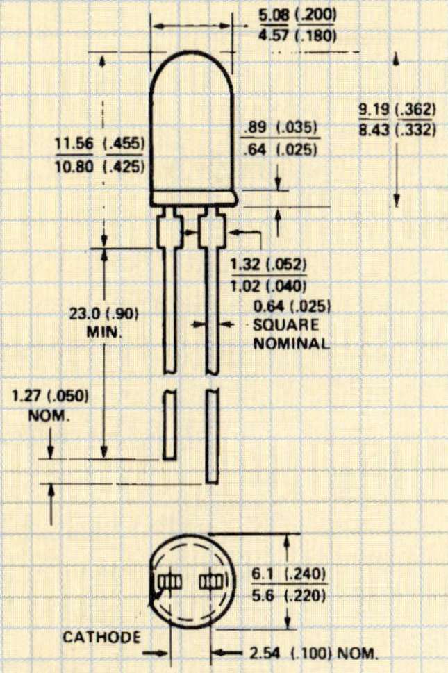

FIGURE 3-11 Device Data Sheet for a Series of Light-Emitting Diodes. (Courtesy of National Semiconductor Corporation.)

that you can isolate component malfunctions within a circuit by determining whether the device is operating to specifications. For example, the MLMP = 3000 series of light-emitting diodes has the following key specifications:

Maximum forward current (I_{FM} or average forward current) = 50 mA

Forward voltage drop (V_F) = 1.4 V to 2.0 V

Reverse breakdown voltage (V_R) = 5 V

EXAMPLE:

What value of resistor would be needed to limit the current to its maximum forward rating if an MLMP = 3000 LED was connected to a 10-V source?

Solution:

Because the MLMP = 3000 LED will drop a maximum of 2 V when forward biased, the series current-limiting resistor will drop the remaining 8 V ($V_S - V_F = V_{RS}$, 10 V − 2 V = 8 V). With V and I known, we can use Ohm's law to calculate the value of R_S.

$$R_S = \frac{V_S - V_F}{I_F} = \frac{10V - 2V}{50 \text{ mA}} = 160 \ \Omega$$

The brightness of an LED is determined by the value of current that passes through the LED. Therefore, more current will mean more light, and less current will mean less light. Doubling the value of R_S will halve the current and halve the LED's brightness, giving it poor visibility. Generally, a value of R_S is chosen that will keep the forward current just below its maximum current rating. In this example, a good choice would be 168 Ω, because this is the next standard-value resistor. Its value would reduce forward current to just below its maximum current rating and ensure a good level of brightness.

$$I_S = \frac{V_S - V_{\text{LED}}}{R_S} = \frac{10V - 2V}{168 \ \Omega} = \frac{8V}{168 \ \Omega} = 47.6 \text{ mA}$$

3-2-5   *LED Applications*

Like the basic P-N junction diode, the light-emitting diode is used in almost every electronic and electrical system. As you proceed through this text, you will see the LED used in a wide variety of circuit applications. For now, let us look at some examples of LED applications.

LED Indicator Circuits

The individual LED is most commonly used as a *power and signal level indicator,* both applications of which can be seen in Figure 3-12.

Figure 3-12(a) shows how an LED can be used as a power indicator on the front panel of a system, such as a CD player. Because the LED is powered by the unit's internal power supply, as shown in the inset, the LED will turn ON and indicate power is present only when the unit is supplying power to its internal circuits.

Figure 3-12(b) shows how three LEDs can be used as signal level indicators to display the code generated by an encoder circuit. For example, when the rotary switch is in position 2, the junction diode D_3 will turn ON and pull line B LOW. This LOW (0 V) signal will not be able to forward bias the LED D_7, and it will not light up. Lines A and C, however, will be pulled HIGH by the resistors R_1 and R_3, and these HIGH outputs will forward bias LEDs D_6 and D_8 and they will light up. Therefore, the code generated by the encoder circuit when it is in position 2 is "HIGH-LOW-HIGH," and this will be displayed on the LEDs as "ON-OFF-ON."

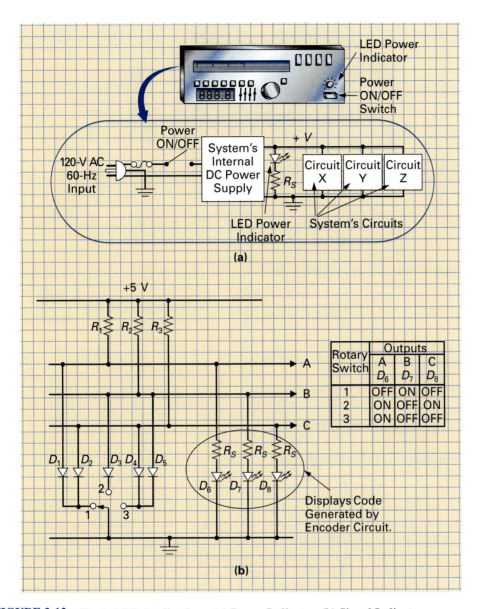

FIGURE 3-12 Single LED Applications. (a) Power Indicator. (b) Signal Indicator.

Bi-Color LEDs

Two-color LEDs were developed so that a single LED package could be used, on an aircraft control panel for example, to indicate whether something was functioning properly (green) or was a problem (red). As you can see in the schematic symbol in Figure 3-13(a), these multicolor LEDs actually contain two *reverse-parallel connected* LEDs. This means that two different-color LEDs are connected in parallel but in the reverse direction inside one package, as seen in the device's data in the inset in Figure 3-13(a). Figure 3-13(b) shows how this device will operate when a 1-Hz (1 cycle per second) 5-V peak square wave is applied to the LED and its current-limiting resistor. When the input voltage is positive, the right or red LED will be forward biased and will turn ON. When the input voltage reverses and is negative, the left or green LED is forward biased and will turn ON. Because a 1-Hz input will have a half-second positive alternation and a half-second negative alternation, the LED package will flash red for half a second, then green for half a second, and then repeat the cycle continuously. You may wish to try this experiment in the lab. If you do, try increasing the frequency of the applied voltage until the switching is so rapid that the two colors blend and appear as yellow.

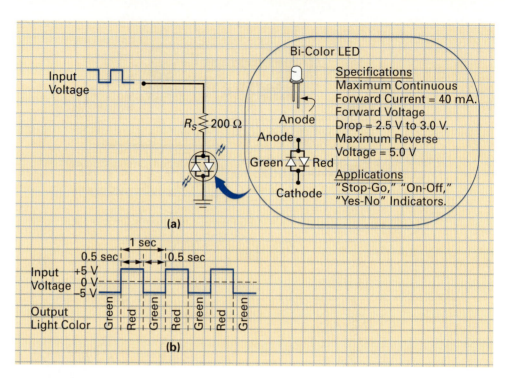

FIGURE 3-13 Bi-Color LED Applications.

Multisegment LED Displays

Seven-Segment Display

Seven LEDs mounted in seven segments. By turning ON a combination of segments, the numbers 0 through 9 can be displayed.

One of the biggest applications of LEDs is in the multisegment display. The **seven-segment display,** which is shown in Figure 3-14(a), is probably the most widely used multisegment LED display. Its seven segments are labeled in a clockwise direction—a, b, c, d, e, f, and g—as shown in the inset in Figure 3-14(a). As can be seen in Figure 3-14(b), the display has seven LEDs mounted in seven segments. Some package types include an eighth LED in a hole, which is used to indicate a decimal-point (dp). As an example, Figure 3-14(c) shows the pin number assignment for a TIL312 seven-segment and two-decimal-point display. The pin numbers are shown inside parentheses. By turning ON a combination of segments, the numbers 0 through 9 can be displayed, as shown in Figure 3-14(d). For example, by turning ON all of the segments, the display will show the number 8. If only segments b and c are turned ON, the display will show the number 1.

Digital Displays

Indicate a quantity using decimal digits.

Analog Displays

Indicate a quantity with the amount of pointer movement.

Generally, in all display applications such as automotive, instrumentation, aircraft, audio, and appliance, more than one LED display is used, as shown in Figure 3-14(e). These **digital displays** indicate a quantity using decimal digits. The older **analog displays** used a moving pointer and scale to indicate a quantity, as shown in Figure 3-14(f), with the amount of pointer movement being an analog, or similar, to the magnitude of the quantity.

These segmented displays are available in either a **common-anode** or a **common-cathode** configuration, as shown in Figure 3-14(g) and (h).

Common-Anode Connection

Common connection between the anodes of LEDs.

The configuration in Figure 3-14(g) is called a common-anode connection because there is a common connection between all the anodes of the LEDs. As you can see in Figure 3-14(g), a common-anode connection will have its "common" connected to a positive supply voltage. To light up the LED, therefore, the cathodes of the LEDs in this configuration will have to be switched LOW. The circuit connections in Figure 3-14(b) and the TIL312 in Figure 3-14(c) are both "common-anode" seven-segment display types.

Common-Cathode Connection

Common connection between the cathodes of LEDs.

The configuration in Figure 3-14(h) is called a common-cathode connection because there is a common connection between all the cathodes of the LEDs. As you can see in Figure 3-14(h), a common-cathode connection will have its "common" connected to ground or 0 V. To light up the LED, the anodes of the LEDs in this configuration will have to be switched HIGH.

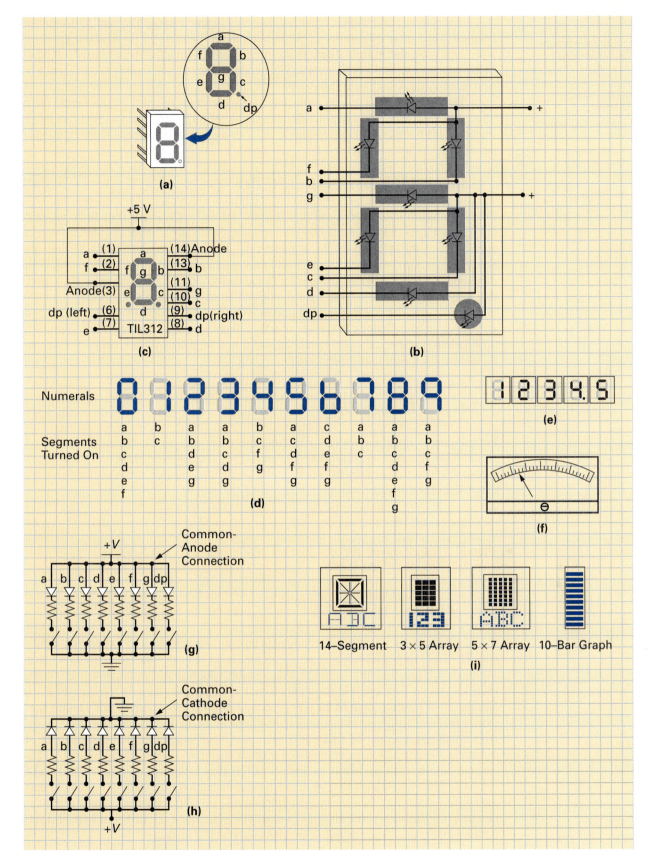

FIGURE 3-14 Multisegmented LED Displays.

Dot Matrix Displays

Displays greater character definition.

Bar Display

Displays the magnitude of a quantity by the number of activated segments.

A variety of LED display types are shown in Figure 3-14(i). The need to display letters resulted in a display with more segments, such as the 14-segment display. For greater character definition, **dot matrix displays** are used. The most popular type consists of 35 small LEDs arranged in a grid of five vertical columns and seven horizontal rows to form a 5×7 matrix.

The **bar display** is rapidly replacing analog meter movements for displaying the magnitude of a quantity. For example, the magnitude of a quantity such as fuel can be represented by the number of activated segments, with no bars lit being zero and all bars lit being maximum.

Infrared-Emitting Diodes (IREDs)

Infrared-emitting diodes are used in a few specialized applications. For example, because the infrared light cannot be seen by the human eye, the IRED is ideally suited for security applications—an intruder would inadvertently break the invisible beam setting off the alarm. The infrared-emitting diode is also used as a light transmitter in fiber-optic communications. Fiber-optic IREDs are similar to the conventional LEDs; however, they are more precisely designed to emit more light in a tighter beam. Fiber optics is a technology in which information, such as voice telephone signals, is converted to light, by an IRED for instance, and then transmitted along the inside of a thin, flexible, glass or plastic fiber. Both of these applications use specially designed IREDs that will be discussed later when you are studying security and communication systems.

3-2-6   *Testing LEDs*

Light-emitting diodes have a very long life expectancy and are much more rugged than their predecessor, the small incandescent light bulb. They do, however, break down and have to be replaced with either exactly the same device or a similar device with the same characteristics. These key characteristics are:

Maximum forward current

Forward voltage drop

Reverse breakdown voltage

To understand how to test an LED, let us first see how a simple LED circuit should operate and then examine what would happen if the circuit developed a malfunction. An example circuit is shown in Figure 3-15(a).

In this circuit, a comparator compares a zero-volts reference to a changing input voltage. The LED in this circuit is functioning as a level indicator and will turn ON whenever the input voltage goes positive. The comparator, which is discussed in a later chapter in more detail, has two inputs labeled "negative ($-$)" and "positive ($+$)." Whenever the input voltage to the positive input of the comparator is below 0 V, the comparator's inputs are "not true" ($+$ is less than $-$), and therefore the output is LOW. In this instance, a LOW or 0-V output will not activate the LED. Whenever the input voltage to the positive input of the comparator is above 0 V, the comparator's inputs are "true" ($+$ is greater than $-$), and therefore the output is HIGH. In this instance, a HIGH or $+5$-V output will turn ON the LED.

Circuits containing LEDs give us a visual indication that there is a problem because the LEDs do not turn ON when the circuit condition is right. As with all circuit malfunctions however, the problem is not always obvious. For example, if the LED in Figure 3-15(a) did not turn ON at any time, the following checks should be made:

1. Check that the input voltage is going positive.
2. Check that the comparator's output is $+5$ V when the input is positive. If the comparator's output does not switch from 0 V to $+5$ V when the input goes positive, you

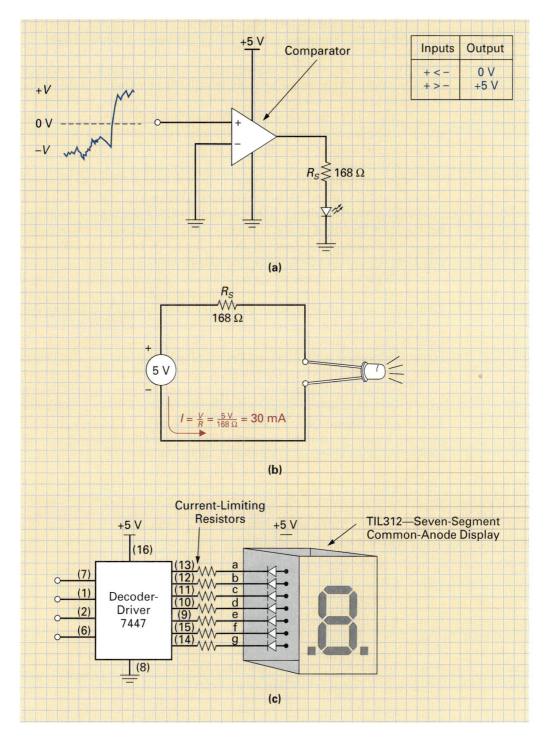

FIGURE 3-15 Testing LEDs. (a) Troubleshooting a Single LED Circuit. (b) LED Test Circuit. (c) Multisegment LED Display Circuit Troubleshooting.

should check the +5-V and 0-V supply voltages to the comparator and the ground at the negative input. After these checks, you should disconnect the comparator's output to see if it functions normally because a shorted output will be pulled LOW constantly.

3. Check the series current-limiting resistor. An open resistor will prevent power from reaching the LED, and like all series opens, will develop all of the source voltage across its terminals. Although a shorted resistor will not limit the forward current, it will probably burn out the LED. Replacing the burnt-out LED will not cure the problem because the shorted resistor will simply burn out the new LED.

As you can see, there are many elements in any circuit that can contribute to a false diagnosis of the problem. However, by systematically checking all of the logical possibilities, you can isolate the circuit fault.

Figure 3-15(b) shows how to construct a simple LED test circuit using a dc power supply and a series current-limiting resistor. If a circuit problem is isolated to the LED, the new LED should be tested with a similar circuit to ensure it is operating correctly before it is inserted into the circuit.

Multisegment LED displays, such as the one seen in Figure 3-15(c), are driven by digital integrated circuits called *decoder-drivers,* which will be discussed in the digital chapter. As with all troubleshooting, it is impossible to isolate a circuit problem unless you fully understand how the circuit should normally operate. At this time, without the full understanding of the decoder-driver circuit, it will be difficult to isolate any faults in this circuit. We can, however, discuss how to check the multisegment LED display.

To begin with, if none of the segments are lighting up, check the supply voltage and ground to all devices in the circuit. To isolate the most common multisegment LED display fault, which is the failure of one or more segments to light up, you must apply the same logic used for the single LED circuit. That is, first determine whether the LED in the multisegment display has failed or whether the *decoder-driver circuit* is not delivering the correct output. For example, a LOW at pin 13 of the decoder-driver should turn ON the "a" segment in the seven-segment display. By following the same steps as the ones we applied to the single LED circuit (the decoder-driver's output voltage, the series current-limiting resistor, and so on), you will be able to diagnose the circuit problem. It is not possible to repair a single LED in a multisegment display. If the fault is isolated to one of the segments, the entire display will have to be replaced. Remember to always check replacements to see that they are compatible (common anode or cathode, I_F rating, and so on).

SELF-TEST EVALUATION POINT FOR SECTION 3-2

Use the following questions to test your understanding of Section 3–2.

1. Light-emitting diodes emit radiation as a result of the recombination of _____ and _____ at the P-N junction.

2. What determines the color of light produced by the LED?

3. What would be the typical voltage drop across a forward biased LED?

   a. 0.7 V   **c.** 0.3 V
   b. 2 V   **d.** 5.6 V

4. A _____ is normally always included to limit I_F to just below its maximum.

5. The amount of light produced by an LED is directly proportional to the amount of _____.

Multiple-Choice Questions

1. The _____ diode is designed to withstand high reverse currents that result when the diode is operated in the reverse breakdown region.
   a. Basic P-N junction   c. Zener
   b. Light-emitting   d. Both (a) and (c) are true

2. When operating in the reverse breakdown region, the _____ the zener will vary over a wide range, while the _____ the zener will vary by only a small amount.
   a. Voltage drop across, forward current through
   b. Forward current through, voltage drop across
   c. Reverse current through, forward current through
   d. Reverse current through, forward drop across

3. The zener diode's symbol is different from all other diode symbols due to its:
   a. Z-shaped cathode bar   c. Straight-bar cathode
   b. Two exiting arrows   d. None of the above

4. A 12-V ± 5% zener diode will have a voltage drop in the _____ to _____ range.
   a. 10.8 V to 13.2 V   c. 5.04 V to 18.96 V
   b. 11.4 V to 12.6 V   d. 11.88 V to 12.12 V

5. A/an _____ circuit maintains the output voltage of a voltage source constant despite variations in the input voltage and the load resistance.
   a. Encoder   c. Comparator
   b. Logic gate   d. Voltage regulator

6. The zener diode is able to maintain the voltage drop across its terminals constant by continually changing its _____ in response to a change in input voltage.
   a. Impedance   c. Power rating
   b. Voltage   d. Both (a) and (c) are true

7. The light-emitting diode is a semiconductor device that converts _____ energy into _____ energy.
   a. Chemical, electrical   c. Electrical, light
   b. Light, electrical   d. Heat, electrical

8. The _____ lead of an LED is distinguished from the other terminal by its longer length, flattening, or close proximity to the flat side of the case.
   a. Anode   b. Cathode

9. The forward voltage drop across a typical LED is usually:
   a. 0.7 V   b. 0.3 V   c. 5 V   d. 2.0 V

10. If all seven cathodes in a seven-segment display are connected to 0 V, the display is referred to as a common _____ configuration and requires a _____ voltage input to turn on an LED segment.
    a. Anode, LOW   c. Anode, HIGH
    b. Cathode, LOW   d. Cathode, HIGH

Practice Problems

11. In reference to the polarity of the applied voltage, which of the circuits in Figure 3-16 are correctly biased for normal zener operation?

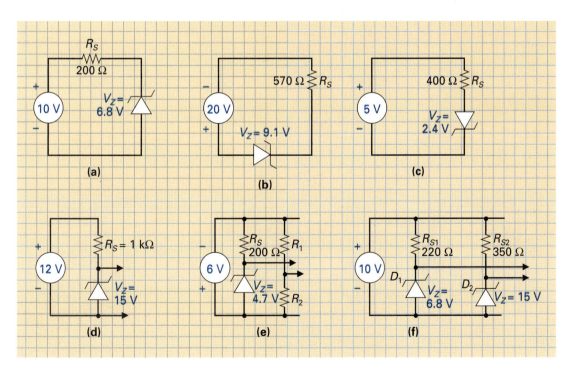

FIGURE 3-16   Biasing Voltage Polarity and Magnitude.

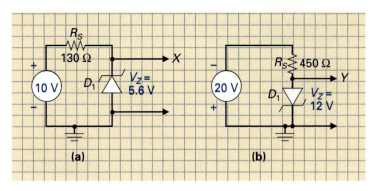

FIGURE 3-17   **Voltage Regulator Circuits (Unloaded).**

12. In reference to the magnitude of the applied voltage, which of the circuits in Figure 3-16 are correctly biased for normal zener operation?

13. Calculate the value of circuit current for each of the zener diode circuits in Figure 3-16.

14. Would a 1-W zener diode have a suitable power dissipation rating for the circuit in Figure 3-16(a)?

15. What wattage or maximum power dissipation rating would you choose for the zener diode in Figure 3-16(e)?

16. What would be the regulated output voltage and polarity at points X and Y for the circuits shown in Figure 3-17 (a) and (b)?

17. Knowing that the zener diode must always be reverse biased, how could we change the circuit connection in Figure 3-17(a) and (b) to obtain opposite polarity supply voltages?

18. Calculate the value of circuit current for both of the circuits shown in Figure 3-17.

19. Which of the light-emitting diodes in Figure 3-18 are biased correctly?

20. Calculate the value of circuit current for each of the LEDs in Figure 3-18.

21. Would a maximum forward current rating of 18 mA be adequate for the LED in Figure 3-18(a)?

22. Would a maximum reverse voltage rating of 5 V be adequate for the LEDs in Figure 3-18(c)?

23. Which of the bi-color LEDs will be ON in Figure 3-19 when the input voltage is $+10$ V, and which will be ON when the input voltage is -10 V?

24. In Figure 3-19, a basic P-N junction diode (D_1) has been included across R_1 so that when the input voltage goes negative this diode will bypass the additional current-limiting resistor R_1. When the input voltage is positive, D_1 is reverse biased and therefore both R_1 and R_2 will limit the value of series current. Which of the colors in the bi-color LED will be brighter?

25. Calculate the value of green and red LED current for the circuit in Figure 3-19.

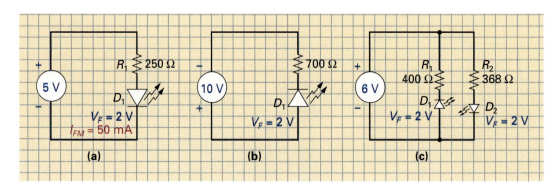

FIGURE 3-18   **Biasing Light-Emitting Diodes.**

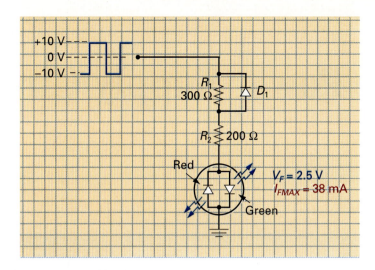

FIGURE 3-19   A Bi-Color Output Display Circuit.

Diode Power Supply Circuits

4

Conducting Achievement

Werner Von Siemens (1816–1892) was born in Lenthe, Germany. He became familiar with the recently developed electric telegraph during military service. Later, in 1847, together with skilled mechanic J. G. Halske, he founded the electrical firm of Siemens and Halske. This firm, under Siemens's guidance, became one of the most important electrical undertakings in the world. Siemens invented cable insulation, an armature for large generators, and the dynamo, or electric generator, which converts mechanical energy to electric energy.

Karl Wilhelm Siemens (1823–1883), Werner's brother, is also well known for his work in the fields of electricity and heat. At the age of 19 he visited England expressly to patent his electroplating invention and never left, making England his home. From 1848 onward, Wilhelm represented his brother's firm in London and became an acknowledged authority. The company installed much of the overland telegraph cable then in existence, as well as an underwater telegraph cable, using submarines. One month before he died, Wilhelm was knighted as Sir William Siemens in acknowledgment of his achievements.

In the family tradition, Alexander Siemens (1847–1928), a nephew of William, went to England in 1867 and worked his way up, beginning in the Siemens's workshops. In 1878, he became manager of the electric department and was responsible for the installation of electric light at Godalming, Surrey, the first English town to be lit with electricity. Like many other members of the family, he patented several inventions and, after the death of Sir William, he became company director.

In honor of the Siemens family's achievements, conductance (G) is measured in the unit siemens (S), with a resistance or impedance of 1 Ω being equal to a conductance or admittance of 1 S.

Introduction

As mentioned previously, *electronic systems are designed to manage the flow of information.* In order for them to achieve this function, all of the electronic circuits within the system require certain constant dc supply voltages. If only a small amount of power is needed, batteries can be used to deliver a dc supply, as is the case with portable equipment such as calculators, watches, cassette radios, multimeters, and so on. With larger electronic systems such as computers, television sets, music systems, and video systems, the dc supply voltages will be obtained from a dc power supply, which is generally a subsystem within the main system. The dc power supply converts a 120-V ac 60-Hz input into the desired dc supply voltages, and it is therefore an electrical system because *electrical systems are designed to manage the flow of power.*

In Chapter 2 we discussed the P-N junction diode, and in Chapter 3 we discussed the zener and light-emitting diodes. In this chapter, we will see how we can use all of these diodes to construct a dc power supply. This chapter is also of great importance to our study of troubleshooting, because more than 90% of all electronic system malfunctions are power related.

4-1 BLOCK DIAGRAM OF A DC POWER SUPPLY

Figure 4-1(a) shows the four basic blocks of a dc power supply. Almost every piece of electronic equipment that makes use of 120 V ac as a source of power will have a built-in dc power supply, as shown in Figure 4-1(b). Stand-alone dc power supplies, such as the one shown in Figure 4-1(c), are also available for use in laboratory experimentation.

Let us now refer to the block diagram of the dc power supply in Figure 4-1(a) and describe the function of each block.

Because the final voltage desired is generally not 120 V, a transformer is usually included to step the ac line voltage up or down to a desired value. Electronic circuits generally require low-voltage supply values such as 12 and 5 V dc, and a step-down transformer would be used. For example, a step-down transformer with a turns ratio of 10:1 would reduce the 120-V ac input to a 12-V ac output. The output current capability of this transformer will be 1:10, which will be ideal because most electronic circuits require low supply voltages with high current capacity.

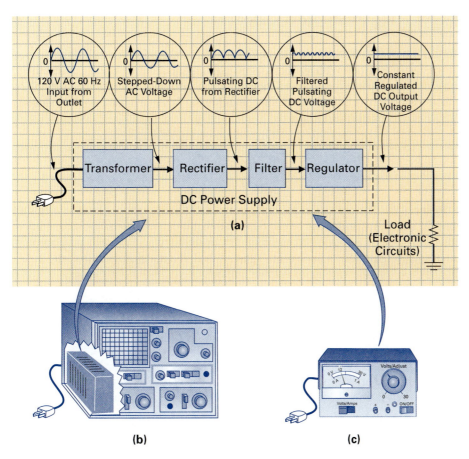

FIGURE 4-1 **DC Power Supply. (a) Block Diagram. (b) Built-in Subsystem. (c) Stand-Alone Unit.**

As shown by the waveforms in Figure 4-1(a), the rectifier converts the stepped-down ac input from the transformer to a pulsating dc output. This pulsating dc output could not be used to power an electronic circuit because of the continuous changes between zero volts and a peak voltage. The filter smoothes out the pulsating dc ripples into an almost constant dc level, as seen in the waveform after the filter.

The final block is called a *regulator*, and although there appears to be no difference between the regulator's input and output waveforms, it serves a very important function. The regulator maintains the dc output voltage from the power supply constant, or stable, despite variations in the ac input voltage or variations in the output load resistance.

Many dc power supplies have several rectifier, filter, and regulator stages, depending on how many dc output voltages are desired for the electronic system. In the following sections, we will examine these electrical circuits in more detail and then combine all of them in a working dc power supply circuit.

SELF-TEST EVALUATION POINT FOR SECTION 4-1

Use the following questions to test your understanding of Section 4–1.

1. List the four basic circuit blocks that make up a dc power supply.
2. Which circuit block converts ac to dc?
3. Which circuit block converts the high ac input voltage into a low ac output voltage?
4. The _____ maintains the dc output voltage constant despite variations in the ac input voltage and the output load resistance.

4-2 TRANSFORMERS

The turns ratio of a transformer in a dc power supply can be selected to either increase or decrease the 120-V ac input. With most electronic equipment, a supply voltage of less than 120 V is required, and therefore a step-down transformer is used. The secondary output voltage, V_S, from the transformer can be calculated with the following formula, which was introduced previously in dc/ac electronic theory:

$$V_S = \frac{N_S}{N_p} \times V_p$$

To apply this formula to an example, refer to the power supply transformer shown in Figure 4-2. This transformer can be connected so that it delivers the same secondary voltage for either a 120-V or a 240-V rms ac input. For example, when the two primary windings are connected in parallel, as shown in Figure 4-2(a), the transformer turns ratio is 6:1 step-down, and therefore the secondary voltage will be

$$V_S = \frac{N_S}{N_p} \times V_p$$

$$= \frac{1}{6} \times 120V_{rms} = 20V_{rms}$$

When the two primary windings are connected in series, as shown in Figure 4-2(b), the transformer turns ratio is 12:1 step-down, and therefore the secondary voltage will be

$$V_S = \frac{N_S}{N_p} \times V_p$$

$$= \frac{1}{12} \times 240V_{rms} = 20V_{rms}$$

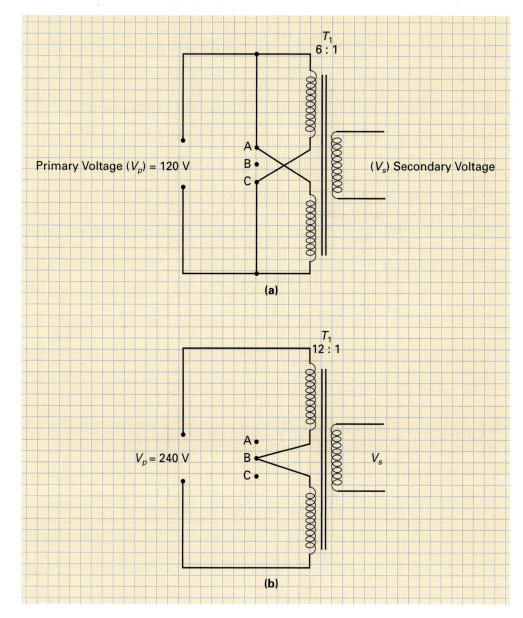

Primary Voltage (V_p) = 120 V

(V_s) Secondary Voltage

T_1
6 : 1

A
B
C

(a)

V_p = 240 V

V_s

T_1
12 : 1

A
B
C

(b)

FIGURE 4-2 **A 120-V/240-V Power Supply Transformer. (a) 120-V Connection.**
(b) 240-V Connection.

By doubling the number of primary turns, we can accept twice the input voltage and deliver the same output voltage.

■ **EXAMPLE:**

Calculate the peak secondary output voltage of the transformer shown in Figure 4-3.

■ *Solution:*

If the rms input is 120 V, the peak input will be

$$V_p = V_{rms} \times 1.414$$
$$= 120 \text{ V} \times 1.414 = 169.7 \text{ V}$$

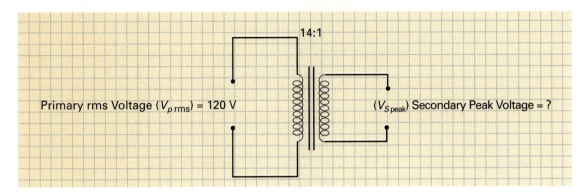

FIGURE 4-3 **Transformer Coupling.**

Now that V_p peak is known, V_S peak can be calculated.

$$V_S = \frac{N_S}{N_p} \times V_p$$

$$= \frac{1}{14} \times 169.7\text{V} = 12.12\text{V}$$

SELF-TEST EVALUATION POINT FOR SECTION 4-2

Use the following questions to test your understanding of Section 4–2.

1. Most dc power supplies employ a step _____ transformer.

2. If a 120-V rms ac input was applied to a 5:1 step-down transformer, what would be the secondary voltage?

4-3 RECTIFIERS

The P-N junction diode's ability to switch current in only one direction makes it ideal for converting two-direction alternating current into one-direction direct current. In this section, we will discuss the three basic diode rectifier circuits: the half-wave rectifier, the full-wave center-tapped rectifier, and the full-wave bridge rectifier.

4-3-1 *Half-Wave Rectifiers*

Half-Wave Rectifier Circuit

A circuit that converts ac to dc by allowing current to flow during only one-half of the ac input cycle.

The **half-wave rectifier circuit** is constructed simply by connecting a diode between the power supply transformer and the load, as shown in Figure 4-4(a). When the secondary ac voltage swings positive, as shown in Figure 4-4(b), the anode of the diode is made positive, causing the diode to turn ON and connect the positive half-cycle of the secondary ac voltage across the load (R_L). When the secondary ac voltage swings negative, as shown in Figure 4-4(c), the anode of the diode is made negative, and therefore the diode will turn OFF. This will prevent any circuit current, and no voltage will be developed across the load (R_L).

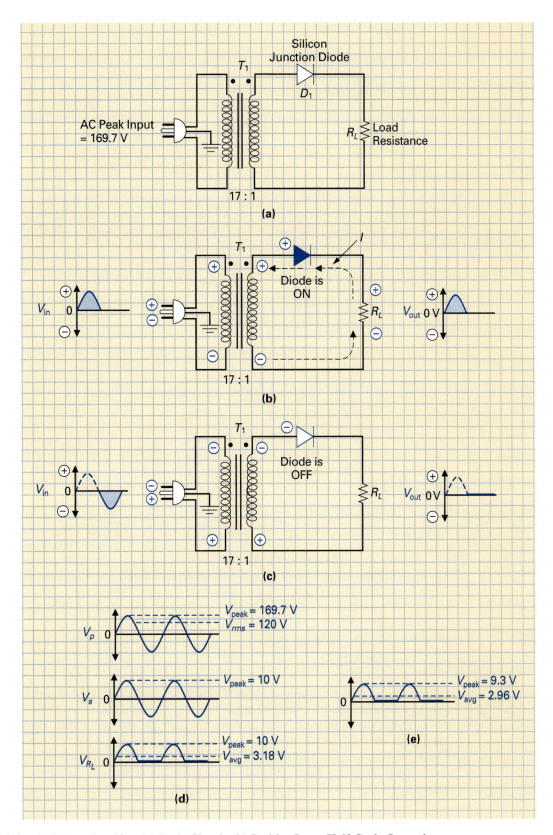

FIGURE 4-4 **Half-Wave Rectifier. (a) Basic Circuit. (b) Positive Input Half-Cycle Operation. (c) Negative Input Half-Cycle Operation. (d) Input/Output Waveforms. (e) Half-Wave Output Minus Diode Barrier Voltage.**

Output Voltage

Figure 4-4(d) illustrates the input and output waveforms for the half-wave rectifier circuit. The 120-V ac (rms) input, or 169.7-V ac peak input, is applied to the 17:1 step-down transformer, which produces an output of

$$V_S = \frac{N_S}{N_p} \times V_p$$

$$= \frac{1}{17} \times 169.7\text{V (peak)} = 10\text{V (peak)}$$

Because the diode will only connect the positive half-cycle of this ac input across the load (R_L), the output voltage (V_{RL}) is a positive pulsating dc waveform of 10 V_{peak}. In this final waveform in Figure 4-4(d), you can see that the circuit is called a half-wave rectifier because only half of the input wave is connected across the output.

The average value of two half-cycles is equal to 0.637 V_{peak}. Therefore, the average value of one half-cycle is equal to 0.318 V_{peak} (0.637/2 = 0.318):

$$V_{av} = 03.18 \times V_{S\,peak}$$

In the example in Figure 4-4, the average voltage of the half-wave output will be

$$V_{av} = 0.318 \times V_{S\,peak}$$
$$V_{av} = 0.318 \times 10\text{ V}$$
$$= 3.18\text{ V}$$

To be more accurate, there will, of course, be a small voltage drop across the diode due to its barrier voltage of 0.7 V for silicon and 0.3 V for germanium. The output from the circuit in Figure 4-4 would actually have a peak of 9.3 V (10 V − 0.7 V), and therefore an average of 2.96 V (0.318 × 9.3 V), as shown in Figure 4-4(e).

$$V_{out} = V_S - V_{diode}$$

Another point to consider is the reverse breakdown voltage of the junction or rectifier diode. When the input swings negative, as illustrated in Figure 4-4(c), the entire negative supply voltage will appear across the open or OFF diode. The maximum reverse breakdown voltage, or peak inverse voltage (PIV) rating of the rectifier diode, must therefore be larger than the peak of the ac voltage at the diode's input.

Output Polarity

The half-wave rectifier circuit can be arranged to produce either a positive pulsating dc output, as shown in Figure 4-5(a), or a negative pulsating dc output, as shown in Figure 4-5(b). By studying the difference between these circuits you can see that in Figure 4-5(a) the rectifier diode is connected to conduct the positive half-cycles of the ac input, while in Figure 4-5(b), the rectifier diode is reversed so that it will conduct the negative half-cycles of the ac input. By changing the direction of the diode in this manner, the rectifier can be made to produce either a positive or a negative dc output.

Ripple Frequency

Referring to the input/output waveforms of the half-wave rectifier circuit shown in Figure 4-6, you can see that one output ripple is produced for every complete cycle of the ac input. Consequently, if a half-wave rectifier is driven by the 120-V ac 60-Hz line voltage, each complete cycle of the ac input and each complete cycle of the output will last for one-

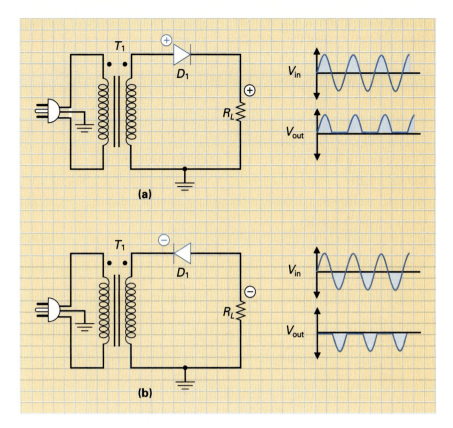

FIGURE 4-5 Changing the Output Polarity of a Half-Wave Rectifier. (a) Positive Pulsating DC. (b) Negative Pulsating DC.

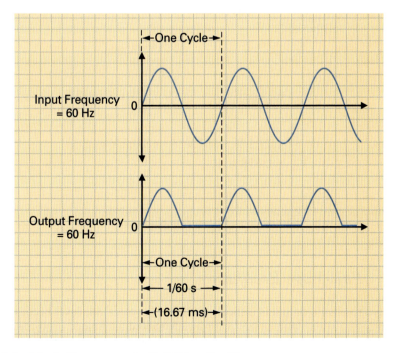

FIGURE 4-6 Ripple Frequency of a Half-Wave Rectifier.

sixtieth or 16.67 msec (1/60 = 1 ÷ 60 = 16.67 msec). The frequency of the pulsating dc output from a rectifier is called the *ripple frequency*, and for half-wave rectifier circuits, the

> Output Pulsating DC Ripple Frequency = Input AC Frequency

■ EXAMPLE:

Calculate the following for the rectifier circuit shown in Figure 4-7:

    a. Output polarity

    b. Peak and average output voltage, taking into account the diode's barrier potential

    c. Output ripple frequency

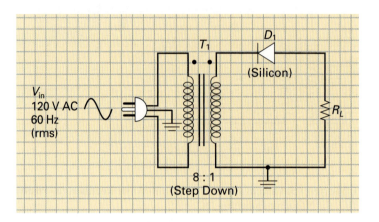

FIGURE 4-7   **Half-Wave Rectifier.**

■ *Solution:*

    a. Negative pulsating dc

    b.

$$V_{\text{in peak}} = 169.7 \text{ V}$$

$$V_S = \frac{N_S}{N_p} \times V_p$$

$$= \frac{1}{8} \times 169.7 \text{ V (peak)} = 21.1 \text{ V (peak)}$$

$$V_{\text{out}} = V_S - V_{\text{diode}}$$

$$= 21.1 \text{ V} - 0.7 \text{ V} = -20.5 \text{ V}$$

$$V_{\text{av}} = 0.318 \times V_{\text{out}}$$

$$= 0.318 \times 20.5 \text{ V} = -6.5 \text{ V}$$

    c. Output Ripple Frequency = Input Frequency = 60 Hz

4-3-2   *Full-Wave Rectifiers*

The half-wave rectifier's output is difficult to filter to a smooth dc level because the output voltage and current are applied to the load for only half of each input cycle. In this section, we will examine two full-wave rectifier circuits, which, as their name implies, switch both half-cycles (or the full ac input wave) of the input through the load in only one direction.

Center-Tapped Rectifier

The basic **full-wave center-tapped rectifier** circuit, which is shown in Figure 4-8(a), contains a center-tapped transformer and two diodes. The center tap of the transformer secondary is grounded (0 V) to create a 180° phase difference between the top and bottom of the secondary winding.

When the ac input (V_{in}) swings positive, the circuit operates in the manner illustrated in Figure 4-8(b). A positive voltage is developed on the top of T_1 secondary turning ON D_1, while a negative voltage is developed on the bottom of T_1 secondary turning OFF D_2. This will permit a flow of electrons up through the load, as indicated by the dashed current (I) line, developing a positive output half-cycle across the load (V_{out}).

When the ac input (V_{in}) swings negative, the circuit operates in the manner illustrated in Figure 4-8(c). A negative voltage is developed on the top of T_1 secondary turning OFF D_1, while a positive voltage is developed on the bottom of T_1 secondary turning ON D_2. This will permit a flow of electrons up through the load as indicated by the dashed current (I) line, again developing a positive output half-cycle across the load (V_{out}).

The two diodes and center-tapped transformer in this circuit switch the two-direction ac current developed across the secondary through the load in only one direction and develop an output voltage across the load of the same polarity. The input/output voltage details of a full-wave center-tapped rectifier are shown in Figure 4-9. A 1:1 turns ratio has been selected in this example to emphasize how the center tap in the secondary of the transformer causes the secondary voltage (V_S) to be divided into two equal halves (V_{S1} and V_{S2}). Referring to the waveforms, you can see that the ac line voltage of 120 V rms, or 169.7 V_{peak} (V_{in}), is applied to the primary of the transformer T_1 (V_p). A 1:1 turns ratio ensures that this same voltage will be developed across the secondary winding ($V_S = V_p = 169.7$ V_{peak}). This secondary voltage will be split in two, as indicated by the waveforms V_{S1} and V_{S2}, which will each have a peak voltage of 1/2 V_S ($V_{S1\,peak} = V_{S2\,peak} = 1/2\,V_{S\,peak} = 1/2 \times 169.7$ V = 84.9 V_{peak}). Because V_{S1} will be connected across the output during one half-cycle and V_{S2} will be connected across the output during the other half-cycle, the output voltage (V_{out}) will also have a peak voltage equal to 1/2 V_S, or 84.9 V. Compared to a half-wave rectifier, therefore, whose peak output voltage is equal to the peak secondary voltage, the full-wave center-tapped rectifier seems to do worse, delivering a peak output of 1/2 V_S. However, the full-wave center-tapped rectifier compensates for the halving of the peak output voltage by doubling the number of half-cycles at the output, compared to a half-wave rectifier. The average output voltage (V_{av}) can be calculated for a full-wave rectifier with the following formula:

$$V_{av} = 0.636 \times \frac{1}{2} V_{S\,peak}$$

Full-Wave Center-Tapped Rectifier

A rectifier circuit that makes use of a center-tapped transformer to cause an output current to flow in the same direction during both half-cycles of the ac input.

■ EXAMPLE:

Calculate the average output voltage from a half-wave and full-wave center-tapped rectifier if $V_{S\,peak} = 169.7$ V.

■ *Solution:*

Half-wave rectifier:

$$V_{av} = 0.318 \times V_{S\,peak}$$
$$= 0.318 \times 169.7\,\text{V} = 54\,\text{V}$$

Full-wave center-tapped rectifier:

If $V_{S\,peak} = 169.7$ V, then $V_{S1\,peak}$ and $V_{S2\,peak} = 84.9$ V

$$V_{av} = 0.636 \times \frac{1}{2} V_{S\,peak}$$

$$= 0.636 \times 84.9\,\text{V} = 54\,\text{V}$$

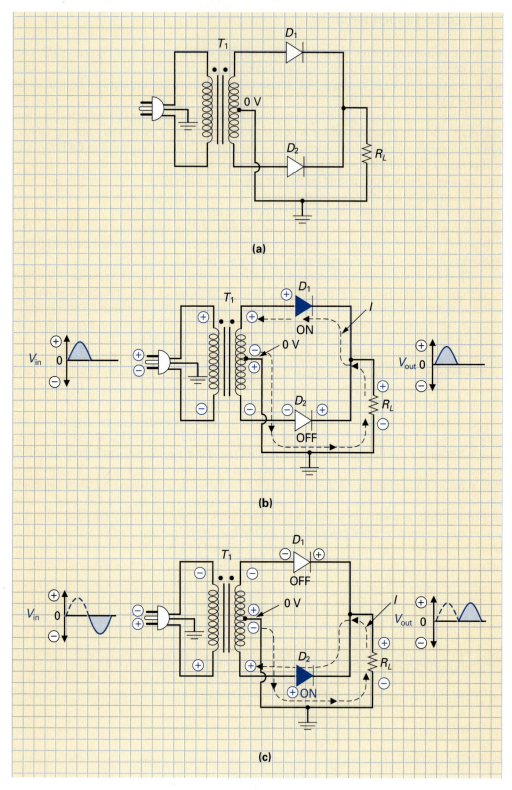

FIGURE 4-8 Full-Wave Center-Tapped Rectifier. (a) Basic Circuit. (b) Positive-Input Half-Cycle Operation. (c) Negative-Input Half-Cycle Operation.

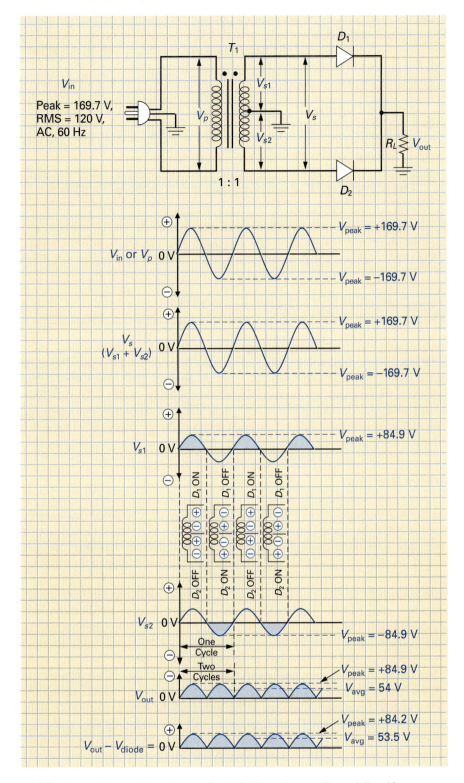

FIGURE 4-9 **Input/Output Waveforms of a Full-Wave Center-Tapped Rectifier.**

As you can see from this example, even though the peak output of the full-wave center-tapped rectifier was half that of the half-wave rectifier, the average output was the same because the full-wave center-tapped rectifier doubles the number of half-cycles at the output, compared to a half-wave rectifier.

Because two half-cycles appear at the output for every one cycle at the input, as shown in Figure 4-9, the ripple frequency will be twice that of the input frequency, and this higher frequency will be easier to filter or smooth of fluctuations.

$$\text{Output Pulsating DC Ripple Frequency} = 2 \times \text{Input AC Frequency}$$

To be completely accurate, we should take into account the barrier voltage drop across the diodes. Because only one diode is ON for each half-cycle, the peak output voltage will only be less 0.7 V (silicon diode), as shown by the last waveform in Figure 4-9. Therefore,

$$V_{\text{out}} = \frac{1}{2} V_{S\,\text{peak}} - 0.7 \text{ V}$$

With regard to the peak inverse voltage, if you refer back to Figure 4-8(b), you will see that the full secondary peak voltage ($V_{S\,\text{peak}}$) appears across the OFF diode D_2 for one half-cycle of the input. Similarly, the full $V_{S\,\text{peak}}$ voltage appears across the OFF D_1 during the other half-cycle of the input, as shown in Figure 4-8(c). Both diodes must therefore have a peak inverse voltage (PIV) rating that is larger than the peak secondary voltage ($V_{S\,\text{peak}}$). For the circuit in Figure 4-9, the diode's maximum reverse voltage rating (PIV) must be greater than 169.7 V.

Bridge Rectifier

By center tapping the secondary of the transformer and having two diodes instead of one, we were able to double the ripple frequency and ease the filtering process. However, that seems to be the only advantage the full-wave center-tapped rectifier has over the half-wave rectifier, and the price we pay is having a more expensive center-tapped transformer and an extra diode.

Bridge Rectifier Circuit
A full-wave rectifier circuit using four diodes that will convert an alternating voltage input into a direct voltage output.

With the **bridge rectifier circuit** shown in Figure 4-10(a), we can have the peak secondary output voltage of the half-wave circuit and the full-wave ripple frequency of the center-tapped circuit. This circuit was originally called a "bridge" rectifier because its shape resembled the framework of a suspension bridge. Figures 4-10(b) and (c) illustrate how the bridge rectifier circuit will behave when an ac input cycle is applied.

When the ac input (V_{in}) swings positive, the circuit operates in the manner illustrated in Figure 4-10(b). A positive potential is applied to the top of the bridge, causing D_2 to turn ON, while a negative potential is applied to the bottom of the bridge, causing D_3 to turn ON. With D_2 and D_3 ON, and D_1 and D_4 OFF, electrons will flow up through the load as indicated by the dashed current line (I), developing a positive output half-cycle across the load (V_{out}).

When the ac input (V_{in}) swings negative, the circuit operates in the manner illustrated in Figure 4-10(c). A negative potential is applied to the top of the bridge, causing D_1 to turn ON, while a positive potential is applied to the bottom of the bridge, causing D_4 to turn ON. With D_1 and D_4 ON and D_2 and D_3 OFF, electrons will flow up through the load as indicated by the dashed current line (I), again developing a positive output half-cycle across the load (V_{out}).

Like the center-tapped rectifier, the bridge rectifier switches two half-cycles of the same polarity through to the load. However, unlike the center-tapped rectifier, the bridge rectifier connects the total peak secondary voltage across the load. A secondary peak voltage of 169.7 V ac would produce a pulsating dc peak across the load of 169.7 V, or the same

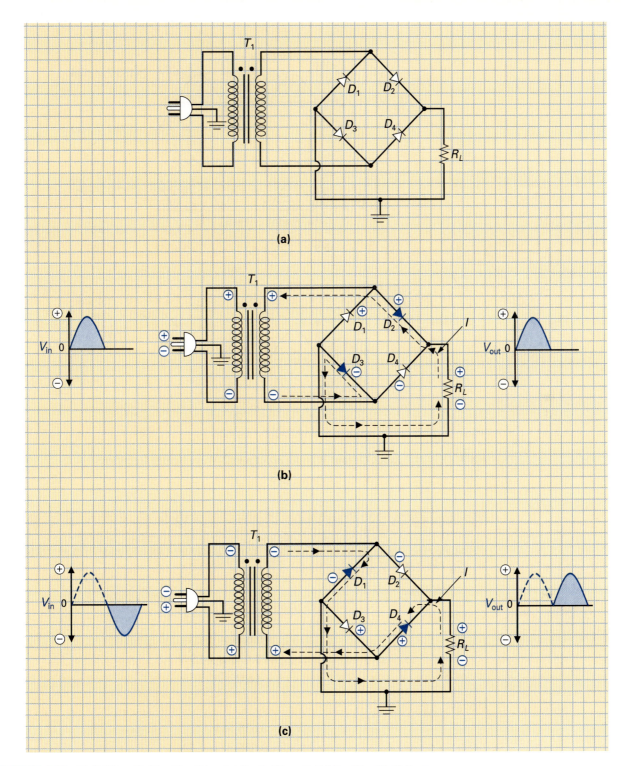

FIGURE 4-10 Full-Wave Bridge Rectifier. (a) Basic Circuit. (b) Positive Half-Cycle
Operation. (c) Negative Half-Cycle Operation.

voltage, as shown in Figure 4-11(a). The average voltage in this case would be larger than
that of a center-tapped rectifier, and equal to

$$V_{av} = 0.36 \times V_{S\,peak}$$

$$= 0.636 \times 169.7\ \text{V} = 107.9\ \text{V}$$

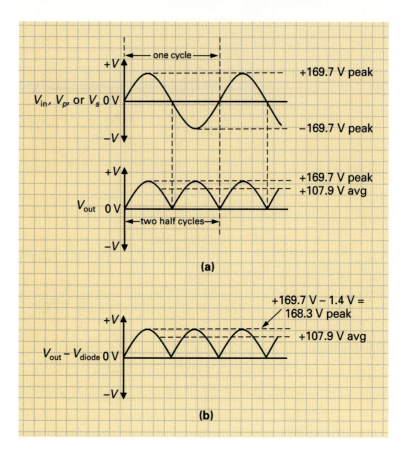

FIGURE 4-11   Input/Output Waveforms of a Full-Wave Bridge Rectifier.

■   EXAMPLE:

Calculate the average output voltage from a center-tapped and bridge rectifier if $V_{S\text{ peak}} = 169.7$ V.

■   *Solution:*

Center-tapped:

$$V_{av} = 0.636 \times 1/2\ V_S$$
$$= 0.636 \times 84.9\ \text{V}$$
$$= 54\ \text{V}$$

Bridge:

$$V_{av} = 0.636 \times V_S$$
$$= 0.636 \times 169.7\ \text{V}$$
$$= 107.9\ \text{V}$$

As you can see from this example, unlike the center-tapped rectifier that only connects half of the peak secondary voltage across the load, the bridge rectifier connects the total peak secondary voltage across the load.

Because two half-cycles appear at the output for every one cycle at the input, as shown in Figure 4-11(a), the ripple frequency will be twice that of the input frequency, and this higher frequency will be easier to filter or smooth of fluctuations.

FIGURE 4-12 Peak Inverse Voltage.

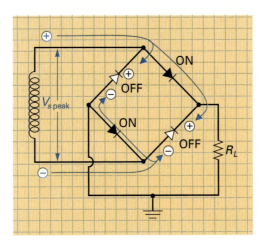

Output Pulsating DC Ripple Frequency = 2 × Input AC Frequency

To be completely accurate, we should take into account the barrier voltage drop across the diodes. Because two diodes are ON for each half-cycle and both diodes are connected in series with the load, a 1.4-V (2 × 0.7 V) drop will occur between the peak secondary voltage (V_S) and the peak output voltage, as shown in Figure 4-11(b).

$$V_{out} = V_{S\,peak} - 1.4\ V$$

With regard to the peak inverse, if you refer to Figure 4-12, you can see that the full $V_{S\,peak}$ voltage appears across the two OFF diodes. Therefore, each diode must have a peak inverse voltage (PIV) rating that is greater than the peak secondary voltage (V_S). For the circuit and values given in Figures 4-10 and 4-11, the diode's maximum reverse voltage rating (PIV) must be greater than 169.7 V.

SELF-TEST EVALUATION POINT FOR SECTION 4-3

Use the following questions to test your understanding of Section 4–3.

1. A rectifier converts an _____ input into a _____ output.
2. A_____-wave rectifier uses only one diode.
3. What would be the ripple frequency out of a half-wave rectifier if a 220-V 50-Hz ac input were applied?
4. Which full-wave rectifier circuit does not need to have a transformer?

4-4 FILTERS

The filter in a dc power supply converts the pulsating dc output from the half-wave or full-wave rectifier into an unvarying dc voltage, as was shown in the waveforms in the basic block diagram in Figure 4-1. In this section, we will examine the three basic types of filters: the **capacitive filter,** the *RC* **filter,** and the *LC* **filter.**

4-4-1 *The Capacitive Filter*

Figure 4-13(a) shows a positive-output, half-wave rectifier and capacitive filter circuit. The waveforms in Figure 4-13(b), (c), and (d) show how the capacitive filter will respond to the

Capacitive Filter
A capacitor used in a power supply filter system to suppress ripple currents while not affecting direct currents.

RC Filter
A selective circuit that makes use of a resistance–capacitance network.

LC Filter
A selective circuit that makes use of an inductance–capacitance network.

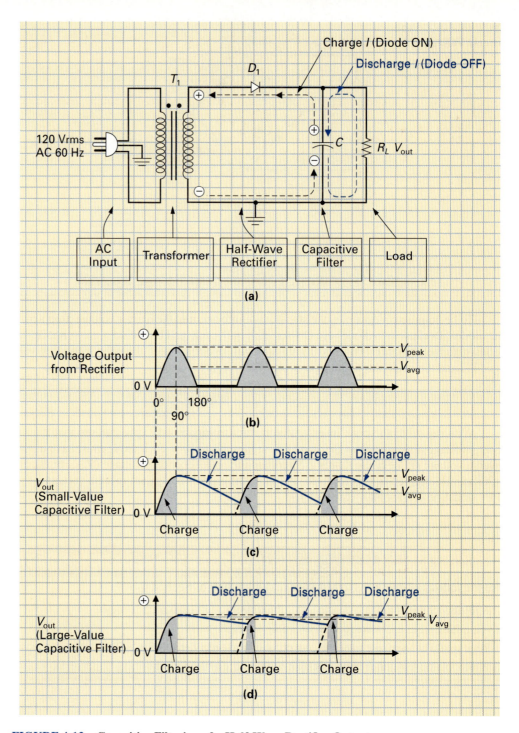

FIGURE 4-13   Capacitive Filtering of a Half-Wave Rectifier Output.

half-wave pulsating dc output from the rectifier. Let us examine how the filter operates by referring to the dashed lines in the circuit in Figure 4-13(a), the output waveform from the rectifier shown in Figure 4-13(b), and the small-value capacitive filter waveform shown in Figure 4-13(c). When the ac input swings positive, the diode is turned ON and the capacitor charges, as indicated by the black dashed charge current line in Figure 4-13(a). The charge time constant will be small because no resistance exists in the charge path except for that of the resistance of the connecting wires (charge $\tau\!\downarrow = R\!\downarrow \times C$). This charge time is indicated by the gray shaded section between 0 and 90° in Figure 4-13(c). When the ac input begins to fall from its positive peak (at 90°), the diode is turned OFF by the large positive potential on the diode's cathode being supplied by the charged capacitor and the decreasing positive potential on the diode's anode being supplied by the input. With the diode OFF, the capacitor begins to discharge, as indicated by the colored dashed discharge current line in Figure 4-13(a). The discharge time constant is a lot longer than the charge time because of the load resistance (discharge $\tau\uparrow = R\uparrow \times C$). The decreasing slope in Figure 4-13(c) illustrates the decreasing voltage across the capacitor, and therefore across the load at the output (V_{out}), as the capacitor discharges. As the ac input and the rectifier's pulsating dc cycle repeat, the output

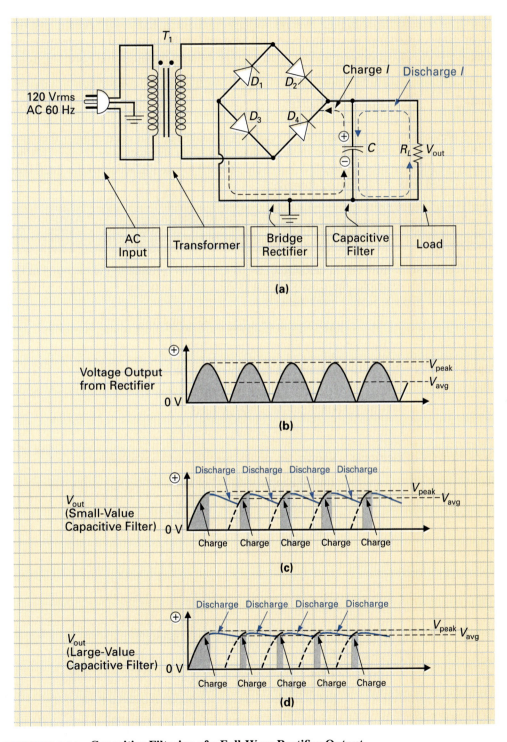

FIGURE 4-14   Capacitive Filtering of a Full-Wave Rectifier Output.

(V_{out}) is, as shown in Figure 4-13(c), an almost constant dc output with a slight variation or ripple above and below the average value.

Figure 4-13(d) shows how a larger-value capacitor will make the charge and discharge time constants longer ($\tau\uparrow = R \times C\uparrow$), therefore decreasing the amount of ripple and increasing the average output voltage.

Figure 4-14(a) shows a positive-output-voltage, full-wave, bridge rectifier and capacitive filter circuit. Figure 4-14(b), (c), and (d) show the output waveforms from the rectifier and the filtered output from a small-value and a large-value capacitor filter. The ripple frequency from a full-wave rectifier is twice that of a half-wave rectifier, and therefore the capacitor does not have too much time to discharge before another positive half-cycle reoccurs. This results in a higher average voltage than in a half-wave circuit, and if a larger-value capacitor is used, the ripple will be even less because of the greater time constant ($\tau\uparrow = R \times C\uparrow$), causing an increased average voltage, as shown in Figure 4-14(d).

4-4-2   *Percent of Ripple*

In the previous two capacitive filter circuits, you could see that even though the output from a filter should be a constant dc level, there is a slight fluctuation or ripple. This fluctuation is called the *percent ripple* and its value is used to rate the action of the filter. It can be calculated with the following formula:

$$\text{Percent Ripple} = \frac{V_{rms} \text{ of Ripple}}{V_{av} \text{ of Ripple}} \times 100$$

To help show how this is calculated, let us apply this formula to the example in Figure 4-15. In this example, the filter's output is fluctuating between $+20$ and $+30$ V, and therefore the ripple's peak-to-peak value is

$$\text{Peak-to-peak of ripple} = 20 \text{ to } 30 \text{ V}$$
$$= 10 \text{ V}_{pk\text{-}pk}$$

A peak-to-peak value of 10 V means that the ripple has a peak value of

$$\text{Peak of ripple} = 1/2 \text{ of pk-pk value}$$
$$= 1/2 \text{ of } 10 \text{ V}$$
$$= 5 \text{ V}_{pk}$$

Once the peak of the ripple is known, the ripple rms value can be calculated:

$$\text{rms of Ripple} = 0.707 \text{ of Peak}$$
$$= 0.707 \text{ of } 5 \text{ V}$$
$$= 3.54 \text{ V}$$

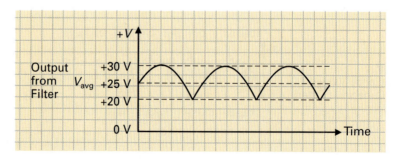

FIGURE 4-15   Percent of Ripple Out of a Filter.

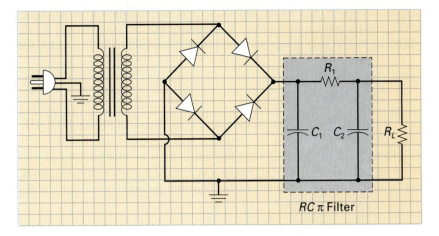

FIGURE 4-16 *RC π* **Filter.**

The average of the ripple is approximately midway between peaks, which in the example in Figure 4-15 is +25 V. Inserting these values in the formula, we can calculate the percent of ripple:

$$\% \text{ Ripple} = \frac{V_{\text{rms}} \text{ of Ripple}}{V_{\text{av}} \text{ of Ripple}} \times 100$$

$$= \frac{3.54 \text{ V}}{25 \text{ V}} \times 100$$

$$= 14\%$$

This means that the average voltage out of the rectifier (+25 V) will fluctuate 14%.

4-4-3   *RC Filters*

The low-pass pi (π) filter shown in Figure 4-16 could be used to further reduce ripple, due to its ability to pass dc but block any fluctuations. It is called a π filter because it contains two vertical parts and one horizontal part and therefore resembles π. The first capacitor, C_1, acts in the same manner as the capacitor in the capacitive filter. The components R_1 and C_2, however, form a voltage divider. Because capacitive reactance increases as frequency decreases ($X_C\uparrow$ is inversely proportional to $f\downarrow$), most of the dc volt-

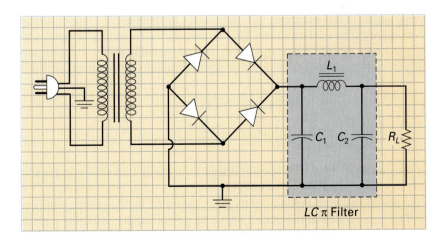

FIGURE 4-17   *LC π* **Filter.**

age will appear across C_2 and be applied to the load, while little of the ac ripple will appear across C_2 and be applied to the load. In summary, the capacitor blocks the dc component and directs it to the load while shunting the ac component away from the load.

4-4-4 *LC Filters*

Replacing R_1 in the RC π filter with an inductor L_1, as shown in Figure 4-17, will increase the efficiency of the filter because the inductor will oppose current without generating heat. The inductor, L_1, will offer a very low opposition or reactance to dc ($X_L\downarrow \propto f\downarrow$) and a very high reactance to the ac ripple ($X_L\uparrow \propto f\uparrow$). Although less power will be wasted, the inductor will be more expensive than the resistor.

SELF-TEST EVALUATION POINT FOR SECTION 4-4

Use the following questions to test your understanding of Section 4–4.

1. Would the filter in a dc power supply be considered:
   a. High pass b. Low pass c. Band pass d. Band stop
2. The percent ripple of a filter is proportional/inversely proportional to the value of the capacitor.
3. Which component would be connected in shunt, and which component would be connected in series to form a low-pass RC filter?
4. The most basic type of power supply filter is a capacitor in _____ with the load. (series/parallel)

4-5 REGULATORS

Voltage Regulator
An electrical circuit or device that maintains the output voltage constant despite variations in load resistance and input voltage.

The **voltage regulator** in a dc power supply maintains the dc output voltage constant despite variations in the ac input voltage and the output load resistance. To help explain why these variations occur, let us first examine the relationship between a source and its load.

4-5-1 *Source and Load*

Ideally, a dc power supply should convert all of its *ac electrical energy input* into a *dc electrical energy output*. However, like all devices, circuits, and systems, a dc power supply is not 100% efficient and, along with the electrical energy output, the dc power supply generates wasted heat. This is why a dc power supply can be represented as a voltage source with an internal resistance (R_{int}), as shown in Figure 4-18(a). The internal resistance, which is normally very small (in this example, 1 Ω), represents the heat energy loss or inefficiency of the dc power supply. The load resistance (R_L) represents the resistance of the electronic circuits in the electronic system. Since R_{int} and R_L are connected in series with one another, the 10-V supply in this example will be developed proportionally across these resistors, resulting in

$$V_{Rint} = 0.1 \text{ V}$$

$$V_{R_L} = \frac{9.9 \text{ V}}{10 \text{ V total}}$$

If the load resistance (R_L) were to decrease dramatically to 1 Ω, as shown in Figure 4-18(b), the decrease in load resistance ($R_L\downarrow$) would cause an increase in load current ($I_L\uparrow$). This current increase would cause an increase in the heat generated by the power supply ($P_{R_{int}}\uparrow = I^2\uparrow \times R$). This loss is seen at the output of the dc power supply, which is now delivering only a 5-V output to the load because the 10-V source voltage is being divided equally across R_{int} and R_L.

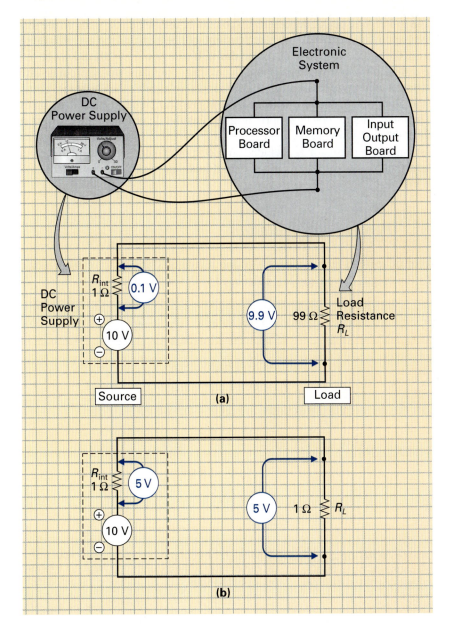

FIGURE 4-18   Source and Load.

$$V_{R_{\text{int}}} = 5 \text{ V}$$

$$V_{R_L} = \frac{5 \text{ V}}{10 \text{ V total}}$$

If the load resistance decreases, the load current will increase, and if too much current is drawn from the source, it will pull down the source voltage. Although a load resistance will generally not change as dramatically as shown in Figure 4-18(b), it will change slightly as different control settings are selected, and this will *load the source*. For example, the load resistance of the electronic circuits in your music system will change as the volume, bass, treble, and other controls are adjusted. This load resistance change would affect the output of a dc power supply if a regulator were not included to maintain the output voltage constant despite variations in load resistance.

To the electric company, each user appliance is a small part of its load, and it is the source. The ac voltage from the electric company can be anywhere between 105 V and 125 V ac rms (148 to 177 V ac peak), depending on consumer use, which depends on the time of day. When many appliances are in use, the load current will be high and the overall load resistance low, causing the source voltage to be pulled down. If a dc power supply did not have a regulator, these different input ac voltages from the electric company would produce different dc output voltages, when we really want the dc supply voltage to the electronic circuits to always be the same value no matter what ac input is present. This is the other reason that a regulator is included in a dc power supply; it maintains the dc output voltage constant despite variations in the ac input voltage.

In summary, a regulator is included in a dc power supply to maintain the dc output voltage constant despite variations in the output load resistance and the ac input voltage, as shown in Figure 4-19.

4-5-2 *Percent of Regulation*

The percent of regulation is a measure of the regulator's ability to regulate—or maintain constant—the output dc voltage. It is calculated with the formula

$$\text{Percent Regulation} = \frac{V_{nl} - V_{fl}}{V_{nl}}$$

$$V_{nl} = \text{No-Load Voltage}$$
$$V_{fl} = \text{Full-Load Voltage}$$

If we apply this formula to the example in Figure 4-20, we get

$$\% \text{ Regulation} = \frac{V_{nl} - V_{fl}}{V_{nl}} \times 100$$

$$= \frac{10 \text{ V} - 5 \text{ V}}{10 \text{ V}} \times 100 = 50\%$$

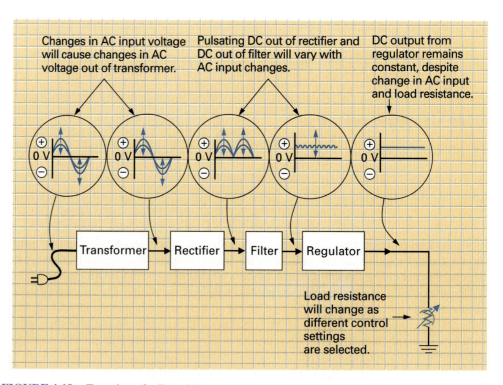

Changes in AC input voltage will cause changes in AC voltage out of transformer.

Pulsating DC out of rectifier and DC out of filter will vary with AC input changes.

DC output from regulator remains constant, despite change in AC input and load resistance.

Transformer → Rectifier → Filter → Regulator

Load resistance will change as different control settings are selected.

FIGURE 4-19 Function of a Regulator.

CHAPTER 4 / DIODE POWER SUPPLY CIRCUITS

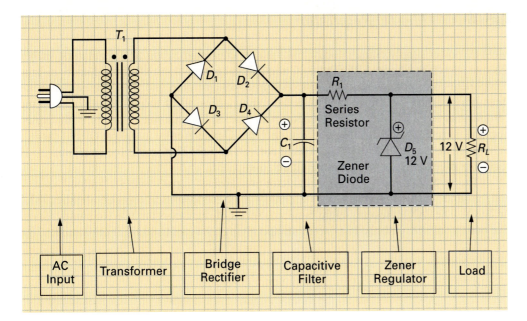

FIGURE 4-20 **Zener Diode Regulator.**

In the ideal situation, a regulator would be included and would maintain the output voltage constant between no-load and full-load, resulting in a percent regulation figure of

$$\% \text{ Regulation } = \frac{V_{nl} - V_{fl}}{V_{nl}} \times 100$$

$$= \frac{10 \text{ V} - 10 \text{ V}}{10 \text{ V}} \times 100$$

$$= 0\%$$

Most regulators achieve a percent regulation figure that is not perfect (0%) but is generally in single digits (2% to 8%).

4-5-3 *Zener Regulator*

In Chapter 3, it was shown how the zener diode could be used as a voltage regulator. Figure 4-20 shows how a zener diode (D_5) and a series resistor (R_1) would be connected in a dc power supply circuit to provide regulation. An increase in the ac input voltage would cause an increase in the dc output from the filter, which would cause an increase in the current through the series resistor (I_S), the zener diode (I_Z), and the load (I_L). The reverse-biased zener diode will, however, maintain a constant voltage across its anode and cathode (in this example, 12 V) despite these input voltage and current variations, with the additional voltage being dropped across the series resistor. For example, if the output from the filter could be between $+15$ and $+20$ V, the zener diode would always drop 12 V, while the series resistor would drop between 3 V (when input is 15 V) and 8 V (when input is 20 V). Consequently, the zener regulator will maintain a constant output voltage despite variation in the ac input voltage.

From the other standpoint, the zener regulator will also maintain a constant output voltage despite variations in load resistance. The zener diode achieves this by increasing and decreasing its current (I_Z) in response to load resistance changes. Despite these changes in zener and load current however, the zener voltage (V_Z), and therefore the output voltage (V_{out} or V_{R_L}), always remains constant. The disadvantage with this regulator is that the series-connected resistor will limit load current and, in addition, generate unwanted heat.

4-5-4   *The IC Regulator*

Most dc power supply circuits today make use of integrated circuit (IC) regulators, such as the one shown in Figure 4-21. These IC regulators contain about 50 individual or discrete components all integrated on one silicon semiconductor chip and then encapsulated in a three-pin package. The inset in Figure 4-21 shows how all of the IC regulator's internal components form circuits, which are shown as blocks. For example, a short-circuit protection and thermal shutdown circuit will protect the IC regulator by turning it OFF if the load current drawn exceeds the regulator's rated current, or if the heat sink is too small and the IC regulator is generating heat faster than it can dissipate it.

The three terminals of the IC regulator are labeled "input" (pin 1), "output" (pin 3), and "ground" (pin 2). The package is generally given the identification code of either 78XX or 79XX. The 78XX (seventy-eight hundred) series of regulators are used to supply a positive output voltage, with the last two digits specifying the output voltage. For example, 7805

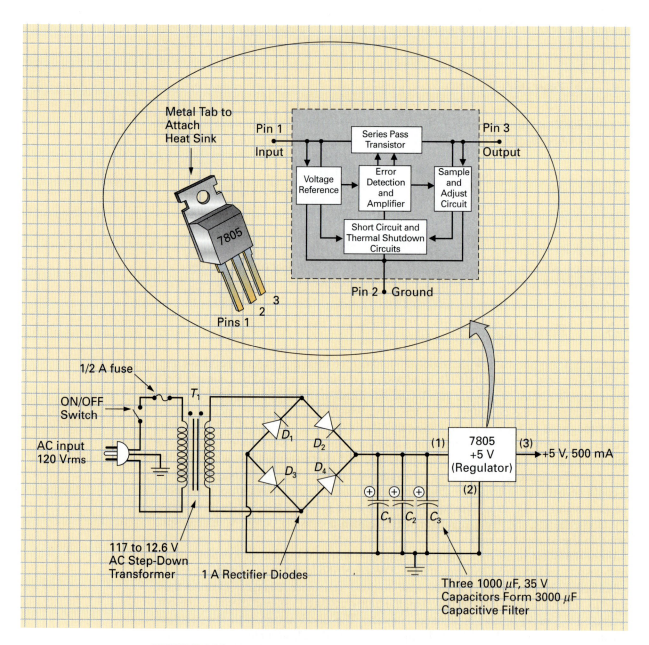

FIGURE 4-21   Integrated Circuit Regulator Within a Power Supply Circuit.

= +5 V, 7812 = +12 V, and so on. The 79XX (seventy-nine hundred) series of regulators are used to supply a negative output voltage, with the last two digits specifying the output voltage. These regulators will deliver a constant regulated output voltage as long as the input voltage is greater than the regulator's rated output voltage and can deliver a maximum output current of up to 1.5 A if properly heat sunk.

Use the following questions to test your understanding of Section 4–5.

1. Why is a voltage regulator needed in a dc power supply?
2. Normally, would a large or small load resistance pull down the source voltage?
3. Is the zener diode in a zener regulator circuit connected in series or in parallel with the load?
4. Indicate the output voltages of the following IC regulators:
   a. 7812 b. 7905 c. 7815 d. 7912

REVIEW QUESTIONS

Multiple-Choice Questions

1. The four main circuit blocks of a dc power supply listed in order from input to output are:
   a. Transformer, rectifier, filter, regulator
   b. Filter, regulator, rectifier, transformer
   c. Transformer, rectifier, regulator, filter
   d. Rectifier, filter, regulator, transformer

2. Which of the four circuit blocks of a dc power supply converts an ac input into a pulsating dc output?
   a. Transformer   c. Filter
   b. Regulator   d. Rectifier

3. Which of the four circuit blocks of a dc power supply converts a high ac voltage into a low ac voltage?
   a. Transformer   c. Filter
   b. Regulator   d. Rectifier

4. Which of the four circuit blocks of a dc power supply maintains the output voltage constant despite variations in the ac input and output load?
   a. Transformer   c. Filter
   b. Regulator   d. Rectifier

5. Which of the four circuit blocks of a dc power supply converts a pulsating dc input into a steady dc output?
   a. Transformer   c. Filter
   b. Regulator   d. Rectifier

6. Which rectifier circuit can be used to generate a negative pulsating dc output?
   a. Half-wave   c. Bridge
   b. Center-tapped   d. All of the above

7. Which rectifier uses four diodes?
   a. Half-wave   c. Bridge
   b. Center-tapped   d. All of the above

8. What rating is used to measure the action of a filter?
   a. Percent rectification   c. Percent regulation
   b. Percent ripple   d. Percent filtration

9. A small load resistance will cause a _____ load current and possibly pull _____ the source voltage.
   a. Small, down   c. Small, up
   b. Large, up   d. Large, down

10. What would be the output voltage of a 7915 IC regulator?
    a. +5 V   b. +15 V   c. −5 V   d. −15 V

Practice Problems

11. If a 240-V rms ac 60-Hz input is applied to the 19:1 step-down transformer shown in Figure 4-22, what would be the peak secondary output voltage?

12. What would be the average voltage out of a positive half-wave rectifier if the transformer secondary in Problem 11 were applied? (Take into account the barrier voltage drop.)

13. What would be the ripple frequency at the output for the circuits described in Problems 11 and 12?

14. Calculate the peak output voltage (taking into account V_{diode}) for the circuit shown in Figure 4-23.

15. What would be the output ripple frequency from the circuit in Figure 4-23?

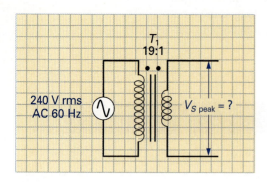

FIGURE 4-22

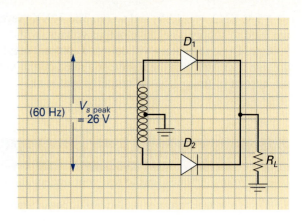

FIGURE 4-23

Diode Multiplier and Clipper Circuits

5

Business Was Dead

Business in 1892 was not going at all well for Almon Strowger, a funeral home director in the small town of La Porte, Indiana. Strangely enough, the telephone, which promoted business for most shopkeepers, was Strowger's downfall. This was because the wife of the owner of a competing funeral parlor served as the town's telephone operator. When people called the operator and asked to be put through to the funeral parlor, they were naturally connected to the operator's husband.

As the old expression states, "necessity is the mother of invention," and for Strowger it became a case of do or die. Making up his mind to cure the problem, he began to study the details of telephone communications. He designed a mechanism that made telephone connections automatic and thus bypassed the town operator. The switch he invented consisted of two parts—a 10-by-10 array of terminals (called the *bank*) arranged in a cylindrical arc and a moveable switch (called the *brush*). The brush stepped along and around a cylinder and could be put in one of 100 positions to connect to one of 100 terminals. The position of the brush was driven by an electromagnet that responded to pulses produced by the telephone dial.

The "Strowger switch" brought about the first automatic telephone exchange system in the world, with improved versions still used extensively in offices up to the 1960s.

Introduction

While rectification is by far the most common use of the diode, there are many other electrical and electronic circuits that make use of the diode's one-way, or unilateral, switching characteristics. In the previous chapters, we have seen the junction diode used in rectifying, encoding, and switching circuit applications. We have also seen the zener diode used in voltage-regulating and voltage-referencing applications and the light-emitting diode used in indicator and displaying applications. These are only a few of the many applications of these diodes. In this chapter, we will examine several other diode-circuit applications, including voltage multiplying, signal detecting, waveform shaping, circuit protecting, and dc restoring.

First, we will continue our discussion from the previous chapter on dc power supply circuits by examining the *voltage-multiplier circuit*. This circuit uses a few diodes to generate a large dc supply voltage that is some multiple of the ac peak input voltage. For example, an oscilloscope's or television's dc power supply would typically use a voltage-multiplier circuit to convert the ac input of 115 V to a 1000-V dc supply for an electrode of the cathode-ray tube.

The second type of circuit is used to change the shape of a waveform and is called a *clipper circuit.* This circuit uses a diode to clip or eliminate an unwanted portion of the ac input signal and converts one type of signal to another. For example, a clipper circuit could be used to clip off the top section and bottom section of a sine wave and change the waveform's shape from sine to almost square.

5-1 VOLTAGE-MULTIPLIER CIRCUITS

As mentioned in the previous chapter, because most electronic circuits require a dc supply voltage that is lower than 115 V, a step-down transformer is normally always included. This step-down transformer will reduce the 115 V ac input voltage from the wall outlet to a lower ac voltage, and then a rectifier, filter, and a regulator will produce a low-value dc supply voltage, as shown in Figure 5-1(a).

In some instances, a circuit or device may require a dc supply voltage that is larger than 115 V and a step-up transformer will have to be used, as shown in Figure 5-1(b). In this example circuit, you can see that the transformer steps the 115 V ac input up to 200 V so that the rectifier, filter, and regulator can produce two dc supply voltages of 178 V and 153 V. Step-up transformers are expensive, and if a wide range of dc supply voltages are needed, it becomes much more economical to use a voltage-multiplier circuit to obtain the higher-value dc supply voltages.

Voltage-multiplier circuits can replace rectifier and filter circuits because they employ diodes and capacitors in almost exactly the same way to convert an ac input to a dc output. The big difference is that, as their name implies, *voltage-multiplier circuits produce a dc output voltage that is some multiple of the peak ac input voltage.* For example, a **voltage-doubler circuit** will produce a dc output voltage that is twice the peak of ac input voltage, whereas a **voltage-tripler circuit** will produce a dc output voltage that is three times the peak of the ac input voltage.

Obtaining a greater output voltage from a multiplier circuit makes it seem as though we get more out than we put in. This is not the case, because voltage multipliers do not internally generate power. If the output voltage of a multiplier is increased by some multiple, the output current of the circuit will be decreased by the same multiple. For example, a voltage doubler will deliver twice the output voltage ($V \times 2$), but only half the output current ($I/2$); and therefore the output power from a voltage-multiplier circuit will never be greater than the input power ($P = V \uparrow I \downarrow$). In fact, because the diodes in a voltage multiplier circuit will dissipate some value of power, there will be slightly less power at the output compared to the input.

5-1-1 *Voltage-Doubler Circuits*

Like the rectifier, there are two types of voltage-doubler circuits: the half-wave voltage doubler and the full-wave voltage doubler. To begin with, let us examine the simplest of the voltage, multiplier circuits, the half-wave voltage doubler.

Half-Wave Voltage-Doubler Circuit

The half-wave voltage-doubler circuit is made up of two diodes and two capacitors and is shown in Figure 5-2(a). This circuit will produce a dc output voltage that is approximately twice the peak value of the ac sine wave input from the secondary of the transformer. In this example, we will assume that the transformer has a 1:1 turns ratio and the 115-V rms, 60-Hz primary voltage will be present at the secondary and applied to the half-wave voltage-doubler circuit. This 115-V rms, secondary voltage will have a peak value of:

$$115 \text{ V} \times 1.414 = 162.6 \text{ V}$$

Voltage-Doubler Circuit

Produces a dc output voltage that is twice the peak of the ac input voltage.

Voltage-Tripler Circuit

Produces a dc output voltage that is three times the peak of the ac input voltage.

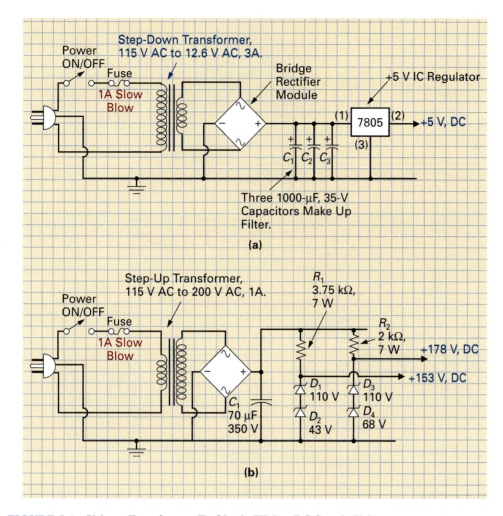

FIGURE 5-1 Using a Transformer To Obtain Higher DC Supply Voltages.

Figure 5-2(b) shows how the circuit will respond to the negative half-cycle of the input sine wave. When the input voltage swings negative, the top of the transformer secondary is made negative and this potential will forward bias D_1 and reverse bias D_2. With D_1 ON, capacitor C_1 will charge to the polarity shown until its plate-to-plate voltage is equal to the peak value of the input sine wave (162.6 V). To simplify the operation of the circuit during this negative half-cycle of the input, the inset in Figure 5-2(b) represents D_1 as a closed switch and D_2 as an open switch. In this equivalent circuit it can be seen how C_1 is connected across the secondary of the transformer by D_1 and will charge to the peak value of the input sine wave. Because there is almost no resistance in the charge path of C_1, the capacitor will charge very quickly to $V_{S\,pk}$.

Figure 5-2(c) shows how the circuit will respond to the positive half-cycle of the input sine wave. When the input voltage swings positive, the top of the transformer secondary is made positive and this potential will forward bias D_2 and reverse bias D_1. Referring to the equivalent circuit in the inset in Figure 5-2(c), you can see what effect diodes D_1 and D_2 will have when they are represented as an open and closed switch. Diode D_2 has now connected the transformer secondary voltage and the charge voltage of C_1 to the output capacitor C_2. Because the transformer's secondary peak voltage ($V_{S\,pk}$) and the capacitor C_1's peak charge voltage ($V_{S\,pk}$) are now *series-aiding voltage sources* ($+/- \rightarrow +/-$), capacitor C_2 will charge to the sum of these two series peak voltages ($C_2 = 2 \times V_{S\,pk}$). Because C_2 is connected across the load R_L, the dc output voltage will equal twice the peak secondary ac voltage, as shown in the simplified diagram in Figure 5-2(d).

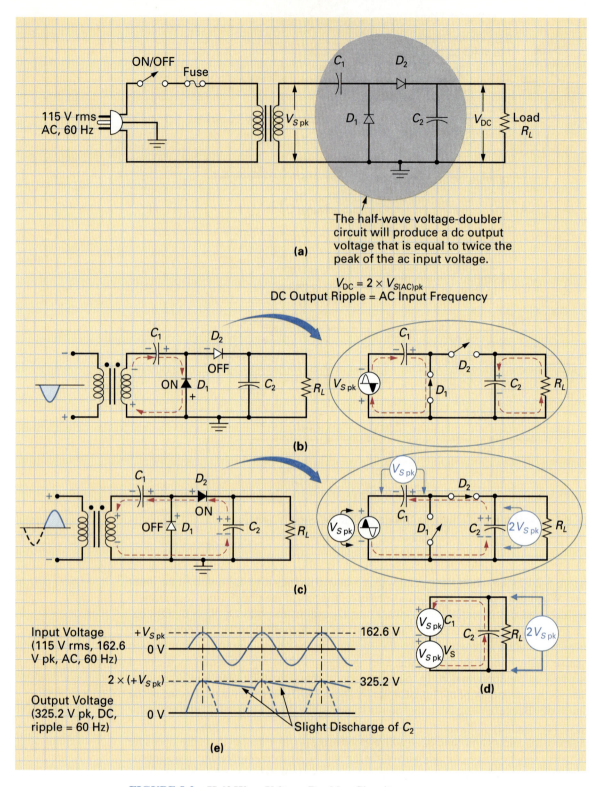

(a)

The half-wave voltage-doubler circuit will produce a dc output voltage that is equal to twice the peak of the ac input voltage.

$$V_{DC} = 2 \times V_{S(AC)pk}$$

DC Output Ripple = AC Input Frequency

(b)

(c)

(d)

Input Voltage (115 V rms, 162.6 V pk, AC, 60 Hz)

$+V_{S\,pk}$ — 162.6 V
0 V

Output Voltage (325.2 V pk, DC, ripple = 60 Hz)

$2 \times (+V_{S\,pk})$ — 325.2 V
0 V

Slight Discharge of C_2

(e)

FIGURE 5-2   **Half-Wave Voltage-Doubler Circuit.**

$$V_{out} = 2 \times V_{S\,pk}$$

$$= 2 \times 162.6 \text{ V}$$
$$= 325.2 \text{ V}$$

When V_S again switches polarity and returns to the condition shown in Figure 5-2(b), the cycle will repeat, with C_1 charging to the peak of the transformer's secondary voltage. During this half-cycle, you may have noticed that the output capacitor C_2 is able to discharge through the load, as shown in the inset in Figure 5-2(b). The time constant formed by the output capacitor and the load resistance is normally very high, and therefore during this negative half-cycle of the input, C_2 only will discharge slightly. This slight discharge of C_2 during the negative alternation of the input voltage can be seen in the output voltage waveform in Figure 5-2(e). Looking closely at this output voltage waveform, you can see that the circuit is called a half-wave voltage doubler because the output capacitor C_2 is charged for only half of the ac input cycle, producing an output that closely resembles that of the filtered half-wave rectifier circuit. Comparing the input voltage waveform to the output voltage waveform in Figure 5-2(e), you can see that for every one cycle of the input (positive-to-positive peak) there is one cycle of the output (positive-to-positive peak). The output dc ripple frequency of a half-wave voltage doubler will therefore equal the input ac frequency:

$$f_{out\ ripple} = f_{in\ ac}$$

Output dc ripple frequency = the input ac frequency

In the circuit in Figure 5-2, the output ripple frequency will equal 60 Hz.

EXAMPLE:

By reversing the directions of the diodes and capacitors in Figure 5-2, you can change the circuit from a *positive half-wave voltage doubler* to a *negative half-wave doubler*. Sketch a negative half-wave doubler circuit and show the C_1 and C_2 charge paths for each alternation of the input. In addition, what would be the peak of the dc output voltage from this circuit, and what would be its ripple frequency?

Solution:

Figure 5-3 shows the component orientation and connection for a negative half-wave doubler circuit. During the positive alternation of the ac input, D_1 will turn ON and allow C_1 to charge, with the polarity shown, to the peak of the ac input. During the negative alternation of the ac input, D_2 will turn ON and connect the two series-aiding peak voltages of V_S and

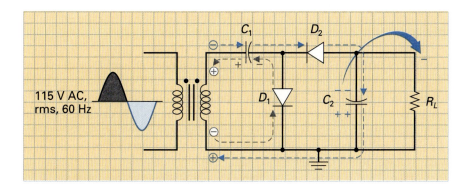

FIGURE 5-3 **Negative Half-Wave Voltage Doubler.**

the charge across C_1, across the output capacitor C_2. The result will be that C_2 will charge to twice the peak of the ac input, and this dc negative voltage will be applied across the load.

$$115 \text{ V ac, rms} \times 1.414 = 162.6 \text{ V ac peak}$$

$$V_{\text{out}} = 2 \times V_{S\,\text{pk}}$$

$$= 2 \times 162.6 \text{ V}$$

$$= 2325.2 \text{ V}$$

$$\text{Output dc Ripple Frequency} = \text{Input ac Frequency} = 60 \text{ Hz}$$

Full-Wave Voltage-Doubler Circuit

The half-wave voltage doubler has two disadvantages. The first is that the ripple frequency of 60 Hz is difficult to filter, and the second is that an expensive output capacitor is required because it must have a voltage rating that is more than twice the peak of the ac input.

The **full-wave voltage-doubler circuit** is shown in Figure 5-4(a). As you can see, this circuit uses the same number of components as the half-wave voltage doubler (two diodes, two capacitors), and yet it overcomes the two disadvantages of difficult filtering and the need for a high-output capacitor voltage rating.

Figure 5-4(b) shows how the circuit will respond to the positive half-cycle of the input sine wave. When the input voltage swings positive, the top of the transformer secondary is made positive and this potential will forward bias D_1 and reverse bias D_2. With D_1 ON, capacitor C_1 will charge to the polarity shown until its plate-to-plate voltage is equal to the peak value of the input sine wave (162.6 V). To simplify the operation of the circuit during this positive half-cycle of the input, the inset in Figure 5-4(b) represents D_1 as a closed switch and D_2 as an open switch. In this equivalent circuit it can be seen how C_1 is connected across the secondary of the transformer by D_1, and therefore it will charge to the peak value of the input sine wave. Because there is almost no resistance in the charge path of C_1, the capacitor will charge very quickly to $V_{S\,\text{pk}}$.

Figure 5-4(c) shows how the circuit will respond to the negative half-cycle of the input sine wave. When the input voltage swings negative, the top of the transformer secondary is made negative and this potential will forward bias D_2 and reverse bias D_1. Referring to the equivalent circuit in the inset in Figure 5-4(c), you can see what effect diodes D_1 and D_2 will have when they are represented as an open and closed switch. In this equivalent circuit it can be seen how C_2 is connected across the secondary of the transformer by D_2 and will charge to the peak value of the input sine wave. As there is almost no resistance in the charge path of C_2, the capacitor will charge very quickly to $V_{S\,\text{pk}}$. Because capacitors C_1 and C_2 are now *series-aiding voltage sources* $(+/- \rightarrow +/-)$, the sum of these two series peak voltages will be applied across the load.

$$V_{\text{out}} = 2 \times V_{S\,\text{pk}}$$

$$= 2 \times 162.6 \text{ V}$$

$$= 325.2 \text{ V}$$

Once the capacitors C_1 and C_2 are charged, the diodes D_1 and D_2 will conduct only during the peaks of the ac input. Between these peaks, C_1 and C_2 will discharge through the load as shown in Figure 5-4(d). The time constant formed by the output capacitors and the load resistance is normally very high, and therefore C_1 and C_2 will discharge only slightly. This slight discharge of C_1 and C_2 can be seen in the output voltage waveform in Figure 5-4(e). Looking closely at this output voltage waveform, you can see that the circuit is called a full-wave voltage doubler because the output capacitors C_1 and C_2 are charged for both half-cycles of the ac input, producing an output that closely resembles that of the filtered full-wave rectifier circuit. Comparing the input voltage waveform to the output voltage waveform in Figure 5-4(e), you can see that for every one cycle of the input (positive-to-positive

CHAPTER 5 / DIODE MULTIPLIER AND CLIPPER CIRCUITS

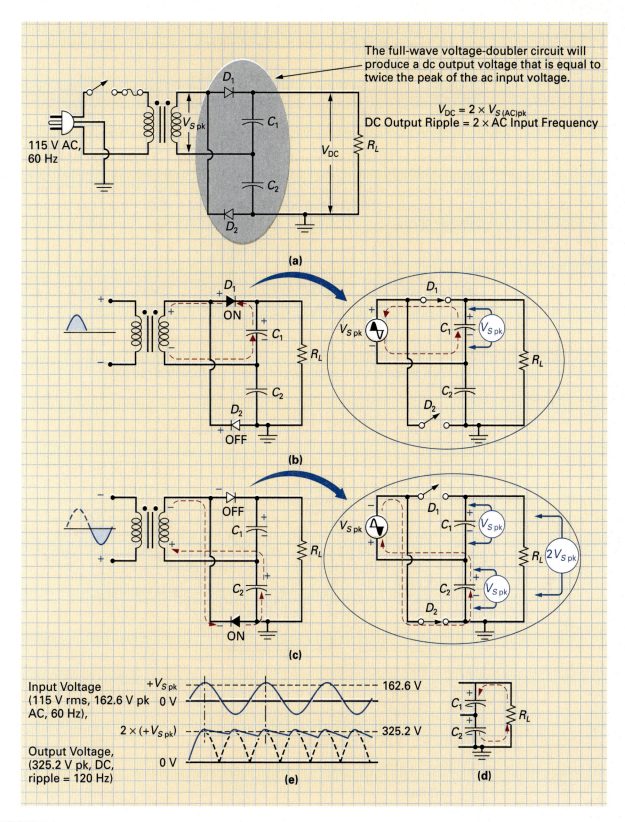

The full-wave voltage-doubler circuit will produce a dc output voltage that is equal to twice the peak of the ac input voltage.

$$V_{DC} = 2 \times V_{S(AC)pk}$$

DC Output Ripple = $2 \times$ AC Input Frequency

(a)

(b)

(c)

Input Voltage
(115 V rms, 162.6 V pk
AC, 60 Hz),

$+V_{S\,pk}$ 162.6 V

0 V

$2 \times (+V_{S\,pk})$ 325.2 V

Output Voltage,
(325.2 V pk, DC,
ripple = 120 Hz)

0 V

(e)

(d)

FIGURE 5-4   Full-Wave Voltage Doubler Circuit.

peak) there are two cycles of the output (positive-to-positive-to-positive peak). The output dc ripple frequency of a full-wave voltage doubler will therefore equal twice the input ac frequency:

$$f_{\text{out ripple}} = 2 \times f_{\text{in ac}}$$

Output dc Ripple Frequency = 2 × Input ac Frequency

In the circuit in Figure 5-4, the output ripple frequency will equal 120 Hz. The full-wave voltage doubler overcomes both disadvantages of the half-wave circuit by having a higher ripple frequency, which is easier to filter. Because the output voltage is split across C_1 and C_2, each capacitor need only have a voltage rating that is slightly more than the peak of the ac input voltage.

■ **EXAMPLE:**

Identify the circuit in Figure 5-5 and then answer the following questions:

  a. What would be the polarity and value of the output voltage at point X and from the inset circuit at point Y?
  b. What would be the output ripple frequency from both circuits?

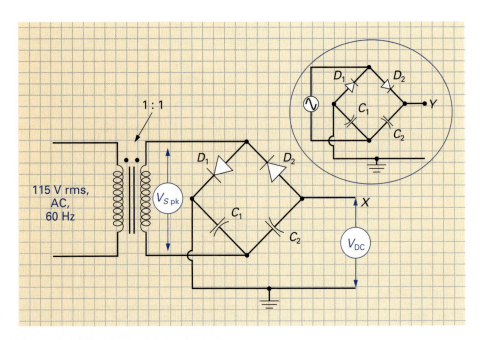

FIGURE 5-5   A DC Power Supply Circuit.

■ *Solution:*

The circuit in Figure 5-5 is a negative full-wave voltage-doubler circuit, and the circuit in the inset in Figure 5-5 is a positive full-wave voltage doubler.

  a. 115 V ac, rms × 1.414 = 162.6 V ac, peak

$$\text{Point } X: \quad \begin{aligned} V_{\text{out}} &= 2 \times V_{S\,\text{pk}} \\ &= 2 \times (-162.6 \text{ V}) \\ &= -325.2 \text{ V} \end{aligned}$$

$$\text{Point } Y: \quad V_{out} = 2 \times V_{S\,pk}$$
$$= 2 \times (+162.6\text{ V})$$
$$= +325.2\text{ V}$$

b. Output dc Ripple Frequency = 2 × Input ac Frequency = 120 Hz

5-1-2   *Voltage-Tripler Circuit*

A voltage-tripler circuit will produce a dc output voltage that is three times the peak of the ac input voltage. Figure 5-6 shows how three diodes and three capacitors can be connected to form a voltage-tripler circuit. To explain how this circuit operates, we will examine how it responds to three half-cycles of the ac input sine wave. These three circuit states are shown in Figure 5-6(a), (b), and (c).

As can be seen in Figure 5-6(a), the first positive half-cycle will turn D_1 ON and allow C_1 to charge to the peak of the ac secondary, voltage. This condition is shown in the simplified diagram in the inset in Figure 5-6(a).

Figure 5-6(b) shows what will occur when the secondary voltage swings negative. Diode D_2 will be forward biased because its anode has been made positive by the charge on C_1 and its cathode has been made negative by the potential at the top of the transformer's secondary. Diode D_2 has now connected the transformer's secondary voltage and the charge voltage of C_1 across the capacitor C_2, as shown in the inset in Figure 5-6(b). Because the transformer's secondary peak voltage ($V_{S\,pk}$), and the capacitor C_1's peak charge voltage

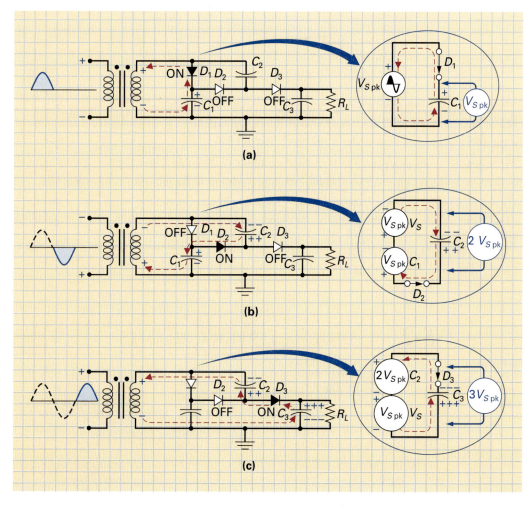

(a)

(b)

(c)

FIGURE 5-6   Voltage-Tripler Circuit.

($V_{S\,pk}$) are now series-aiding voltage sources ($+/- \rightarrow +/-$), capacitor C_2 will charge to the sum of these two series peak voltages ($C_2 = 2 \times V_{S\,pk}$).

Figure 5-6(c) shows what will happen at the peak of the next positive half-cycle of the input. Diode D_3 will be forward biased because its anode has been made positive by the charge on C_2 and its cathode has been made negative by the potential at the bottom of the transformer's secondary. Diode D_3 has now connected the transformer's secondary voltage and the charge voltage of C_2 across the capacitor C_3, as shown in the inset in Figure 5-6(c). Because the transformer's secondary peak voltage ($V_{S\,pk}$) and C_2's double peak-charge voltage ($2 \times V_{S\,pk}$) are now series-aiding voltage sources ($+/- \rightarrow +/-$), capacitor C_3 will charge to the sum of these two series peak voltages, which is three times the secondary peak voltage:

$$C_3 = V_{S\,pk} + (2 \times V_{S\,pk}) = 3 \times V_{S\,pk}$$

■ **EXAMPLE:**

What would be the peak output voltage across the load from the voltage-tripler circuit in Figure 5-6, if the ac input was 115 V rms?

■ *Solution:*

$$115 \text{ V ac, rms} \times 1.414 = 162.6 \text{ V ac, peak}$$
$$V_{out} = 3 \times V_{S\,pk}$$
$$= 3 \times 162.6 \text{ V}$$
$$= 487.8 \text{ V}$$

SELF-TEST EVALUATION POINT FOR SECTION 5-1

Use the following questions to test your understanding of Section 5–1.

1. A half-wave voltage doubler whose input is 115 V rms will produce an output that is about _____ volts, peak.

2. What are the two disadvantages of the half-wave voltage doubler compared to the full-wave voltage doubler?

3. What would be the peak output from a voltage-tripler and voltage-quadrupler circuit for a 115-V rms, ac input?

4. Assuming a 115-V rms, 60-Hz input, what would be the output ripple frequency from a half-wave voltage-doubler circuit and a full-wave voltage-doubler circuit?

5-2   DIODE CLIPPER CIRCUITS

Clipper Circuit or Limiter

Used to eliminate an unwanted section of waveform.

A **clipper circuit** is used to cut off or eliminate an unwanted section of a waveform. It is used in basically one of two applications:

    a. In many applications, *it is used to remove a natural part of the waveform.* For example, the half-wave rectifier is a basic clipper circuit because it eliminates either the positive or negative alternation of the ac input signal.

    b. In other applications, *it is used to prevent a voltage from exceeding a certain value.* In these instances, the clipper circuit is often called a **limiter** because it will limit a high-amplitude voltage pulse or other signal.

5-2-1   *Series Clipper Circuits*

The diode is an ideal clipper because it can be connected in one of two ways and biased at a certain reference level to clip off a certain part of the input waveform. There are two types of clipper circuits: the **series clipper,** which contains a diode that is in series with the load, and the **shunt clipper,** which has a diode that is in shunt, or in parallel, with the load. In this section, we will examine the different series clipper circuits and their applications.

Basic Series Clipper Circuits

Figure 5-7 shows the two basic series clipper circuits along with their input and output waveforms. Comparing these circuits to the previously discussed half-wave rectifier circuits, you will notice that they are identical.

The circuit shown in Figure 5-7(a) is called a **negative series clipper** because this circuit has a diode connected in *series* with the load or output, and its orientation is such that it will *clip* off the *negative* alternation of the ac input. Referring to the waveforms in Figure 5-7(b), you can see that during the positive alternation of the ac input, the diode is forward biased and equivalent to a closed switch, as shown in the inset. The positive alternation is connected to the output; however, there will be a 0.7-V loss in voltage between input and output due to the series-connected diode.

Series Clipper

A circuit that will clip off part of the input signal. Also known as a limiter because the circuit will limit the ac input. A series clipper circuit has a clipping or limiting device in series with the load.

Shunt Clipper

A circuit that will clip off part of the input signal. Also known as a limiter because the circuit will limit the ac input. A shunt clipper circuit has a clipping or limiting device in shunt with the load.

Negative Series Clipper

A circuit that has a diode connected in series with the load or output; its orientation is such that it will clip off the negative alternation of the ac input.

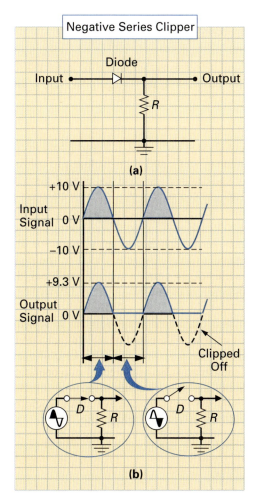

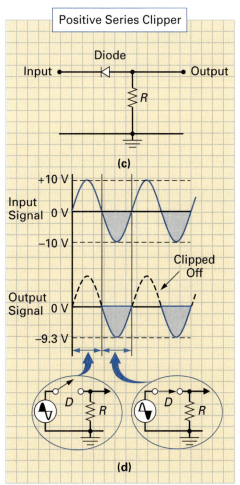

FIGURE 5-7   Basic Series Clipper Circuits.

$$V_{out} = V_{in} - 0.7 \text{ V}$$

$$V_{out} = 10 \text{ V} - 0.7 \text{ V} = 9.3 \text{ V}$$

During the negative alternation of the ac input, the diode is reverse biased and equivalent to an open switch, as shown in the inset in Figure 5-7(b). The negative alternation will be blocked or clipped by the series diode, and the output voltage will be 0 V.

The circuit shown in Figure 5-7(c) is called a **positive series clipper** because this circuit has a diode connected in *series* with the load or output and its orientation is such that it will *clip* off the *positive* alternation of the ac input. Comparing this circuit to the one in Figure 5-7(a), you can see that the only difference is the direction of the diode. Here, the positive alternation of the ac input will be clipped off, and the negative alternation of the ac input will be passed to the output. Referring to the waveforms in Figure 5-7(d), you can clearly see this result. Once again the amplitude of the output voltage takes into account the 0.7-V drop due to the series-connected diode.

As mentioned earlier, the construction of these circuits is no different than that of the half-wave rectifier circuits discussed in detail in the previous chapter. The difference between the two is their application: a half-wave rectifier circuit is used in a dc power supply to convert ac to dc, whereas a series clipper circuit is used to clip off or eliminate a certain part of an ac signal. To draw a parallel, this "same construction/different application" situation could be compared to the electromagnet and inductor. They are also constructed in exactly the same way but used in two completely different applications.

■ **EXAMPLE:**

Referring to the circuit in Figure 5-8(a), sketch the waveforms that would be obtained for each of the test points 1, 2, and 3. Label each waveform with its voltage peak values and time periods.

■ *Solution:*

Figure 5-8(b) shows the waveforms that will be obtained at test points 1, 2, and 3 for the circuit in Figure 5-8(a). The differentiator circuit (made up of C_1 and R_1) is included to convert the square-wave input to the positive and negative spike signal shown at test point 2. As with all differentiator circuits, the values of R and C are chosen so that the RC time constant is short compared to the half-cycle period of the input square wave. In this example, the time constant is equal to 12 msec ($\tau = R \times C = 1 \text{ k}\Omega \times 12 \text{ }\mu\text{F} = 12$ msec). Even taking into account 5 time constants (which is the time it takes for a capacitor to fully charge or discharge), the time of 60 msec is still short compared to the half-cycle period of the input square wave, which is 100 msec. The negative series clipper circuit will clip or block the negative spike from the differentiator circuit, and therefore the output signal at TP_3 will be a positive 5-V spike.

A circuit such as the one shown in Figure 5-8 is often used in applications in which a positive pulse is needed from a square wave. This narrow pulse could be used to activate or trigger a circuit into operation only for the duration of the positive spike. This kind of operation is called **positive-edge triggering** because the circuit is being triggered into operation only on the positive edge of the square-wave input. If a positive series clipper circuit was used in place of the negative series clipper in Figure 5-8(a), the positive spike would be clipped and the negative spike output could provide **negative-edge triggering.**

Biased Series Clipper Circuits

With the basic series clippers discussed in the previous section, the clipping level was at 0 V, with the direction of the diode determining whether everything above or below 0 V was eliminated. With **biased series clipper circuits,** the clipping level can be changed by apply-

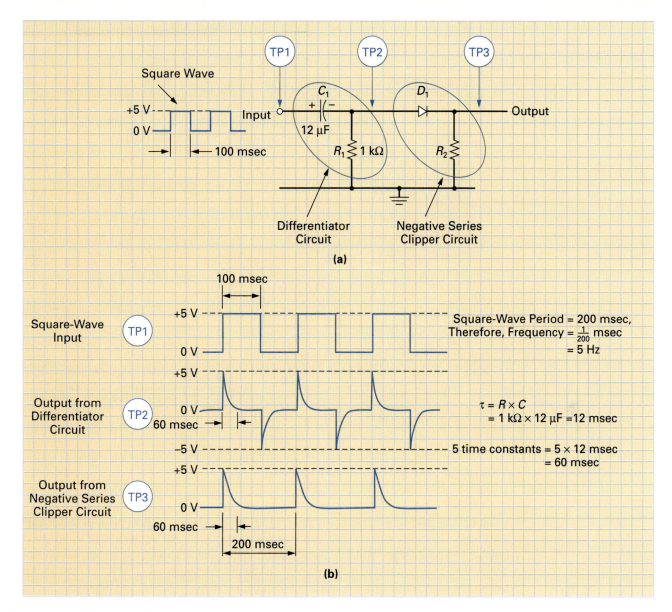

FIGURE 5-8 *Positive Pulse Generator Circuit.*

ing a bias voltage to the series-connected diode. The two basic types of biased series clipper circuits are shown in Figure 5-9.

In Figure 5-9(a), the cathode of the diode is set at +5 V by a dc source. This +5-V dc supply will set the clipping level because the diode cannot conduct until the input signal overcomes the +5-V supply and the diode's 0.7-V barrier voltage. Referring to the input/output waveforms in Figure 5-9(b), you can see that the output remains at +5-V except when the positive ac input exceeds this voltage level. At this positive peak of the ac input waveform, the diode will conduct and switch this portion of the input signal through to the output. The peak output will, of course, be 0.7 V less than the input due to the voltage drop across the series-connected diode. As you can see in the output waveform in Figure 5-9(b), part of the positive alternation and all of the negative alternation will be clipped off.

Figure 5-9(c) shows how the circuit can be converted to a *biased positive series clipper* by reversing the direction of the diode and the polarity of the bias voltage. The −5 V-dc supply will set the clipping level because the diode cannot conduct until the input signal overcomes the −5-V supply and the diode's 0.7-V barrier voltage. Referring to the input/output waveforms in Figure 5-9(d), you can see that the output remains at −5 V except

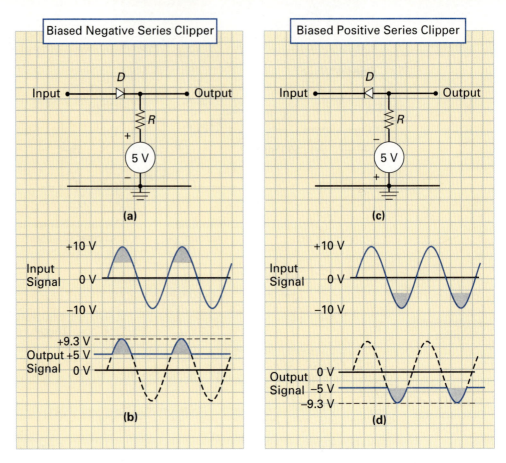

FIGURE 5-9 Biased Series Clipper Circuits.

when the negative ac input exceeds this voltage level. At this negative peak of the ac input waveform, the diode will conduct and switch this portion of the input signal through to the output. The peak output will, of course, be 0.7 V less than the input due to the voltage drop across the series-connected diode. As you can see in the output waveform in Figure 5-9(d), part of the negative alternation and all of the positive alternation will be clipped off.

5-2-2   *Shunt Clipper Circuits*

As previously mentioned, there are two types of clipper circuits: the series clipper, which contains a diode that is in series with the load, and the shunt clipper, which has a diode that is in shunt, or in parallel, with the load. In this section, we will examine the different shunt clipper circuits and their applications.

Basic Shunt Clipper Circuits

Figure 5-10 shows the two basic shunt clipper circuits along with their input and output waveforms. These shunt clipper circuits achieve exactly the same results as the series clipper circuits but operate in almost an opposite way. To explain this, the series clipper produced an output when the diode was forward biased and no output when the diode was reverse biased. As you will see in the operation of the shunt clipper, it will short the input signal to ground when it is forward biased and produce an output when the diode is reverse biased. Let us now examine the two basic shunt clipper circuits in more detail.

The circuit shown in Figure 5-10(a) is called a **negative shunt clipper** because this circuit has a diode connected in *shunt* with the load or output and its orientation is such that

Negative Shunt Clipper

A circuit that has a diode connected in shunt with the load or output; its orientation is such that it will clip off the negative alternation of the ac input.

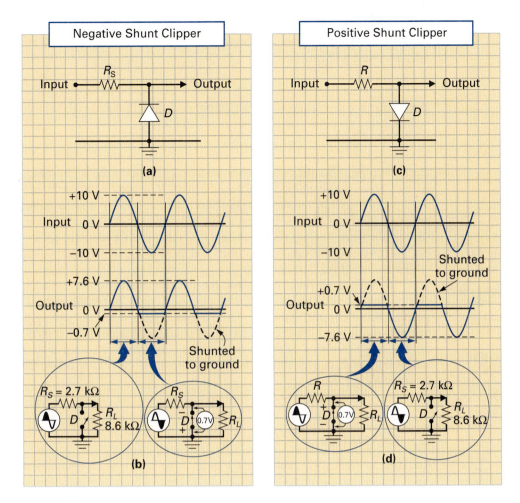

FIGURE 5-10 Basic Shunt Clipper Circuits.

it will *clip* off the *negative* alternation of the ac input by shunting it to ground. Referring to the waveforms in Figure 5-10(b), you can see that during the positive alternation of the ac input, the diode is reverse biased and equivalent to an open switch, as seen in the inset. With the diode effectively removed from the circuit, the output voltage developed across the load can be calculated using the standard voltage-divider formula.

$$V_{\text{out}} = \frac{R_L}{R_L + R_S} = V_{\text{in}}$$

Applying this formula to the example values given in the inset for the positive alternation in Figure 5-10(b), the peak output voltage will be as follows:

$$V_{\text{out}} = \frac{R_L}{R_L + R_S} \times V_{\text{in}}$$

$$V_{\text{out}} = \frac{8.6 \text{ k}\Omega}{8.6 \text{ k}\Omega + 2.7 \text{ k}\Omega} \times 10 \text{ V} = 7.6 \text{ V}$$

As you can see from this example, although the output signal will resemble the positive alternation of the input in shape, the peak output will be less than the peak input voltage due to the voltage division of R_S and R_L.

During the negative alternation of the ac input signal, the shunt clipper diode is forward biased, as seen in the inset in Figure 5-10(b). The forward-biased diode will develop its usual drop of 0.7 V and because the diode is connected in parallel with the load, the load voltage will equal the forward voltage drop across the diode. To state this with a formula, the output voltage will equal

$$V_{out} = -0.7 \text{ V}$$

During the negative alternation, the output voltage will equal -0.7 V, with the remaining voltage being dropped across the series resistor, R_S.

$$V_{R_S} = -10 \text{ V}_{peak} - (-0.7 \text{ V}) = -9.3 \text{ V}_{peak}$$

During this negative alternation of the ac input, the reason the series resistor R_S is included is clear. When the shunt diode is forward biased, R_S acts as a current-limiting resistor. Without R_S in the circuit, the diode would short the ac input voltage to ground, which would probably cause the diode's maximum forward current rating to be exceeded. In most applications, the value of R_S is chosen to be a low value so that when the diode is reverse biased the voltage drop across R_S will be small and most of the input signal voltage will be developed across the load.

To clip off, or shunt to ground, the positive alternation of the ac input, we simply change the diode's direction, as shown in Figure 5-10(c). Referring to the waveforms for the **positive shunt clipper**, shown in Figure 5-10(d), you can see that during the positive alternation of the ac input, the diode is forward biased and therefore the output is $+0.7$ V. During the negative alternation of the ac input, the diode is reverse biased and, if we assume the same values of R_S and R_L, the negative peak output will be -7.6 V due to the voltage-divider action of R_S and R_L.

■ **EXAMPLE:**

Referring to Figure 5-11, calculate the peak of the square wave's output voltage.

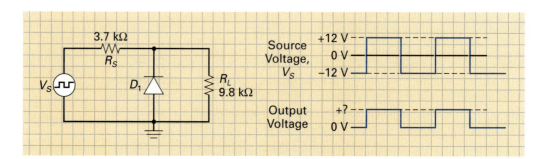

FIGURE 5-11 **Positive Square Wave Generator Circuit.**

■ *Solution:*

When the input signal goes negative, the shunt clipper diode will conduct and shunt the signal to ground. When the input signal is positive, however, the diode is reverse biased and the series resistor and the load form a voltage divider, with the following voltage being developed across the output or load:

$$V_{out} = \frac{R_L}{R_L + R_S} \times V_{in}$$

$$V_{out} = \frac{9.8 \text{ k}\Omega}{9.8 \text{ k}\Omega + 3.7 \text{ k}\Omega} \times 12 \text{ V} = \frac{9.8 \text{ k}\Omega}{13.5 \text{ k}\Omega} \times 12 \text{ V} = 8.7 \text{ V}$$

Biased Shunt Clipper Circuits

Like the series clipper circuit, shunt clipper circuits can have their clipping level adjusted by introducing a bias voltage. Figure 5-12 illustrates how a bias voltage can be applied to the shunt-connected diodes to adjust the point at which a certain section of the input signal is eliminated.

Figure 5-12(a) shows how a shunt-connected diode and a bias voltage can be connected to clip off, or shunt to ground, a section of the negative alternation. Referring to the waveforms in Figure 5-12(b), you can see that during the positive alternation of the ac input the diode is reverse biased and therefore equivalent to an open switch, as shown in the inset. Assuming the values given in the inset for the positive alternation in Figure 5-12(b), the output voltage developed across the load will be:

$$V_{\text{out}} = \frac{R_L}{R_L + R_S} \times V_{\text{in}}$$

$$V_{\text{out}} = \frac{9.3 \text{ k}\Omega}{9.3 \text{ k}\Omega + 1.3 \text{ k}\Omega} \times 10 \text{ V} = \frac{9.3}{10.6 \text{ k}\Omega} \times 10\text{V} = 8.8 \text{ V}$$

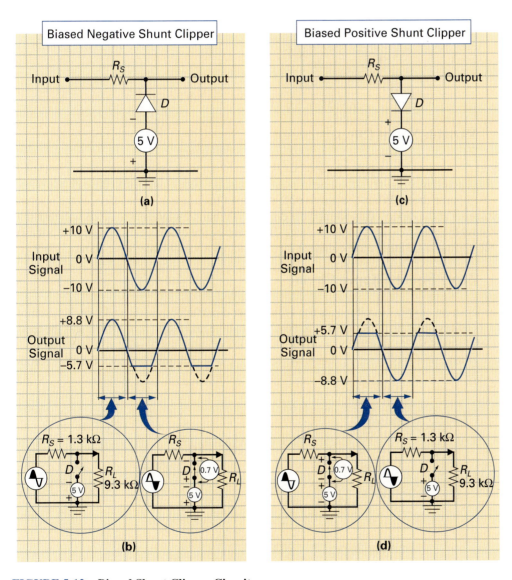

FIGURE 5-12 **Biased Shunt Clipper Circuits.**

Referring back to Figure 5-12(a), you can see that the anode of the diode is connected to −5 V. This means that the diode can conduct only when the input voltage exceeds the bias voltage of −5 V and the diode's barrier voltage of 0.7 V, a total of −5.7 V. Referring to the waveforms in Figure 5-12(b), you can see that when the negative alternation of the ac input exceeds −5.7 V, the diode is forward biased and the input signal is clipped at this voltage.

$$V_{out} = V_{bias} + V_F$$

$$= -5 V + (-0.7 V) = -5.7 V$$

Figure 5-12(c) shows how the positive peak of the ac input signal can be clipped by reversing the direction of the diode and the polarity of the bias voltage from Figure 5-12(a). Referring to the waveforms for this biased positive shunt clipper circuit in Figure 5-12(d), you can see that when the input exceeds +5.7 V, the diode is forward biased and the input signal will be clipped at this voltage.

$$V_{out} = V_{bias} + V_F = +5 V + (+0.7 V) = +5.7 V$$

During the negative alternation of the ac input signal, the diode is reverse biased and therefore equivalent to an open switch, as shown in the inset in Figure 5-12(d). Assuming the values given in this inset, the output voltage developed across the load will be

$$V_{out} = \frac{R_L}{R_L + R_S} \times V_{in}$$

$$V_{out} = \frac{9.3 \text{ k}\Omega}{9.3 \text{ k}\Omega + 1.3 \text{ k}\Omega} \times (-10 V) = -8.8 V$$

Variable Shunt Clipper Circuit

Figure 5-13 shows how a potentiometer can be used to adjust the positive shunt clipping level. In this circuit, R_1 will vary the value of positive biasing voltage applied to the cathode of the shunt clipping diode. When the positive alternation of the ac input signal exceeds the positive voltage set at the diode's cathode and the diode 0.7-V barrier voltage, the diode will turn ON and limit the positive output to this voltage. By reversing the diode's direction and the polarity of the bias voltage, the circuit could be made to act as a variable negative shunt clipper.

Zener Shunt Clipper Circuit

A zener shunt clipper circuit makes use of the zener diode's forward-biased switching action and its reverse-bias zener action to clip both the positive and negative alternation of the input signal. Figure 5-14(a) shows how differently a clipper circuit will react if a zener diode is used instead of a P-N junction diode. When the input signal goes positive, the zener diode

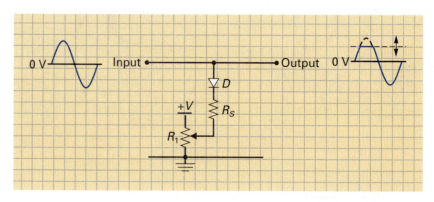

FIGURE 5-13 **Variable Shunt Clipper Circuit.**

CHAPTER 5 / DIODE MULTIPLIER AND CLIPPER CIRCUITS

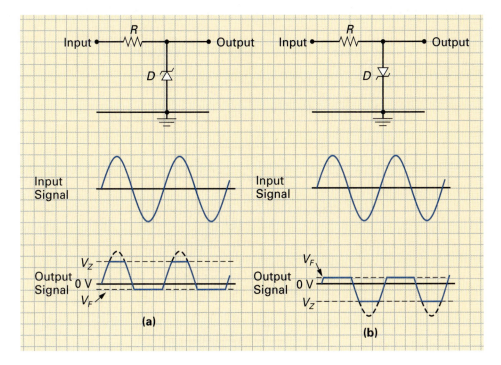

FIGURE 5-14 **Zener Shunt Clipper Circuits.**

will remain OFF until the input signal reaches the diode's zener voltage (V_Z). At this time, the zener diode will conduct and hold the output voltage at this level. When the input signal goes negative, the zener diode will be forward biased and clip the input signal's negative alternation at -0.7 V (V_F).

If we were to reverse the direction of the zener diode, the output signal clipping would be reversed, as shown in Figure 5-14(b). In this instance, the positive alternation of the input signal would be limited to $+0.7$ V (V_F), and the negative alternation of the input would be clipped at the zener voltage (V_Z).

Symmetrical Zener Shunt Clipper Circuit

Figure 5-15(a) shows how two back-to-back zener diodes can be used to limit both input signal peaks to produce a clipped symmetrical output. Referring to the input/output waveforms in Figure 5-15(b), you can see that both the positive and negative alternation have a simplified equivalent circuit, as shown in the insets.

When the input signal swings positive and exceeds $+6.7$ V, both zener diodes will conduct. Diode D_1 will conduct because it is forward biased, and D_2 will conduct because it has been sent into its reverse breakdown region—its zener voltage rating is 6 V. The output will be limited or clipped in this case to

$$V_{out} = V_Z + V_F$$

$$= 6 \text{ V} + 0.7 \text{ V} = +6.7 \text{ V}$$

When the negative alternation of the input exceeds -6.7 V, diode D_1 will drop its zener voltage of 6 V because it is operating in its reverse breakdown region and D_2 will drop 0.7 V because it is forward biased. The output will therefore be clipped to

$$V_{out} = -V_Z + (-0.7 \text{ V})$$

$$= -6 \text{ V} + (-0.7 \text{ V}) = -6.7 \text{ V}$$

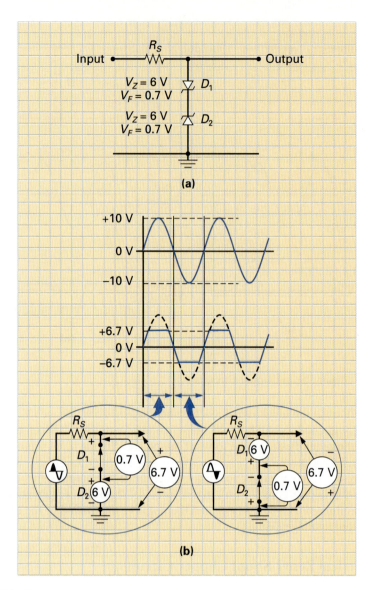

FIGURE 5-15 **Symmetrical Zener Shunt Clipper Circuit.**

In summary, if the input signal exceeds these two extremes of $+6.7$ V and -6.7 V, the diodes will conduct and shunt the peaks to ground. Between these two voltages, however, neither of the diodes can conduct, and the input signal is passed through to the output.

Shunt Clipper Circuit Applications

In most applications, shunt clipper circuits are used to prevent a voltage from exceeding a certain value. In these instances, the shunt clipper circuit is often called a *limiter*, because it will limit an unwanted high-amplitude voltage or current pulse or spike. These voltage spikes or current surges are often referred to as **transient voltages or currents.** By definition, these are pulses, or sudden momentary cycles, that occur in a circuit because of a sudden change in voltage or load. Figure 5-16 shows three examples of how a shunt clipper or limiter circuit can be used in a circuit to eliminate transient voltages or currents.

In Figure 5-16(a), a symmetrical zener shunt circuit is being used to protect a dc power supply from the transients that can occur in the ac voltage from the wall outlet, as seen in the waveform in the inset. These zeners, which would typically have a V_Z rating of 200 V, will

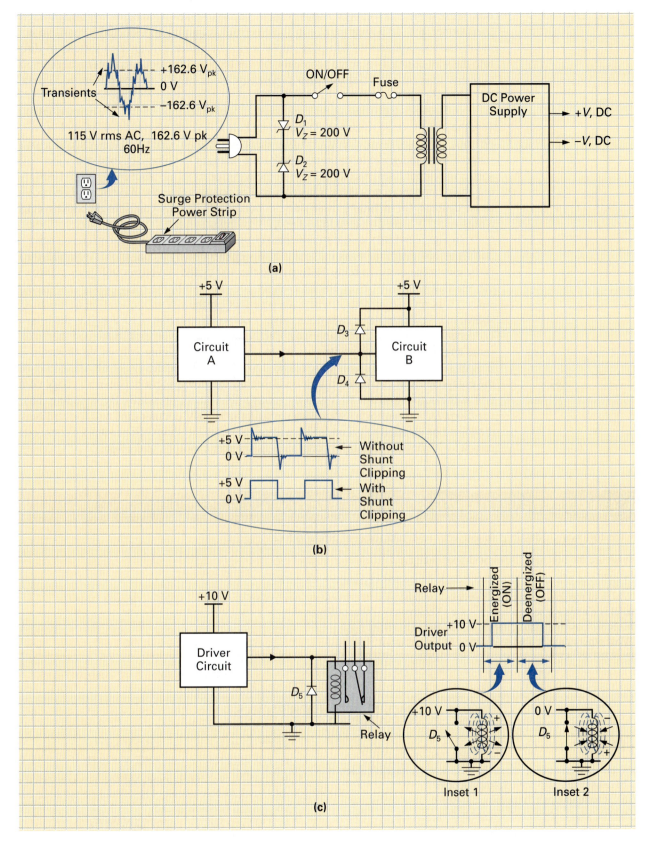

FIGURE 5-16 Transient Protection Circuits.

limit these transients to 200 V and prevent these large spikes from being coupled through the transformer and dc power supply to the electronic equipment. Generally, sensitive equipment, such as computers, will plug into a *surge protection power strip* instead of directly into the wall outlet. This surge protection power strip will include a symmetrical zener shunt clipper circuit, or a special *transient suppressor diode,* which will be discussed in detail in the following chapter.

In Figure 5-16(b), the shunt-connected clipper diodes D_3 and D_4 are being used to prevent the input voltage to circuit B from going below 0 V and above +5 V. Many low-voltage circuits, such as digital circuits, will be damaged if their input voltages exceed a voltage range. Electromagnetic interference (EMI) and changes in voltage or load can inject voltage spikes into input lines. As can be seen in the waveforms in the inset in Figure 5-16(b), without shunt clipper circuits these transients will be coupled directly into circuit B and possibly cause damage. By including these shunt-connected diodes, we can limit the input voltage to a high of 5.7 V with D_3 and a low of −0.7 V with D_4.

In Figure 5-16(c), a shunt-connected clipper diode is being used to protect a driver circuit from a counter emf transient that is always generated by a coil. In this example, the coil is the electromagnet of a relay. Referring to the associated waveform in Figure 5-16(c), you can see that when the driver circuit's output goes to +10 V, the relay coil is energized. Because D_5 is reverse biased, it will have no effect on the circuit's operation at this time, as shown in inset 1. However, when the driver's output drops from +10 V to 0 V, the relay coil is deenergized and the collapsing magnetic field will induce an emf in the coil that is of the same amplitude but of opposite polarity to that of the energizing voltage, as shown in inset 2. If this −10-V counter emf were allowed to go back to the driver circuit, it could damage the components in the driver circuit because they are now caught between a −10-V counter emf at the output and a +10-V supply voltage (20 volts of pressure). By including the clipper diode D_5 in shunt or parallel with the coil, the −10-V counter emf will forward bias the diode and be shunted to ground. You will often find these shunt clipper diodes connected in parallel with devices that have coils, such as speakers, motors, and so on.

SELF-TEST EVALUATION POINT FOR SECTION 5-2

Use the following questions to test your understanding of Section 5–2.

1. What are the two basic types of clipper circuits?
2. Which type of clipper circuit has a diode connected in parallel with the load?
3. Which of the following clipper circuits could be used to remove the positive spike of a differentiator circuit's output?
   a. Basic negative series clipper
   b. Basic positive series clipper
   c. Basic positive shunt clipper
   d. Both (b) and (c)
4. What other name is used to describe a clipper circuit?

REVIEW QUESTIONS

Multiple-Choice Questions

1. What key advantage does the full-wave voltage doubler have over the half-wave voltage doubler?
   a. Its output ripple frequency is lower
   b. Its output ripple frequency is higher
   c. It uses fewer components
   d. Both (a) and (b) are true

2. If a 115-V rms, 60-Hz input voltage was connected to a half-wave doubler circuit, what would be the peak output voltage and ripple frequency?
   a. 230 V, 60 Hz   c. 325.2 V, 60 Hz
   b. 230 V, 120 Hz   d. 325.2 V, 120 Hz

3. If a 115-V rms, 60-Hz input voltage was connected to a full-wave doubler circuit, what would be the peak output voltage and ripple frequency?
   a. 230 V, 60 Hz   c. 325.2 V, 60 Hz
   b. 230 V, 120 Hz   d. 325.2 V, 120 Hz

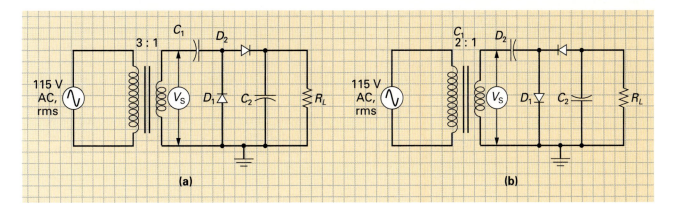

FIGURE 5-17   **Multiplier Circuits.**

4. If a 115-V rms, 60-Hz input voltage was connected to a volt-age-tripler circuit, what would be the peak output voltage?
   a. 487.8 V   c. 406 V
   b. 345 V   d. 325.2 V

5. The _____ clipper contains a diode that is connected end-to-end with the load, whereas the _____ clipper has a diode that is connected in parallel with the load.
   a. Series, biased   c. Series, shunt
   b. Shunt, series   d. Shunt, biased

6. A positive series clipper circuit will pass the _____ alternation of the ac input.
   a. Positive   b. Negative

7. A negative series clipper circuit will eliminate the _____ alternation of the ac input.
   a. Positive   b. Negative

8. Which of the following circuits would be used in modulator communication circuits?
   a. Shunt clipper   c. Positive clamper
   b. Series clipper   d. Symmetrical shunt clipper

9. Which of the following shunt clipper circuits is generally used at the front end of a dc power supply circuit to protect against transients from the 115-V ac input?
   a. Positive shunt clipper
   b. Negative shunt clipper
   c. Variable shunt clipper
   d. Symmetrical zener shunt clipper

10. A shunt clipper diode is generally always connected across a coil to protect the drive circuit from _____ .
   a. Electromagnetic interference
   b. Modulation
   c. The energizing voltage
   d. The coil's counter emf

Practice Problems

11. Identify the circuits shown in Figure 5-17, and then calculate the output voltage from each circuit.

12. Identify the circuit shown in Figure 5-18, and then calculate the voltage measured by voltmeters *A*, *B*, and *C*.

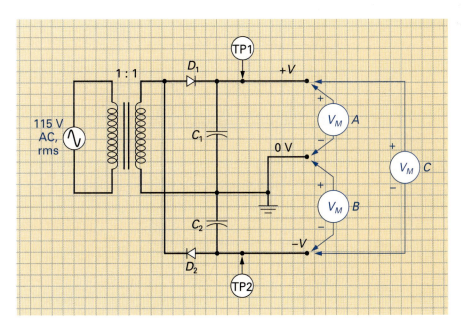

FIGURE 5-18   **A Dual-Polarity Power Supply Circuit.**

13. Why would the circuit in Figure 5-18 be called a dual-output power supply circuit?

14. Identify the circuit shown in Figure 5-19, and indicate which of the output waveforms shown in Figure 5-19(c), (d), or (e) will be present on the oscilloscope for the function generator input signal shown in Figure 5-19(b).

15. Calculate the positive and negative peak of the output from the circuit in Figure 5-19.

16. What circuits would generate the outputs shown in Figure 5-19(c), (d), and (e), for the input shown in Figure 5-19(b)?

17. Identify the circuit shown in Figure 5-20, and calculate the positive and negative peak of the output signal.

18. Some computers save programs (which are instructions and data) that do not presently need to be used on the tape of an audio cassette. These programs are saved in exactly the same way that you save or record a piece of music on an audio cassette. When a computer wants to use this program, the cassette will have to be played and the program loaded back into the computer's memory. Figure 5-21 shows how a retrieved computer program that has been

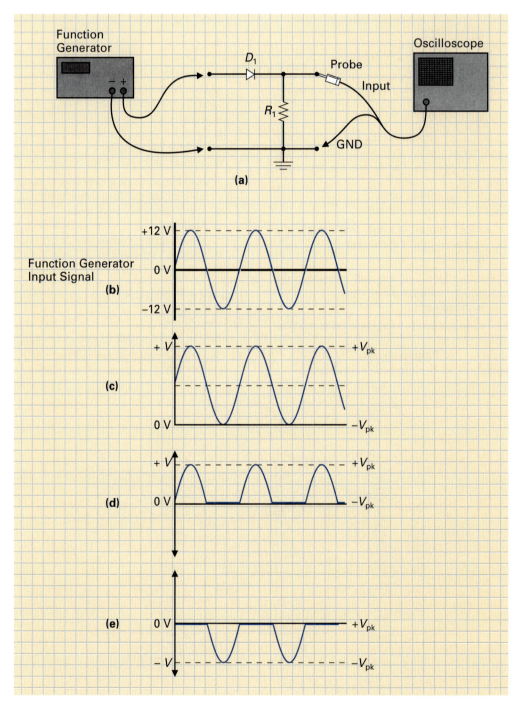

FIGURE 5-19 A Circuit's Input/Output Signals.

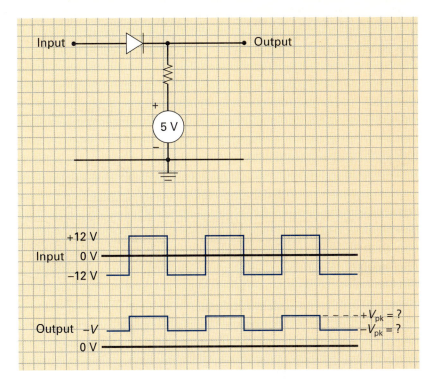

FIGURE 5-20 A Biased Negative Series Clipper Circuit.

saved on an audio cassette is limited and amplified before being put back into the computer's memory. Identify the circuit formed by D_1 and D_2, and describe why you think this circuit is included.

19. The four drive circuits in Figure 5-22 are used to produce a sequence of 0-V outputs to the four windings of a stepper motor. For example, first drive circuit A will switch 0 V to its output and cause winding A to energize, then drive cir-

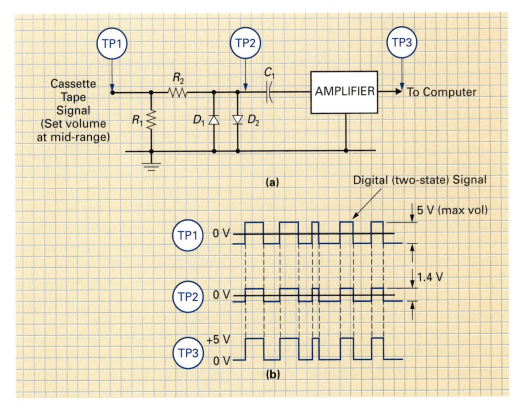

FIGURE 5-21 Digital Cassette Tape Signal Processing.

cuit B will produce a 0-V output and energize winding B, and then C, and then D, and then the cycle will repeat. The result is that the rotor of the motor will step around in a clockwise direction. Why are diodes D_1 through D_4 included in this circuit, and what forward current rating should these diodes have?

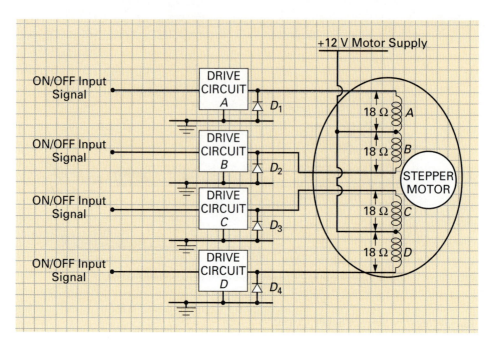

FIGURE 5-22   A Stepper Motor Drive Circuit.

Special Application Diodes

Magnetic Attraction

Andre Ampère was born near Lyon, France, in 1775. His father was a rich silk merchant who tutored him privately; however, by the time the boy had reached his teens he had read the works of many of the great mathematicians. Ampère possessed a photographic memory, which, in conjunction with a precocious ability in mathematics, made him a challenge for any professor of science.

In 1796, he began giving private lessons in mathematics, chemistry, and languages. It was in this capacity that he met his wife and was married in 1799. In 1801, he was offered a position as the professor of physics at Bourg, and because his wife was sick at the time, he traveled on ahead. His wife died a few days later and he never recovered from the blow. In fact, in later life he confided to a friend that he realized at the time of his wife's death that he could love nothing else except his work. In 1809, he became professor of mathematics at École Polytechnique in Paris, a position he held for the remainder of his life.

In 1820, Ampère witnessed Hans Christian Oersted's discovery that a compass needle could be deflected by a current-carrying wire. Inspired by this first basic step in electromagnetism, Ampère began to experiment with characteristic industry and care. In only a few weeks he advanced Oersted's discovery by leaps and bounds, developing several mathematical laws of electromagnetism. He also discovered that a coil of wire carrying a current would act like a magnet, and if an iron bar were placed in its center, it would become magnetized. He called this device a solenoid, a name that is still given to electromagnets containing a movable element. In honor of his achievements in electricity, the basic unit of electric current, the ampere, is named after him. On June 10, 1836, Ampère died in Marseille of what he told a friend just moments before his death was "a broken heart."

Introduction

Up to this point we have seen how the semiconductor P-N junction can be used to construct a junction diode, zener diode, and light-emitting diode and how these devices are used in a variety of circuit applications. In this chapter, we will study some other types of diodes that have unique characteristics that make them well suited for special circuit application. These diodes include the photodiode, varactor diode, transient suppressor diode, and the constant-current diode.

6-1   THE PHOTODIODE

In Chapter 3 we discussed in detail the operation and characteristics of the light-emitting diode or LED. The LED is a photoemitting or light-transmitting device. In this section we will be discussing in detail the operation and characteristics of its counterpart, the photodiode, which is a photodetecting or light-receiving device. The LED and photodiode are referred to as "optoelectronic devices" because their operation combines both optics and electronics. Like most areas of science, optoelectronics has its own set of terms and units. Before discussing the photodiode, we will briefly examine the basic principles of light.

6-1-1   *The Optical Electromagnetic Spectrum*

Optical Electromagnetic Spectrum

That part of the electromagnetic spectrum that encompasses infrared, visible, and ultraviolet light.

Infrared Band

The range of frequencies below that which the human eye can detect.

Visible Light Band

The range of frequencies the human eye can detect.

Ultraviolet Band

The range of frequencies above that which the human eye can detect.

Like radio waves, light is electromagnetic radiation—but in a different range of frequencies. The light, or **optical electromagnetic spectrum,** is midway between the microwave frequency band and the X-ray frequency band, as shown in Figure 6-1(a). This optical light spectrum extends from 300 GHz to about 9,300,000 GHz and contains three bands called the **infrared band, visible light band,** and **ultraviolet band.** The human eye can only detect the narrow band of frequencies in the visible light band between about 428,000 GHz and 750,000 GHz. Light waves cannot be seen below this visible region in the infrared band and above this visible region in the ultraviolet band. The frequencies in the visible light band can be divided into the colors of the rainbow: red, orange, yellow, green, blue, and violet, with each color corresponding to a narrow range of frequencies.

As you can see in Figure 6-1(a), the frequency values in the optical spectrum are large and difficult to work with. To make these values easier to handle, most optical frequencies are converted to their wavelength equivalent. To review, wavelength is symbolized by the Greek letter lambda (λ) and is measured in meters. As can be seen in Figure 6-1(b), wavelength is, as its name states, the physical length of one cycle of a transmitted electromagnetic wave. Like radio waves, the wavelength of a light wave is calculated by dividing the light wave's velocity by its frequency, as shown in the following formula:

$$\lambda = \frac{c}{f}$$

where λ = the length of one cycle of the electromagnetic wave in meters;

           c = the speed of the electromagnetic wave, which is the speed of light or 3×10^8 meters/second; and

           f = the frequency of the electromagnetic wave in hertz.

■ **EXAMPLE:**

Calculate the wavelengths of the two frequencies at either end of the microwave frequency band, which extends from 300 MHz to 300 GHz.

■ *Solution:*

For 300 MHz:

$$\lambda = \frac{c}{f} = \frac{3 \times 10^8 \text{ m/sec}}{300 \times 10^6 \text{ Hz}} = 1 \text{ meter}$$

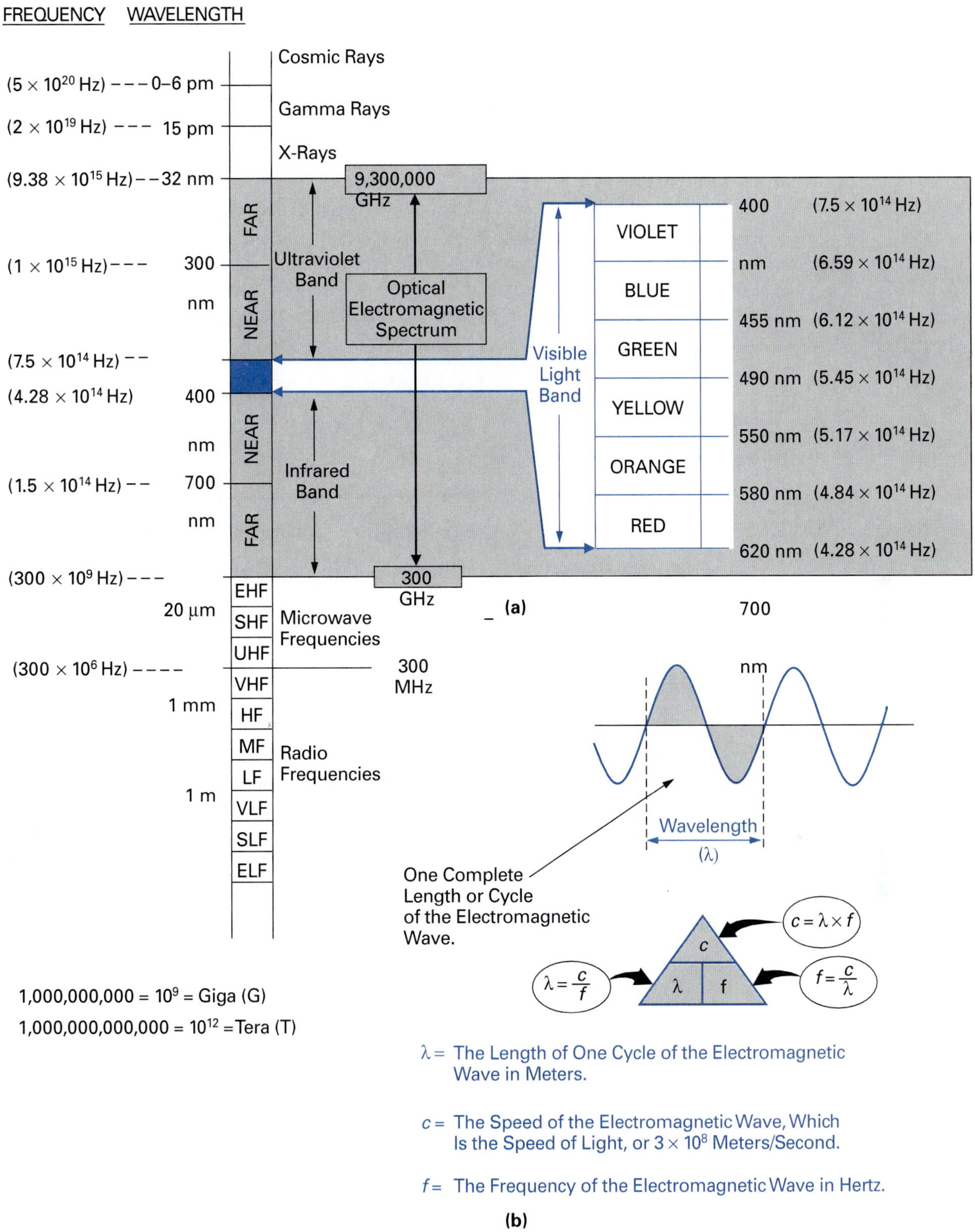

FREQUENCY   WAVELENGTH

$(5 \times 10^{20}$ Hz$)$ – – –0–6 pm — Cosmic Rays

$(2 \times 10^{19}$ Hz$)$ – – – 15 pm — Gamma Rays

X-Rays

$(9.38 \times 10^{15}$ Hz$)$ – –32 nm —

FAR

9,300,000 GHz

$(1 \times 10^{15}$ Hz$)$ – – – 300 nm

Ultraviolet Band

NEAR

Optical Electromagnetic Spectrum

$(7.5 \times 10^{14}$ Hz$)$ – –

$(4.28 \times 10^{14}$ Hz$)$ 400 nm

NEAR

$(1.5 \times 10^{14}$ Hz$)$ – –   700 nm

Infrared Band

FAR

$(300 \times 10^9$ Hz$)$ – – –

20 μm

EHF

$(300 \times 10^6$ Hz$)$ – – – –

SHF   Microwave Frequencies

UHF

VHF

1 mm

HF

MF   Radio Frequencies

LF

1 m

VLF

SLF

ELF

300 GHz

300 MHz

Visible Light Band

400         $(7.5 \times 10^{14}$ Hz$)$

VIOLET

nm $(6.59 \times 10^{14}$ Hz$)$

BLUE

455 nm $(6.12 \times 10^{14}$ Hz$)$

GREEN

490 nm $(5.45 \times 10^{14}$ Hz$)$

YELLOW

550 nm $(5.17 \times 10^{14}$ Hz$)$

ORANGE

580 nm $(4.84 \times 10^{14}$ Hz$)$

RED

620 nm $(4.28 \times 10^{14}$ Hz$)$

(a)   700

1,000,000,000 = 10^9 = Giga (G)
1,000,000,000,000 = 10^{12} =Tera (T)

One Complete Length or Cycle of the Electromagnetic Wave.

Wavelength (λ)

nm

$c = \lambda \times f$

$\lambda = \dfrac{c}{f}$ λ f $f = \dfrac{c}{\lambda}$

c

$\lambda =$ The Length of One Cycle of the Electromagnetic Wave in Meters.

$c =$ The Speed of the Electromagnetic Wave, Which Is the Speed of Light, or 3×10^8 Meters/Second.

$f =$ The Frequency of the Electromagnetic Wave in Hertz.

(b)

FIGURE 6-1   The Optical Electromagnetic Spectrum.

For 300 GHz:

$$\lambda = \frac{c}{f} = \frac{3 \times 10^8 \text{ m/sec}}{300 \times 10^9 \text{ Hz}} = 0.001 \text{ meter}$$

These wavelengths are shown alongside their frequency equivalents in Figure 6-1(a).

As you can see from the previous example, because wavelength is inversely proportional to frequency, the higher the frequency the smaller the wavelength ($f\uparrow$, $\lambda\downarrow$). In fact, at light frequencies the calculated wavelengths are so small we have to continually convert the answer in meters to a smaller unit such as micrometers or nanometers. For example, the visible light band begins at 4.28×10^{14} Hz. This frequency is equivalent to the following wavelength:

$$\lambda = \frac{c}{f} = \frac{3 \times 10^8 \text{ m/sec}}{4.28 \times 10^{14} \text{ Hz}} = 0.0000007 \text{ meter}$$

To add a prefix to meter:

$$0.0000007 \text{ m} = 0.7 \times 10^{-6} \text{ m} = 0.7 \text{ micrometers or } 0.7 \text{ microns}$$

$$\text{or} \quad 0.0000007 \text{ m} = 700 \times 10^{-9} \text{ m} = 700 \text{ nanometers}$$

To save a step, we could use a smaller light-speed unit and convert frequency directly to micrometers (more commonly called *microns*) or nanometers. For example:

Speed of Light ($\uparrow$)	*Wavelength* ($\downarrow$)
3×10^8 meters/second	= meters
3×10^{10} centimeters/second	= centimeters
3×10^{11} millimeters/second	= millimeters
3×10^{14} micrometers/second	= micrometers or microns
3×10^{17} nanometers/second	= nanometers

Do not think that these speed of light values are all different. Light will simply travel one centimeter 100 times faster than it will travel one meter.

To apply these to the visible light band starting frequency of 4.28×10^{14} Hz, the wavelength in microns and nanometers would be:

$$\lambda = \frac{c}{f} = \frac{3 \times 10^{14} \text{ }\mu\text{m/sec}}{4.28 \times 10^{14} \text{ Hz}} = 0.7 \text{ micrometers or } 0.7 \text{ microns}$$

$$\lambda = \frac{c}{f} = \frac{3 \times 10^{17} \text{nm/sec}}{4.28 \times 10^{14} \text{ Hz}} = 700.9 \text{ nanometers}$$

Angstrom (Å)

1×10^{-10} meters

Another unit that is commonly used is the **angstrom,** which is symbolized by Å and is equal to 1×10^{-10} meters. Therefore, our example of 0.0000007009 meters would be equal to 7009×10^{-10} meters, or 7009 angstroms (7009 Å). To convert to angstroms, simply use 3×10^{18} for the speed of light:

$$\lambda = \frac{c}{f} = \frac{3 \times 10^{18}}{4.28 \times 10^{14}} = 7,009 \text{ angstroms}$$

It is important to understand the relationship between frequency and wavelength, because these two terms are used interchangeably when we discuss electromagnetic radio or light waves. In fact, most manufacturer's data sheets will only list a LED's output frequency or a photodiode's best input frequency by wavelength.

A manufacturer's data sheet lists a LED as producing the highest output power at a wavelength of 800 nm or 8,000 Å. In what light wave band is this wavelength?

■ *Solution:*

Referring to Figure 6-1, you can see that a wavelength of 600 nanometers is within the near infrared light wave band. This band, which is close to the visible band (hence the name "near infrared"), extends in wavelength from 20 microns to 700 nanometers. In frequency, these wavelengths are equivalent to 150 terahertz (150×10^{12} Hz or 1.5×10^{14} Hz) to 428 terahertz (4.28×10^{14} Hz).

Before we begin our discussion on the photodiode, there is one other aspect of light that we should address. The **wave theory** of light assumes that light propagates or travels as electromagnetic waves. This theory is ideal for explaining why light travels at the speed of light through a vacuum (186,282.4 miles/sec or 3×10^8 meters/sec), at a slightly slower velocity in air, at even slower speeds in other materials such as glass and water and the light-bending action that occurs as light passes through these materials. The wave theory, however, cannot explain the interaction that occurs between light and semiconductor materials. To explain this action, we must use the **quantum theory** or **particle theory** of light. The quantum theory states that light consists of tiny particles, and each of these discrete quanta or individual packets of energy is called a **photon.** These photons are uncharged particles that have wavelike characteristics. The photon's energy is determined by its frequency, with a higher-frequency photon having more energy than a lower-frequency photon. Therefore, ultraviolet photons will have a higher energy content than visible light-wave photons.

To understand and explain the behavior of light waves, we will have to think of them as electromagnetic waves that contain many tiny particles.

6-1-2   *Photodiode Construction and Symbol*

A **photodiode** is a photodetecting or light-receiving device that contains a semiconductor P-N junction.

Figure 6-2(a) shows a typical photodiode package. A glass window or convex lens allows light to enter the case and strike the semiconductor photodiode that is mounted within the metal case.

Figure 6-2(b) shows how the photodiode is constructed in basically one of two ways. The **P-N photodiode** contains a *p*-type region that is diffused into an *n*-type substrate. A metal base makes the connection between the cathode terminal and the *n*-type region, while a metal ring makes contact between the anode terminal and the *p*-type region. Light enters the photodiode through the hole in the metal ring. The **PIN photodiode** is constructed in almost exactly the same way, except that the device has an intrinsic layer between the *p* and *n* regions, hence the name PIN (*p*-type layer, intrinsic layer, *n*-type layer). The addition of the intrinsic layer, which is a pure semiconductor having no impurities, makes the photodiode respond better to low-frequency (infrared) photons, which tend to penetrate deeper into the diode's regions. The intrinsic layer also creates a larger depletion region, which causes the photodiode to produce a more linear change in current in response to light-intensity changes.

Figure 6-2(c) shows the two commonly used photodiode schematic symbols. With the LED, the two arrows pointed away from the diode to indicate that it generated a light output. With the photodiode, the two arrows point toward the diode to indicate that it responds to a light input.

Wave Theory of Light

Assumes that light propagates or travels as electromagnetic waves.

Quantum Theory or Particle Theory

States that light consists of tiny particles.

Photon

A discrete particle or quantum of light.

Photodiode

A photodetecting or light-receiving device that contains a semiconductor P-N junction.

P-N Photodiode

Contains a *p*-type region that is diffused into an *n*-type substrate.

PIN Photodiode

Has an intrinsic layer between the *p* and *n* regions.

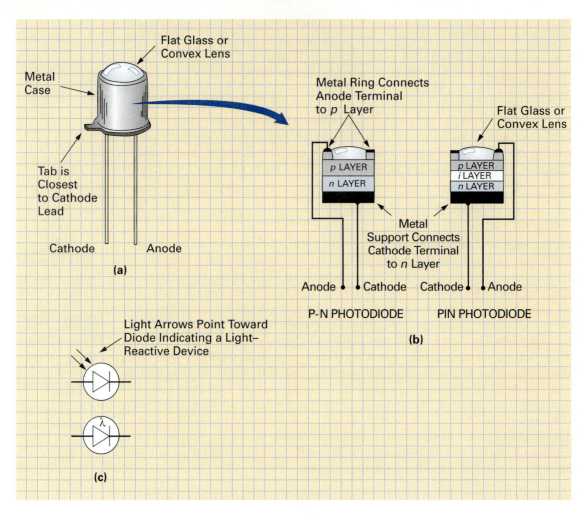

FIGURE 6-2 **Photodiodes. (a) Typical Package. (b) P-N and PIN Photodiode Construction. (c) Schematic Symbols.**

6-1-3 *Photodiode Operation*

Photovoltaic Mode

When the photodiode generates an output voltage in response to a light input.

Photodiodes can be operated in one of two modes, as shown in Figure 6-3. When used in the **photovoltaic mode,** the photodiode will generate an output voltage (voltaic) in response to a light (photo) input. Figure 6-3(a) shows how the photovoltaic photodiode cell, or solar cell, will operate in this mode. When light passes through the photodiode's window, the light's photons are absorbed at different depths in the semiconductor material, depending on their energy content (wavelength). High-energy photons will collide with semiconductor atoms and transfer energy to the atoms. If enough energy is transferred from photon to atom, a valence electron will be released from the valence band to the conduction band. This will result in a free electron (negative charge) and a positively charged atom (electron–hole pair). Negative free electrons in the depletion region will be attracted to the positive ions in the *n*-type region, while positive holes in the depletion region will be attracted to the negative ions in the *p*-type region. This separation of charges will generate a potential difference or small voltage across the P-N junction that is typically about 0.45 V. If a load were connected across the photodiode, a small electron current would flow from the *n*-type region (cathode) to the *p*-type region (anode). Photovoltaic photodiodes can be used as a light meter in a camera. They can also be arranged into banks or arrays, where they charge batteries in remote locations such as communications satellites, freeway emergency telephones, and portable equipment such as calculators.

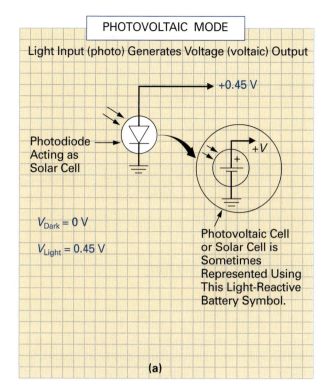

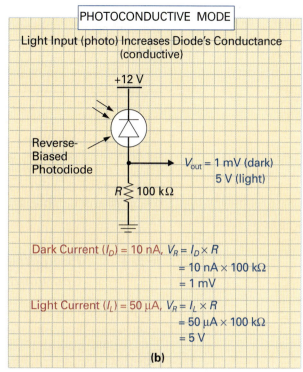

FIGURE 6-3 Photodiode Operation.

Photodiodes are most widely used in the **photoconductive mode,** in which they will change their conductance (conductive) when light (photo) is applied. In this mode, the photodiode is reverse biased (*n*-type region is made positive; *p*-type region is made negative), as shown in the example circuit in Figure 6-3(b). The reverse-biased photodiode will have a wide depletion region, and only a small reverse current will pass through the diode. The reverse current that passes through the photodiode when no light is being applied is called the **dark current (I_D),** and in the example in Figure 6-3(b) this is equal to 10 nA. When light is applied, photons enter the depletion region and create electron–hole pairs. The electrons are attracted to the positive bias voltage ($+12$ V), and the holes are attracted to the negative bias voltage (ground). This movement of separated electrons and holes makes up a reverse current through the photodiode. An increase in the light intensity will result in an increase in the reverse current and in the photodiode's conductivity. The reverse current that passes through the photodiode when light is being applied is called the **light current (I_L),** and in the example in Figure 6-3(b) this is equal to 50 µA.

Photoconductive Mode

When the photodiode changes conductance in response to a light input.

Dark Current (I_D)

The reverse current that passes through the photodiode when no light is being applied.

Light Current (I_L)

The reverse current that passes through the photodiode when light is being applied.

■ EXAMPLE:

Referring to Figure 6-3(b), calculate

  a. The ratio of light current to dark current.

  b. The output voltage when no light is applied.

  c. The output voltage when light is applied.

■ Solution:

a.
$$\text{Ratio} = \frac{I_L}{I_D} = \frac{50\ \mu A}{10\ nA} = 5{,}000$$

Therefore, the photodiode's light current is 5,000 times larger than the photodiode's dark current.

b.
$$V_R = I_D \times R = 10 \text{ nA} \times 100 \text{ k}\Omega = 1 \text{ mV}$$

c.
$$V_R = I_L \times R = 50 \text{ }\mu\text{A} \times 100 \text{ k}\Omega = 5 \text{ V}$$

The PIN photodiode is more widely used than the P-N photodiode due to its greater sensitivity. This is because the intrinsic layer adds to the photodiode's depletion region and makes a much wider depletion region for a given reverse bias voltage. This wider depletion region increases the chance that electron-hole pairs will be generated by photons and increases the conductance of the photodiode.

6-1-4 *Photodiode Application: Fiber-Optic Communications*

The PIN photodiode is one of the most frequently used light detectors in fiber optic communication links. Fiber optics is a technology in which light is transmitted along the inside of a thin, flexible, glass or plastic fiber. The light signal transmitted down an optical fiber is equivalent to an electrical signal passing down a copper wire.

SELF-TEST EVALUATION POINT FOR SECTION 6-1

Use the following questions to test your understanding of Section 6–1.

1. What is the difference between an LED and a photodiode?
2. Define the difference between photoconduction and photovoltaic action.
3. A photodiode has a peak spectral response at 615 nm. This wavelength is in what optical electromagnetic band?
4. Give the full names of the following abbreviations:
   a. PIN   **b.** IDP   **c.** IRED
5. An increase in light will cause a/an _____ in the conduction of a photodiode.

6-2   THE VARACTOR DIODE

When any diode is reverse biased, a depletion region is formed, as shown in Figure 6-4(a). Increasing the reverse bias voltage applied across the diode increases the width of the depletion layer. Conversely, when the reverse bias voltage is decreased, the depletion region width becomes narrower. This depletion region has an absence of majority carriers and acts like an insulator, preventing conduction between the *n* and *p* regions of the diode—just as a dielectric separates the two plates of a capacitor. In fact, the similarities between a reverse-biased diode and a capacitor are many. When reverse biased, a diode exhibits a small value of capacitance that can be varied by varying the reverse bias voltage and therefore the width of the depletion region. The basic P-N junction diode has only a very small amount of internal junction capacitance; however, special diodes can be constructed to have a larger value of internal capacitance. These special diodes are called **varactor diodes** because, by varying the reverse bias voltage, they can be made to operate as *voltage-controlled variable capacitors.*

Varactor Diodes
Diodes that operate as voltage-controlled variable capacitors.

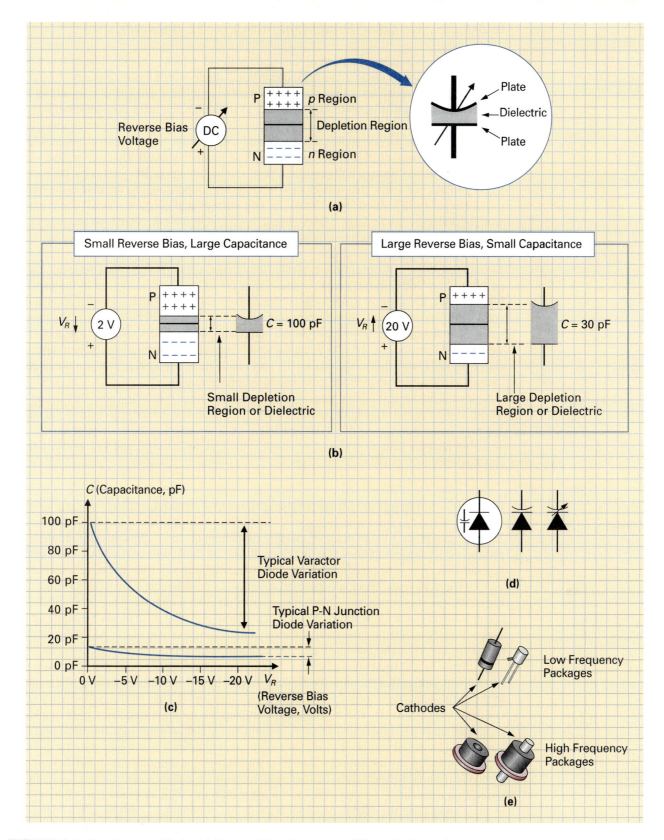

FIGURE 6-4 The Varactor Diode. (a) Reverse Bias Capacitance Effect. (b) Operation.
(c) Typical Capacitance versus Reverse Bias Voltage Characteristics.
(d) Schematic Symbols. (e) Low-Frequency and High-Frequency Packages.

6-2-1   *Varactor Diode Operation and Characteristics*

Figure 6-4(b) shows the operation of the varactor diode, which is operated in its reverse region. The varactor diode is a specially constructed diode with a small impurity dose at its junction. The impurity level increases as you travel away from the junction, resulting in a greater "capacitance to reverse bias voltage" change than that of a conventional-junction diode. The reverse bias voltage (V_R) is varied to control the width of the depletion region and is always less than the reverse breakdown voltage rating of the diode. As an example, Figure 6-4(b) shows how a small reverse bias voltage will produce a small depletion region, while a large reverse bias voltage will produce a large depletion region. Remembering the capacitance formula, we know that capacitance is inversely proportional to the distance between the plates ($C \propto 1/d$). Therefore, a small reverse voltage will produce a small depletion region, or dielectric, and a large capacitance ($d\downarrow$, $C\uparrow$). On the other hand, an increase in reverse bias voltage will result in a large depletion region or dielectric and a small capacitance ($d\uparrow$, $C\downarrow$).

Figure 6-4(c) illustrates the typical "capacitance versus reverse bias voltage" curves produced by a conventional junction and a varactor diode. As you can see in the curve in Figure 6-4(c), the varactor diode's capacitance varies inversely with the applied reverse bias voltage. For example, at −2 V, the varactor's depletion region will be small and the capacitance large at approximately 80 pF. On the other hand, as the reverse bias voltage across the varactor is increased toward −20 V, the varactor's capacitance decreases rapidly to approximately 25 pF. Figure 6-4(c) also shows the very small change in a basic P-N junction diode's internal capacitance over the same reverse bias voltage range.

The schematic symbols used to represent a varactor (variable-capacitor) diode are shown in Figure 6-4(d). Figure 6-4(e) shows the typical low-frequency (below 500 mHz) and high-frequency (above 500 mHz) packages. A wide assortment of varactor diodes are available with capacitance values that range from 1 pF to 2,000 pF, and with power ratings that range from 500 mW to 35 W.

6-2-2   *Varactor Diode Application: Tuned Circuit*

Varactor diodes are used in place of variable capacitors in many applications. One such application is shown in Figure 6-5(a), in which a varactor diode will be used in a tuned circuit in a radio receiver. The tuned circuit in a radio receiver acts as a filter to pass a selected station, or frequency, and block all other radio stations, or frequencies. To explain this further, Figure 6-5(a) shows the basic block diagram of an AM (amplitude-modulated) radio receiver. As you can see, the tuned circuit is at the front end of a radio receiver, and its operation will be controlled by the radio's tuning control to select one of the AM radio stations present in the AM radio band between 535 kHz and 1605 kHz.

Figure 6-5(b) shows the schematic diagram of a basic tuned circuit that contains a varactor diode. Notice that the varactor diode is reverse biased because its cathode is connected to the positive source voltage ($+V$) via the tuning resistor (R_T). As R_T is adjusted, the capacitance of the varactor D_1 will be changed, changing the resonant frequency of the tank circuit made up of D_1's capacitance and L_1's inductance. Because the varactor diode acts as a variable capacitor, it can be used in place of the more expensive, mechanically variable air capacitors. Let us review how this parallel resonant band-pass tuning circuit will operate. At resonance (which is determined by the capacitance of D_1 and the inductance of L_1), the tank has a high impedance, so very little of the input current at this resonant frequency will be shunted away from the output. At frequencies above resonance, X_C will be low, and most of the input signal at frequencies above resonance will be shunted away from the output by the capacitance of D_1. At frequencies below resonance, X_L will be low, and the shunting action of L_1 will again prevent any frequencies below resonance from appearing at the output.

This band-pass filter circuit will therefore tune in (select or pass) one frequency that contains the information we desire and allow this signal to proceed to the other circuits in the radio or television receiver. All of the other millions of information-carrying frequen-

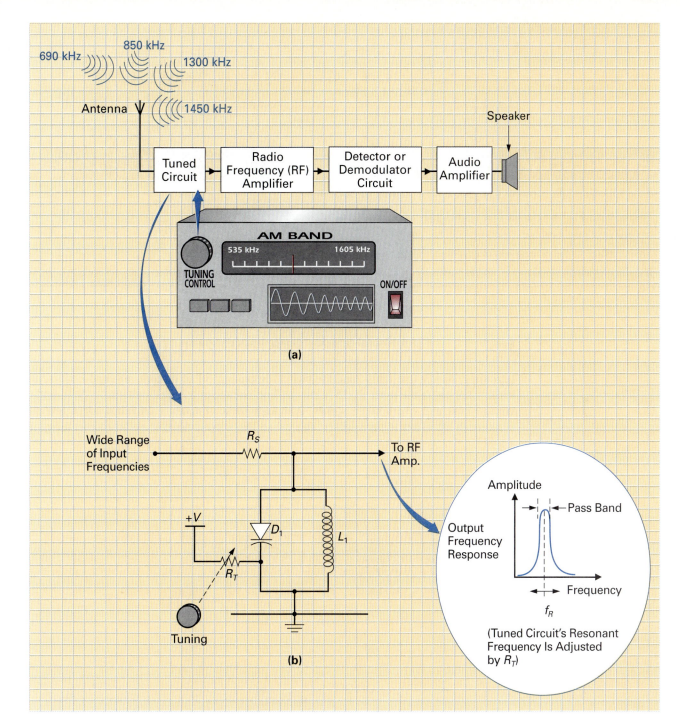

FIGURE 6-5 **The Varactor Diode in a Tuned Circuit.**

cies, however, will be blocked by the resonant action of the filter. Varactor diodes that are used in these circuits are often called *varicap diodes* or *tuning diodes*.

EXAMPLE:

Calculate the resonant frequency of the tuned circuit in Figure 6-5(b), if

D_1 Capacitance $= 14$ pF

L_1 Inductance $= 5$ mH

$$f_R = \frac{1}{2\pi \sqrt{L \times C_{VD1}}} = \frac{1}{2\pi \sqrt{(5 \times 10^{-3}) \times (14 \times 10^{-12})}}$$
$$= 601.5 \text{ kHz}$$

SELF-TEST EVALUATION POINT FOR SECTION 6-2

Use the following questions to test your understanding of Section 6–2.

1. True or false: The varactor diode is normally always operated in its forward region.
2. In what application can varactors normally be found?
3. True or false: As the reverse bias is increased, the capacitance of the varactor will decrease.
4. If a silicon varactor diode was forward biased by an applied voltage of +0.7 V, what would be its value of capacitance?

6-3   TRANSIENT SUPPRESSOR DIODE

Lightning, power line faults, and the switching on and off of motors, air-conditioners, and heaters can cause the normal 115 V rms ac line voltage at the wall outlet to contain under-voltage dips and overvoltage spikes. Although these *transients* only last for a few microseconds, the overvoltage spikes can cause the input line voltage to momentarily in-

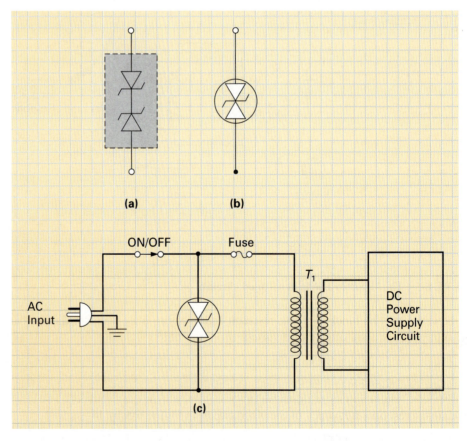

FIGURE 6-6   **Transient Suppressor Diodes. (a) Construction. (b) Schematic Symbol. (c) Bidirectional Circuit Application (AC Line Voltage).**

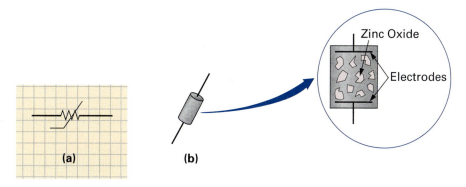

FIGURE 6-7 Metal Oxide Varistors (MOVs). (a) Schematic Symbol. (b) Physical Appearance and Construction.

crease by 1,000 V or more. In sensitive equipment, such as televisions and computers, shunt filtering devices are connected between the ac line input and the primary of the dc power supply's transformer to eliminate these transients before they get into, and possibly damage, the system.

One such device that can be used to filter the ac line voltage is the **transient suppressor diode.** Transient suppressor diodes are also called **transorbs** because they "absorb transients." Referring to Figure 6-6(a), you can see that this diode contains two zener diodes that are connected back-to-back. The schematic symbol for this diode is shown in Figure 6-6(b).

Figure 6-6(c) shows how a transorb would be connected across the ac power line input to a dc power supply. Because the zeners within the transient suppressor diode are connected back-to-back, they will operate in either direction (the device is "bidirectional") and monitor both alternations of the ac input. If a voltage surge occurs that exceeds the V_Z (zener voltage) of the diodes, they will break down and shunt the surge away from the power supply.

Most manufacturers' transorbs have a high power-dissipation rating because they may have to handle momentary power line surges in the hundreds of watts. For example, the Motorola 1N5908-1N6389 series of transorbs can dissipate 1.5 kW for a period of approximately 10 msec (most surges last for a few milliseconds). The devices must also have a fast turn-on time so that they can limit or clamp any voltage spikes. For example, the Motorola P6KE6.8 series has a response time of less than 1 nsec.

In dc applications, a single unidirectional (one-direction) transient suppressor can be used instead of a bidirectional (two-direction) transient suppressor. These single transorbs have the same schematic symbol as a zener.

Metal-oxide varistors (MOVs) are currently replacing zener diode and transient diode suppressors because they are able to shunt a much higher current surge and are cheaper. These are not semiconductor devices (in fact, they contain a zinc-oxide and bismuth-oxide compound in a ceramic body) but are connected in the same way as a transient suppressor diode. They are called **varistors** because they operate as a "voltage-dependent resistor" that will have a very low resistance at a certain breakdown voltage. The MOV's schematic symbol, typical appearance, and construction are shown in Figure 6-7.

Transient Suppressor Diode

A device used to protect voltage-sensitive electronic devices in danger of destruction by high-energy voltage transients.

Transorb

Absorb transients. Another name for transient suppressor diode.

Metal-Oxide Varistors (MOVs)

Devices that are replacing zener diode and transient diode suppressors because they are able to shunt a much higher current surge and are cheaper.

Varistor

Voltage-dependent resistor.

SELF-TEST EVALUATION POINT FOR SECTION 6-3

Use the following questions to test your understanding of Section 6–3.

1. True or false: A bidirectional transient suppressor diode would be used to suppress ac power surges.

2. True or false: A unidirectional transient suppressor diode would be used to suppress dc power surges.

3. Besides the ability to dissipate the large burst of power in a surge, what other important feature should a transorb have?

4. What device is largely replacing the transorb in transient protection applications?

Multiple-Choice Questions

1. Which of the following wavelengths is equivalent to the frequency 545 THz?
   a 580 nm   **b.** 550 nm   **c.** 400 nm   **d.** 700 nm

2. When operated in the photoconductive mode, the photodiode is normally
   a. Forward biased   **c.** Acting as a voltage source
   b. Reverse biased   **d.** None of the above

3. When light increases, the conductance of a photodiode will _____.
   a. Increase   **c.** Remain the same
   b. Decrease   **d.** Be unaffected

4. The LED is classified as a photo _____ device, while the photodiode is classified as a photo _____ device.
   a. Transmitting, emitting   **c.** Receiving, transmitting
   b. Detecting, emitting   **d.** Emitting, detecting

5. A varactor diode is generally operated in its _____ biased mode.
   a. Forward   **b.** Reverse   **c.** Photo   **d.** Un-

6. As the varactor diode's reverse bias voltage increases, the width of a varactor's depletion region _____, and therefore the varactor's capacitance _____.
   a. Increases, decreases   **c.** Decreases, increases
   b. Increases, increases   **d.** Both (a) and (c) are true

7. In what application are varactor diodes typically used?
   a. Voltage Regulator Circuits
   b. Constant-Current Circuits
   c. Transient Suppressor Applications
   d. Tuned Circuits

8. Which diode is used to shunt high voltage surges?
   a. P-N junction diode   **c.** Transorb diode
   b. Constant-current diode   **d.** Varactor diode

9. Which of the following diodes are normally always reverse biased?
   a. Varicap   **c.** PIN photodiode
   b. Transorb   **d.** All of the above

10. Which of the following devices could be used in place of a unidirectional transorb?
    a. Tuning diode   **c.** Metal-oxide varistor
    b. Zener diode   **d.** Both (b) and (c) are true

Practice Problems

11. Convert the following optical frequencies to their wavelength equivalents:
    a. 6×10^{14} Hz to a wavelength in nanometers.
    b. 1×10^{15} Hz to a wavelength in angstroms.
    c. 1×10^{14} Hz to a wavelength in microns.

12. Convert the following wavelengths to their frequency equivalents:
    a. 500 nm   **b.** 4550 Å   **c.** 850 μm

13. Calculate the ratio of light current to dark current for the photodiodes shown in Figure 6-8.

14. Which of the photoconductive photodiodes in Figure 6-8 are correctly biased?

15. Calculate the dark and light output voltage developed across the photodiode in Figure 6-8(d).

16. In Figure 6-9, a small circular disk containing 12 holes is attached to a dc (toy) motor. The electronic circuit surrounding this motor is designed to function as a "tachometer," which is an instrument that measures angular speed in revolutions per minute (rpm). The input control resistor R_1 is used to adjust the speed of the motor, while the output LED D_2 and the oscilloscope are used to calculate the motor's angular speed in rpm. This circuit operates in the following way. LED D_1 is permanently ON, and its light is

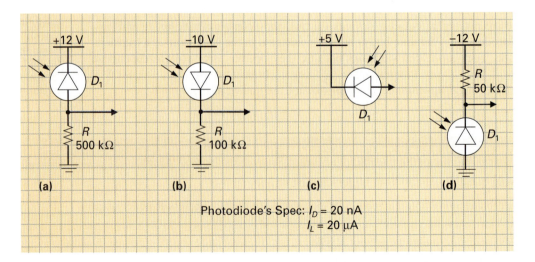

FIGURE 6-8   **Photodiode Biasing.**

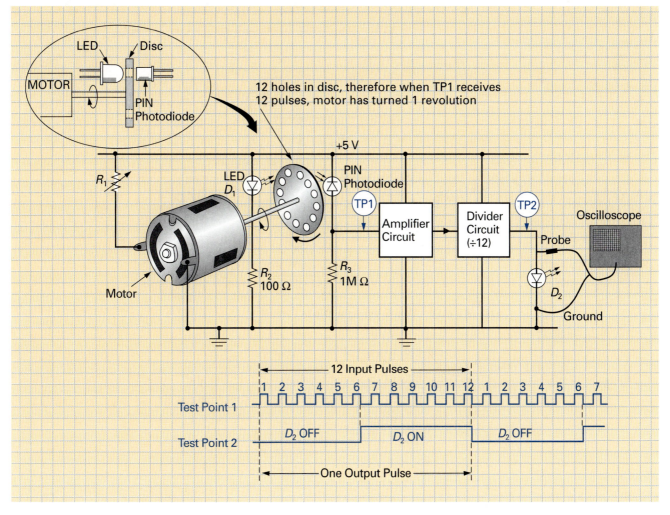

FIGURE 6-9 Tachometer Circuit.

applied to the 12-hole rotating disc that is attached to the motor. As the motor rotates, light passes through the holes and is applied to a PIN photodiode. The photodiode will conduct more and less to develop a series of pulses across the resistor R_3 at test point 1, as seen in the TP1 waveform. Because there are 12 holes in the disc, 12 pulses will be generated for each revolution of the disc and therefore each revolution of the motor. These pulses are amplified. Then a divide-by-12 circuit will generate 1 pulse for every 12 input pulses at TP2, as shown in the TP2 waveform. Because the disc generates 12 pulses/revolution of the motor ($\times 12$), and the divider circuit generates 1 pulse for every 12 input pulses ($\div 12$), the LED (D_2) and the oscilloscope will receive 1 pulse/revolution of the motor. Measuring the pulses/second received at D_2 and the oscilloscope will enable us to calculate the motor's angular speed in revolutions/minute.

a. If the motor is turning at a speed of 3,300 rpm, what will be the frequency of the pulses at TP1 and TP2?

b. If the oscilloscope is measuring 24 pps at TP2, what is the motor's rpm and the pps rate at TP1?

17. Are the varactor diodes in Figure 6-10 correctly or incorrectly biased?

18. Calculate the frequency range of the tuned circuit shown in Figure 6-10.

19. Which of the transient suppressor diodes in Figure 6-11(c), (d), (e), and (f) should be used to protect the circuits in Figure 6-11(a) and (b)?

20. In Figure 6-12, the four special diodes are being used in this low-voltage bridge rectifier circuit due to their smaller voltage drop. If the input ac is 6 V rms, what will be the peak of the pulsating dc from the rectifier (taking into account an individual diode drop of 0.4 V)?

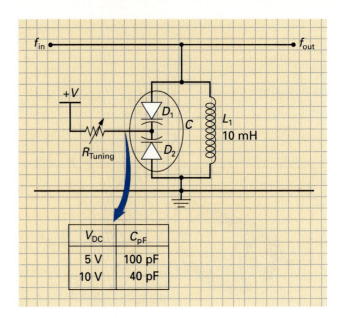

V_{DC}	C_{pF}
5 V	100 pF
10 V	40 pF

FIGURE 6-10   Tuned Varactor Diode Circuit.

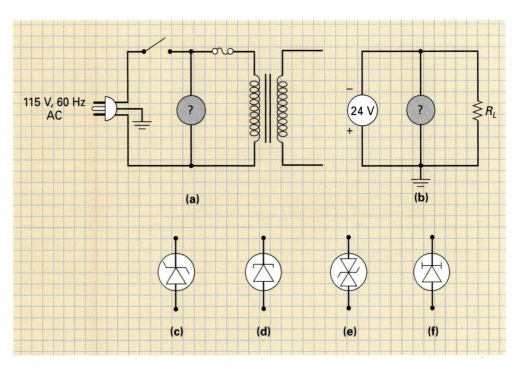

FIGURE 6-11   Transient Voltage Protection.

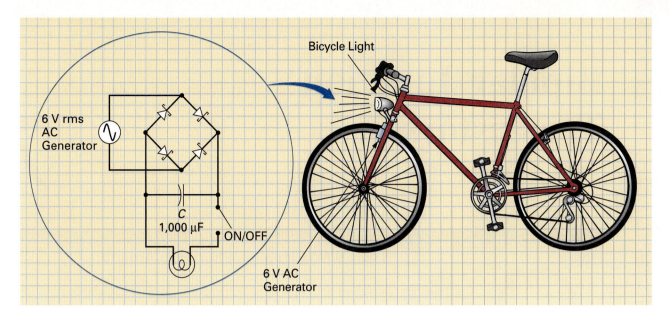

FIGURE 6-12 Low-Voltage Diode Application.

Bipolar Junction Transistors

7

There's No Sleeping When He's Around!

Carl Friedrich Gauss was born April 30, 1777, to poor, uneducated parents in Brunswick, Germany. He was a child of precocious abilities, particularly in mental computation. In elementary school he soon impressed his teachers, who said that mathematical ability came easier to Gauss than speech.

In secondary school he rapidly distinguished himself in ancient languages and mathematics. At 14, Gauss was presented to the court of the duke of Brunswick, where he displayed his computing skill. Until his death in 1806, the duke generously supported Gauss and his family, encouraging the boy with textbooks and a laboratory.

In the early years of the nineteenth century, Gauss's interest was in astronomy, and his accumulated work on celestial mechanics was published in 1809. In 1828, at a conference in Berlin, Gauss met physicist Wilhelm Weber, who would eventually become famous for his work on electricity. They worked together for many years and became close friends, investigating electromagnetism and the use of a magnetic needle for current measurement. In 1833, they constructed an electric telegraph system that could communicate across Göttingen from Gauss's observatory to Weber's physics laboratory. (This telegraph system of communication was later developed independently by U.S. inventor Samuel Morse.)

Gauss conceived almost all of his fundamental mathematical discoveries between the ages of 14 and 17. There are many stories of his genius in his early years, one of which involved a sarcastic teacher who liked giving his students long-winded problems and then resting, or on some occasions sleeping, in class. On his first day with Gauss, who was eight years old, the teacher began, as usual, by telling the students to find the sum of all the numbers from 1 to 100. The teacher barely had a chance to sit down before Gauss raised his hand and said "5,050." The dumbfounded teacher, who believed Gauss must have heard the problem before and memorized the answer, asked Gauss to explain how he had solved the problem. He replied: "The numbers 1, 2, 3, 4, 5, and so on to 100 can be paired as 1 and 100, 2 and 99, 3 and 98, and so on. Since each pair has a sum of 101, and there are 50 pairs, the total is 5,050."

Introduction

In 1948, a component known as a transistor sparked a whole new era in electronics, the effects of which have not been fully realized even to this day. A transistor is a three-element device made of semiconductor materials used to control electron flow by varying the voltages applied to its three elements. Having the ability to control the amount of current through the transistor allows us to achieve two very important applications: switching and amplification.

Like the diode, transistors are formed by *p* and *n* regions and, as we are already aware, the point at which a *p* and an *n* region join is known as a *junction*. Transistors in general are classified as being either the *bipolar* or *unipolar* type. The bipolar type has two P-N junctions, while unipolar transistors have only one P-N junction. In this chapter, we will study all of the details relating to the *bipolar* transistor, or as it is also known, the *bipolar junction transistor* or *BJT*.

7-1 FIRST APPROXIMATION DESCRIPTION OF A BIPOLAR TRANSISTOR

In most cases it is easier to build a jigsaw puzzle when you can refer to the completed picture on the box. The same is true whenever anyone is trying to learn anything new, especially a science that contains many small pieces. These first approximation descriptions are a means for you to quickly see the complete picture without having to wait until you connect all of the pieces. Like the diode's first approximation description, this general overview will cover the transistor's basic construction, schematic symbol, physical appearance, basic operation, and main applications.

7-1-1 *Transistor Types (NPN and PNP)*

Like the diode, a bipolar transistor is constructed from a semiconductor material. However, unlike the diode, which has two oppositely doped regions and one P-N junction, the transistor has three alternately doped semiconductor regions and two P-N junctions. These three alternately doped regions are arranged in one of two different ways, as shown in Figure 7-1.

With the **NPN transistor** shown in Figure 7-1(a), a thin, lightly doped *p*-type region known as the **base** (symbolized *B*) is sandwiched between two *n*-type regions called the **emitter** (symbolized *E*) and the **collector** (symbolized *C*). Looking at the NPN transistor's schematic symbol in Figure 7-1(b), you can see that an arrow is used to indicate the emitter lead. As a memory aid for the NPN transistor's schematic symbol, you may want to remember that when the emitter arrow is "**N**ot **P**ointing i**N**" to the base, the transistor is an **"NPN."** An easier method is to think of the arrow as a diode, with the tip of the arrow or cathode pointing to an *n* terminal and the back of the arrow or anode pointing to a *p* terminal, as seen in the inset in Figure 7-1(b).

The **PNP transistor** can be seen in Figure 7-1(c). With this transistor type, a thin, lightly doped *n*-type region (base) is placed between two *p*-type regions (emitter and collector). Figure 7-1(d) illustrates the PNP transistor's schematic symbol. Once again, if you think of the emitter arrow as a diode, as shown in the inset in Figure 7-1(d), the tip of the arrow or cathode is pointing to an *n* terminal and the back of the arrow or anode is pointing to a *p* terminal.

NPN Transistor

A thin, lightly doped *p*-type region (base) is sandwiched between two *n*-type regions (emitter and collector).

Base

The region that lies between an emitter and a collector of a transistor and into which minority carriers are injected.

Emitter

A transistor region from which charge carriers are injected into the base.

Collector

A semiconductor region through which a flow of charge carriers leaves the base of the transistor.

PNP Transistor

A thin, lightly doped *n*-type region (base) is placed between two *p*-type regions (emitter and collector).

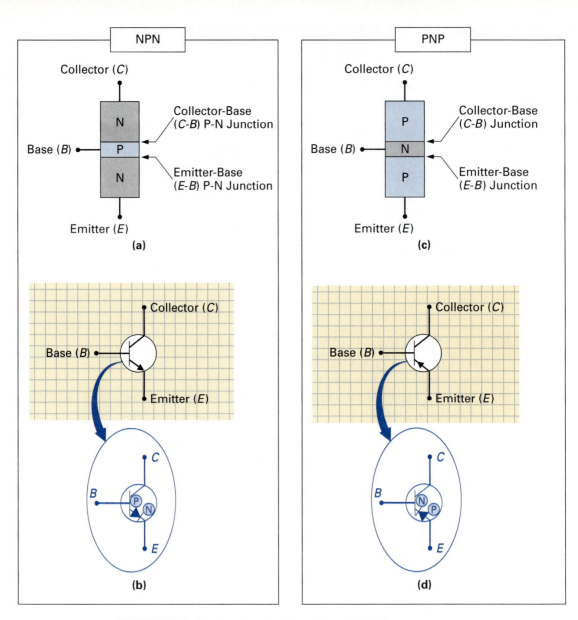

FIGURE 7-1 Bipolar Junction Transistor (BJT) Types.

7-1-2   *Transistor Construction and Packaging*

Like the diode, the three layers of an NPN or PNP transistor are not formed by joining three alternately doped regions. These three layers are formed by a "diffusion process," which first melts the base region into the collector region, and then melts the emitter region into the base region. For example, with the NPN transistor shown in Figure 7-2(a), the construction process would begin by diffusing or melting a *p*-type base region into the *n*-type collector region. Once this *p*-type base region is formed, an *n*-type emitter region is diffused or melted into the newly diffused *p*-type base region to form an NPN transistor. Keep in mind that manufacturers will generally construct thousands of these transistors simultaneously on a thin semiconductor wafer or disc, as shown in Figure 7-2(a). Once tested, these discs, which are about 3 inches in diameter, are cut to separate the individual transistors. Each transistor is placed in a package, as shown in Figure 7-2(b). The package will protect the transistor from humidity and dust, provide a means for electrical connection between the three semiconductor regions and the three transistor terminals, and serve as a heat sink to conduct away any heat generated by the transistor.

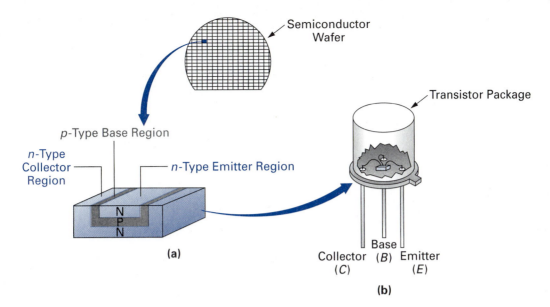

FIGURE 7-2 Bipolar Junction Transistor Construction and Packaging.

Figure 7-3 illustrates some of the typical low-power and high-power transistor packages. Most low-power, small-signal transistors are hermetically sealed in a metal, plastic, or epoxy package. Four of the low-power packages shown in Figure 7-3(a) have their three leads protruding from the bottom of the package because these package types are usually inserted and soldered into holes in printed circuit boards (PCBs). The surface mount technology (SMT) low-power transistor package, on the other hand, has flat metal legs that mount directly onto the surface of the PCB. These transistor packages are generally used in high-component-density PCBs because they use less space than a "through-hole" package. To explain this in more detail, a through-hole transistor package needs a hole through the PCB and a connecting pad around the hole to make a connection to the circuit. With an SMT package, however, no holes are needed, only a small connecting pad. Without the need for holes, pads on printed circuit boards can be smaller and placed closer together, resulting in considerable space saving.

The high-power packages, shown in Figure 7-3(b), are designed to be mounted onto the equipment's metal frame or chassis so that the additional metal will act as a heat sink and conduct the heat away from the transistor. With these high-power transistor packages, two or three leads may protrude from the package. If only two leads are present, the metal case will serve as a collector connection, and the two pins will be the base and emitter.

Transistor package types are normally given a reference number. These designations begin with the letters "TO," which stands for transistor outline, and are followed by a number. Figure 7-3 includes some examples of TO reference designators.

7-1-3   *Transistor Operation*

Figure 7-4 shows an NPN bipolar transistor, and the inset shows how a transistor can be thought of as containing two diodes: a *base-to-collector diode* and a *base-to-emitter diode*. With an NPN transistor, both diodes will be back-to-back and "**N**ot be **P**ointing i**N**" (NPN) to the base, as shown in the inset in Figure 7-4. For a PNP transistor, the base–collector and base–emitter diodes will both be pointing into the base.

Transistors are basically controlled to operate as a switch, or they are controlled to operate as a variable resistor. Let us now examine each of these operating modes.

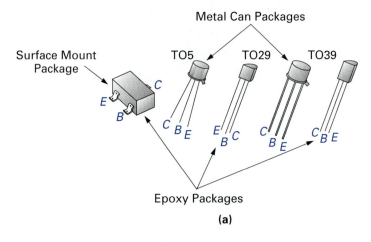

Metal Can Packages

Surface Mount Package

TO5   TO29   TO39

Epoxy Packages

(a)

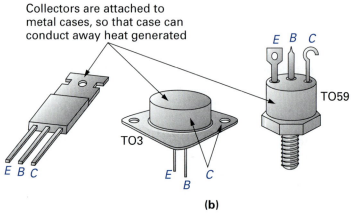

Collectors are attached to metal cases, so that case can conduct away heat generated

E B C

TO59

TO3

E B C

E   C
   B

(b)

FIGURE 7-3   **Bipolar Junction Transistor Package Types. (a) Low Power. (b) High Power.**

FIGURE 7-4   **The Base–Collector and Base–Emitter Diodes Within a Bipolar Transistor.**

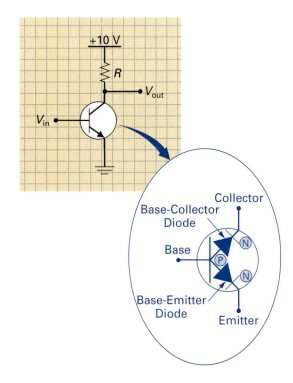

+10 V

R

V_{out}

V_{in}

Collector

Base-Collector Diode

Base

N

P

N

Base-Emitter Diode

Emitter

The Transistor's ON/OFF Switching Action

Figure 7-5 illustrates how the transistor can be made to operate as a switch. This ON/OFF switching action of the transistor is controlled by the transistor's base-to-emitter (B–E) diode. If the B–E diode of the transistor is forward biased, the transistor will turn ON; if the B–E diode of the transistor is reverse biased, the transistor will turn OFF.

To begin with, let us see how the transistor can be switched ON. In Figure 7-5(a), the B–E diode of the transistor is forward biased (anode at base is +5 V, cathode at emitter is 0 V), and the transistor will turn ON. Its collector and emitter output terminals will be equivalent to a closed switch, as shown in Figure 7-5(b). This low resistance between the transistor's collector and emitter will cause a current (I), as shown in Figure 7-5(b). The output voltage in this condition will be zero volts because all of the +10 V supply voltage will be dropped across R. Another way to describe this would be to say that the low resistance path between the transistor's emitter and collector connects the zero-volt emitter potential through to the output.

Now let us see how the transistor can be switched OFF. In Figure 7-5(c), the transistor has 0 V being applied to its base input. In this condition, the B–E diode of the transistor is reverse biased (anode at base is 0 V, cathode at emitter is 0 V), and so the transistor will turn OFF and its collector and emitter output terminals will be equivalent to an open switch, as shown in Figure 7-5(d). This high resistance between the transistor's collector and emitter will prevent any current and any voltage drop, resulting in the full +10 V supply voltage being applied to the output, as shown in Figure 7-5(d).

The Transistor's Variable-Resistor Action

In the previous section we saw how the transistor can be biased to operate in one of two states: ON or OFF. When operated in this two-state way, the transistor is being switched ON and OFF in almost the same way as a junction diode. The transistor, however, has another ability that the diode does not have—it can also function as a variable resistor, as shown in the equivalent circuit in Figure 7-6(a). In Figure 7-5 we saw how +5-V base input bias voltage would result in a low resistance between emitter and collector (closed switch) and how a 0-V base input bias would result in a high resistance between emitter and collector (open switch). The table in Figure 7-6(b) shows an example of the relationship between base input bias voltage (V_B) and emitter-to-collector resistance (R_{CE}). In this table, you can

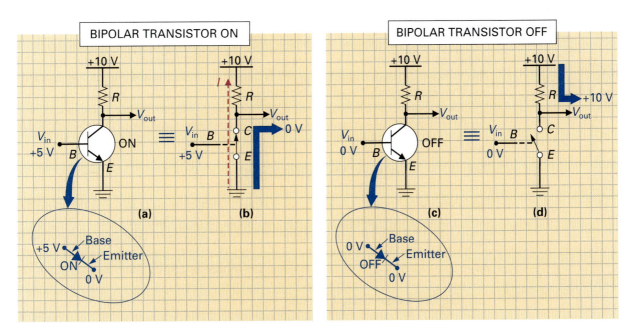

FIGURE 7-5 The Bipolar Transistor's ON/OFF Switching Action.

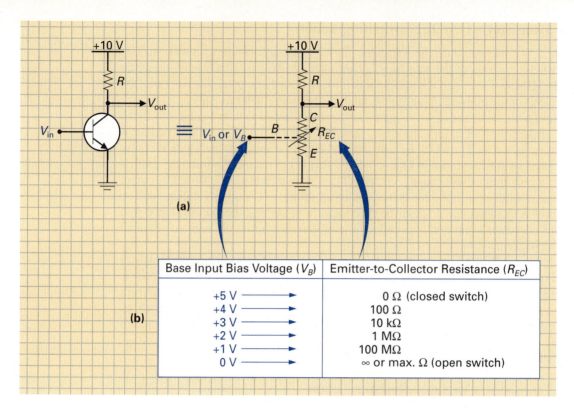

Base Input Bias Voltage (V_B)	Emitter-to-Collector Resistance (R_{EC})
+5 V ⟶	0 Ω (closed switch)
+4 V ⟶	100 Ω
+3 V ⟶	10 kΩ
+2 V ⟶	1 MΩ
+1 V ⟶	100 MΩ
0 V ⟶	∞ or max. Ω (open switch)

(b)

FIGURE 7-6   The Bipolar Transistor's Variable-Resistor Action.

see that the transistor is not only going to be driven between the two extremes of fully ON and fully OFF. When the base input voltage is at some voltage level between +5 V and 0 V, the transistor is partially ON; therefore, the transistor's emitter-to-collector resistance is somewhere between 0 Ω and maximum Ω. For example, when $V_B = +4$ V, the transistor is not fully ON and its emitter-to-collector resistance will be slightly higher, at 100 Ω. If the base input bias voltage is further reduced to +3 V, for example, you can see in the table that the emitter-to-collector resistance will further increase to 10 kΩ. Further decreases in base input voltage ($V_B\downarrow$) will cause further increases in emitter-to-collector resistance ($R_{CE}\uparrow$) until $V_B = 0$ V and $R_{CE} =$ maximum Ω.

As a matter of interest, the name *transistor* was derived from the fact that through base control we can "transfer" different values of "resistance" between the emitter and collector. This effect of "transferring resistance" is known as **transistance** and the component that functions in this manner is called the *transistor*.

Now that we have seen how the transistor can be made to operate as either a switch or a variable resistor, let us see how these characteristics can be made use of in circuit applications.

Transistance

The effect of transferring resistance.

7-1-4   *Transistor Applications*

The transistor's impact on electronics has been phenomenal. It initiated the multibillion-dollar semiconductor industry and was the key element behind many other inventions, such as integrated circuits (ICs), optoelectronic devices, and digital computer electronics. In all of these applications, however, the transistor is basically made to operate in one of two ways: as a switch or as a variable resistor. Let us now briefly examine an example of each.

Digital Logic Gate Circuit

A digital logic gate circuit makes use of the transistor's ON/OFF switching action. Digital circuits are often referred to as "switching" or "two-state" circuits because their main control device (the transistor) is switched between the two states of ON and OFF. The transistor is at the very heart of all digital electronic circuits. For example, transistors are used to construct logic gate circuits, gates are used to construct flip-flop circuits, flip-flops are used to construct register and counter circuits, and these circuits are used to construct microprocessor, memory, and input/output circuits—the three basic blocks of a digital computer.

Figure 7-7(a) shows how the transistor can be used to construct a NOT gate or INVERTER gate. The basic NOT gate circuit is constructed using one NPN transistor and two resistors. This logic gate has only one input (A) and one output (Y), and its schematic symbol is shown in Figure 7-7(b). Figure 7-7(c) shows how this logic gate will react to the two different input possibilities. When the input is 0 V (logic 0), the transistor's base–emitter P-N diode will be reverse biased and so the transistor will turn OFF. Referring to the inset for this circuit condition in Figure 7-7(c), you can see that the OFF transistor is equivalent to an open switch between emitter and collector, and therefore the +5-V supply voltage will be connected to the output. In summary, a logic 0 input (0 V) will be converted to a logic 1 output (+5 V). On the other hand, when the input is +5 V (logic 1), the transistor's base–emitter P-N diode will be forward biased and so the transistor will turn ON. Referring to the inset for this circuit condition in Figure 7-7(c), you can see that the

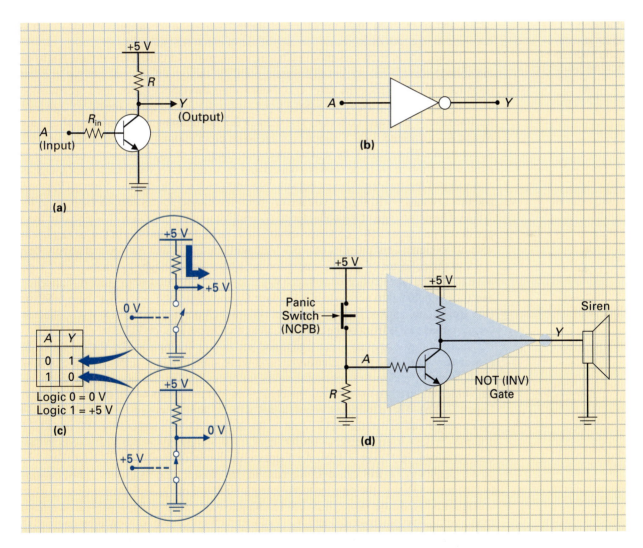

FIGURE 7-7   **The Transistor Being Used in a Digital Electronic Circuit. (a) Basic NOT or INVERTER Gate Circuit. (b) NOT Gate Schematic Symbol. (c) NOT Gate Function Table. (d) NOT Gate Security System Application.**

ON transistor is equivalent to a closed switch between emitter and collector, and therefore 0 V will be connected to the output. In summary, a logic 1 input (+5 V) will be converted to a logic 0 output (0 V).

Referring to the function table in Figure 7-7(c), you can see that the output logic level is "NOT" the same as the input logic level—hence the name NOT gate.

As an application, Figure 7-7(d) shows how a NOT or INV gate can be used to invert an input control signal. In this circuit, you can see that a normally closed push-button (NCPB) switch is used as a panic switch to activate a siren in a security system. Because the push button is normally closed, it will produce +5 V at A when it is not in alarm. If this voltage were connected directly to the siren, the siren would be activated incorrectly. By including the NOT gate between the switch circuit and the siren, the normally HIGH output of the NCPB will be inverted to a LOW and not activate the siren when we are not in alarm. When the panic switch is pressed, however, the NCPB contacts will open, producing a LOW input voltage to the NOT gate. This LOW input will be inverted to a HIGH output and activate the siren.

Analog Amplifier Circuit

When used as a variable resistor, the transistor is the controlling element in many analog or linear circuit applications such as amplifiers, oscillators, modulators, detectors, regulators, and so on. The most important of these applications is **amplification,** which is the boosting in strength or increasing in amplitude of electronic signals.

Figure 7-8(a) shows a simplified transistor amplifier circuit, while Figure 7-8(b) shows the voltage waveforms present at different points in the circuit. As you can see, the transistor is labeled Q_1 because the letter Q is the standard letter designation used for transistors.

Before applying an ac sine-wave input signal, let us determine the dc voltage levels at the transistor's base and collector. The 6.3 kΩ/3.7 kΩ resistance ratio of the voltage divider R_1 and R_2 causes the +10-V supply voltage to be proportionally divided, producing +3.7 V dc across R_2. This +3.7 V dc will be applied to the base of the transistor, causing the base–emitter junction of Q_1 to be forward biased and Q_1 to turn ON. With transistor Q_1 ON, a certain value of resistance will exist between the transistor's collector and emitter (R_{CE} or R_{EC}), and this resistance will form a voltage divider with R_E and R_C, as seen in the inset in Figure 7-8(a). The dc voltage at the collector of Q_1 relative to ground (V_C) will be equal to the voltage developed across Q_1's collector–emitter resistance (R_{CE}) and R_E. In this example circuit, with no ac signal applied, a V_B of +3.7 V dc will cause R_{CE} and R_E to cumulatively develop +8 V at the collector of Q_1. The transistor has a base bias voltage (V_B) that is +3.7 V dc relative to ground and a collector voltage (V_C) that is +8 V dc relative to ground. Capacitors C_1 and C_2 are included to act as dc blocks, with C_1 preventing the +3.7-V base bias voltage (V_B) from being applied back to the input (V_{in}) and C_2 preventing the +8-V dc collector reference voltage (V_C) from being applied across the output (V_{R_L} or V_{out}).

Let us now apply an input signal and see how it is amplified by the amplifier circuit in Figure 7-8(a). The alternating-input sine-wave signal (V_{in}) is applied to the base of Q_1 via C_1, which, like most capacitors, offers no opposition to this ac signal. This input signal, which has a peak-to-peak voltage change of 200 mV, is shown in the first waveform in Figure 7-8(b). The alternating input signal will be superimposed on the +3.7-V dc base bias voltage and cause the +3.7 V dc at the base of Q_1 to increase by 100 mv (3.7 V + 100 mV = 3.8 V), and decrease by 100 mv (3.7 V − 100 mV = 3.6 V), as seen in the second waveform in Figure 7-8(b). An increase in the input signal ($V_{in}\uparrow$), and therefore the base voltage ($V_B\uparrow$), will cause an increase in the emitter diode's forward bias, causing Q_1 to turn more ON and the emitter-to-collector resistance of Q_1 to decrease ($R_{EC}\downarrow$). Because voltage drop is always proportional to resistance, a decrease in $R_{EC}\downarrow$ will cause a decrease in the voltage drop across R_{CE} and R_E ($V_C\downarrow$), and this decrease in V_C will be coupled to the output via C_2, causing a decrease in the output voltage developed across the load ($V_{out}\downarrow$).

Now let us examine what will happen when the sine-wave input signal decreases. A decrease in the input signal ($V_{in}\downarrow$), and therefore the base voltage ($V_B\downarrow$), will cause a decrease in the emitter diode's forward bias, causing Q_1 to turn less ON and the emitter-to-

Amplification

Boosting in strength or increasing amplitude of electronic signals.

CHAPTER 7 / BIPOLAR JUNCTION TRANSISTORS

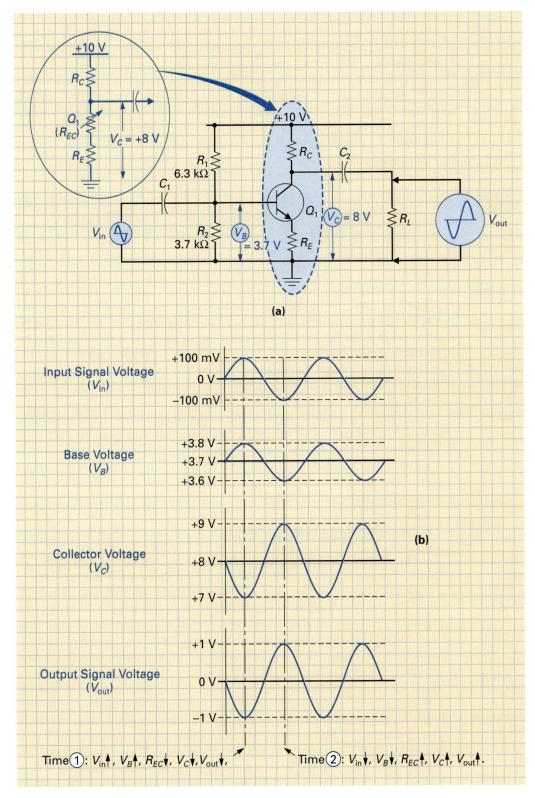

(a)

(b)

FIGURE 7-8   The Transistor Being Used in an Analog Electronic Circuit. (a) Basic Amplifier Circuit. (b) Input/Output Voltage Waveforms.

collector resistance of Q_1 to increase ($R_{EC}\uparrow$). Because voltage drop is always proportional to resistance, an increase in $R_{EC}\uparrow$ will cause an increase in the voltage drop across R_{EC} and R_E ($V_C\uparrow$). This increase in V_C will be coupled to the output via C_2, causing an increase in the output voltage developed across the load ($V_{out}\uparrow$).

Comparing the input signal voltage (V_{in}) to the output signal voltage (V_{out}) in Figure 7-8(b), you can see that a change in the input signal voltage produces a corresponding greater change in the output signal voltage. The ratio (comparison) of output signal voltage change to input signal voltage change is a measure of this circuit's **voltage gain (A_V)**. In this example, the **output signal voltage change (ΔV_{out})** is between $+1$ V and -1 V, and the **input signal voltage change (ΔV_{in})** is between $+100$ mV and -100 mV. The circuit's voltage gain between input and output will therefore be

$$\text{Voltage Gain } (A_V) = \frac{\text{Output Voltage Change } (\Delta V_{out})}{\text{Input Voltage Change } (\Delta V_{in})}$$

$$A_V = \frac{+1 \text{ to } -1 \text{ V}}{+100 \text{ mV to } -100 \text{ mV}} = \frac{2 \text{ V}}{200 \text{ mV}} = 10$$

A voltage gain of 10 means that the output voltage is 10 times larger than the input voltage. The transistor does not produce this gain magically within its NPN semiconductor structure. The gain or amplification is achieved by the input signal controlling the conduction of the transistor, which takes energy from the collector supply voltage and develops this energy across the load resistor. Amplification is achieved by having a small input voltage control a transistor and its large collector supply voltage, so that a small input voltage change results in a similar but larger output voltage change.

Comparing the input signal to the output signal at time 1 and time 2 in Figure 7-8(b), you can see that this circuit will invert the input signal voltage in the same way that the NOT gate inverts its input voltage (positive input voltage swing produces a negative output voltage swing, and vice versa). This inversion always occurs with this particular transistor circuit arrangement; however, it is not a problem because the shape of the input signal is still preserved at the output (both input and output signals are sinusoidal).

A Switching Regulator Circuit

In the previous chapter, it was shown how a series resistor and zener diode could be used in a dc power supply to function as a voltage regulator. These regulator types maintain a constant output voltage because variations in input voltage or load current are dissipated as heat. These **series dissipative regulators** generally have a low "conversion efficiency" of typically 60% to 70% and should be used only in low- to medium-load current applications.

Series switching regulators, on the other hand, have a conversion efficiency of typically 90%. To explain the operation of these regulator types, refer to the simplified circuit in Figure 7-9(a). To improve efficiency, a series-pass transistor (Q_1) is operated as a switch rather than as a variable resistor. This means that Q_1 is switched ON and OFF, and therefore either switches the $+12$-V input at its collector through to its emitter, or blocks the $+12$ V from passing through to the emitter. These $+12$-V pulses at the emitter of Q_1 charge capacitor C_1 to an average voltage (which in this example is $+5$ V) and this voltage is applied to the load (R_L). To explain this in more detail, when Q_1 is turned ON by a HIGH base voltage from the switching regulator IC, the unregulated $+12$ V at Q_1's collector is switched through to Q_1's emitter, where it reverse biases D_1, and is applied to the series-connected inductor L_1 and parallel capacitor C_1. Inductor L_1 and capacitor C_1 act as a low-pass filter because series-connected L_1 opposes the ON/OFF changes in current and passes a relatively constant current to the load: shunt- or parallel-connected C_1 opposes the ON/OFF changes in voltage and holds the output voltage relatively constant at $+5$ V. When Q_1 is turned OFF by a LOW base voltage from the switching regulator IC, the unregulated $+12$-V input is disconnected from the LC filter, the inductor's magnetic field will collapse and produce a currrent through the load, and the $+5$-V charge held by C_1

Voltage Gain (A_V)
The ratio of the output signal voltage change to input signal voltage change.

Output Signal Voltage Change
Change in output signal voltage in response to a change in the input signal voltage.

Input Signal Voltage Change
The input voltage change that causes a corresponding change in the output voltage.

Series Dissipative Regulators
Voltage regulators that maintain a constant output voltage by causing variations in input voltage or load current to be dissipated as heat.

Series Switching Regulators
A regulator circuit containing a power transistor in series with the load that is switched ON and OFF to regulate the dc output voltage delivered to the load.

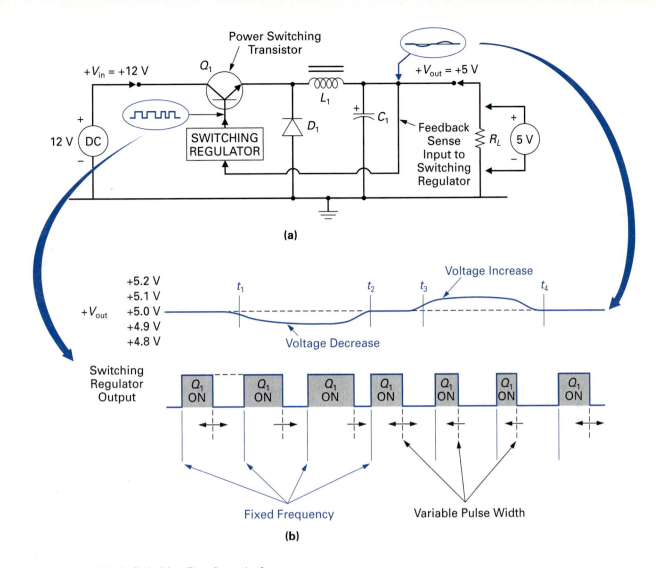

FIGURE 7-9 Basic Switching Regulator Action.

will still be applied across the load. Inductor L_1, therefore, smoothes out the current changes, while capacitor C_1 smoothes out the voltage changes caused by the ON/OFF switching of transistor Q_1. The next, and most important, question is: how does this circuit regulate, or maintain constant, the output voltage? The answer is: through a closed-loop "sense and adjust" system controlled by a switching regulator IC. The switching regulator IC operates by comparing an internal fixed reference voltage to a sense input, which is taken from the +5-V output, as is shown in Figure 7-9(a). Referring to the waveforms shown in Figure 7-9(b), you can see that whenever the output voltage falls below +5 V (from time t_1 to t_2), the switching regulator responds by increasing the width of the positive output pulse applied to the base of Q_1. This increases the ON time of Q_1, which raises the average output voltage, bringing the output back up to +5 V. On the other hand, whenever the output voltage rises above +5 V (between time t_3 and t_4), the switching regulator responds by decreasing the width of the positive output pulse applied to the base of Q_1. This decreases the ON time of Q_1, which lowers the average output voltage, bringing the output back down to +5 V. The net result is that the output voltage will remain locked at +5 V despite variations in the input voltage and variations in the load.

A switching regulator, therefore, is a voltage regulator that chops up, or switches ON and OFF (at typically a 20-kHz rate), a dc input voltage to efficiently produce a regulated dc output voltage. A **switching power supply** uses switching regulators and is generally small

Switching Power Supply

A dc power supply that makes use of a series switching regulator controlled by a pulse-width modulator to regulate the output voltage.

in size and very efficient. The only disadvantage is that the circuitry is generally a little more complex, and therefore a little more costly.

Use the following questions to test your understanding of Section 7-1.

1. What are the two basic types of bipolar transistor?
2. Name the three terminals of a bipolar transistor.
3. What are the two basic ways in which a transistor is made to operate?
4. Which of the modes of operation mentioned in question 3 is made use of in digital circuits and which is made use of in analog circuits?

7-2   SECOND APPROXIMATION DESCRIPTION OF A BIPOLAR TRANSISTOR

Now that we have a good understanding of the bipolar junction transistor's (BJT's) general characteristics, operation, and applications, let us examine all of these aspects in a little more detail.

7-2-1   *Basic Bipolar Transistor Action*

When describing diodes previously, we saw how the P-N junction of a diode could be either forward or reverse biased to either permit or block the flow of current through the device. The transistor must also be biased correctly; however, in this case, two P-N junctions rather than one must have the correct external supply voltages applied.

A Correctly Biased NPN Transistor Circuit

Figure 7-10(a) shows how an NPN transistor should be biased for normal operation. In this circuit, a +10-V supply voltage is connected to the transistor's collector (C) via a 1-kΩ collector resistor (R_C). The emitter (E) of the transistor is connected to ground via a 1.5-kΩ emitter resistor (R_E), and, as an example, an input voltage of +3.7 V is being applied to the base (B). The output voltage (V_{out}) is taken from the collector, and this collector voltage (V_C) will be equal to the voltage developed across the transistor's collector-to-emitter and the emitter resistor R_E.

As previously mentioned in the first approximation description of the transistor, the transistor can be thought of as containing two diodes, as shown in Figure 7-10(b). In normal operation, *the transistor's emitter diode or junction is forward biased, while the transistor's collector diode or junction is reverse biased.* To explain how these junctions are biased ON and OFF simultaneously, let us see how the input voltage of +3.7 V will affect this transistor circuit. An input voltage of +3.7 V is large enough to overcome the barrier voltage of the emitter diode (base–emitter junction), and so it will turn ON (base or anode is +, emitter or cathode is connected to ground or 0 V). Like any forward-biased silicon diode, the emitter diode will drop 0.7 V between base and emitter, and so the +3.7 V at the base will produce +3.0 V at the emitter. Knowing the voltage drop across the emitter resistor ($V_{R_E} = 3$ V) and the resistance of the emitter resistor ($R_E = 1.5$ kΩ), we can calculate the value of current through the emitter resistor.

$$I_{R_E} = \frac{V_{R_E}}{R_E} = \frac{3\text{ V}}{1.5\text{ k}\Omega} = 2\text{ mA}$$

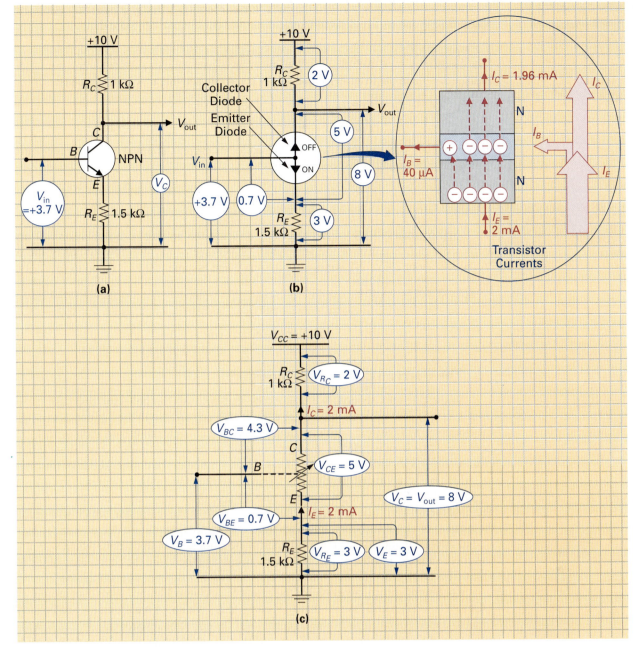

FIGURE 7-10 A Correctly Biased NPN Transistor Circuit.

This emitter resistor current of 2 mA will leave ground, travel through R_E, and then enter the transistor's n-type emitter region. This current at the transistor's emitter terminal is called the **emitter current (I_E).** The forward-biased emitter diode will cause the steady stream of electrons entering the emitter to head toward the base region, as shown in the inset in Figure 7-10(b). The base is a very thin, lightly doped region with very few holes in relation to the number of electrons entering the transistor from the emitter. Consequently, only a few electrons combine with the holes in the base region and flow out of the base region. This relatively small current at the transistor's base terminal is called the **base current (I_B).** Because only a few electrons combine with holes in the base region, there is an accumulation of electrons in the base's p layer. These free electrons, feeling the attraction of the large, positive collector supply voltage ($+10$ V), will travel through the n-type collector junction and out of the transistor to the positive external collector supply voltage. The current emerging out of

Emitter Current (I_E)

The current at the transistor's emitter terminal.

Base Current (I_B)

The relatively small current at the transistor's base terminal.

Collector Current (I_C)
The current emerging out
of the transistor's collector.

the transistor's collector is called the **collector current (I_C).** Because both the collector current and base current are derived from the emitter current, we can state that

$$I_E = I_B + I_C$$

In the example in the inset in Figure 7-10(b), you can see that this is true because

$$I_E = I_B + I_C$$
$$I_E = 40 \ \mu A + 1.96 \ mA = 2 \ mA \qquad (40 \ \mu A = 0.04 \ mA)$$

Stated another way, we can say that the collector current is equal to the emitter current minus the current that is lost out of the base.

$$I_C = I_E - I_B$$
$$I_C = 2 \ mA - 40 \ \mu A = 1.96 \ mA$$

Approximately 98% of the electrons entering the emitter of a transistor will arrive at the collector. Because of the very small percentage of current flowing out of the base (I_B equals about 2% of I_E), we can approximate and assume that I_C is equal to I_E.

$$I_C \cong I_E$$

(I_C approximately equals I_E)

The Current-Controlled Transistor

In the previous section, we discovered that because the collector and base currents (I_C and I_B) are derived from the emitter current (I_E), an increase in the emitter current ($I_E \uparrow$), for example, will cause a corresponding increase in collector and base current ($I_C \uparrow$, $I_B \uparrow$). Looking at this from a different angle, an increase in the applied base voltage (base input increases to +3.8 V) will increase the forward bias applied to the emitter diode of the transistor, which will draw more electrons up from the emitter and cause an increase in I_E, I_B, and I_C. Similarly, a decrease in the applied base voltage (base input decreases to +3.6 V) will decrease the forward bias applied to the emitter diode of the transistor, which will decrease the number of electrons being drawn up from the emitter and cause a decrease in I_E, I_B, and I_C. The applied input base voltage will control the amount of base current, which will in turn control the amount of emitter and collector current, and therefore the conduction of the transistor. This is why *the bipolar transistor is known as a current-controlled device.*

Continuing our calculations for the example circuit in Figure 7-10(b), let us apply this current relationship and assume that I_C is equal to I_E, which, as we previously calculated, is equal to 2 mA. Knowing the value of current for the collector resistor ($I_{R_C} = 2$ mA) and the resistance of the collector resistor ($R_C = 1$ kΩ), we can calculate the voltage drop across the collector resistor.

$$V_{R_C} = I_{R_C} \times R_C = 2 \ mA \times 1 \ k\Omega = 2 \ V$$

With 2 V being dropped across R_C, the voltage at the transistor's collector (V_C) will be

$$V_C = +10 \ V - V_{R_C} = 10 \ V - 2 \ V = 8 \ V$$

Because the voltage at the transistor's collector relative to ground is applied to the output, the output voltage will also be equal to 8 V.

$$V_C = V_{out} = 8 \ V$$

At this stage, we can determine a very important point about any correctly biased NPN transistor circuit. *A properly biased transistor will have a forward-biased base–emitter junction (emitter diode is ON), and a reverse-biased base–collector junction (collector diode is OFF).* We can confirm this with our example circuit in Figure 7-10(b), because we now know the voltages at each of the transistor's terminals.

Emitter diode (base–emitter junction) is forward biased (ON) because
Anode (base) is connected to $+3.7$ V (V_{in})
Cathode (emitter) is connected to 0 V via R_E

Collector diode (base–collector junction) is reverse biased (OFF) because
Anode (base) is connected to $+3.7$ V (V_{in})
Cathode (collector) is at $+8$ V (due to 2 V drop across R_C)

Keep in mind that even though the collector diode (base–collector junction) is reverse biased, current will still flow through the collector region. This is because most of the electrons traveling from emitter-to-base (through the forward-biased emitter diode) do not find many holes in the thin, lightly doped base region, and therefore the base current is always very small. Almost 98% of the electrons accumulating in the base region feel the strong attraction of the positive collector supply voltage and flow up into the collector region and then out of the collector as collector current.

With the example circuit in Figure 7-10(a) and (b), the emitter diode is ON and the collector diode is OFF, and the transistor is said to be operating in its normal, or *active*, region.

Operating a Transistor in the Active Region

A transistor is said to be in **active operation,** or in the **active region,** when its base–emitter junction is forward biased (emitter diode is ON), and the base–collector junction is reverse biased (collector diode is OFF). In this mode, the transistor is equivalent to a variable resistor between collector and emitter.

In Figure 7-10(c), our transistor circuit example has been redrawn with the transistor this time being shown as a variable resistor between collector and emitter and with all of our calculated voltage and current values inserted. Before we go any further with this circuit, let us discuss some of the letter abbreviations used in transistor circuits. To begin with, the term V_{CC} is used to denote the "stable collector voltage" and this dc supply voltage will typically be positive for an NPN transistor. Two Cs are used in this abbreviation ($+V_{CC}$) because V_C (V sub single C) is used to describe the voltage at the transistor's collector relative to ground. The doubling up of letters such as V_{CC}, V_{EE}, or V_{BB} is used to denote a constant dc bias voltage for the collector (V_{CC}), emitter (V_{EE}), and base (V_{BB}). A single sub letter abbreviation such as V_C, V_E, or V_B is used to denote a transistor terminal voltage relative to ground. The other voltage abbreviations, V_{CE}, V_{BE}, and V_{CB}, are used for the voltage difference between two terminals of the transistor. For example, V_{CE} is used to denote the potential difference between the transistor's collector and emitter terminals. Finally, I_E, I_B, and I_C are, as previously stated, used to denote the transistor's emitter current (I_E), base current (I_B), and collector current (I_C).

Because the transistor's resistance between emitter and collector in Figure 7-10(c) is in series with R_C and R_E, we can calculate the voltage drop between collector and emitter (V_{CE}) because V_{RC} and V_{RE} are known.

$$V_{CE} = V_{CC} - (V_{R_E} + V_{R_C})$$
$$V_{CE} = 10\text{ V} - (3\text{ V} + 2\text{ V}) = 10\text{ V} - 5\text{ V} = 5\text{ V}$$

Now that we know the voltage drop between the transistor's collector and emitter (V_{CE}), we can calculate the transistor's equivalent resistance between collector and emitter (R_{CE}) because we know that the current through the transistor is 2 mA.

$$R_{CE} = \frac{V_{CE}}{I_C} = \frac{5\text{ V}}{2\text{ mA}} = 2.5\text{ k}\Omega$$

Operating the Transistor in Cutoff and Saturation

Figure 7-11 shows the three basic ways in which a transistor can be operated. As we have already discovered, the bias voltages applied to a transistor control the transistor's op-

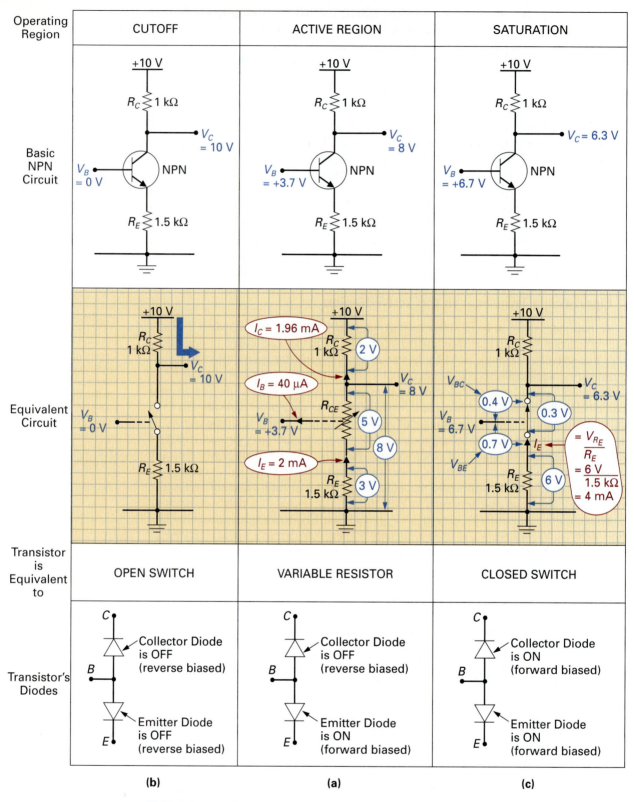

FIGURE 7-11 The Three Bipolar Transistor Operating Regions.

eration by controlling the two P-N junctions (or diodes) in a bipolar transistor. For example, the center column reviews how a transistor will operate in the active region. As you can see, our previous circuit example with all of its values has been used. To summarize: *When a transistor is operated in the active region, its emitter diode is biased ON, its collector diode is biased OFF, and the transistor is equivalent to a variable resistor between the collector and the emitter.*

The left column in Figure 7-11 shows how the same transistor circuit can be driven into **cutoff.** A transistor is in cutoff when the bias voltage is reduced to a point that it stops current in the transistor. In this example circuit, you can see that when the base input bias voltage (V_B) is reduced to 0 V, the transistor is cut off. *In cutoff, both the emitter and the collector diode of the transistor will be biased OFF, the transistor is equivalent to an open switch between the collector and the emitter, and the transistor current is zero.*

The right column in Figure 7-11(c) shows how the same transistor circuit can be driven into **saturation.** A transistor is in saturation when the bias voltage is increased to such a point that any further increase in bias voltage will not cause any further increase in current through the transistor. In the equivalent circuit in Figure 7-11(c), you can see that when the base input bias voltage (V_B) is increased to +6.7 V, the emitter diode of the transistor will be heavily forward biased and the emitter current will be large.

Cutoff
A transistor is in cutoff when the bias voltage is reduced to a point that it stops current in the transistor.

Saturation
A transistor is in saturation when the bias voltage is increased to such a point that further increase will not cause any increase in current through the transistor.

$$I_E = \frac{V_B - V_{BE}}{R_E} = \frac{6.7 \text{ V} - 0.7 \text{ V}}{1.5 \text{ k}\Omega} = 4.\text{mA}$$

Because I_B and I_C are both derived from I_E, an increase in I_E will cause a corresponding increase in both I_B and I_C. These high values of current through the transistor account for why a transistor operating in saturation is said to be equivalent to a closed switch (high conductance, low resistance). Although the transistor's resistance between the collector and emitter (R_{CE}) is assumed to be 0 Ω, there is still some small value of R_{CE}. Typically, a saturated transistor will have a 0.3-V drop between the collector and emitter ($V_{CE} = 0.3$ V), as shown in the equivalent circuit in Figure 7-11(c). If $V_{BE} = 0.7$ V and $V_{CE} = 0.3$ V, then the voltage drop across the collector diode of a saturated transistor (V_{BC}) will be 0.4 V. This means that the base of the transistor (anode) is now +0.4 V relative to the collector (cathode), and there is not enough reverse bias voltage to turn OFF the collector diode. *In saturation, both the emitter and collector diodes are said to be forward biased, the transistor is equivalent to a closed switch, and any further increase in bias voltage will not cause any further increase in current through the transistor.*

Biasing PNP Bipolar Transistors

Generally the PNP transistor is not employed as much as the NPN transistor in most circuit applications. The only difference that occurs with PNP transistor circuits is that the polarity of V_{CC} and the base bias voltage (V_B) need to be reversed to a negative voltage, as shown in Figure 7-12. The PNP transistor has the same basic operating characteristics as the NPN transistor, and all of the previously discussed equations still apply. Referring to the inset in Figure 7-12, you will see that the −3.7-V base bias voltage will forward bias the emitter diode and the −10-V V_{CC} will reverse bias the collector diode, so that the transistor is operating in the active region. Also, the electron transistor currents are in the opposite direction. This, however, makes no difference because the sum of the collector current entering the collector and base current entering the base is equal to the value of emitter current leaving the emitter, so $I_E = I_B + I_C$ still applies.

7-2-2   *Bipolar Transistor Circuit Configurations and Characteristics*

In the previous sections, we have seen how the bipolar junction transistor can be used in digital two-state switching circuits and analog or linear circuits such as the amplifier. In all of

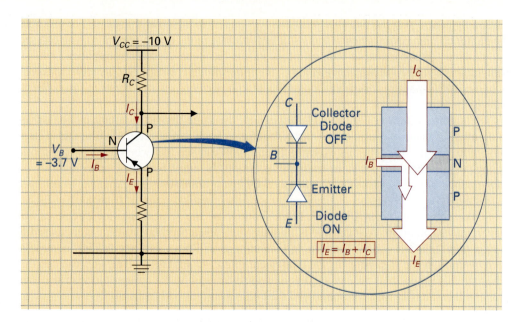

FIGURE 7-12   **A Correctly Biased PNP Transistor Circuit.**

Configurations

Different circuit interconnections.

Common

Shared by two or more services, circuits, or devices. Although the term *common ground* is frequently used to describe two or more connections sharing a common ground, the term *common* alone does not indicate a ground connection, only a shared connection.

Common–Emitter (C–E) Circuit

Configuration in which the input signal is applied between the base and the emitter, while the output signal appears between the transistor's collector and emitter.

these different circuit interconnections or **configurations,** the bipolar transistor was used as the main controlling element, with one of its three leads being used as a common reference and the other two leads being used as an input and an output. Although there are many thousands of different bipolar transistor circuit applications, all of these circuits can be classified in one of three groups based on which of the transistor's leads is used as the **common** reference. These three different circuit configurations are shown in Figure 7-13. With the *common–emitter (C–E)* bipolar transistor circuit configuration, shown in Figure 7-13(a), the input signal is applied between the base and emitter, while the output signal appears between the transistor's collector and emitter. With this circuit arrangement, the input signal controls the transistor's base current, which in turn controls the transistor's output collector current, and the emitter lead is common to both the input and output. Similarly, with the *common–base (C–B)* circuit configuration shown in Figure 7-13(b), the input signal is applied between the transistor's emitter and base, the output signal is developed across the transistor's collector and base, and the base is common to both input and output. Finally, with the *common–collector (C–C)* circuit configuration shown in Figure 7-13(c), the input is applied between the base and collector, the output is developed across the emitter and collector, and the collector is common to both the input and output.

To begin with, we will discuss the common–emitter circuit configuration characteristics because it has been this circuit arrangement that we have been using in all of the circuit examples in this chapter.

Common–Emitter Circuits

With the **common–emitter circuit,** the transistor's emitter lead is common to both the input and output signals. In this circuit configuration, the base serves as the input lead and the collector serves as the output lead. Figure 7-14 contains a basic common–emitter (C–E) circuit, its associated input/output voltage and current waveforms, characteristic curves, and table of typical characteristics. Using this illustration, we will examine the operation and characteristics of the C–E circuit configuration.

DC Current Gain   Referring to the C–E circuit in Figure 7-14(a), and its associated waveforms in Figure 7-14(b), let us now examine this circuit's basic operation.

Before applying the ac sine-wave input signal (V_{in}), let us assume that $V_{in} = 0$ V and examine the transistor's dc operating characteristics, or the "no-input-signal" condition. The voltage divider R_1 and R_2 will divide the V_{CC} supply voltage, producing a positive dc base

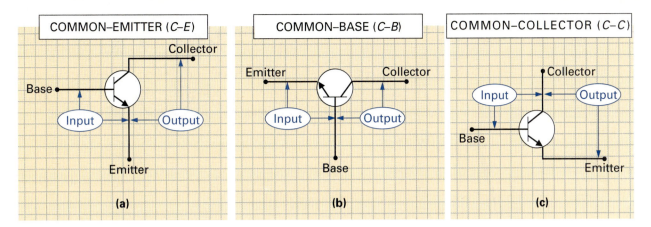

FIGURE 7-13 Bipolar Transistor Circuit Configurations.

bias voltage across R_2. This base bias voltage will be applied to the base of Q_1, and because it is generally greater than 0.7 V, it will forward bias the transistor's base–emitter junction, turning Q_1 ON (in the example in Figure 7-14, $V_B = 3$ V dc). Capacitor C_1 is included to act as a dc block, preventing the base bias voltage (V_B) from being applied back to the input (V_{in}). The value of dc base bias voltage will determine the value of base current (I_B) flowing out of the transistor's base, and this value of I_B will in turn determine the value of collector current (I_C) flowing out of the transistor's collector and through R_C. Because the transistor's output current (I_C) is so much larger than the transistor's very small input current (I_B), the circuit produces an increase in current, or a **current gain.** The current gain in a common–emitter configuration is called the transistor's **beta** (symbolized β). A transistor's **dc beta** ($β_{DC}$) indicates a common-emitter transistor's "dc current gain," and it is the ratio of its output current (I_C) to its input current (I_B). This ratio can be expressed mathematically as

$$\beta_{DC} = \frac{I_C}{I_B}$$

> **Current Gain**
> The increase in current produced by the transistor circuit.
>
> **Beta (β)**
> The transistor's current gain in a common–emitter configuration.
>
> **DC Beta ($β_{DC}$)**
> The ratio of a transistor's dc output current to its input current.

■ **EXAMPLE:**

As can be seen in Figure 7-14(b), the "no input signal" level of $I_C = 1$ mA and $I_B = 30$ μA. What is the transistor's dc beta?

■ *Solution:*

$$\beta_{DC} = \frac{I_C}{I_B}$$
$$= \frac{1\,\text{mA}}{30\,\mu\text{A}}$$
$$= 33.3$$

This value indicates that I_C is 33.3 times greater than I_B; therefore, the dc current gain between input and output is 33.3.

Another form of system analysis, known as "hybrid parameters," uses the term *hfe* instead of $β_{DC}$ to indicate a transistor's dc current gain.

AC Current Gain   When amplifying an ac waveform, a transistor has applied to it both the dc voltage to make it operational and the ac signal voltage to be amplified that

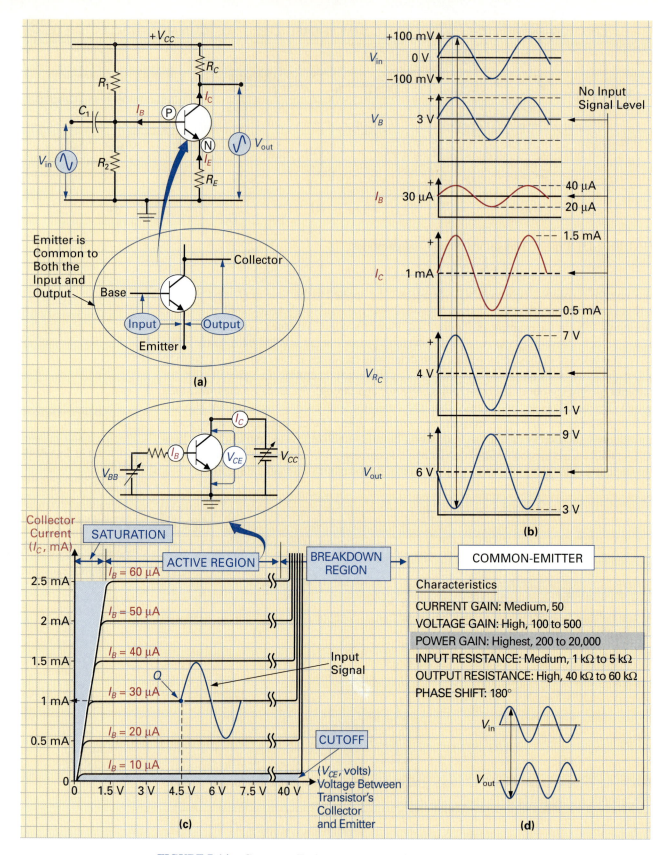

FIGURE 7-14   Common–Emitter *(C–E)* Circuit Configuration Characteristics.

varies the base bias voltage and the base current. Referring to the first four waveforms in Figure 7-14(b), let us now examine what will happen when we apply a sine-wave input signal. As V_{in} increases above 0 V to a peak positive voltage ($V_{in}\uparrow$), it will cause the forward base bias voltage applied to the transistor to increase ($V_B\uparrow$) above its dc reference or "no input signal level." This increase in V_B will increase the forward conduction of the transistor's emitter diode, resulting in an increase in the input base current ($I_B\uparrow$) and a corresponding larger increase in the output collector current ($I_C\uparrow$). The ratio of output collector current change (ΔI_C) to input base current change (ΔI_B) is the transistor's ac current gain or **ac beta** (β_{AC}). The formula for calculating a transistor's ac current gain is

$$\beta_{AC} = \frac{\Delta I_C}{\Delta I_B}$$

EXAMPLE:

Calculate the ac current gain of the example in Figure 7-14(b).

Solution:

$$\beta_{AC} = \frac{\Delta I_C}{\Delta I_B}$$
$$= \frac{1.5 \text{ to } 0.5 \text{ mA}}{40 \text{ to } 20 \text{ }\mu\text{A}}$$
$$= \frac{1 \text{ mA}}{20 \text{ }\mu\text{A}}$$
$$= 50$$

This means that the alternating collector current at the output is 50 times greater than the alternating base current at the input.

Common–emitter transistors will typically have beta, or current gain values, of 50.

Voltage Gain The common–emitter circuit is not only used to increase the level of current between input and output. It can also be used to increase the amplitude of the input signal voltage, or produce a voltage gain. This action can be seen by examining the C–E circuit in Figure 7-14(a) and by following the changes in the associated waveforms in Figure 7-14(b). An input signal voltage increase from 0 V to a positive peak ($V_{in}\uparrow$) causes an increase in the dc base bias voltage ($V_B\uparrow$), causing the emitter diode of Q_1 to turn more ON and result in an increase in both I_B and I_C. Because the I_C flows through R_C, and because voltage drop is proportional to current, an increase in I_C will cause an increase in the voltage drop across R_C ($V_{R_C}\uparrow$). The output voltage is equal to the voltage developed across Q_1's collector-to-emitter resistance (R_{CE}) and R_E. Because R_C is in series with Q_1's collector-to-emitter resistance (R_{CE}) and R_E, an increase in $V_{R_C}\uparrow$ will cause a decrease in the voltage developed across R_{CE} and R_E, which is $V_{out}\downarrow$. This action is summarized with Kirchhoff's voltage law, which states that the sum of the voltages in a series circuit is equal to the voltage applied ($V_{R_C} + V_{out} = V_{CC}$). Using the example in Figure 7-14(b), you can see that when

$V_{R_C}\uparrow$ to 7 V, $V_{out}\downarrow$ to 3 V.   ($V_{R_C} + V_{out} = V_{CC}$, 7 V + 3 V = 10 V)

$V_{R_C}\downarrow$ to 1 V, $V_{out}\uparrow$ to 9 V.   ($V_{R_C} + V_{out} = V_{CC}$, 1 V + 9 V = 10 V)

Although the input voltage (V_{in}) and output voltage (V_{out}) are out of phase with one another, you can see from the example values in Figure 7-14(b) that there is an increase in the signal voltage between input and output. This voltage gain between input and output is possible because the output current (I_C) is so much larger than the input current (I_B). The amount of voltage gain (which is symbolized A_V) can be calculated by comparing the output voltage change (ΔV_{out}) to the input voltage change (ΔV_{in}).

$$A_V = \frac{\Delta V_{out}}{\Delta V_{in}}$$

■ EXAMPLE:

Calculate the voltage gain of the circuit and its associated waveforms in Figure 7-14(a) and (b).

■ *Solution:*

$$A_V = \frac{\Delta V_{out}}{\Delta V_{in}} = \frac{+9 \text{ V to} +3 \text{ V}}{+100 \text{ mV to} -100 \text{ mV}} = \frac{6 \text{ V}}{200 \text{ mV}} = 30$$

This value indicates that the output ac signal voltage is 30 times larger than the input ac signal voltage.

Most common–emitter transistor circuits have high voltage gains between 100 to 500.

Power Gain   As we have seen so far, the common–emitter circuit provides both current gain and voltage gain. Because power is equal to the product of current and voltage ($P = V \times I$), it is not surprising that the C–E circuit configuration also provides **power gain** (A_P). The power gain of a circuit can be calculated by dividing the output signal power (P_{out}) by the input signal power (P_{in}).

Power Gain (A_P)
The ratio of the output signal power to the input signal power.

$$A_P = \frac{P_{out}}{P_{in}}$$

To calculate the amount of input power (P_{in}) applied to the C–E circuit, we will have to multiply the change in input signal voltage (ΔV_{in}) by the accompanying change in input signal current (ΔI_{in} or ΔI_B).

$$P_{in} = \Delta V_{in} \times \Delta I_{in}$$

To calculate the amount of output power (P_{out}) delivered by the C–E circuit, we will have to multiply the change in output signal voltage (ΔV_{out}) produced by the change in output signal current (ΔI_{out} or ΔI_C).

$$P_{out} = \Delta V_{out} \times \Delta I_{out}$$

The power gain of a common–emitter circuit is therefore calculated with the formula

$$A_P = \frac{P_{out}}{P_{in}} = \frac{\Delta V_{out} \times \Delta I_{out}}{\Delta V_{in} \times \Delta I_{in}}$$

■ EXAMPLE:

Calculate the power gain of the example circuit in Figure 7-14.

Solution:

$$A_P = \frac{P_{out}}{P_{in}} = \frac{\Delta V_{out} \times \Delta I_{out}}{\Delta V_{in} \times \Delta I_{in}}$$

$$= \frac{(9 \text{ V to } 3 \text{ V}) \times (1.5 \text{ mA to } 0.5 \text{ mA})}{(+100 \text{ mV to } -100 \text{ mV}) \times (40 \text{ } \mu\text{A to } 20 \text{ } \mu\text{A})}$$

$$= \frac{6 \text{ V} \times 1 \text{ mA}}{200 \text{ mV} \times 20 \text{ } \mu\text{A}} = \frac{6 \text{ mW}}{4 \text{ } \mu\text{W}} = 1{,}500$$

In this example, the common–emitter circuit has increased the input signal power from 4 μW to 6 mW—a power gain of 1,500.

The power gain of the circuit in Figure 7-14 can also be calculated by multiplying the previously calculated C–E circuit voltage gain (A_V) by the previously calculated C–E circuit current gain (β_{AC}). Since

$$A_V = \frac{\Delta V_{out}}{\Delta V_{in}}, \quad \text{and} \quad \beta_{AC} = \frac{\Delta I_{out} \text{ (or } \Delta I_C)}{\Delta I_{in} \text{ (or } \Delta I_B)}$$

$$A_P = V \times I = \frac{\Delta V_{out}}{\Delta V_{in}} \times \frac{\Delta I_{out} \text{ (or } \Delta I_C)}{\Delta I_{in} \text{ (or } \Delta I_B)} \quad \text{or } A_P = A_V \times \beta_{AC}$$

$$\boxed{A_P = A_V \times \beta_{AC}}$$

For the example circuit in Figure 7-14, this will be

$$A_P = A_V \times \beta_{AC} = 30 \times 50 = 1{,}500$$

indicating that the common–emitter circuit's output power in Figure 7-14 is 1,500 times larger than the input power.

The power gain of common–emitter transistor circuits can be as high as 20,000, making this characteristic the circuit's key advantage.

Collector Characteristic Curves   One of the easiest ways to compare several variables is to combine all of the values in a graph. Figure 7-14(c) shows a special graph, called the **collector characteristic curves,** for a typical common–emitter transistor circuit. The data for this graph are obtained by using the transistor test circuit, shown in the inset in Figure 7-14(c), which will apply different values of base bias voltage (V_{BB}) and collector bias voltage (V_{CC}) to an NPN transistor. The two ammeters and one voltmeter in this test circuit are used to measure the circuit's I_B, I_C, and V_{CE} response to each different circuit condition. The values obtained from this test circuit are then used to plot the transistor's collector current (I_C in mA) on the vertical axis against the transistor's collector–emitter voltage (V_{CE}) drop on the horizontal axis for various values of base current (I_B in μA). This graph shows the relationship between a transistor's input base current, output collector current, and collector-to-emitter voltage drop.

Let us now examine the typical set of collector characteristic curves shown in Figure 7-14(c). When V_{CE} is increased from zero, by increasing V_{CC}, the collector current rises very rapidly, as indicated by the rapid vertical rise in any of the curves. When the collector diode of the transistor is reverse biased by the voltage V_{CE}, the collector current levels off. At this point, any one of the curves can be followed based on the amount of base current, which is determined by the value of base bias voltage (V_{BB}) applied. This flat part of the curve is known as the transistor's **active region.** The transistor is normally operated in this region, where it is equivalent to a variable resistor between the collector and emitter.

As an example, let us use these curves in Figure 7-14(c) to calculate the value of output current (I_C) for a given value of input current (I_B) and collector-to-emitter voltage (V_{CE}).

Collector Characteristic Curves

Graph for a typical common–emitter transistor circuit.

Active Region

Flat part of the collector characteristic curve. A transistor is normally operated in this region, where it is equivalent to a variable resistor between the collector and emitter.

If V_{BB} is adjusted to produce a base current of 30 μA and V_{CC} is adjusted until the voltage between the transistor's collector and emitter (V_{CE}) is 4.5 V, the output collector current (I_C) will be equal to approximately 1 mA. This is determined by first locating 4.5 V on the horizontal axis (V_{CE} = 4.5 V), following this point directly up to the I_B = 30 μA curve, and then moving directly to the left to determine the value of output current on the vertical axis (I_C = 1 mA). When these values of V_{CE} and I_B are present, the transistor is said to be operating at point Q. This dc operating point is often referred to as a **quiescent operating point (Q point),** which means a dc steady-state or no-input-signal operating point. The Q point of a transistor is set by the circuit's dc bias components and supply voltages. For instance, in the circuit in Figure 7-14(a), R_1 and R_2 were used to set the dc base bias voltage (V_B, and therefore I_B), and the values of V_{CC}, R_C, and R_E were chosen to set the transistor's dc collector–emitter voltage (V_{CE}). At this dc operating point, we can calculate the transistor's dc current gain (β_{DC}), because both I_B and I_C are known

$$\beta_{DC} = \frac{I_C}{I_B} = \frac{1\text{ mA}}{30\text{ μA}} = 33.3$$

If a sine-wave signal was applied to the circuit, as shown in the waveforms in Figure 7-14(b) and the characteristic curves in Figure 7-14(c), it would cause the transistor's input base current, and therefore output collector current, to alternate above and below the transistor's Q point (dc operating point). This ac input signal voltage will cause the input base current (I_B) to increase between 20 μA and 40 μA, and this input base current change will generate an output collector current change of 0.5 mA to 1.5 mA. The transistor's ac current gain (β_{AC}) will be

$$\beta_{AC} = \frac{\Delta I_C}{\Delta I_B}$$
$$= \frac{1.5\text{ to }0.5\text{ mA}}{40\text{ to }20\text{ μA}}$$
$$= \frac{1\text{ mA}}{20\text{ μA}}$$
$$= 50$$

Returning to the collector characteristic curves in Figure 7-14(c), you can see that if the collector supply voltage (V_{CC}) is increased to an extreme, a point will be reached where the V_{CE} voltage across the transistor will cause the transistor to break down, as indicated by the rapid rise in I_C. This section of the curve is called the **breakdown region** of the graph, and the damaging value of current through the transistor will generally burn out and destroy the device. As an example, for the 2N3904 bipolar transistor, breakdown will occur at a V_{CE} voltage of 40 V.

There are two shaded sections shown in the set of collector characteristic curves in Figure 7-14(c). These two shaded sections represent the other two operating regions of the transistor. To begin with, let us examine the vertically shaded **saturation region.** If the base bias voltage (V_{BB}) is increased to a large positive value, the emitter diode of the transistor will turn ON heavily, I_B will be a large value, and the transistor will be operating in saturation. In this operating region, the transistor is equivalent to a closed switch between its collector and emitter (both the emitter and collector diode are forward biased), and therefore the voltage drop between collector and emitter will be almost zero (V_{CE} = typically 0.3 V, when transistor is saturated), and I_C will be a large value that is limited only by the externally connected components. The horizontally shaded section represents the **cutoff region** of the transistor. If the base bias voltage (V_{BB}) is decreased to zero, the emitter diode of the transistor will turn OFF, I_B will be zero, and the transistor will be operating in cutoff. In this operating region, the transistor is equivalent to an open switch between its collector and emitter (both the emitter and collector diode are reverse biased), and the voltage drop between collector and emitter will be equal to V_{CC} and I_C will be zero.

Quiescent Operating Point (Q Point)

The voltage or current value that sets up the no-input-signal or operating point bias voltage.

Breakdown Region

The point at which the collector supply voltage will cause a damaging value of current through the transistor.

Saturation Region

The point at which the collector supply voltage has the transistor operating at saturation.

Cutoff Region

The point at which the collector supply voltage has the transistor operating in cutoff.

CHAPTER 7 / BIPOLAR JUNCTION TRANSISTORS

A set of collector characteristic curves are therefore generally included in a manufacturer's device data sheet and can be used to determine the values of I_B, I_C, and V_{CE} at any operating point.

Input Resistance   The **input resistance (R_{in})** of a common–emitter transistor is the amount of opposition offered to an input signal by the input base–emitter junction (emitter diode). Because the base–emitter junction is normally forward biased when the transistor is operating in the active region, the opposition to input current is relatively small. However, the extremely small base region will only support a very small input base current. On average, if no additional components are connected in series with the transistor's base–emitter junction, the input resistance of a C–E transistor circuit is typically a medium value between 1 kΩ and 5 kΩ. This typical value is an average because the transistor's input resistance is a "dynamic or changing quantity" that will vary slightly as the input signal changes the conduction of the C–E transistor's emitter diode, and this changes I_B ($R \updownarrow = V/I \updownarrow$).

Because the transistor has a small value of input P-N junction capacitance and input terminal inductance, the opposition to the input signal is not only resistive but, to a small extent, reactive. For this reason, the total opposition offered by the transistor to an input signal is often referred to as the **input impedance (Z_{in})** because impedance is the total combined resistive and reactive input opposition.

Output Resistance   The **output resistance (R_{out})** of a common–emitter transistor is the amount of opposition offered to an output signal by the output base–collector junction (collector diode). This junction is normally reverse biased when the transistor is operating in the active region, and therefore the C–E transistor's output resistance is relatively high. However, because a unique action occurs within the transistor and allows current to flow through this reverse-biased junction (electron accumulation at the base and then conduction through collector diode due to attraction of $+V_{CC}$), the output current (I_C) is normally large, and so the output resistance is not an extremely large value. On average, if no load resistor is connected in series with the transistor's collector diode, the output resistance of a C–E transistor circuit is typically a high value between 40 kΩ and 60 kΩ.

Because the transistor has a small value of output P-N junction capacitance and output terminal inductance, the opposition to the output signal is not only resistive but also reactive. For this reason, the total opposition offered by the transistor to the output signal is often referred to as the **output impedance (Z_{out}).**

Common–Base Circuits

With the **common–base circuit,** the transistor's base lead is common to both the input and output signal. In this circuit configuration, the emitter serves as the input lead and the collector serves as the output lead. Figure 7-15 contains a basic common–base (C–B) circuit, its associated input/output voltage and current waveforms, and table of typical characteristics. Using this illustration, we will examine the operation and characteristics of the C–B circuit configuration.

DC Current Gain   Referring to the C–B circuit in Figure 7-15(a), and its associated waveforms in Figure 7-15(b), let us now examine this circuit's basic operation.

Before applying the ac sine-wave input signal (V_{in}), let us assume that $V_{in} = 0$ V and examine the transistor's dc operating characteristics, or the "no-input-signal" condition. The voltage divider R_1 and R_2 will divide the V_{CC} supply voltage, producing a positive dc base bias voltage across R_2. This base bias voltage will be applied to the base of Q_1. Because it is generally greater than 0.7 V, it will forward bias the transistor's base–emitter junction, turning Q_1 ON. Because the common–base circuit's input current is I_E and its output current is I_C, the current gain between input and output will be determined by the ratio of I_C to I_E. This ratio for calculating a C–B transistor's dc current gain is called the transistor's **dc alpha (α_{DC})** and is equal to

Input Resistance (R_{in})
The amount of opposition offered to an input signal by the input base–emitter junction (emitter diode).

Input Impedance (Z_{in})
The total opposition offered by the transistor to an input signal.

Output Resistance (R_{out})
The amount of opposition offered to an output signal by the output base–collector junction (collector diode).

Output Impedance (Z_{out})
The total opposition offered by the transistor to the output signal.

Common–Base (C–B) Circuit
Configuration in which the input signal is applied between the transistor's emitter and base, while the output is developed across the transistor's collector and base.

DC Alpha (α_{DC})
The ratio for calculating a C–B transistor's dc current gain.

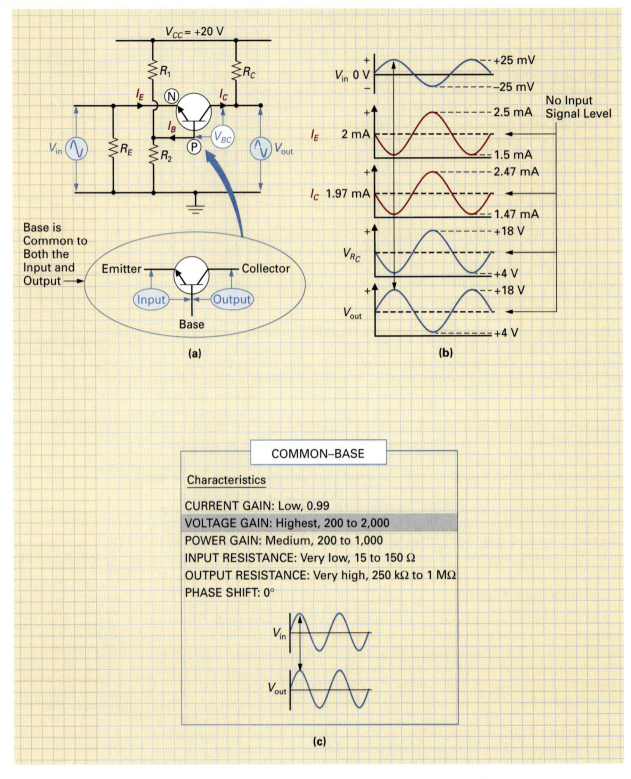

(a)

(b)

COMMON–BASE

Characteristics

CURRENT GAIN: Low, 0.99
VOLTAGE GAIN: Highest, 200 to 2,000
POWER GAIN: Medium, 200 to 1,000
INPUT RESISTANCE: Very low, 15 to 150 Ω
OUTPUT RESISTANCE: Very high, 250 kΩ to 1 MΩ
PHASE SHIFT: 0°

(c)

FIGURE 7-15 **Common–Base (*C–B*) Circuit Configuration Characteristics.**

$$\alpha_{DC} = \frac{I_C}{I_E}$$

The "no-input-signal" or "steady-state" dc levels of I_E and I_C are determined by the value of voltage developed across R_2, which is controlling the conduction of the transistor's forward-biased base–emitter junction (emitter diode). Because the output current I_C is always slightly lower than the input current I_E (due to the small I_B current flow out of the base), the C–B transistor circuit does not increase current between input and output. In fact, there is a slight loss in current between input and output, which is why the C–B circuit configuration is said to have a current gain that is less than 1.

EXAMPLE:

Calculate the dc alpha of the circuit in Figure 7-15(a) if $I_C = 1.97$ mA and $I_E = 2$ mA.

Solution:

$$\alpha_{DC} = \frac{I_C}{I_E} = \frac{1.97 \text{ mA}}{2 \text{ mA}} = 0.985$$

This value of 0.985 indicates that I_C is 98.5% of I_E (0.985 $\times$ 100 = 98.5).

As you can see from this example, the difference between I_C and I_E is generally so small that we always assume that the dc alpha is 1, which means that $I_C = I_E$.

AC Current Gain   When amplifying an ac waveform, a transistor has applied to it both the dc voltage to make it operational and the ac signal voltage to be amplified that varies the base–emitter bias and the input emitter current. Referring to the waveforms in Figure 7-15(b), let us now examine what will happen when we apply a sine-wave input signal. As mentioned previously, the positive voltage developed across R_2 will make the NPN transistor's base positive with respect to the emitter and forward bias the P-N base–emitter junction.

As the input voltage swings positive ($V_{in}\uparrow$), it will reduce the forward bias across the transistor's P-N base–emitter junction. For example, if $V_B = +5$ V and the transistor's n-type emitter is made positive, the P-N base-emitter diode will be turned more OFF. Turning the transistor's emitter diode less ON will cause a decrease in emitter current ($I_E\downarrow$), a decrease in collector current ($I_C\downarrow$), and a decrease in the voltage developed across R_C ($V_{R_C}\downarrow$). Because V_{R_C} and V_{out} are connected in series across V_{CC}, a decrease in $V_{R_C}\downarrow$ must be accompanied by an increase in $V_{out}\uparrow$. To explain this another way, the decrease in I_E and the subsequent decrease in both I_C and I_B means that the conduction of the transistor has decreased. This decrease in conduction means that the normally forward-biased base–emitter junction has turned less ON and the normally reverse-biased base–collector junction has turned more OFF. Because the transistor's base–collector junction is in series with R_C and R_2 across V_{CC}, an increase in the transistor's base–collector resistance ($R_{BC}\uparrow$) will cause an increase in the voltage developed across the transistor's base–collector junction ($V_{BC}\uparrow$), which will cause an increase in $V_{out}\uparrow$.

Similarly, as the input voltage swings negative ($V_{in}\downarrow$), it will increase the forward bias across the transistor's P-N base–emitter junction. For example, if $V_B = -5$ V and the transistor's n-type emitter is made negative, the P-N base–emitter diode will be turned more ON. Turning the transistor's emitter diode more ON will cause an increase in emitter current ($I_E\uparrow$), an increase in collector current ($I_C\uparrow$), and an increase in the voltage developed across R_C ($V_{R_C}\uparrow$). Because V_{R_C} and V_{out} are connected in series across V_{CC}, an increase in $V_{R_C}\uparrow$ must be accompanied by a decrease in $V_{out}\downarrow$.

AC Alpha (α_{AC})

The ratio of input emitter current change to output collector current change.

Now that we have seen how the input voltage causes a change in input current (I_E) and output current (I_C), let us examine the *C–B* circuit's ac current gain. The ratio of input emitter current change (ΔI_E) to output collector current change (ΔI_C) is the *C–B* transistor's ac current gain or **ac alpha (α_{AC})**. The formula for calculating a *C–B* transistor's ac current gain is

$$\alpha_{AC} = \frac{\Delta I_C}{\Delta I_E}$$

EXAMPLE:

Calculate the ac current gain of the example in Figure 7-15(b).

Solution:

$$\alpha_{AC} = \frac{\Delta I_C}{\Delta I_E}$$

$$= \frac{2.47 \text{ to } 1.47 \text{ mA}}{2.5 \text{ to } 1.5 \text{ mA}}$$

$$= \frac{1 \text{ mA}}{1 \text{ mA}}$$

$$= 1$$

This means that the change in output collector current is equal to the change in input emitter current, and therefore the ac current gain is 1. (The output is 1 times larger than the input, 1 mA $\times$ 1 = 1 mA.)

Common–base transistors will typically have an ac alpha, or ac current gain, of 0.99.

Voltage Gain   Although the common–base circuit does not achieve any current gain, it does make up for this disadvantage by achieving a very large voltage gain between input and output. Returning to the *C–B* circuit in Figure 7-15(a) and its waveforms in Figure 7-15(b), let us see how this very high voltage gain is obtained. Only a small input voltage (V_{in}) is needed to control the conduction of the transistor's emitter diode, and therefore the input emitter current (I_E) and output collector current (I_C). Even though I_C is slightly lower than I_E, it is still a relatively large value of current and will develop a large voltage change across R_C for a very small change in V_{in}. Because V_{R_C} and V_{out} are in series and connected across V_{CC}, a large change in voltage across R_C will cause a large change in the voltage developed across the transistor output (V_{BC}) and V_{out}. As before, the amount of voltage gain (which is symbolized A_V) can be calculated by comparing the output voltage change (ΔV_{out}) to the input voltage change (ΔV_{in}).

$$A_V = \frac{\Delta V_{out}}{\Delta V_{in}}$$

EXAMPLE:

Calculate the voltage gain of the circuit and its associated waveforms in Figure 7-15(a) and (b).

$$A_V = \frac{\Delta V_{out}}{\Delta V_{in}} = \frac{+18 \text{ V to } +4 \text{ V}}{+25 \text{ mV to } -25 \text{ mV}} = \frac{14 \text{ V}}{50 \text{ mV}} = 280$$

This value indicates that the output ac signal voltage is 280 times larger than the ac input signal voltage.

Most common–base transistor circuits have very high voltage gains between 200 and 2,000. While on the topic of comparing the input voltage to output voltage, you can see by looking at Figure 7-15(b) that, unlike the *C–E* circuit, the common–base circuit has no phase shift between input and output (V_{out} is in phase with V_{in}).

Power Gain Although the common–base circuit achieves no current gain, it does have a very high voltage gain and therefore can provide a medium amount of power gain ($P\uparrow = V\uparrow \times I$). The power gain of a circuit can be calculated by dividing the output signal power (P_{out}) by the input signal power (P_{in}).

$$A_P = \frac{P_{out}}{P_{in}}$$

To calculate the amount of input power (P_{in}) applied to the *C–B* circuit, we will have to multiply the change in input signal voltage (ΔV_{in}) by the accompanying change in input signal current (ΔI_{in} or ΔI_E).

$$P_{in} = \Delta V_{in} \times \Delta I_{in}$$
$$P_{in} = 50 \text{ mV} \times 1 \text{ mA} = 50 \text{ }\mu\text{W}$$

To calculate the amount of output power (P_{out}) delivered by the *C–B* circuit, we will have to multiply the change in output signal voltage (ΔV_{out}) produced by the change in output signal current (ΔI_{out} or ΔI_C).

$$P_{out} = \Delta V_{out} \times \Delta I_{out}$$
$$P_{out} = 14 \text{ V} \times 1 \text{ mA} = 14 \text{ mW}$$

The power gain of a common–base circuit is calculated with the formula

$$A_P = \frac{P_{out}}{P_{in}} = \frac{\Delta V_{out} \times \Delta I_{out}}{\Delta V_{in} \times \Delta I_{in}}$$

EXAMPLE:

Calculate the power gain of the example circuit in Figure 7-15.

Solution:

$$A_P = \frac{P_{out}}{P_{in}} = \frac{\Delta V_{out} \times \Delta I_{out}}{\Delta V_{in} \times \Delta I_{in}}$$

$$= \frac{14 \text{ V} \times 1 \text{ mA}}{50 \text{ mV} \times 1 \text{ mA}} = \frac{14 \text{ mW}}{50 \text{ }\mu\text{W}} = 280$$

In this example, the common–base circuit has increased the input signal power from 50 μW to 14 mW—a power gain of 280.

The power gain of the circuit in Figure 7-15 can also be calculated by multiplying the previously calculated *C–B* circuit voltage gain (A_V) by the previously calculated *C–B* circuit current gain (α_{AC}).

$$A_P = A_V \times \alpha_{AC}$$

For the example circuit in Figure 7-15, this will be

$$A_P = A_V \times \alpha_{AC} = 280 \times 1 = 280$$

indicating that the common–base circuit's output power in Figure 7-15 is 280 times larger than the input power.

Typical common–base circuits will have power gains from 200 to 1,000.

Input Resistance   The input resistance (R_{in}) of a common–base transistor is the amount of opposition offered to an input signal by the input base–emitter junction (emitter diode). Because the base–emitter junction is normally forward biased and the input emitter current (I_E) is relatively large, the input signal sees a very low input resistance. On average, if no additional components are connected in series with the transistor's base–emitter junction, the input resistance of a *C–B* transistor circuit is typically a low value between 15 Ω and 150 Ω. This typical value is an average because the transistor's input resistance is a dynamic or changing quantity that will vary slightly as the input signal changes the conduction of the *C–B* transistor's emitter diode, and this changes I_E ($R\updownarrow = V/I\updownarrow$).

Output Resistance   The output resistance (R_{out}) of a common–base transistor is the amount of opposition offered to an output signal by the output base–collector junction (collector diode). This junction is normally reverse biased when the transistor is operating in the active region, and therefore the *C–B* transistor's output resistance is relatively high. On average, if no load resistor is connected in series with the transistor's collector diode, the output resistance of a *C–B* transistor circuit is typically a very high value between 250 kΩ and 1 MΩ.

Common–Collector Circuits

With the **common–collector circuit,** the transistor's collector lead is common to both the input and output signal. In this circuit configuration therefore, the base serves as the input lead and the emitter serves as the output lead. Figure 7-15 contains a basic common–collector (*C–C*) circuit, its associated input/output voltage and current waveforms, and table of typical characteristics. Using this illustration, we will examine the operation and characteristics of the *C–C* circuit configuration.

DC Current Gain   Referring to the *C–C* circuit in Figure 7-16(a) and its associated waveforms in Figure 7-16(b), let us now examine this circuit's basic operation.

Before applying the ac sine-wave input signal (V_{in}), let us assume that $V_{in} = 0$ V and examine the transistor's dc operating characteristics, or the "no-input-signal" condition. The voltage divider R_1 and R_2 will divide the V_{CC} supply voltage, producing a positive dc base bias voltage across R_2. This base bias voltage will be applied to the base of Q_1. Because it is generally greater than 0.7 V, it will forward bias the transistor's base–emitter junction, turning Q_1 ON. Because the common–collector (*C–C*) circuit's input current I_B is much smaller than the output current I_E, the circuit provides a high current gain. In fact, the common–collector circuit provides a slightly higher gain than the *C–E* circuit because the common–collector's output current (I_E) is slightly higher than the *C–E*'s output current (I_C).

Like any circuit configuration, the dc current gain is equal to the ratio of output current to input current. For the *C–C* circuit, this is equal to the ratio of I_E to I_B:

$$\text{DC Current Gain} = \frac{I_E}{I_B}$$

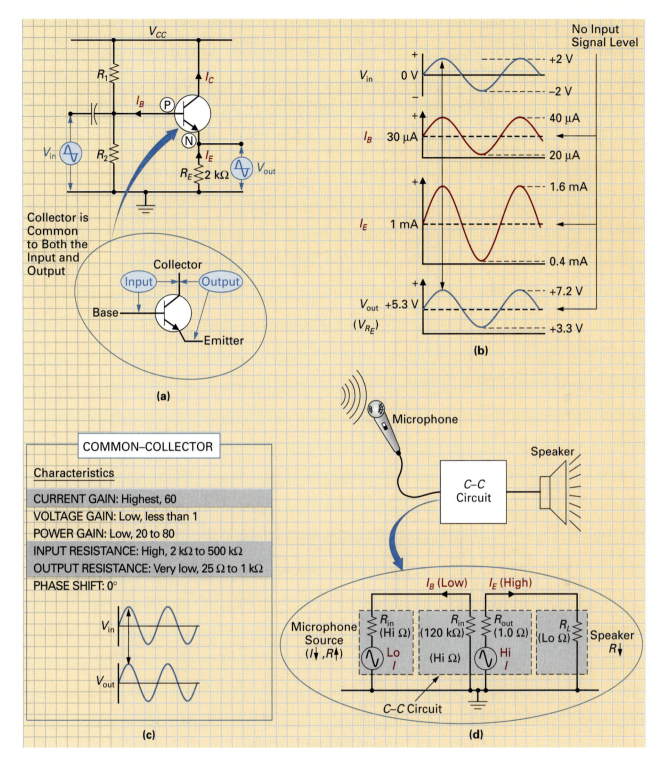

FIGURE 7-16   Common–Collector *(C–C)* Circuit Configuration Characteristics.

Transistor manufacturers will generally not provide specifications for all three circuit configurations. In most cases, because the common–emitter (C–E) circuit configuration is most frequently used, manufacturers will give the transistor's characteristics for only the C–E circuit configuration. In these instances, we will have to convert this C–E circuit data to equivalent specifications for other configurations. For example, in most data sheets the transistor's dc current gain will be listed as β_{DC}. As we know, dc beta is the measure of a C–E circuit's current gain because it compares input current I_B to output current I_C. How, then, can we convert this value so that it indicates the dc current gain of a common–collector circuit? The answer is as follows:

$$\text{Common–Collector DC Current Gain} = \frac{\text{Output Current}}{\text{Input Current}} = \frac{I_E}{I_B}$$

Since $I_E = I_B + I_C$,

$$\text{DC Current Gain} = \frac{I_E}{I_B} = \frac{(I_B + I_C)}{I_B}$$

Since $I_B \div I_B = 1$,

$$\text{DC Current Gain} = 1 + \frac{I_C}{I_B}$$

Since $\dfrac{I_C}{I_B} = \beta_{DC}$,

$$\text{DC Current Gain} = 1 + \frac{I_C}{I_B} = 1 + \beta_{DC}$$

$$\boxed{\text{DC Current Gain} = 1 + \beta_{DC}}$$

■ **EXAMPLE:**

Calculate the dc current gain of the C–C circuit in Figure 7-16 if the transistor's $\beta_{DC} = 32.33$.

■ *Solution:*

$$\text{DC Current Gain} = 1 + \frac{I_C}{I_B} = 1 + \frac{(I_E - I_B)}{I_B} = 1 + \frac{1\ mA - 30\ \mu A}{30\ \mu A}$$

$$= 1 + \frac{970\ \mu A}{30\ \mu A} = 1 + 32.33 = 33.33$$

or

$$\text{DC Current Gain} = 1 + \beta_{DC} = 1 + 32.33 = 33.33$$

As you can see in this example, the dc current gain of a common–collector circuit ($\beta_{DC} + 1$) is slightly higher than the dc current gain of a C–E circuit (β_{DC}). In most instances, the extra 1 makes so little difference when the transistor's dc current gain is a large value of about 30, as in this example, that we assume that the current gain of a common–collector circuit is equal to the current gain of a C–E circuit.

$$\boxed{C\text{–}C \text{ DC Current Gain} \cong C\text{–}E \text{ DC Current Gain } (\beta_{DC})}$$

AC Current Gain When amplifying an ac waveform, a transistor has applied to it both the dc voltage to make it operational and the ac signal voltage that varies the base–emitter bias and the input base current. Referring to the waveforms in Figure 7-16(b), let us now examine what will happen when we apply a sine-wave input signal. As mentioned previously, the positive voltage developed across R_2 will make the NPN transistor's base positive with respect to the emitter and therefore forward bias the P-N base–emitter junction.

As the input voltage swings positive ($V_{in}\uparrow$), it will add to the forward bias applied across the transistor's P-N base–emitter junction. This means that the transistor's emitter diode will turn more ON and cause an increase in the $I_B\uparrow$ and therefore a proportional but much larger increase in the output current $I_E\uparrow$.

Similarly, as the input voltage swings negative ($V_{in}\downarrow$), it will subtract from the forward bias applied across the transistor's P-N base–emitter junction. This means that the transistor's emitter diode will turn less ON and cause a decrease in the $I_B\downarrow$ and therefore a proportional but larger decrease in the output current $I_E\downarrow$.

The ac current gain of a common–collector transistor is calculated using the same formula as dc current gain. However, with an ac current, we will compare the output current change (ΔI_E) to the input current change (ΔI_B).

$$\text{AC Current Gain} = \frac{\Delta I_E}{\Delta I_B}$$

■ **EXAMPLE:**

Calculate the ac current gain of the circuit in Figure 7-16(a), using the values in Figure 7-16(b).

■ *Solution:*

$$\text{AC Current Gain} = \frac{\Delta I_E}{\Delta I_B} = \frac{1.6\text{ mA} - 0.4\text{ mA}}{40\text{ μA} - 20\text{ μA}} = 60$$

Like the common–collector's dc current gain, because there is so little difference between I_E and I_C, we can assume that the ac current gain is equivalent to β_{AC}.

$$C\text{–}C \text{ AC Current Gain} \cong C\text{–}E \text{ AC Current Gain } (\beta_{AC})$$

Common–collector transistor circuit configurations can have current gains as high as 60, indicating that I_E is 60 times larger than I_B.

Voltage Gain Although the common–collector circuit has a very high current gain rating, it cannot increase voltage between input and output. Returning to the C–C circuit in Figure 7-16(a) and its waveforms in Figure 7-16(b), let us see why this circuit has a very low voltage gain.

As the input voltage swings positive ($V_{in}\uparrow$), it will add to the forward bias applied across the transistor's P-N base–emitter junction ($V_{BE}\uparrow$). As the transistor's emitter diode turns more ON, it will cause an increase in $I_B\uparrow$, a proportional but larger increase in $I_E\uparrow$, and therefore an increase in the voltage developed across R_E (V_{R_E}, V_{out}, or $V_E\uparrow$). This increase in the voltage developed across R_E has a **degenerative effect** because an increase in the emitter voltage ($V_E\uparrow$) will counter the initial increase in base voltage ($V_B\uparrow$), and therefore the voltage difference between the transistor's base and emitter will remain almost constant (V_{BE} is almost constant). In other words, if the base goes positive and then the emitter goes positive, there is almost no increase in the potential difference between the base and the emitter and so the change in forward bias is almost zero. There is, in fact, a very small

Degenerative Effect

An effect that causes a reduction in amplification due to negative feedback.

change in forward bias between base and emitter, and this will cause a small change in I_B and I_E, and therefore a small output voltage will be developed across R_E. Comparing the input and output voltage signals in Figure 7-15(b), you can see that both are about 4 V pk-pk and both are in phase with one another. The common–collector circuit is often referred to as an **emitter-follower** or **voltage-follower** because the emitter output voltage seems to track or follow the phase and amplitude of the input voltage.

As with all circuit configurations, the amount of voltage gain (A_V) can be calculated by comparing the output voltage change (ΔV_{out}) to the input voltage change (ΔV_{in}).

$$A_V = \frac{\Delta V_{out}}{\Delta V_{in}}$$

EXAMPLE:

Calculate the voltage gain of the circuit and its associated waveforms in Figure 7-16(a) and (b).

Solution:

$$A_V = \frac{\Delta V_{out}}{\Delta V_{in}} = \frac{+7.2 \text{ V to} +3.3 \text{ V}}{+2 \text{ V to} -2 \text{ V}} = \frac{3.9 \text{ V}}{4 \text{ V}} = 0.975$$

This value indicates that the output ac signal voltage is 0.975 or 97.5% of the ac input signal voltage ($0.975 \times 4 \text{ V} = 3.9 \text{ V}$).

Most common–collector transistor circuits have a voltage gain that is less than 1. However, in most circuit examples it is assumed that output voltage change equals input voltage change.

Power Gain Although the common–collector circuit achieves no voltage gain, it does have a very high current gain and therefore can provide a small amount of power gain ($P\uparrow = V \times I\uparrow$). As before, the power gain of a circuit can be calculated by dividing the output signal power (P_{out}) by the input signal power (P_{in}).

$$A_P = \frac{P_{out}}{P_{in}}$$

To calculate the amount of input power (P_{in}) applied to the C–C circuit, we will have to multiply the change in input signal voltage (ΔV_{in}) by the accompanying change in input signal current (ΔI_{in} or ΔI_B).

$$P_{in} = \Delta V_{in} \times \Delta I_{in}$$
$$P_{in} = 4 \text{ V} \times 20 \text{ }\mu\text{A} = 80 \text{ }\mu\text{W}$$

To calculate the amount of output power (P_{out}) delivered by the C–C circuit, we will have to multiply the change in output signal voltage (ΔV_{out}) produced by the change in output signal current (ΔI_{out} or ΔI_E).

$$P_{out} = \Delta V_{out} \times \Delta I_{out}$$
$$P_{out} = 3.9 \text{ V} \times 1.2 \text{ mA} = 4.68 \text{ mW}$$

The power gain of a common–collector circuit is therefore calculated with the formula

$$A_P = \frac{P_{out}}{P_{in}} = \frac{\Delta V_{out} \times \Delta I_{out}}{\Delta V_{in} \times \Delta I_{in}}$$

EXAMPLE:

Calculate the power gain of the example circuit in Figure 7-16.

Solution:

$$A_P = \frac{P_{out}}{P_{in}} = \frac{\Delta V_{out} \times \Delta I_{out}}{\Delta V_{in} \times \Delta I_{in}}$$

$$= \frac{3.9 \text{ V} \times 1.2 \text{ mA}}{4 \text{ V} \times 20 \text{ } \mu\text{A}} = \frac{4.68 \text{ mW}}{80 \text{ } \mu\text{W}} = 58.5$$

In this example, the common–collector circuit has increased the input signal power from 80 μW to 4.68 mW—a power gain of 58.5.

The power gain of the circuit in Figure 7-16 can also be calculated by multiplying the previously calculated C–C circuit voltage gain (A_V) by the previously calculated C–C circuit current gain.

$$A_P = A_V \times \text{AC Current Gain}$$

For the example circuit in Figure 7-16, this will be

$$A_P = A_V \times \text{AC Current Gain} = 0.975 \times 60 = 58.5$$

indicating that the common–collector circuit's output power in Figure 7-16 is 58.5 times larger than the input power.

Typical common–collector circuits will have power gains from 20 to 80.

Input Resistance An input signal voltage will see a very large input resistance when it is applied to a common–collector circuit. This is because the input signal sees the very large emitter-connected resistor ($R_E \uparrow\uparrow$) and, to a smaller extent, the resistance of the forward-biased base–emitter junction ($R_{in} \uparrow = V_{in}/I_B \downarrow$: R_{in} is large because I_{in} or I_B is small). Using these two elements, we can derive a formula for calculating the input resistance of a C–C transistor circuit.

$$R_{in} = R_E \times \text{AC Current Gain}$$

Since

$$C\text{–}C \text{ AC Current Gain} \cong C\text{–}E \text{ AC Current Gain} (\beta_{AC})$$

the input resistance can also be calculated with the formula

$$R_{in} = R_E \times \beta_{AC}$$

EXAMPLE:

Calculate the input resistance of the circuit in Figure 7-16, assuming $\beta_{AC} = 60$.

Solution:

$$R_{in} = R_E \times \beta_{AC} = 2 \text{ k}\Omega \times 60 = 120 \text{ k}\Omega$$

This means that an input voltage signal will see this C–C circuit as a resistance of 120 kΩ.

The input resistance of a C–C transistor circuit is typically a very large value between 2 kΩ and 500 kΩ.

Output Resistance   The output signal from a common–collector circuit sees a very low output resistance, as proved by this circuit's very high output current gain. Like this circuit's input resistance, the output resistance is largely dependent on the value of the emitter resistor R_E.

The output resistance of a typical C–C transistor circuit is a very low value between 25 Ω and 1 kΩ.

Impedance or Resistance Matching   Do not be misled into thinking that the very high input resistance and low output resistance of the common–collector transistor circuit are disadvantages. On the contrary, the very high input resistance and low output resistance of this configuration are made use of in many circuit applications, along with the C–C circuit's other advantage of high current gain.

To explain why a high input resistance and a low output resistance are good circuit characteristics, refer to the application circuit in Figure 7-16(d). In this example, a microphone is connected to the input of a C–C amplifier and the output of this circuit is applied to a speaker. As we know, the sound wave input to the microphone will physically move a magnet within the microphone, which will in turn interact to induce a signal voltage into a stationary coil. This voice signal voltage from the microphone, which is our source, is then applied across the input resistance of our example C–C circuit, which is our load.

In the inset in Figure 7-16(d), you can see that the microphone has been represented as a low-current ac source with a high internal resistance and the input resistance of the C–C circuit is shown as a high-value (in the previous example, 120 kΩ) resistor. Remembering our previous discussion on sources and loads, we know that a small load resistance will cause a large current to be drawn from the source, and this large current will drain or pull down the source voltage. Many signal sources, such as microphones, can only generate a small signal source voltage because they have a high internal resistance. If this small signal source voltage is applied across an amplifier with a small input resistance, a large current will be drawn from the source. This heavy load will pull the signal voltage down to such a small value that it will not be large enough to control the amplifier. A large amplifier input resistance ($R_{in}\uparrow$), on the other hand, will not load the source. Therefore, the input voltage applied to the amplifier will be large enough to control the amplifier circuit, to vary its transistor currents, and to achieve the gain between amplifier input and output. In summary, the high input resistance of the C–C circuit can be connected to a high resistance source because it will not draw an excessive current and pull down the source voltage.

Referring again to the inset in Figure 7-16(d), you can see that at the output end, the C–C's output circuit has been represented as a high current source with a low value internal output resistor and the speaker has been represented as a low resistance load. The low output resistance of the C–C circuit means that this circuit can deliver the high current output that is needed to drive the low resistance load.

As you will see later in application circuits, most C–C circuits are used as a resistance or **impedance matching circuit** that can match, or isolate, a high resistance (low current) source, such as a microphone, to a low resistance (high current) load, such as the speaker. By acting as a **buffer current amplifier,** the C–C circuit can ensure that power is efficiently transferred from source to load.

Impedance Matching Circuit

A circuit that can match, or isolate, a high resistance (low current) source.

Buffer Current Amplifier

The C–C circuit that can ensure that power is efficiently transferred from source to load.

7-2-3   *Bipolar Junction Transistor Data Sheet*

Like the diode's data sheets, manufacturer's bipolar transistor data sheets list the typical dc and ac operating characteristics of the device. Figure 7-17 shows the data sheet of a typical *general-purpose switching or amplifying bipolar transistor,* with annotations describing the key characteristics.

As before, notes are included in these data sheets to call out important characteristics and to explain some of the terms that have not been previously used.

DEVICE: 2N3903 and 2N3904—NPN Silicon Switching and Amplifier Transistors

Maximum continuous collector current (I_C) = 200 mA.

MAXIMUM RATINGS

Rating	Symbol	Value	Unit
Collector-Emitter Voltage	V_{CEO}	40	Vdc
Collector-Base Voltge	V_{CBO}	60	Vdc
Emitter-Base Voltage	V_{EBO}	6.0	Vdc
Collector Current — Continuous	I_C	200	mAdc
Total Device Dissipation @ T_A = 25°C Derate above 25°C	P_D	625 5.0	mW mW/°C
*Total Device Dissipation @ T_C = 25°C Derate above 25°C	P_D	1.5 12	Watts mW/°C
Operating and Storage Junction Temperature Range	T_J, T_{stg}	-55 to $+150$	°C

*THERMAL CHARACTERISTICS

Characteristic	Symbol	Max	Unit
Thermal Resistance, Junction to Ambient	$R_{\theta JA}$	200	°C/W
Thermal Resistance, Junction to Case	$R_{\theta JC}$	83.3	°C/W

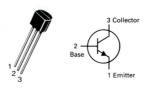

2N3903
2N3904★

CASE 29-04, STYLE 1
TO-92 (TO-226AA)

3 Collector

2 Base

1 Emitter

GENERAL PURPOSE TRANSISTORS

NPN SILICON

★This is a Motorola
designated preferred device.

ELECTRICAL CHARACTERISTICS (T_A = 25°C unless otherwise noted.)

OFF Characteristic (operated in cutoff)	Symbol	Min	Max	Unit
Collector-Emitter Breakdown Voltage(1) (I_C = 1.0 mAdc, I_B = 0)	$V_{(BR)CEO}$	40	—	Vdc
Collector-Base Breakdown Voltage (I_C = 10 μAdc, I_E = 0)	$V_{(BR)CBO}$	60	—	Vdc
Emitter-Base Breakdown Voltage (I_E = 10 μAdc, I_C = 0)	$V_{(BR)EBO}$	6.0	—	Vdc
Base Cutoff Current (V_{CE} = 30 Vdc, V_{EB} = 3.0 Vdc)	I_{BL}	—	50	nAdc
Collector Cutoff Current (V_{CE} = 30 Vdc, V_{EB} = 3.0 Vdc)	I_{CEX}	—	50	nAdc

NOTE: The "O" following CBO, CEO, EBO indicates the third terminal is "open." For example, $V_{(BR)CEO}$ means the breakdown voltage between collector and emitter with the base open.

ON Characteristic (operated in active and saturation region)		Symbol	Min	Max	Unit
DC Current Gain(1) (I_C = 0.1 mAdc, V_{CE} = 1.0 Vdc)	2N3903 2N3904	h_{FE}	20 40	— —	—
(I_C = 1.0 mAdc, V_{CE} = 1.0 Vdc)	2N3903 2N3904		35 70	— —	
(I_C = 10 mAdc, V_{CE} = 1.0 Vdc)	2N3903 2N3904		50 100	150 300	
(I_C = 50 mAdc, V_{CE} = 1.0 Vdc)	2N3903 2N3904		30 60	— —	
(I_C = 100 mAdc, V_{CE} = 1.0 Vdc)	2N3903 2N3904		15 30	— —	
Collector-Emitter Saturation Voltage(1) (I_C = 10 mAdc, I_B = 1.0 mAdc) (I_C = 50 mAdc, I_B = 5.0 mAdc)		$V_{CE(sat)}$	— —	0.2 0.3	Vdc
Base-Emitter Saturation Voltage(1) (I_C = 10 mAdc, I_B = 1.0 mAdc) (I_C = 50 mAdc, I_B = 5.0 mAdc)		$V_{BE(sat)}$	0.65 —	0.85 0.95	Vdc

h_{FE} = β_{DC}, dc current gain is measured at different values of I_C.

Maximum base–emitter voltage (V_{BE}) when transistor is saturated

Maximum value of voltage between collector and emitter (V_{CE}) when transistor is in saturation.

FIGURE 7-17 A General-Purpose NPN Silicon Transistor. (Copyright of Motorola. Used by permission.)

7-2-4   *Testing Bipolar Junction Transistors*

Although transistors are exceptionally more reliable than their counterpart, the vacuum tube, they still will malfunction. These failures are normally the result of excessive temperature, current, or mechanical abuse and generally result in one of three problems:

1. An open between two or three of the transistor's leads
2. A short between two or three of the transistor's leads
3. A change in the transistor's characteristics

Transistor Tester

The **transistor tester** shown in Figure 7-18 is a special test instrument that can be used to test both NPN and PNP bipolar transistors. This special meter can be used to determine whether an open or short exists between any of the transistor's three terminals, the transistor's dc current gain (β_{DC}), and whether an undesirable value of leakage current is present through one of the transistor's junctions.

Ohmmeter Transistor Test

If the transistor tester is not available, the ohmmeter can be used to detect open and shorted junctions, which are the most common transistor failures. Figure 7-19 shows the step-by-step procedure for testing an NPN transistor. Following through this test procedure, we begin by reverse biasing the collector diode and then the emitter diode, and then forward biasing the collector diode and then the emitter diode. The table in Figure 7-19 shows the order and action to be performed for each step and the reading that should result if the NPN transistor junction is operating correctly.

7-2-5   *Bipolar Transistor Biasing Circuits*

As we discovered in the previous discussion on transistor circuit configurations, the ac operation of a transistor is determined by the "dc bias level," or "no input signal level." This

FIGURE 7-18   **Transistor Tester. (Courtesy of Sencore, Inc.—Test Equipment for the Professional Servicer. 1-800-Sencore.)**

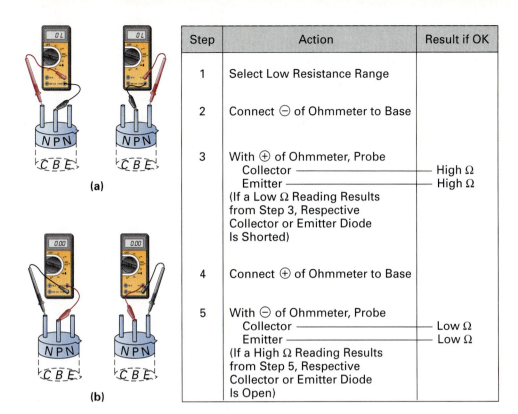

Step	Action	Result if OK
1	Select Low Resistance Range	
2	Connect ⊖ of Ohmmeter to Base	
3	With ⊕ of Ohmmeter, Probe Collector ———————— Emitter ———————— (If a Low Ω Reading Results from Step 3, Respective Collector or Emitter Diode Is Shorted)	High Ω High Ω
4	Connect ⊕ of Ohmmeter to Base	
5	With ⊖ of Ohmmeter, Probe Collector ———————— Emitter ———————— (If a High Ω Reading Results from Step 5, Respective Collector or Emitter Diode Is Open)	Low Ω Low Ω

FIGURE 7-19 **NPN Ohmmeter Test Procedure. (a) Reverse-Biasing Collector, Then Emitter, Diode. (b) Forward-Biasing Collector, Then Emitter, Diode.**

steady-state or dc operating level is set by the value of the circuit's dc supply voltage (V_{CC}) and the value of the circuit's biasing resistors. This single supply voltage and the one or more biasing resistors set up the initial dc values of transistor current (I_B, I_E and I_C) and transistor voltage (V_{BE}, V_{CE} and V_{BC}).

In this section, we will examine some of the more commonly used methods for setting the "initial dc operating point" of a bipolar transistor circuit. As you encounter different circuit applications, you will see that many of these circuits include combinations of these basic biasing techniques and additional special-purpose components for specific functions. Because the common–emitter (C–E) circuit configuration is used more extensively than the C–B and the C–C, we will use this configuration in all of the following basic biasing circuit examples.

Base Biasing

Figure 7-20(a) shows how a common–emitter transistor circuit could be base biased. With **base biasing,** the emitter diode of the transistor is forward biased by applying a positive base bias voltage ($+V_{BB}$) via a current-limiting resistor (R_B) to the base of Q_1. In Figure 7-20(b), the transistor circuit from Figure 7-20(a) has been redrawn so as to simplify the analysis of the circuit. The transistor is now represented as a diode between base and emitter (emitter diode), and the transistor's emitter-to-collector has been represented as a variable resistor. Assuming Q_1 is a silicon bipolar transistor, the forward-biased emitter diode will have a standard base–emitter voltage drop of 0.7 V (emitter diode drop = 0.7 V).

Base biasing
A transistor biasing method in which the dc supply voltage is applied to the base of the transistor via a base bias resistor.

$$V_{BE} = 0.7 \text{ V}$$

The base bias resistor (R_B) and the transistor's emitter diode form a series circuit across V_{BB}, as seen in Figure 17-20(b). Therefo re, the voltage drop across R_B (V_{RB}) will be equal to the difference between V_{BB} and V_{BE}.

FIGURE 7-20　A Base-Biased Common Emitter Circuit. (a) Basic Circuit. (b) Simplified Equivalent Circuit.

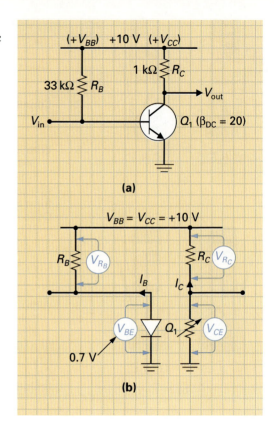

(a)

(b)

$$V_{R_B} = V_{BB} - V_{BE}$$
$$= V_{BB} - 0.7 \text{ V}$$

$$V_{R_B} = V_{BB} - V_{BE} = 10 \text{ V} - 0.7 \text{ V} = 9.3 \text{ V}$$

Now that the resistance and voltage drop across R_B are known, we can calculate the current through R_B (I_{R_B}). Because a series circuit is involved, the current through R_B (I_{R_B}) will also be equal to the transistor base current I_B.

$$I_B = \frac{V_{R_B}}{R_B}$$

$$I_B = \frac{V_{R_B}}{R_B} = \frac{9.3 \text{ V}}{33 \text{ k}\Omega} = 282 \text{ }\mu\text{A}$$

Because the transistor's dc current gain (β_{DC}) is given in Figure 7-20(a), we can calculate I_C because β_{DC} tells us how much greater the output current I_C is compared to the input current I_B.

$$I_C = I_B \times \beta_{DC}$$

$$I_C = I_B \times \beta_{DC} = 282 \text{ }\mu\text{A} \times 20 = 5.6 \text{ mA}$$

Because the current through R_C is I_C, we can now calculate the voltage drop across R_C (V_{R_C}).

$$V_{R_C} = I_C \times R_C$$

$$V_{R_C} = I_C \times R_C = 5.6 \text{ mA} \times 1 \text{ k}\Omega = 5.6 \text{ V}$$

CHAPTER 7 / BIPOLAR JUNCTION TRANSISTORS

Now that V_{R_C} is known, we can calculate the voltage drop across the transistor's collector-to-emitter because V_{CE} and V_{R_C} are in series and will be equal to the applied voltage V_{CC}.

$$V_{CE} = V_{CC} - V_{R_C}$$

$$V_{CE} = V_{CC} - V_{R_C} = 10\,\text{V} - 5.6\,\text{V} = 4.4\,\text{V}$$

Combining the previous two equations, we can obtain the following V_{CE} formula:

$$V_{CE} = V_{CC} - V_{R_C}$$

Since

$$V_{R_C} = I_C \times R_C$$

$$V_{CE} = V_{CC} - (I_C \times R_C)$$

$$V_{CE} = V_{CC} - (I_C \times R_C) = 10\,\text{V} - (5.6\,\text{mA} \times 1\,\text{k}\Omega) = 4.4\,\text{V}$$

Using the preceding formulas, which are all basically Ohm's law, you can calculate the current and voltage values in a base biased circuit.

DC Load Line In a transistor circuit, such as the example in Figure 7-20, V_{CC} and V_{R_C} are constants. On the other hand, the input current I_B and the output current I_C are variables. Using the example circuit in Figure 7-20, let us calculate what collector-to-emitter voltage drops (V_{CE}) will result for different values of I_C.

a. When Q_1 is OFF, $I_C = 0$ mA, and therefore V_{CE} equals

$$V_{CE} = V_{CC} - (I_C \times R_C) = 10\,\text{V} - (0\,\text{mA} \times 1\,\text{k}\Omega) = 10\,\text{V} - 0\,\text{V} = 10\,\text{V}$$

This would make sense because Q_1 would be equivalent to an open switch between collector and emitter when it is OFF, and therefore all of the 10-V V_{CC} supply voltage would appear across the open. Figure 7-21 shows how this point would be plotted on a graph (point A).

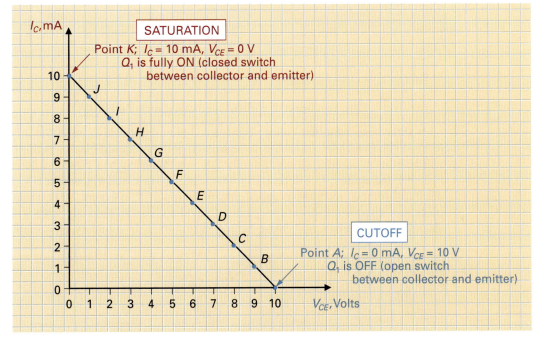

FIGURE 7-21 **A Transistor DC Load Line with Cutoff and Saturation Points.**

b. When $I_C = 1$ mA,

$$V_{CE} = 10 \text{ V} - (1 \text{ mA} \times 1 \text{ k}\Omega) = 10 \text{ V} - 1 \text{ V} = 9 \text{ V (point } B)$$

c. When $I_C = 2$ mA,

$$V_{CE} = 10 \text{ V} - (2 \text{ mA} \times 1 \text{ k}\Omega) = 10 \text{ V} - 2 \text{ V} = 8 \text{ V (point } C)$$

d. When $I_C = 3$ mA,

$$V_{CE} = 10 \text{ V} - (3 \text{ mA} \times 1 \text{ k}\Omega) = 10 \text{ V} - 3 \text{ V} = 7 \text{ V (point } D)$$

e. When $I_C = 4$ mA, $V_{CE} = 6$ V (point E)
f. When $I_C = 5$ mA, $V_{CE} = 5$ V (point F)
g. When $I_C = 6$ mA, $V_{CE} = 4$ V (point G)
h. When $I_C = 7$ mA, $V_{CE} = 3$ V (point H)
i. When $I_C = 8$ mA, $V_{CE} = 2$ V (point I)
j. When $I_C = 9$ mA, $V_{CE} = 1$ V (point J)
k. When $I_C = 10$ mA, the only resistance is that of R_C because Q_1 is fully ON and is equivalent to a closed switch between collector and emitter. It is not a surprise that the voltage drop across Q_1's collector-to-emitter is almost 0 V.

$$V_{CE} = 10 \text{ V} - (10 \text{ mA} \times 1 \text{ k}\Omega) = 10 \text{ V} - 10 \text{ V} = 0 \text{ V (point } K)$$

The line drawn in the graph in Figure 7-21 is called the **dc load line** because it is a line representing all the dc operating points of the transistor for a given load resistance. In this example, the transistor's load was the 1 kΩ collector-connected resistor R_C in Figure 7-20.

Cutoff and Saturation Points Let us now examine the two extreme points in a transistor's dc load line, which in the example in Figure 7-21 were points A and K. If a transistor's base input bias voltage is reduced to zero, its input current I_B will be zero, Q_1 will turn OFF and be equivalent to an open switch between the collector and emitter, the output current I_C will be 0 mA, and a V_{CE} will be 10 V. This point in the transistor dc load line is called *cutoff* (point A in Figure 7-21) because the output collector current is reduced to zero, or cut off. In summary, at cutoff:

$$I_{C(Cutoff)} = 0 \text{ mA}$$

$$V_{CE(Cutoff)} = V_{CC}$$

In the example circuit in Figure 7-20 and its dc load line in Figure 7-21, with Q_1 cut OFF:

$$I_{C(Cutoff)} = 0 \text{ mA}, \qquad V_{CE(Cutoff)} = V_{CC} = 10 \text{ V}$$

If the base input bias voltage is increased to a large positive value, the transistor's collector diode (which is normally reverse biased) will be forward biased. In this condition, I_B will be at its maximum, Q_1 will be fully ON and equivalent to a closed switch between the collector and emitter, I_C will be at its maximum of 10 mA, and V_{CE} will be 0 V. This point in the transistor's dc load line is called *saturation* (point K in Figure 7-21) because, just as a point is reached where a wet sponge is saturated and cannot hold any more water, the transistor at saturation cannot increase I_C beyond this point. In summary, at saturation:

$$I_{C(Sat.)} = \frac{V_{CC}}{R_C}$$

$$V_{CE(Sat.)} = 0.\text{V}$$

In the example circuit in Figure 7-20 and its dc load line in Figure 7-21, with Q_1 saturated:

CHAPTER 7 / BIPOLAR JUNCTION TRANSISTORS

$$I_{C(\text{Sat.})} = \frac{V_{CC}}{R_C} = \frac{10\text{ V}}{1\text{ k}\Omega} = 10\text{ mA}$$

$$V_{CE(\text{Sat.})} = 0\text{ V}$$

Rearranging the formula $\beta_{DC} = I_C/I_B$, we can calculate the value of input base current that causes the output saturation current:

$$\beta_{DC} = \frac{I_C}{I_B}$$

Therefore,

$$I_{B(\text{Sat.})} = \frac{I_{C(\text{Sat.})}}{\beta_{DC}}$$

In the example circuit in Figure 7-20 and its dc load line in Figure 7-21, the input current that will cause saturation will be

$$I_{B(\text{Sat.})} = \frac{I_{C(\text{Sat.})}}{\beta_{DC}} = \frac{10\text{ mA}}{20} = 500\text{ }\mu\text{A}$$

Figure 7-22 summarizes all of our base bias circuit calculations so far by including the dc load line from Figure 7-21 in a set of collector characteristic curves for the transistor circuit example in Figure 7-20. As you can see in the graph in Figure 7-22, at cutoff $I_B = 0$ μA, $I_C = 0$ mA, and $V_{CE} = V_{CC}$, which is 10 V. On the other hand, at saturation $I_B = 500$ μA, $I_C = 10$ mA, and $V_{CE} = 0$ V.

Quiescent Point   Generally, the value of the base bias resistor (R_B) is chosen so that the value of base current (I_B) is near the middle of the dc load line. For example, if a base bias resistance of 37.2 kΩ was used in the example circuit in Figure 7-20 ($R_B = 37.2$ kΩ), it would produce a base current of 250 mA ($I_B = 9.3$ V/37.2 kΩ = 250 μA). Referring to the dc load line in Figure 7-22, you can see that this value of base current is halfway between cutoff at 0 μA, and saturation at 500 μA. This point is called the *quiescent* (at rest) or *Q point* and is defined as *the dc bias point at which the circuit rests when no ac input signal*

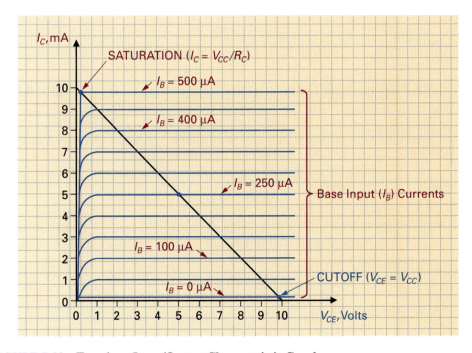

FIGURE 7-22   **Transistor Input/Output Characteristic Graph.**

is applied. An ac input signal voltage will vary I_B above and below this Q point, resulting in a corresponding but larger change in I_C.

■ **EXAMPLE:**

Complete the following for the circuit shown in Figure 7-23.

FIGURE 7-23 Bipolar Transistor Example.

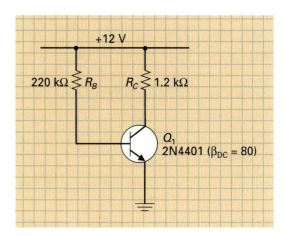

a. Calculate I_B.
b. Calculate I_C.
c. Calculate V_{CE}.
d. Sketch the circuit's dc load line with saturation and cutoff points.
e. Indicate where the Q point is on the circuit's dc load line.

■ *Solution:*

a. Since $V_{BE} = 0.7$ V and $V_{BB} = 12$ V,

$$V_{R_B} = 12\ \text{V} - 0.7\ \text{V} = 11.3\ \text{V}$$

$$I_B = \frac{V_{R_B}}{R_B} = \frac{11.3\ \text{V}}{220\ \text{k}\Omega} = 51.4\ \mu\text{A}$$

b. $$I_C = I_B \times \beta_{DC} = 51.4\ \mu\text{A} \times 80 = 4.1\ \text{mA}$$

c. $$V_{R_C} = I_C \times R_C = 4.1\ \text{mA} \times 1.2\ \text{k}\Omega = 4.92\ \text{V}$$
$$V_{CE} = V_{CC} - V_{R_C} = 12\ \text{V} - 4.92\ \text{V} = 7.08\ \text{V}$$

d. At cutoff, the transistor is OFF and therefore equivalent to an open switch between collector and emitter. All of the V_{CC} supply voltage will therefore be across Q_1.
At cutoff, $V_{CE} = V_{CC} = 12$ V (see cutoff in the dc load line in Figure 7-24).
At saturation, the transistor is fully ON and therefore equivalent to a closed switch between the collector and emitter. The only resistance is that of R_C, and so at saturation,

$$I_{C(Sat.)} = \frac{V_{CC}}{R_C} = \frac{12\ \text{V}}{1.2\ \text{k}\Omega} = 10\ \text{mA}$$

(see saturation in the dc load line in Figure 7-24).

e. The operating point or Q point of this circuit is set by the base bias resistor R_B. This Q point will be at

$$I_C = 4.1\ \text{mA}$$

which produces a

$$V_{CE} = 7.08\ \text{V}$$

This quiescent (Q) point is also shown on Figure 7-24.

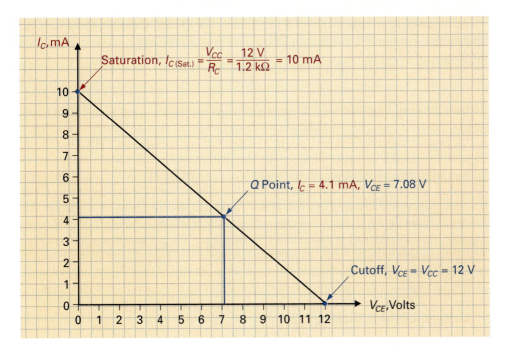

FIGURE 7-24   The DC Load Line for the Circuit in Figure 7-23.

■ **EXAMPLE:**

Calculate the current through the lamp in Figure 7-25.

FIGURE 7-25   Two-State Lamp Circuit.

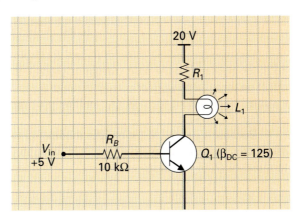

■ **Solution:**

$$V_{BE} = 0.7 \text{ V}, \qquad V_{in} = +5 \text{ V}$$

Therefore,

$$V_{R_B} = V_{in} - 0.7 \text{ V}$$
$$= 5 \text{ V} - 0.7 \text{ V} = 4.3 \text{ V}$$

$$I_B = \frac{V_{R_B}}{R_B} = \frac{4.3 \text{ V}}{10 \text{ k}\Omega} = 430 \text{ }\mu\text{A}$$

$$I_C = I_B \times \beta_{DC}$$
$$= 430 \text{ }\mu\text{A} \times 125 = 53.75 \text{ mA}$$

An input of zero volts ($V_{in} = 0$ V) will turn OFF Q_1 and therefore lamp L_1. On the other hand, an input of +5 V will turn ON Q_1 and permit a collector current, and therefore lamp current, of 53.75 mA.

Base Biasing Applications   Base bias circuits are used in switching circuit applications like the two-state ON/OFF lamp circuit discussed in the previous example. In these circuits, the bipolar transistor is equivalent to a switch and is controlled by a HIGH/LOW input voltage that drives the transistor between the two extremes of cutoff and saturation.

The advantage of this biasing technique is circuit simplicity because only one resistor is needed to set the base bias voltage. The disadvantage of the base-biased circuit is that it cannot compensate for changes in its dc bias current due to changes in temperature. To explain this in more detail, a change in temperature will result in a change in the internal resistance of the transistor (all semiconductor devices have a negative temperature coefficient of resistance—temperature $\uparrow$ causes internal resistance $\downarrow$). This change in the transistor's internal resistance will change the transistor's dc bias currents (I_B and I_C), which will change or shift the transistor's dc operating point or Q point away from the desired midpoint.

Voltage-Divider Biasing

Voltage-Divider Biasing

A biasing method used with amplifiers in which a series arrangement of two fixed-value resistors is connected across the voltage source. The result is that a desired fraction of the total voltage is obtained at the center of the two resistors and is used to bias the amplifier.

Figure 7-26(a) shows how a common–emitter transistor circuit could be **voltage-divider biased.** The name of this biasing method comes from the two- resistor series voltage divider (R_1 and R_2) connected to the transistor's base. In this most widely used biasing method, the emitter diode of Q_1 is forward biased by the voltage developed across R_2 (V_{R_2}), as seen in the simplified equivalent circuit in Figure 7-26(b). To calculate the voltage developed across R_2, and therefore the voltage applied to Q_1's base, we can use the voltage-divider formula.

$$V_{R_2} \text{ or } V_B = \frac{R_2}{R_1 + R_2} \times V_{CC}$$

$$V_{R_2} \text{ or } V_B = \frac{R_2}{R_1 + R_2} \times V_{CC} = \frac{10 \text{ k}\Omega}{20 \text{ k}\Omega + 10 \text{ k}\Omega} \times 20 \text{ V} = 0.333 \times 20 \text{ V} = 6.7 \text{ V}$$

Because the current through R_1 and R_2 (from ground to $+V_{CC}$) is generally more than 10 times greater than the base current of Q_1 (I_B), it is normally assumed that I_B will have no effect on the voltage-divider current through R_1 and R_2. The R_1 and R_2 voltage divider can be assumed to be independent of Q_1, and the previous voltage-divider formula can be used to calculate V_{R_2} or V_B.

Because $V_B = 6.7$ V, the emitter diode of Q_1 will be forward biased. Assuming a 0.7-V drop across the transistor's base–emitter junction ($V_{BE} = 0.7$ V), the voltage at the emitter terminal of Q_1 (V_E) will be

$$V_{R_E} \text{ or } V_E = V_B - 0.7 \text{ V}$$

$$V_{R_E} \text{ or } V_E = V_B - 0.7 \text{ V} = 6.7 \text{ V} - 0.7 \text{ V} = 6 \text{ V}$$

Now that the voltage drop across R_E (V_{R_E}) is known, along with its resistance, we can calculate the current through R_E and the value of current being injected into the transistor's emitter.

$$I_{R_E} = I_E = \frac{V_{R_E}}{R_E}$$

$$I_{R_E} = I_E = \frac{V_{R_E}}{R_E} = \frac{6 \text{ V}}{5 \text{ k}\Omega} = 1.2 \text{ mA}$$

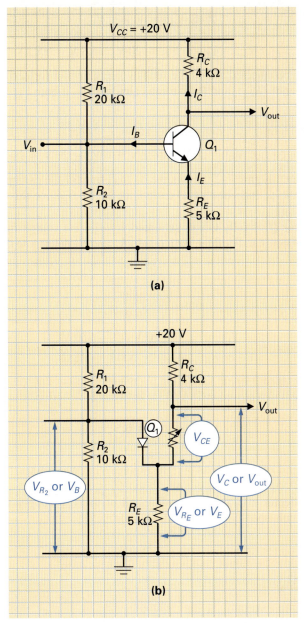

FIGURE 7-26 **A Voltage-Divider-Biased Common–Emitter Circuit. (a) Basic Circuit.**
(b) Simplified Equivalent Circuit.

Because we know that a transistor collector current (I_C) is approximately equal to the emitter current (I_E), we can state that

$$I_E \cong I_C$$

$$I_E \cong I_C = 1.2 \text{ mA}$$

Now that I_C is known, we can calculate the voltage drop across R_C (V_{R_C}) because both its resistance and current are known.

$$V_{R_C} = I_C \times R_C$$

$$V_{R_C} = I_C \times R_C = 1.2 \text{ mA} \times 4 \text{ k}\Omega = 4.8 \text{ V}$$

The dc quiescent voltage at the collector of Q_1 with respect to ground (V_C), which is also V_{out}, will be equal to the dc supply voltage (V_{CC}) minus the voltage drop across R_C.

$$V_C \text{ or } V_{\text{out}} = V_{CC} - V_{R_C}$$

$$V_C \text{ or } V_{\text{out}} = V_{CC} - V_{R_C} = 20\text{ V} - 4.8\text{ V} = 15.2\text{ V}$$

Because V_{CC} is connected across the series voltage divider formed by R_C, Q_1's collector-to-emitter resistance (R_{CE}), and R_E, we can calculate V_{CE} if both V_{R_C} and V_E are known:

$$V_{CE} = V_{CC} - (V_{R_C} + V_E)$$

$$V_{CE} = V_{CC} - (V_{R_C} + V_E) = 20\text{ V} - (4.8\text{ V} + 6\text{ V}) = 20\text{ V} - 10.8\text{ V} = 9.2\text{ V}$$

DC Load Line Figure 7-27 shows the dc load line for the example circuit in Figure 7-26. Referring to the dc load line's two endpoints, let us examine this circuit's saturation and cutoff points.

When transistor Q_1 is fully ON or saturated, it will have approximately 0 Ω of resistance between its collector and emitter. As a result, R_C and R_E determine the value of I_C when Q_1 is saturated.

$$I_{C(\text{Sat.})} = \frac{V_{CC}}{R_C + R_E}$$

$$I_{C(\text{Sat.})} = \frac{V_{CC}}{R_C + R_E} = \frac{20\text{ V}}{4\text{ k}\Omega + 5\text{ k}\Omega} = \frac{20\text{ V}}{9\text{ k}\Omega} = 2.2\text{ mA}$$

As you can see in Figure 7-27, at saturation, I_C is maximum at 2.2 mA and V_{CE} is 0 V because Q_1 is equivalent to a closed switch (0 Ω) between Q_1's collector and emitter.

$$V_{CE(\text{Sat.})} = 0\text{ V}$$

At the other end of the dc load line in Figure 7-27, we can see how the transistor's characteristics are plotted when it is cut off. When Q_1 is cut OFF, it is equivalent to an open switch between collector-to-emitter. Therefore, all of the V_{CC} supply voltage will appear across the series circuit open.

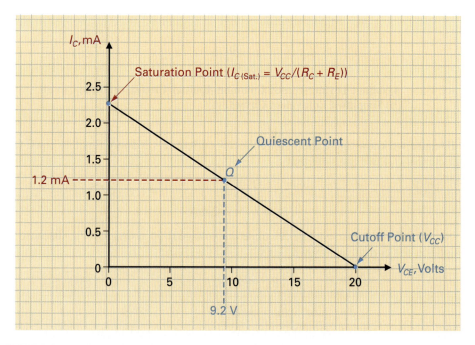

FIGURE 7-27 The DC Load Line for the Circuit in Figure 7-25.

CHAPTER 7 / BIPOLAR JUNCTION TRANSISTORS

$$V_{CE(\text{Cutoff})} = V_{CC}$$

$$V_{CE(\text{Cutoff})} = V_{CC} = 20 \text{ V}$$

As you can see in Figure 7-27, when Q_1 is cut OFF, all of the V_{CC} supply voltage will appear across Q_1's collector-to-emitter terminals, and I_C will be blocked and equal to zero.

$$I_{C(\text{Cutoff})} = 0 \text{ mA}$$

Generally, the value of the voltage-divider resistors R_1 and R_2 is chosen so that the value of base current (I_B) is near the middle of the dc load line. Referring to Figure 7-27, you can see that by plotting our previously calculated values of I_C (which at rest was 1.2 mA) and V_{CE} ($V_{CE} = 9.2$ V), we obtain a Q point that is near the middle of the dc load line.

■ **EXAMPLE:**

Calculate the following for the circuit shown in Figure 7-28.

a. V_B and V_E.

b. Determine whether C_E will have any effect on the dc operating voltages.

c. I_C.

d. V_C and V_{CE}.

e. Sketch the circuit's dc load line and include the saturation, cutoff, and Q points.

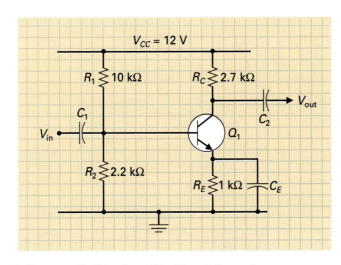

FIGURE 7-28   A Common-Emitter Amplifier Circuit Example.

■ *Solution:*

a. $$V_B = \frac{R_2}{R_1 + R_2} \times V_{CC} = \frac{2.2 \text{ k}\Omega}{10 \text{ k}\Omega + 2.2 \text{ k}\Omega} \times 12 \text{ V} = 2.16 \text{ V}$$

$$V_E = V_B - 0.7 \text{ V} = 2.16 \text{ V} - 0.7 \text{ V} = 1.46 \text{ V}$$

b. Since all capacitors can be thought of as a dc block, C_E will have no effect on the circuit's dc operating voltages.

c. $$I_E = \frac{V_E}{R_E} = \frac{1.46 \text{ V}}{1 \text{ k}\Omega} = 1.46 \text{ mA}$$

$$I_C \cong I_E = 1.46 \text{ mA}$$

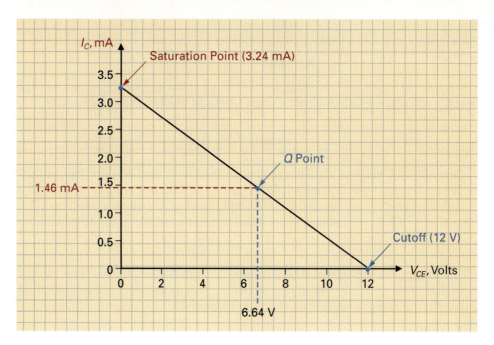

FIGURE 7-29   **The DC Load Line for the Circuit in Figure 7-28.**

d. $V_{R_C} = I_C \times R_C = 1.46 \text{ mA} \times 2.7 \text{ k}\Omega = 3.9 \text{ V}$

V_{out} or $V_C = V_{CC} - V_{R_C} = 12 \text{ V} - 3.9 \text{ V} = 8.1 \text{ V}$

$V_{CE} = V_{CC} - (V_{R_C} + V_E) = 12 \text{ V} - (3.9 \text{ V} + 1.46 \text{ V}) = 12 \text{ V} - 5.36 \text{ V} = 6.64 \text{ V}$

e. $I_{C(\text{Sat.})} = \dfrac{V_{CC}}{R_C + R_E} = \dfrac{12 \text{ V}}{2.7 \text{ k}\Omega + 1 \text{ k}\Omega} = \dfrac{12 \text{ V}}{3.7 \text{ k}\Omega} = 3.24 \text{ mA}$

$V_{CE(\text{Cutoff})} = V_{CC} = 12 \text{ V}$

Q point, $I_C = 1.46 \text{ mA}$ and $V_{CE} = 6.64 \text{ V}$

(This information is plotted on the graph in Figure 7-29.)

Voltage-Divider Bias Applications   Voltage-divider biased circuits are used in analog or linear circuit applications such as the amplifier circuit discussed in the previous example. In these circuits, the bipolar transistor is equivalent to a variable resistor and is controlled by an alternating input signal voltage.

Unlike the base-biased circuit, the voltage-divider biased circuit has very good temperature stability due to the emitter resistor R_E. To explain this in more detail, let us assume that there is an increase in the temperature surrounding a voltage-divider circuit, such as the example circuit in Figure 7-28. As temperature increases, it causes an increase in the transistor's internal currents ($I_B\uparrow, I_E\uparrow, I_C\uparrow$) because all semiconductor devices have a negative temperature coefficient of resistance (temperature $\uparrow, R\downarrow, I\uparrow$). An increase in $I_E\uparrow$ will cause an increase in the voltage drop across $R_E\uparrow$, which will decrease the voltage difference between the transistor's base and emitter ($V_{BE}\downarrow$). Decreasing the forward bias applied to the transistor's emitter diode will decrease all of the transistor's internal currents ($I_B\downarrow, I_E\downarrow, I_C\downarrow$) and return them to their original values. Therefore, a change in output current (I_C) due to temperature will effectively be fed back to the input and change the input current (I_B), which is why a circuit containing an emitter resistor is said to have **emitter feedback** for temperature stability.

Emitter Feedback

The coupling from the emitter output to the base input in a transistor amplifier.

CHAPTER 7 / BIPOLAR JUNCTION TRANSISTORS

Use the following questions to test your understanding of Section 7-2.

1. The bipolar transistor is a _____ (voltage/current) controlled device.
2. When a bipolar transistor is being operated in the active region, its emitter diode is _____ biased and its collector diode is _____ biased.
3. Which of the following is correct:
   a. $I_E = I_C - I_B$ b. $I_C = I_E - I_B$ c. $I_B = I_C - I_E$
4. When a transistor is in cutoff, it is equivalent to a/an _____ between its collector and emitter.
5. When a transistor is in saturation, it is equivalent to a/an _____ between its collector and emitter.
6. Which of the bipolar transistor circuit configurations has the best
   a. Voltage gain   b. Current gain   c. Power gain
7. Which biasing method makes use of two series-connected resistors across the V_{CC} supply voltage?
8. Which biasing technique has a single resistor connected in series with the base of the transistor?

REVIEW QUESTIONS

Multiple-Choice Questions

1. The bipolar junction transistor has three terminals called the
   a. Drain, source, gate
   b. Anode, cathode, gate
   c. Main terminal 1, main terminal 2, gate
   d. Emitter, base, collector

2. The name bipolar junction transistor was given to the device because it has
   a. Two P-N junctions
   b. Two magnetic poles
   c. One p region and one n region
   d. Two magnetic junctions

3. An NPN transistor is normally biased so that its base is _____.
   a. Positive   b. Negative

4. Which is considered the most common bipolar junction transistor configuration?
   a. Common–base   c. Common–emitter
   b. Common–collector   d. None of the above

5. A common–collector circuit is often called a/an _____.
   a. Base follower   c. Collector follower
   b. Emitter follower   d. None of the above

6. With the NPN transistor schematic symbol, the emitter arrow will point _____ the base, whereas with the PNP transistor schematic symbol, the emitter arrow will point _____ the base.
   a. Toward, away from   b. Away from, toward

7. The transistor's ON/OFF switching action is made use of in _____ circuits.
   a. Analog   c. Linear
   b. Digital   d. Both (a) and (c) are true

8. The transistor's variable resistor action is made use of in _____ circuits.
   a. Analog   c. Linear
   b. Digital   d. Both (a) and (c) are true

9. Approximately 98% of the electrons entering the _____ of a bipolar transistor will arrive at the _____, and the remainder will flow out of the _____.
   a. Emitter, collector, base   c. Collector, emitter, base
   b. Base, collector, emitter   d. Emitter, base, collector

10. The common–base circuit configuration achieves the highest _____ gain, the common–emitter achieves the highest _____ gain, and the common–collector achieves the highest _____ gain.
    a. Voltage, current, power   c. Voltage, power, current
    b. Current, power, voltage   d. Power, voltage, current

11. Which of the following abbreviations is used to denote the voltage drop between a transistor's base and emitter?
    a. I_{BE} b. V_{CC} c. V_{CE} d. V_{BE}

12. Which of the following abbreviations is used to denote the voltage drop between a transistor's collector and emitter?
    a. V_C b. V_{CE} c. V_E d. V_{CC}

13. A transistor's _____ specification indicates the gain in dc current between the input and output of a common–emitter circuit.
    a. α_{AC} b. α_{DC} c. β_{AC} d. β_{DC}

14. Consider the following for a base-biased bipolar transistor circuit: $R_B = 33$ kΩ, $R_C = 560$ Ω, Q_1 (β_{DC}) = 25, $V_{CC} = +10$ V. What is V_{BE}?
    a. 1.43 mV
    b. 25×33 kΩ
    c. 0.7 V
    d. Not enough information given to calculate.

15. Which point on the dc load line results in an $I_C = V_{CC}/R_C$ and a $V_{CE} = 0$ V?
   a. Saturation point
   b. Cutoff point
   c. Q point
   d. None of the above

16. Which point on the dc load line results in a $V_{CE} = V_{CC}$ and an $I_C = 0$?
   a. Saturation point   **c.** Q point
   b. Cutoff point   **d.** None of the above

17. The midway point on the dc load line at which a transistor is biased with dc voltages when no signal input is applied is called the
   a. Saturation point
   b. Cutoff point
   c. Q point
   d. None of the above

18. Which transistor biasing method makes use of one current-limiting resistor in the base circuit?
   a. Base bias
   b. Voltage-divider bias
   c. Emitter-follower bias
   d. Current-divider bias

19. A forward-biased transistor emitter or collector diode should have a _____ resistance, while a reverse-biased emitter and collector diode should have a _____ resistance.
   a. Low, low   **c.** High, high
   b. High, low   **d.** Low, high

20. A transistor tester will check a transistor's _____.
   a. Opens or shorts between any of the terminals
   b. Gain
   c. Reverse leakage current value
   d. All of the above

Practice Problems

21. Identify the type and terminals of the transistors shown in Figure 7-30.

22. A bipolar transistor is correctly biased for operation in the active region when its emitter diode is forward biased and its collector diode is reverse biased. Referring to Figure 7-31, which of the bipolar transistor circuits is correctly biased?

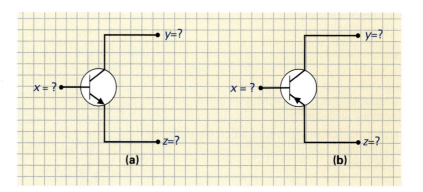

FIGURE 7-30   Identify the Transistor Type and Terminals.

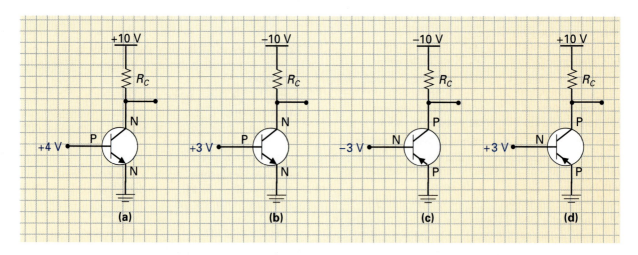

FIGURE 7-31   Identifying the Correctly Biased (Active Region) Bipolar Transistors.

23. Calculate the value of the missing current in the following examples:
   a. $I_E = 25$ mA, $I_C = 24.6$ mA, $I_B = ?$
   b. $I_B = 600$ μA, $I_C = 14$ mA, $I_E = ?$
   c. $I_E = 4.1$ mA, $I_B = 56.7$ μA, $I_C = ?$

24. Calculate the voltage gain (AV) of the transistor amplifier whose input/output waveforms are shown in Figure 7-32.

25. Identify the configuration of the actual bipolar transistor electronic system circuits shown in Figure 7-33.

26. Identify the bipolar transistor type and the biasing technique used in Figure 7-33.

27. Calculate the following for the base-biased transistor circuit shown in Figure 7-34:
   a. I_B b. I_C c. V_{CE}

28. Calculate the following for the voltage-divider-biased transistor circuit shown in Figure 7-35:
   a. V_B and V_E b. I_C c. V_C d. V_{CE}

29. Sketch the dc load line for the circuit in Figure 7-35, showing the saturation, cutoff, and Q points.

Troubleshooting Questions

30. Which of the transistors being tested in Figure 7-36 are good or bad? If bad, state the suspected problem.

FIGURE 7-32   **Transistor Amplifier Input/Output Waveforms.**

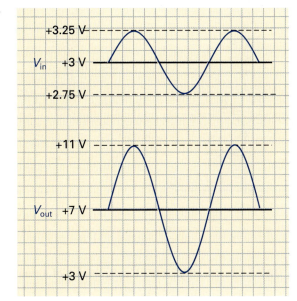

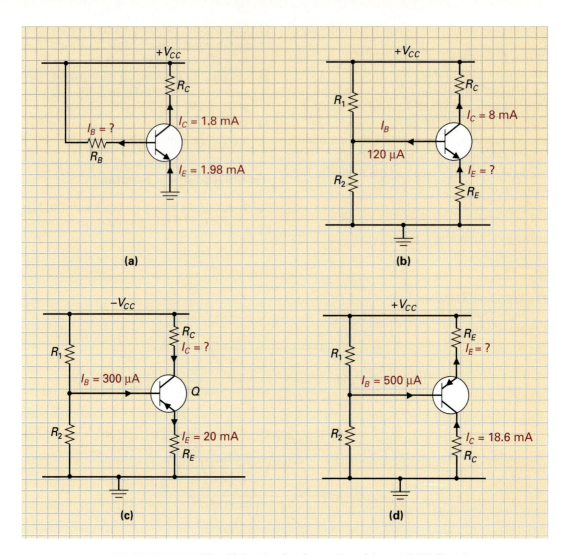

FIGURE 7-33   Identifying the Configuration of Actual BJT Circuits.

FIGURE 7-34   Base-Biased Transistor Circuit.

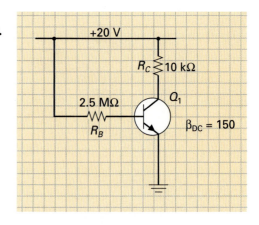

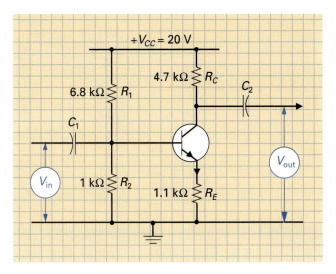

FIGURE 7-35   Voltage-Divider-Biased Transistor Circuit.

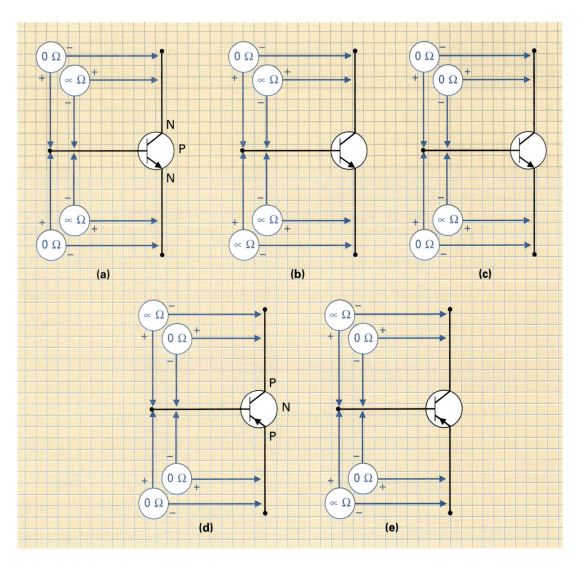

FIGURE 7-36   Testing Transistors with the Ohmmeter.

Amplifier Circuits

8

The First Pocket Calculator

During the seventeenth century, European thinkers were obsessed with any device that could help them with mathematical calculation. Scottish mathematician John Napier decided to meet this need, and in 1614 he published his new discovery of logarithms. In this book, consisting mostly of tediously computed tables, Napier stated that a logarithm is the exponent of a base number. For example, the common logarithm (base 10) of 100 is 2 ($100 = 10^2$), the common logarithm of 10 is 1 ($10 = 10^1$), the common logarithm of 27 is 1.43136 ($27 = 10^{1.43136}$), the common logarithm of 6 is 0.77815 ($6 = 10^{0.77815}$). Any number, no matter how large or small, can be represented by or converted to a logarithm. He also outlined how the multiplication of two numbers could be achieved by simply adding the numbers' logarithms. For example, if the logarithm of 2 (which is 0.30103) is added to the logarithm of 4 (which is 0.60206), the result will be 0.90309, which is the logarithm of the number 8 ($0.30103 + 0.60206 = 0.90309$, $2 \times 4 = 8$). Therefore, the multiplication of two large numbers can be achieved by looking up the logarithms of the two numbers in a log table, adding them together, and then finding the number that corresponds to the sum in an antilog (reverse log) table. In this example, the antilog of 0.90309 is 8.

Napier's table of logarithms was used by William Oughtred, who, just 10 years after Napier's death in 1617, developed a handy mechanical device that could be used for rapid calculation. This device, considered the first pocket calculator, was the slide rule.

As well as being a brilliant mathematician, Napier was also interested in designing military weapons. One such unfinished project was a death ray system consisting of an arrangement of mirrors and lenses that would produce a concentrated lethal beam of sunlight.

Introduction

The amplifier is probably one of the most widely used electronic circuits. Its function is to increase the amplitude of either an ac or dc input signal. Although this function is quite basic, it is vital and needed in almost every electronic system. Amplification is achieved by having a small signal at the input control a large amount of power at the output. The output signal should be a direct copy of the input signal, except larger in amplitude.

As an electronics technician, it is important for you to be able to recognize the appearance, understand the operation, and know the characteristics of various bipolar transistor amplifier circuits. The amplifier circuits in this chapter have been organized by the frequency of the signal they will amplify. In this chapter, we will examine *direct-current (dc) amplifiers,* which amplify dc and low-frequency ac signals, and *audio-frequency (AF) amplifiers,* which amplify ac signals between 20 Hz and 20,000 Hz. In the following chapter, we will continue the coverage started in this chapter, examining bipolar transistor amplifier circuits that operate at progressively higher frequencies, such as *radio-frequency (RF) amplifiers, intermediate-frequency (IF) amplifiers,* and *video-frequency (VF) amplifiers.*

8-1   AMPLIFIER PRINCIPLES

Before we examine how the bipolar transistor can be used as the controlling element in a direct-current (dc) or audio-frequency (AF) amplifier, let us begin by discussing a few basic amplifier principles.

8-1-1   *Basic Amplifier Types and Gain*

The amplifier's main objective is to produce a **gain,** which is symbolized *A.* Gain is the ratio of the amplitude of the output signal to the amplitude of the input signal. This can be stated mathematically as:

The ratio of the amplitude of the output signal to the amplitude of the input signal.

$$\text{Gain } (A) = \frac{\text{Amplitude of Output Signal}}{\text{Amplitude of Input Signal}}$$

There are three basic amplifier types, and these are shown in Figure 8-1. As you can see from this illustration, and as shown with the buffer amplifier in the previous chapter, an amplifier is often symbolized as a triangle with an input terminal on the left and an output terminal on the right.

Voltage Gain

Figure 8-1(a) shows the **voltage amplifier,** which is designed to produce an output signal voltage that is greater than the input signal voltage. This type of amplifier will typically receive an input signal (V_{in}) measured in millivolts and produce an output signal (V_{out}) normally measured in volts. The amount of voltage amplification, or **voltage gain (A_V),** is a ratio of the output signal voltage to the input signal voltage. If a dc voltage signal is applied, the following formula can be used:

Voltage Amplifier
An amplifier designed to produce an output signal voltage that is greater than the input signal voltage.

Voltage Gain (A_V)
The ratio of the output signal voltage to the input signal voltage.

$$A_V = \frac{V_{\text{out}}}{V_{\text{in}}}$$

DC Voltage Amplifier Gain

▪ **EXAMPLE:**

A photodiode is being used to sense the light level in a room. Based on the amount of light present, the photodiode will produce a dc signal voltage of between 0.2 V and 5 V. This dc voltage signal is applied to a direct-current (dc) voltage amplifier, which is included to boost the dc voltage signal to a more usable voltage level. If an input voltage signal of 200 mV is applied to the voltage amplifier, and the amplifier produces an output voltage of 2 V, what is the amplifier's voltage gain?

SECTION 8-1 / AMPLIFIER PRINCIPLES   **197**

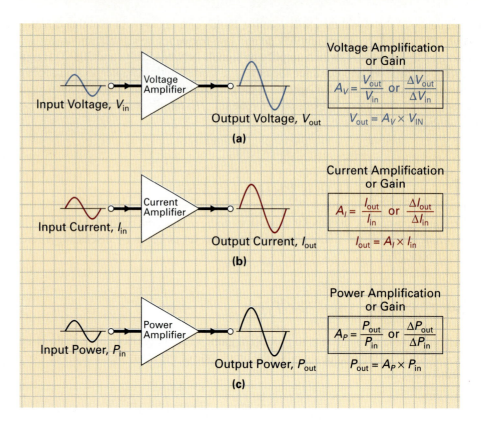

FIGURE 8-1 **Basic Amplifier Types and Gain. (a) Voltage Amplifier. (b) Current Amplifier. (c) Power Amplifier.**

■ *Solution:*

$$A_V = \frac{V_{out}}{V_{in}} = \frac{2 \text{ V}}{200 \text{ mV}} = 10$$

If an ac signal is applied to a voltage amplifier, the Greek capital letter delta (Δ) precedes the V_{out} and V_{in} in the gain formula to indicate the ac signal voltage change.

$$A_V = \frac{\Delta V_{out}}{\Delta V_{in}}$$

AC Voltage Amplifier Gain

■ **EXAMPLE:**

A microphone is being used to sense the loudness level in a room. The ac signal voltage produced by the microphone is applied to an audio-frequency (AF) voltage amplifier. If an input voltage signal of 400 mV peak–peak is applied to the voltage amplifier, and the amplifier produces an output voltage of 8 V peak–peak, what is the amplifier's voltage gain?

■ *Solution:*

$$A_V = \frac{V_{out}}{V_{in}} = \frac{8 \text{ V}}{400 \text{ mV}} = 20$$

By transposing the gain formula, we can obtain a formula for calculating the output of a voltage amplifier if the voltage gain and input voltage are known.

$$V_{out} = A_V \times V_{in}$$

■ **EXAMPLE:**

What would be the output from a common–base voltage amplifier with a gain of 50 if a multimeter on the ac volt's setting measures an input signal of 18 mV?

■ *Solution:*

A multimeter on the ac volt's setting is calibrated to display rms voltage. If the input voltage is an rms value, the output voltage value will also be an rms value. Therefore,

$$V_{out(rms)} = A_V \times V_{in(rms)} = 50 \times 18 \text{ mV} = 900 \text{ mV}_{rms}$$

Current Gain

Figure 8-1(b) shows the **current amplifier,** which is designed to produce an output signal current (I_{out}) that is greater than the input signal current (I_{in}). The amount of current amplification, or **current gain (A_I),** can be calculated by using the same gain ratio formula of output over input.

$$A_I = \frac{I_{out}}{I_{in}} \qquad\qquad A_I = \frac{\Delta I_{out}}{\Delta I_{in}}$$

DC Current Amplifier Gain   AC Current Amplifier Gain

■ **EXAMPLE:**

A common–collector amplifier has an input of 0.25 mA and an output of 0.37 mA. What is its current gain?

■ *Solution:*

$$A_I = \frac{I_{out}}{I_{in}} = \frac{0.37\text{mA}}{0.25\text{mA}} = 1.5$$

By transposing these gain formulas, we can calculate the output of a current amplifier by multiplying the amplifier's current gain by the input current.

$$I_{out} = A_I \times I_{in}$$

■ **EXAMPLE:**

If a 150-μA input signal is applied to a current amplifier with a gain of 35, what will the output signal current be?

■ *Solution:*

$$I_{out} = A_I \times I_{in} = 35 \times 150 \text{ μA} = 5.25 \text{ mA}$$

Power Gain

Figure 8-1(c) shows the **power amplifier,** which is designed to produce an output signal power that is greater than the input signal power. This type of amplifier will typically receive an input signal (P_{in}) measured in milliwatts and produce an output signal (P_{out}) normally measured in watts. The amount of power amplification, or **power gain (A_P),** can be calculated by using the same gain-ratio formula of output over input.

$$A_P = \frac{P_{out}}{P_{in}} \qquad\qquad A_P = \frac{\Delta P_{out}}{\Delta P_{in}}$$

DC Power Amplifier Gain   AC Power Amplifier Gain

As before, we can transpose the power gain formula to calculate the output from a power amplifier when the input power is known.

$$P_{out} = A_P \times P_{in}$$

Because power is equal to the product of voltage and current ($P = V \times I$), the power gain of an amplifier can also be calculated by multiplying the amount of voltage gain by the amount of current gain.

$$A_P = A_V \times A_I$$

■ **EXAMPLE:**

Calculate the gain of a common–emitter power amplifier if it has a voltage gain of 43 and a current gain of 10. Also, determine the output power from the amplifier when an input of 53 µW is applied.

■ *Solution:*

$$A_P = A_V \times A_I = 43 \times 10 = 430$$
$$P_{out} = A_P \times P_{in} = 430 \times 53 \text{ µW} = 22.8 \text{ mW}$$

8-1-2   *Bipolar Transistor Circuit Configurations*

The voltage, current, and power amplifier block symbols shown in Figure 8-1 are used to represent a complete electronic amplifier circuit, as shown in Figure 8-2. This amplifier circuit generally contains one or more transistors, which act as the circuit's controlling element, and several associated components, such as resistors and capacitors, which control the characteristics and stability of the transistor. As we discussed in Chapter 7, by connecting the bipolar junction transistor in one of three different configurations, we could have it function as a power amplifier, voltage amplifier, or current amplifier. Let us briefly review these three circuit configurations and their characteristics, which are shown in Figure 8-3.

1. The common–emitter circuit configuration, shown in Figure 8-3(a), provides a high current gain and a high voltage gain, and therefore the combined power gain is high. It also has a medium input impedance, a high output impedance, and a 180° phase shift between input and output. All of these generally good characteristics make the common–emitter transistor circuit configuration the most widely used.

2. The common–base circuit configuration, shown in Figure 8-3(b), provides a high voltage gain. However, its current output is slightly less than the current input, so the com-

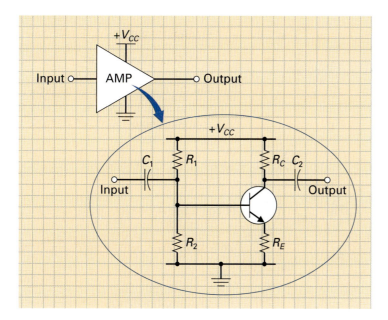

FIGURE 8-2 **The Amplifier Circuit's Components.**

bined power gain is medium. It also has a very low input impedance, which will load a source, and a high output impedance, which accounts for the low output current. This configuration does not invert the signal between input and output. All of these generally poor characteristics make the common–base transistor circuit configuration the most infrequently used.

3. The common–collector circuit configuration, shown in Figure 8-3(c), provides a high current gain; its voltage output, however, is slightly less than the voltage input, so the combined power gain is low. The circuit's key advantage is its high input impedance and very low output impedance, making it ideal as an impedance-matching device between a low-current (high-impedance) source and a high-current (low-impedance) load. This circuit is also referred to as an *emitter follower*.

All of the direct-current, audio-frequency amplifiers discussed in this chapter, and all of the radio-frequency, intermediate-frequency, and video-frequency amplifiers discussed in the next chapter, will use one of these three circuit configurations in order to take advantage of that configuration's characteristics.

8-1-3   *Amplifier Class of Operation*

Before we discuss the four basic amplifier classes of operation, or modes of operation, let us briefly review the basic dc biasing of a transistor amplifier by referring to Figure 8-4. This group of characteristic curves was discussed previously in Chapter 7. The key point to remember with regard to this section is that the transistor amplifier is generally biased at a dc operating point, which is called the quiescent or Q point. The input signal will vary the transistor's dc bias voltage above and below its Q point, causing a change in base current (I_B), collector current (I_C), and output voltage (V_{CE}), as shown in Figure 8-5(a). All of the bipolar transistor amplifier circuits discussed in Chapter 7 were operated in this way. However, as you will see in the different application amplifier circuits discussed in this chapter, by biasing an amplifier at different points on the dc load line, we can obtain some circuit characteristics that are ideal for certain applications.

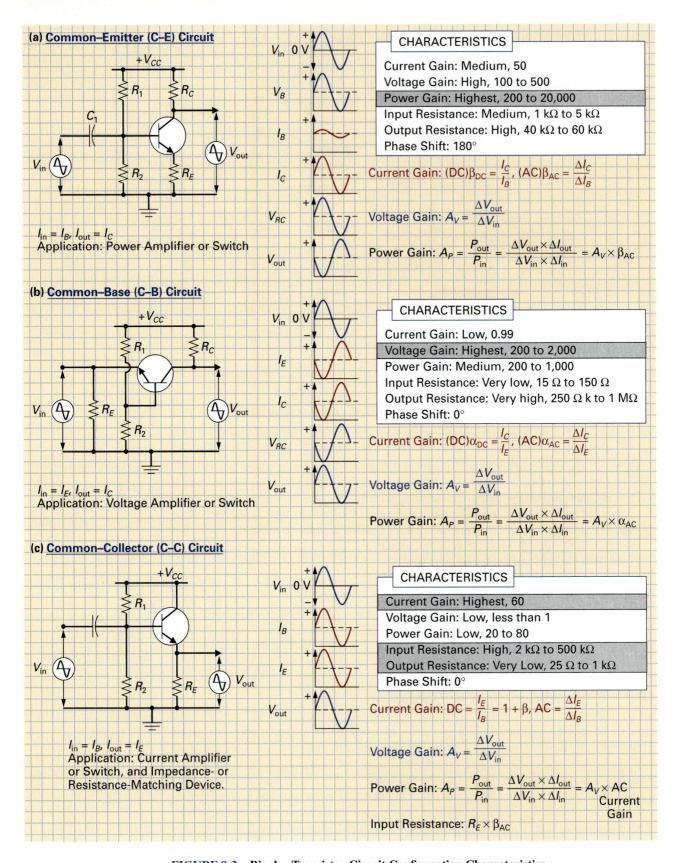

(a) Common–Emitter (C–E) Circuit

$I_{in} = I_B$, $I_{out} = I_C$
Application: Power Amplifier or Switch

CHARACTERISTICS

Current Gain: Medium, 50
Voltage Gain: High, 100 to 500
Power Gain: Highest, 200 to 20,000
Input Resistance: Medium, 1 kΩ to 5 kΩ
Output Resistance: High, 40 kΩ to 60 kΩ
Phase Shift: 180°

Current Gain: (DC)$\beta_{DC} = \dfrac{I_C}{I_B}$, (AC)$\beta_{AC} = \dfrac{\Delta I_C}{\Delta I_B}$

Voltage Gain: $A_V = \dfrac{\Delta V_{out}}{\Delta V_{in}}$

Power Gain: $A_P = \dfrac{P_{out}}{P_{in}} = \dfrac{\Delta V_{out} \times \Delta I_{out}}{\Delta V_{in} \times \Delta I_{in}} = A_V \times \beta_{AC}$

(b) Common–Base (C–B) Circuit

$I_{in} = I_E$, $I_{out} = I_C$
Application: Voltage Amplifier or Switch

CHARACTERISTICS

Current Gain: Low, 0.99
Voltage Gain: Highest, 200 to 2,000
Power Gain: Medium, 200 to 1,000
Input Resistance: Very low, 15 Ω to 150 Ω
Output Resistance: Very high, 250 Ω k to 1 MΩ
Phase Shift: 0°

Current Gain: (DC)$\alpha_{DC} = \dfrac{I_C}{I_E}$, (AC)$\alpha_{AC} = \dfrac{\Delta I_C}{\Delta I_E}$

Voltage Gain: $A_V = \dfrac{\Delta V_{out}}{\Delta V_{in}}$

Power Gain: $A_P = \dfrac{P_{out}}{P_{in}} = \dfrac{\Delta V_{out} \times \Delta I_{out}}{\Delta V_{in} \times \Delta I_{in}} = A_V \times \alpha_{AC}$

(c) Common–Collector (C–C) Circuit

$I_{in} = I_B$, $I_{out} = I_E$
Application: Current Amplifier
or Switch, and Impedance- or
Resistance-Matching Device.

CHARACTERISTICS

Current Gain: Highest, 60
Voltage Gain: Low, less than 1
Power Gain: Low, 20 to 80
Input Resistance: High, 2 kΩ to 500 kΩ
Output Resistance: Very Low, 25 Ω to 1 kΩ
Phase Shift: 0°

Current Gain: DC $= \dfrac{I_E}{I_B} = 1 + \beta$, AC $= \dfrac{\Delta I_E}{\Delta I_B}$

Voltage Gain: $A_V = \dfrac{\Delta V_{out}}{\Delta V_{in}}$

Power Gain: $A_P = \dfrac{P_{out}}{P_{in}} = \dfrac{\Delta V_{out} \times \Delta I_{out}}{\Delta V_{in} \times \Delta I_{in}} = A_V \times$ AC Current Gain

Input Resistance: $R_E \times \beta_{AC}$

FIGURE 8-3   Bipolar Transistor Circuit Configuration Characteristics.

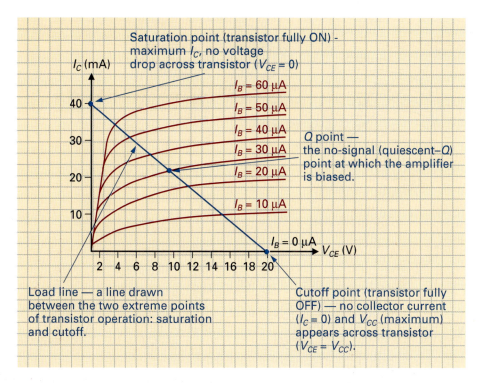

FIGURE 8-4 The Bipolar Transistor's Load Line and Q Point in its Family of Characteristic Curves.

Class-A Amplifier Operation

Any transistor amplifier that has its dc operating point set near the center of the load line, so that the output current (I_C) flows during the entire cycle of the ac input signal, is said to be operating as a **class-A amplifier.** All of the amplifier circuits discussed in Chapter 7 were of the class A-type because they were biased at a midpoint on the dc load line so that any change in the input current would produce a proportional but amplified change in the output current, as shown in Figure 8-5(a). Class-A amplifiers are said to operate in a linear manner because a change in the input signal produces a proportional change in the output signal.

Class-B Amplifier Operation

When an amplifier is biased so that its output current will flow for only half of the complete ac input signal cycle, the circuit is said to be operating as a **class-B amplifier,** as can be seen in Figure 8-5(b). To achieve this mode of operation, the amplifier is biased so that its Q point is at cutoff; therefore, the output current will flow only during the time that the input alternation forward biases the transistor. The advantage of this amplifier class of operation is that the circuit will consume only half the power of an equivalent class-A amplifier because the amplifier is turned ON for only 50% of the time. Stated another way, a class-B amplifier works for half of the input cycle and then rests for the other half of the input cycle, as opposed to a class-A amplifier, which is working all of the time. Class-B amplifiers can be pushed to deliver a lot more power than a class-A amplifier because of their 50% work ratio. A simple analogy is that if a person who normally works a full eight-hour day was instructed to work for only half that time, he or she could be pushed to deliver a lot more in four hours than in the usual eight-hour day.

Class-C Amplifier Operation

When an amplifier is biased so that its output current will flow for less than one half of the ac input cycle, it is said to be operating as a **class-C amplifier,** as seen in Figure 8-5(c). This is achieved by biasing the transistor amplifier operating point below the cutoff point so

Class-A Amplifier
A transistor amplifier that has its dc operating point set near the center of the load line, so that the output current (IC) flows during the entire cycle of the ac input signal.

Class-B Amplifier
An amplifier biased so that its output current will flow for only half of the complete ac input signal cycle.

Class-C Amplifier
An amplifier biased so that its output current will flow for less than one half of the ac input cycle.

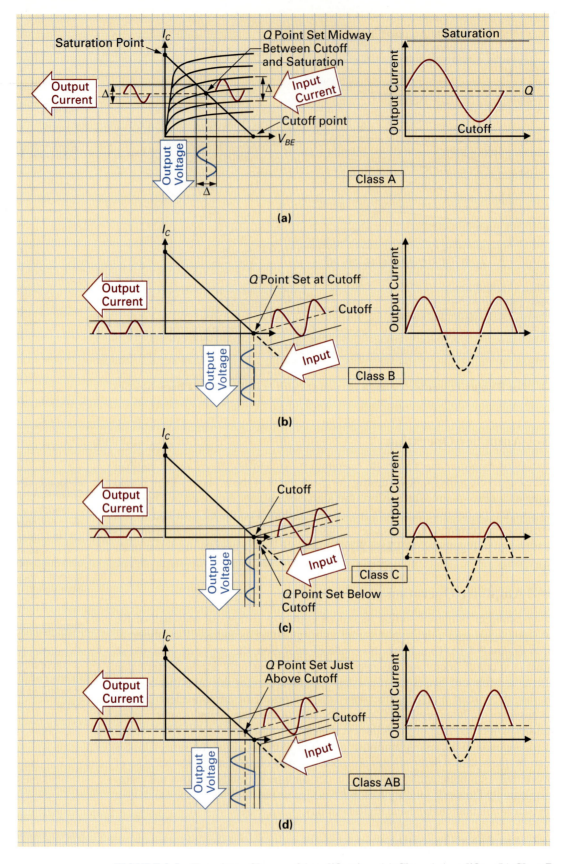

FIGURE 8-5 Transistor Classes of Amplification. (a) Class-A Amplifier. (b) Class-B Amplifier. (c) Class-C Amplifier. (d) Class-AB Amplifier.

that for most of the time the amplifier is reverse biased or cutoff. For only a small period of time will it produce an output pulse of current. This type of amplifier is therefore very efficient because it only consumes power for a very short time, compared to the complete cycle of the ac input signal.

Class-AB Amplifier Operation

When an amplifier is biased so that its output current will flow for slightly less than one full cycle of the ac input (class A), but more than one-half of the ac input cycle (class B), the amplifier is said to be operating as a **class-AB amplifier,** as shown in Figure 8-5(d). This type of amplifier is a compromise between a class A and a class B, and its operating point is set at slightly above cutoff, causing an output current to be produced for slightly more than one-half of the input cycle.

Amplifier Efficiency

The Greek letter *eta,* η, (pronounced "eat-eh"), is used to designate an amplifier's efficiency, which is the ratio of the power delivered to the load by the amplifier to the power consumed by the amplifier. As a formula, it would appear as follows:

$$\eta = \frac{\text{Power Delivered to Load by Amplifier } (P_L)}{\text{DC Power Consumed by Amplifier } (P_S)} \times 100\%$$

Class-A amplifiers are constantly drawing power from the $+V_{CC}$ dc power supply the whole time that they are supplying power to the load. As a result, class-A amplifiers tend to be fairly inefficient. On the other hand, class-B amplifiers are OFF for 50% of the input cycle, and, because they consume no power during this time, they tend to be more efficient than their class-A counterparts. Class-C amplifiers are OFF, and therefore not consuming power, for more than 50% of the input cycle. A class-C amplifier is therefore more efficient than a class-B amplifier. The class-AB amplifier's power consumption is midway between the class A and the class B and so is its efficiency.

As we proceed through this chapter, you will see how the efficiency of these different classes of amplifiers can be calculated and how their efficiency is generally at best as follows:

AMPLIFIER CLASS	EFFICIENCY
A	35%
AB	50%
B	75%
C	99%

8-1-4   *Amplifier Frequency Response*

By changing the design of an amplifier circuit, you can change its characteristics. One of the key characteristics of an amplifier circuit is its response to different input signal frequencies. Some amplifier circuits are best suited to amplify high-frequency signals, while other amplifier circuits respond better to low-frequency signals. For example, audio frequency amplifiers provide a high gain to any audio-frequency signals between 20 Hz and 20 kHz, whereas a radio-frequency amplifier would provide almost no gain to any input signal in this range. A **frequency response curve** is used to show the response of an amplifier circuit to different signal input frequencies by plotting gain against frequency. As an example, Figure 8-6 shows the frequency response curve for an audio-frequency power amplifier. The amplifier's **bandwidth** is the group or band of frequencies between the half-power points. These half-power points are the points at which the gain has fallen below 50% of the maximum power gain (−3 dB). Referring to the example in Figure 8-6, you can see that the audio-frequency power amplifier's gain increases above half power at 20 Hz and falls below half power at 20,000 Hz. The bandwidth for this amplifier will be 19,980 Hz (19.98 kHz), which is the difference between 20,000 Hz and 20 Hz.

$$BW = f_{HI} - f_{LO}$$

BW = Bandwidth, f_{HI} = Upper Frequency Limit, f_{LO} = Lower Frequency Limit
$$BW = f_{HI} - f_{LO} = 20,000 \text{ Hz} - 20 \text{ Hz} = 19,980 \text{ Hz}$$

The upper frequency limit at which the gain of the amplifier falls below half power is also called the **cutoff frequency.**

Figure 8-6(b) shows how the bandwidth of a current amplifier is determined. In this diagram, you may have noticed that 70.7%, or 0.707 of the maximum gain, is used to determine the bandwidth. This is because 70.7% of the maximum current gain is equivalent to the half-power points. To prove this, let us use an example.

▪ EXAMPLE:

An amplifier is delivering 100-mA to a 2-kΩ load. Therefore, the power delivered is

$$P = I^2 \times R = 100 \text{ mA} \times 2 \text{ k}\Omega = 20 \text{ W}$$

If the input signal frequency were to increase and cause the amplifier's current output to drop to 70.7 mA, what would be the power delivered to the same load?

▪ *Solution:*

$$P = I^2 \times R = 70.7 \text{ mA} \times 2 \text{ k}\Omega = 10 \text{ W}$$

As you can see from this example, the 70.7% current points are equal to the half-power points; an amplifier's bandwidth exists between the 70.7% current points, or half-power points.

The same is true for a voltage amplifier, as shown in Figure 8-6(c), because 70.7% of the maximum voltage gain is equivalent to the half-power points. Let us again prove this with an example.

▪ EXAMPLE:

An amplifier is delivering 10 V to a 1-kΩ load. Therefore, the power delivered is

$$P = \frac{V^2}{R} = \frac{10 \text{ V}^2}{1 \text{ k}\Omega} = 100 \text{ mW}$$

If the input signal frequency were to increase and cause the amplifier's voltage output to drop to 7.07 V, what would be the power delivered to the same load?

▪ *Solution:*

$$P = \frac{V^2}{R} = \frac{7.07 \text{ V}^2}{1 \text{ k}\Omega} = 50 \text{ mW}$$

CHAPTER 8 / AMPLIFIER CIRCUITS

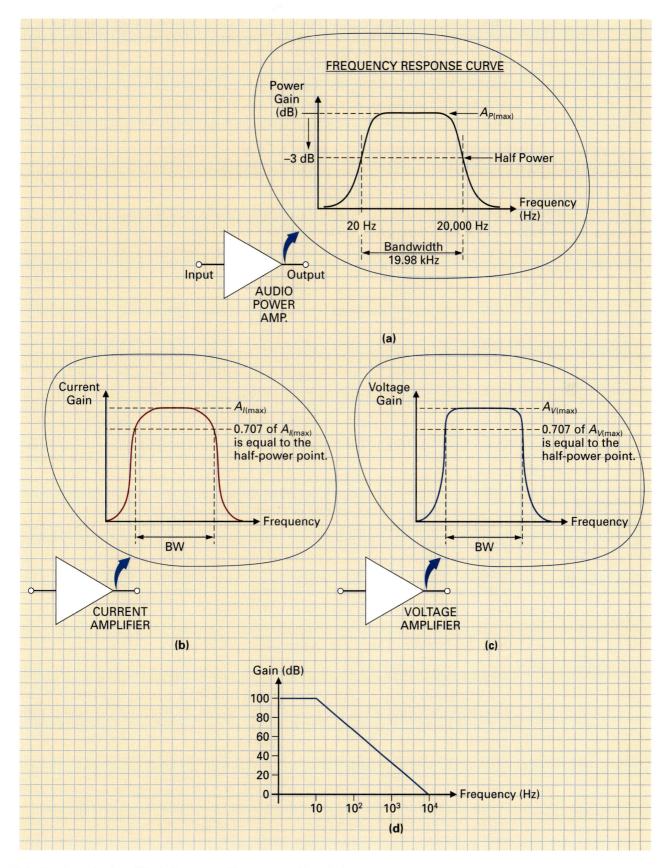

FIGURE 8-6   An Amplifier's Frequency Response and Bandwidth.

Once again, as you can see from this example, the 70.7% voltage points are equal to the half-power points. An amplifier's bandwidth therefore exists between the 70.7% voltage or current points, or half-power points.

A **Bode diagram** is another form of frequency response curve. As you can see from the example in Figure 8-6(d), this diagram is an approximation using straight lines and plots a circuit's gain in decibels against frequency.

Bode Diagram

A frequency response curve that plots gain in decibels versus frequency.

Use the following questions to test your understanding of Section 8–1.

1. How is the gain of an amplifier calculated?
2. If a voltage amplifier has a gain of 75 and an input of 0.5 mV, what will be the output voltage?
3. The common–base configuration provides a good _____ gain, the common–emitter provides a good _____ gain, and the common–collector provides a good _____ gain.
4. Define cutoff frequency.
5. What is a frequency response curve?
6. What class of operation causes no signal distortion?
7. Which amplifier class of operation has its Q point set near the center of the load line so that output current is present for the entire cycle of the ac input signal?
8. A _____ is used to show the response of an amplifier circuit to different input signal frequencies.

8-2 DIRECT-CURRENT (DC) AMPLIFIERS

Now that the principles of amplifier circuits have been discussed, let us discuss some specific amplifier types and their applications. In this section we will examine the operation and characteristics of **direct-current or dc amplifiers,** which are normally always class-A biased (Q point is midway between saturation and cutoff); therefore, the transistor is always conducting. Direct-current amplifiers are used to amplify dc voltages or currents, or low-frequency ac voltages or currents, produced by transducers or sensors. These transducers will sense heat, light, pressure, or vibration and produce a corresponding electrical signal that is usually extremely weak in amplitude and must be amplified up to a more usable level. To begin with, we will study a basic single-stage dc amplifier.

Direct-Current (DC) Amplifiers

Amplifiers for dc and low-frequency ac signals.

8-2-1 *Single-Stage DC Amplifier Circuits*

Figure 8-7(a) shows how a direct-current (dc) amplifier can be used to amplify a +0.2-V to +5-V dc signal from a light-dependent resistor (LDR), which is being used to sense the ambient light level. Resistors R_1 and R_2 provide voltage-divider bias for the common–emitter transistor amplifier. Collector resistor R_3 operates in conjunction with the transistor to develop an output signal voltage. Resistor R_E provides "emitter feedback" for temperature stability. To review, resistor R_E achieves circuit thermal stability through emitter feedback. With R_E included in the circuit, V_{CC} is developed across R_C, Q_1's collector–emitter, and R_E. An increase in temperature will cause an increase in I_C, I_E, and therefore the voltage developed across R_E. An increase in V_{RE} will decrease the base–emitter bias voltage applied, causing Q_1's internal currents to decrease back to their original values. A change in the output current (I_C) due to temperature will cause the emitter resistor to feed back a control voltage to the input.

Because the input signal is a dc voltage, capacitors cannot be connected in the signal path (a capacitor will act as a block to any dc signals). The lack of an input coupling capacitor means that the base bias of Q_1 is set not only by the resistance values of R_1 and R_2 but also by the internal resistance of the LDR source in parallel with R_1. Similarly, the lack of an output coupling capacitor means that the resistance of the load is connected in parallel with the resistance of transistor Q_1 and R_4. When designing a dc amplifier, the resistance of the input source and resistance of the load will have to be taken into account because these values will affect the bias network. The bias network will govern base voltage and base current and, in turn, control collector current and output voltage. As you know, the load resistance shown in Figure 8-7(a) merely represents the resistance of the device connected to the output of this amplifier. In most instances, this load resistance will represent the input resistance of the next amplifier stage.

The common–emitter dc amplifier shown in Figure 8-7(a) will provide both voltage and current amplification; however, its main purpose is as a voltage amplifier. As you can see in Figure 8-7(b), which shows the typical frequency response of a dc amplifier, the voltage gain of a dc amplifier remains almost constant from 0 Hz (dc) to about 2,000 Hz (ac). Beyond 2,000 Hz or 2 kHz, the voltage gain of a dc amplifier drops off rapidly. The frequency at which a transistor's gain falls below 70.7% is called the transistor's *cutoff frequency*. Consider this frequency cutoff action in more detail: As the ac input signal frequency increases, its cycle time or period decreases, and eventually this cycle time is faster than the transit time needed for a charge carrier to pass through the transistor. At this time, the gain of the transistor will fall to almost zero because the transistor's internal currents do not have enough time to respond to the high-frequency input signal changes. Different types of bipolar transistors will have different cutoff frequency ratings, which will be specified in the manufacturer's data sheet.

From a dc input signal of 0 Hz up to an ac input signal frequency of about 2 kHz, the voltage gain of a dc amplifier is approximately equal to the ratio of the transistor's collector resistance R_C to the emitter resistor R_E.

$$A_V = \frac{R_C}{R_E}$$

No Load Connected

With no load connected, the voltage gain of the dc amplifier in Figure 8-7(a) will be equal to R_3 divided by R_4.

However, if a load is connected to the output, the collector current out of the transistor will split and pass through R_3 and R_L; therefore, these two resistors are effectively in parallel. The voltage gain of an amplifier in this instance is equal to the equivalent parallel resistance of R_3 and R_L divided by the emitter resistance (R_4 or R_E).

$$A_V = \frac{R_3 \| R_L}{R_E}$$

Load Connected

To calculate the transistor's equivalent parallel collector resistance, we will use the standard parallel resistance formula: product over sum.

$$R_C = R_3 \| R_L = \frac{R_3 \times R_L}{R_3 + R_L}$$

Let us now practice using these formulas with an example.

■ EXAMPLE:

Calculate the voltage gain of the circuit in Figure 8-7(a). For the dc input signal voltage range given, what would be the output dc signal voltage range?

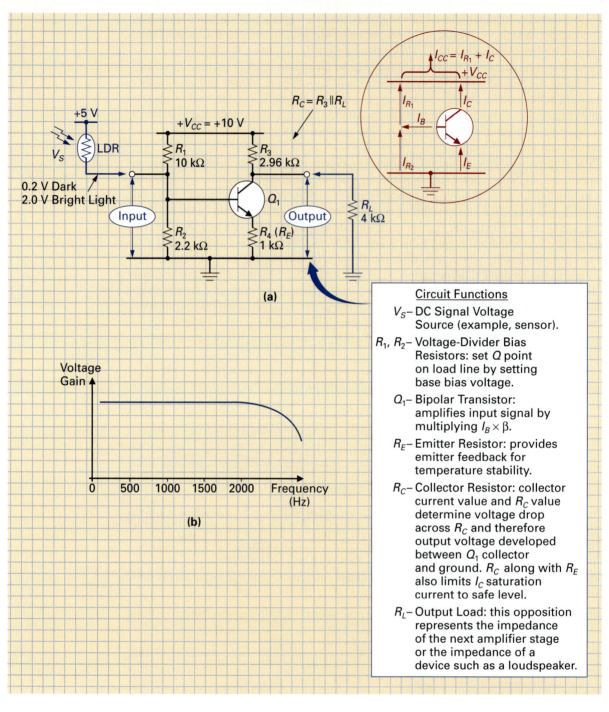

FIGURE 8-7   Single-Stage Direct-Current (DC) Amplifier. (a) Basic Circuit.
(b) Frequency Response.

■ *Solution:*

$$R_C = \frac{R_3 \times R_L}{R_3 + R_L} = \frac{2.96 \text{ k}\Omega \times 4 \text{ k}\Omega}{2.96 \text{ k}\Omega + 4 \text{ k}\Omega} = \frac{11.84 \text{ k}\Omega}{6.96 \text{ k}\Omega} = 1.7 \text{ k}\Omega$$

$$A_V = \frac{R_C}{R_E} = \frac{1.7 \text{ k}\Omega}{1 \text{ k}\Omega} = 1.7$$

The output signal voltage will be 1.7 times greater than the input signal voltage. Therefore, because the dc input signal voltage can be anywhere from 0.2 V to 2 V, the output signal voltage will be anywhere from

$$V_{out} = V_{in} \times A_V = 0.2 \text{ V} \times 1.7 \text{ V} = 0.34 \text{ V}$$

to

$$V_{out} = V_{in} \times A_V = 2 \text{ V} \times 1.7 \text{ V} = 3.4 \text{ V}$$

As mentioned previously, an amplifier's efficiency is equal to the ratio of the power delivered to the load by the amplifier to the power consumed by the amplifier.

$$\eta = \frac{\text{Power Delivered to Load by Amplifier } (P_L)}{\text{DC Power Consumed by Amplifier } (P_S)} \times 100\%$$

To calculate the efficiency of the dc amplifier in Figure 8-7, we will have to know how much power is being delivered to the load and how much dc power is being consumed by the amplifier. To calculate the amount of power delivered to the load (P_L), we can use the standard V^2/R power formula.

$$P_L = \frac{V_L^2}{R_L}$$

P_L = Power delivered to the load

V_L = Voltage developed across load

R_L = Load resistance

To calculate the amount of power consumed by the amplifier circuit, or the power supplied to the amplifier by the dc power supply (P_S), we will use the standard VI power formula.

$$P_S = V_{CC} \times I_{CC}$$

P_S = Power supplied to the amplifier by the dc power supply

V_{CC} = Circuit's dc supply voltage

I_{CC} = Total current drawn by the amplifier circuit

To see how the efficiency of an amplifier is calculated, let us apply these formulas to the example circuit in Figure 8-7(a).

■ **EXAMPLE:**

Calculate the efficiency of the class-A amplifier in Figure 8-7(a).

■ *Solution:*

Because we know from the previous example that the output voltage is between 0.34 V and 3.4 V, the average output voltage will be

$$V_{avg} = \frac{V_1 + V_2}{2} = \frac{0.34 \text{ V} + 3.4 \text{ V}}{2} = 1.87$$

This voltage will be applied across a 4-kΩ load, so the power delivered to the load will be

$$P_L = \frac{V_L^2}{R_L} = \frac{1.87 \text{ V}^2}{4 \text{ k}\Omega} = 874.2 \text{ μW}$$

To calculate the amount of power consumed by the amplifier, we will need to know the total amount of current being drawn by the amplifier circuit (I_{CC}). This total value of current is equal to the value of current flowing through the voltage-divider resistor R_1 (I_{R_1}) and the value of collector current (I_C), as shown in the inset in Figure 8-7(a). To determine these values, we will have to do the following calculations:

$$V_{R_2} = \frac{R_2}{R_1 + R_2} \times V_{CC} = 1.8 \text{ V}$$

$$I_{R_2} = \frac{V_{R_2}}{R_2} = \frac{1.8 \text{ V}}{2.2 \text{ k}\Omega} = 818 \ \mu\text{A}$$

$I_{R_2} \cong I_{R_1}$ (I_B can be ignored); therefore $I_{R_1} = 818 \ \mu\text{A}$

$V_E = V_B - 0.7 \text{ V} = 1.8 \text{ V} - 0.7 \text{ V} = 1.1 \text{ V}$

$$I_E = \frac{V_E}{R_E} = \frac{1.1 \text{ V}}{1 \text{ k}\Omega} = 1.1 \text{ mA}$$

$I_E \cong I_C$; therefore $I_C = 1.1 \text{ mA}$

Now that I_{R_1} and I_C are known, we can calculate I_{CC}.

$$I_{CC} = I_{R_1} + I_C = 818 \ \mu\text{A} + 1.1 \text{ mA} = 1.92 \text{ mA}$$

Now that I_{CC} is known, we can calculate the power supplied to the amplifier by the dc power supply.

$$P_S = V_{CC} \times I_{CC} = 10 \text{ V} \times 1.92 \text{ mA} = 19.2 \text{ mW}$$

If the power delivered to the load is equal to 874.2 μW, and the power supplied to the amplifier by the dc power supply is equal to 19.2 mW, the class-A amplifier's efficiency in Figure 8-7(a) will be

$$\eta = \frac{P_L}{P_S} \times 100\% = \frac{874.2 \ \mu\text{W}}{19.2 \text{ mW}} \times 100\% = 4.5\%$$

With the common–emitter circuit configuration used in Figure 8-7(a), there will be a 180° phase difference between the input signal voltage and the output signal voltage. For example, when the input signal voltage goes more positive, the output voltage will go less positive, and when the input signal voltage goes less positive, the output voltage will go more positive. This inverting action of the amplifier is not a problem because the output voltage signal still changes in accordance with the input signal change. For example, if the light signal increased in Figure 8-7(a), the positive voltage change at the input of Q_1 would also increase, causing the voltage signal at the collector of Q_1 to decrease proportionally, so that the drop in voltage accurately represents the signal increase.

8-2-2   *Multiple-Stage DC Amplifier Circuits*

In some applications, a single dc amplifier stage may not provide enough gain. In these instances, several dc amplifier stages may be needed to increase the input signal up to a desired amplitude.

Direct-Coupled DC Amplifier Circuits

Figure 8-8(a) shows how we can couple the output of one dc amplifier stage into the input of another dc amplifier stage so that we can achieve a higher overall gain. This two-stage dc amplifier circuit uses **direct coupling,** which means that the output of the first stage is coupled directly into the input of the second stage. The first stage of this multiple-stage dc amplifier is identical to the single-stage dc amplifier discussed previously in Fig-

Direct Coupling

The output of the first stage is coupled directly into the input of the second stage.

CHAPTER 8 / AMPLIFIER CIRCUITS

ure 8-7(a), in that it contains four resistors (R_1 to R_4) and a transistor (Q_1). The second-stage dc amplifier circuit is made up of transistor Q_2 and resistors R_5 and R_6. The base bias for Q_2 is provided by the dc collector voltage of Q_1. The input signal is applied to the base of Q_1 and, in controlling Q_1, this signal will also control Q_2. Because the input signal may be a dc voltage or current, dc blocking components such as capacitors and transformers cannot be connected in the signal path.

The dc input signal, or ac input signal, is amplified by the first stage and then further amplified by the second stage. When two or more dc amplifiers are connected end-to-end, or **cascaded,** in this way, the overall dc amplifier voltage gain is equal to the product of all the individual stage gains.

$$A_{\text{Total}} = A_1 \times A_2 \times A_3 \times \cdots$$

The disadvantage of this direct-coupled dc amplifier is its lack of isolation between stages. As the number of direct-coupled stages increases, the collector bias voltage applied to the next stage becomes progressively larger. Because it is this collector voltage that controls the dc bias of the next stage, only a few direct-coupled stages can be cascaded before the collector voltage becomes so large that it will drive the next amplifier stage's output beyond the $+V_{CC}$ supply voltage. Referring to the inset in Figure 8-8(a), you can see that the Q point (dc base bias point) on the load line is continually increased by each stage until the signal input drives the transistor into saturation and the upper part of the input signal is clipped. The output signal of an amplifier should in most cases be a direct copy of the input signal, only larger in amplitude. Any differences or irregularities introduced unintentionally into the signal are unwanted and are called **signal distortion.** To prevent this distortion from occuring with direct-coupled dc amplifiers, it is generally necessary to decrease the gain of the first stage so that we can control the gain of the second stage. This compromise of sacrificing gain for circuit control is typical with most amplifier circuits.

Cascaded
Two or more dc amplifiers connected end-to-end.

Signal Distortion
Any unwanted differences or irregularities introduced unintentionally into the signal.

EXAMPLE:

Both of the stages of the dc amplifier in Figure 8-8(a) have a voltage gain of 10. If the gain of the first stage is decreased to 80%, what is the overall gain of this two-stage amplifier?

Solution:
$$80\% \text{ of } 10 = 8$$
$$A_{V(\text{Total})} = A_{V_1} \times A_{V_2} = 8 \times 10 = 80$$

The output signal voltage of the amplifier in Figure 8-8(a) will be 80 times greater than the input signal voltage.

Resistive-Coupled DC Amplifier Circuits

Figure 8-8(b) shows how we can overcome the lack of isolation (and therefore lack of control) of a direct-coupled dc amplifier by connecting a large-value resistor (R_5) between two complete dc amplifier stages. This **resistive-coupling** technique means that the bias voltages and currents of each stage are somewhat independent of one another, so we can control the gain of each stage to produce a higher gain output than a direct-coupled amplifier. The disadvantage of this circuit is that, because the resistor is connected in the signal path, there is a voltage drop and therefore signal voltage loss across the coupling resistor.

Resistive Coupling
A connection made by a large-value resistor.

Zener-Coupled DC Amplifier Circuits

Figure 8-8(c) shows how we can achieve circuit control (through stage isolation) and reduce coupling loss by **zener coupling** two dc amplifier stages. The zener diode (D_1) is reverse biased and operating in its zener region because the voltage at the collector of Q_1 is generally several volts more positive than the voltage at the base of Q_2.

Zener Coupling
A connection made by a zener diode.

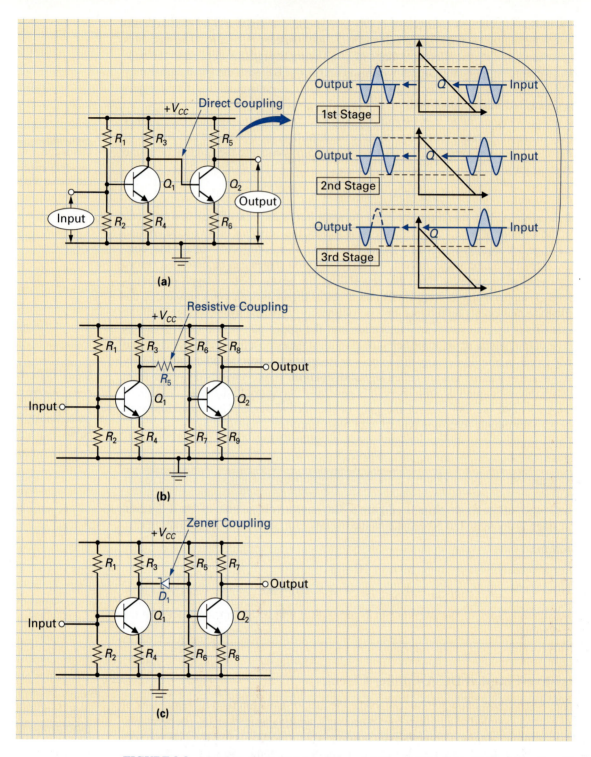

FIGURE 8-8 Multiple-Stage DC Amplifiers. (a) Direct Coupling. (b) Resistive Coupling. (c) Zener Coupling.

An applied input signal will be amplified by the first stage as before and then be applied to the zener. This changing signal voltage at the collector of Q_1 will cause the zener diode's internal resistance to continually change, resulting in a change in reverse current through the zener but an almost constant voltage drop across the zener. Because the signal voltage at the collector of Q_1 will be developed across the zener diode and R_6, the voltage across the zener diode will remain almost constant and the changing signal voltage will be developed across R_6. The changing signal voltage developed across R_6 will be applied to the base of Q_2, controlling the second-stage dc amplifier and therefore causing additional gain. Zener coupling two dc amplifier stages in this way causes the overall voltage gain to be almost equal to the calculated voltage gain of stage-1 gain times stage-2 gain.

Complementary DC Amplifier Circuits

Figure 8-9(a) shows the basic **complementary dc amplifier,** which makes use of an NPN transistor and a PNP transistor. This circuit operates in almost the same way as a direct-coupled dc amplifier, except for the use of complementary or opposite bipolar transistor types. An applied input signal will be amplified by the first stage as before. To properly bias the second PNP stage, Q_2's emitter and collector leads are connected in reverse so that its emitter is connected to $+V_{CC}$ via R_5 and its collector lead is connected to ground via R_6. As we know, the base of a PNP transistor must be more positive than its collector, while the opposite is true for the NPN transistor, whose base must be more negative than its collector. Because the input signal is applied to the base of Q_2 and the output signal is taken from the collector, this second-stage dc amplifier is also a common–emitter amplifier. It is the voltage developed across R_3 (between Q_1's collector and $+V_{CC}$) that is applied to Q_2's base–emitter and controls Q_2. It is the voltage developed across R_6 (between Q_2's collector and ground) that is applied to the output and is used to control the following stage.

Complementary DC Amplifier

An amplifier in which NPN and PNP transistors are used in an alternating sequence.

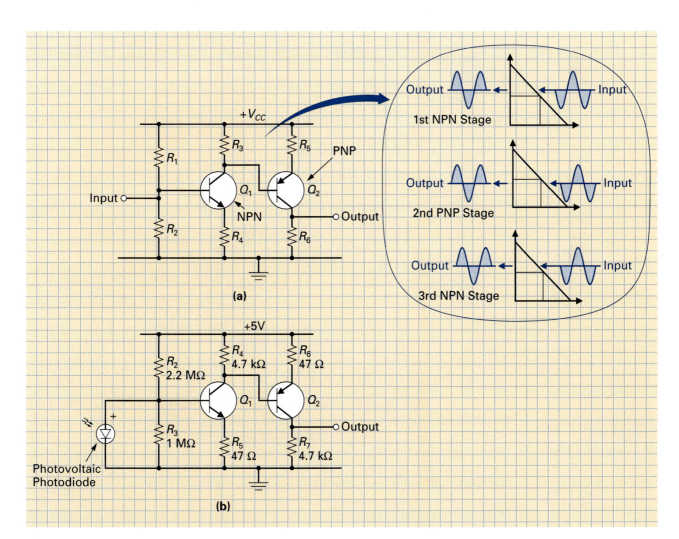

FIGURE 8-9 Complementary DC Amplifier Circuits. (a) Basic Circuit. (b) Application: Light-Sensing Amplifier.

As mentioned previously, with a standard direct-coupled dc amplifier, as the number of direct-coupled stages increases, the collector voltage of each stage becomes progressively larger. As a result, only a few stages can be cascaded before the collector bias voltage for the next stage becomes so large that the input signal drives the transistor into saturation and signal distortion occurs. A solution to this problem is to cascade complementary transistors (an NPN followed by a PNP, followed by an NPN, followed by a PNP, and so on). Because the collector of an NPN stage is connected to $+V_{CC}$ via a resistor, its output collector bias voltage will be more positive. On the other hand, because the collector of a PNP transistor is connected to ground via a resistor, its output collector bias voltage will be less positive. As can be seen in the inset in Figure 8-9(a), this complementary dc amplifier arrangement enables us to cascade several stages to achieve high signal gain without distortion because the Q point, which is set by the previous stage, continually alternates between being either more positive (slightly above the midpoint) or less positive (slightly below the midpoint).

Figure 8-9(b) shows a typical two-stage complementary dc amplifier circuit. This circuit is being used to amplify a small voltage input signal being generated by a photovoltaic cell. The voltage gain of each stage can be calculated with the previously discussed formula and is approximately equal to the ratio of the collector resistance to the emitter resistance.

$$A_V = \frac{R_C}{R_E}$$

■ EXAMPLE:

Calculate the approximate individual-stage gain and overall dc amplifier gain of the circuit in Figure 8-9(b).

■ Solution:

The gain of the first NPN stage will be approximately equal to the ratio of R_4 to R_5 (ignoring the input resistance of Q_2).

$$A_V = \frac{R_C}{R_E} = \frac{R_4}{R_5} = \frac{4.7 \text{ k}\Omega}{47 \text{ }\Omega} = 100$$

The gain of the second PNP stage will be equal to the ratio of R_7 to R_6.

$$A_V = \frac{R_C}{R_E} = \frac{R_6}{R_7} = \frac{4.7 \text{ k}\Omega}{47 \text{ }\Omega} = 100$$

The overall gain of the circuit is equal to the product of the first-stage gain and the second-stage gain.

$$A_{\text{Total}} = A_1 \times A_2 = 100 \times 100 = 10,000$$

Darlington Pair DC Amplifier Circuits

Figure 8-10(a) shows how the collectors of two transistors can be tied together and the emitter of one transistor can be direct coupled to the base of the other to form a very useful dc amplifier arrangement called a **darlington pair.** Named after its creator, this pair of bipolar transistors function as one and are packaged no differently than an ordinary transistor, as shown in the inset in Figure 8-10(a). Referring to the circuit in Figure 8-10(a), you can see that resistors R_1 and R_2 provide voltage-divider bias for the first transistor, Q_1, while resistor R_3 serves as a collector resistor for both transistors. An emitter resistor is generally not included in the circuit because it would reduce the amount of current gain (I_E and therefore I_C), which in turn would reduce voltage gain (V_{out} or V_{CE}). An applied input signal voltage will cause a variation in the base current of Q_1, a corresponding change in Q_1's emitter current, and therefore a variation in Q_2's base current and collector current. By controlling Q_1,

Darlington Pair

A dc amplifier arrangement wherein the collectors of two transistors are tied together, with the emitter of one transistor direct coupled to the base of the other.

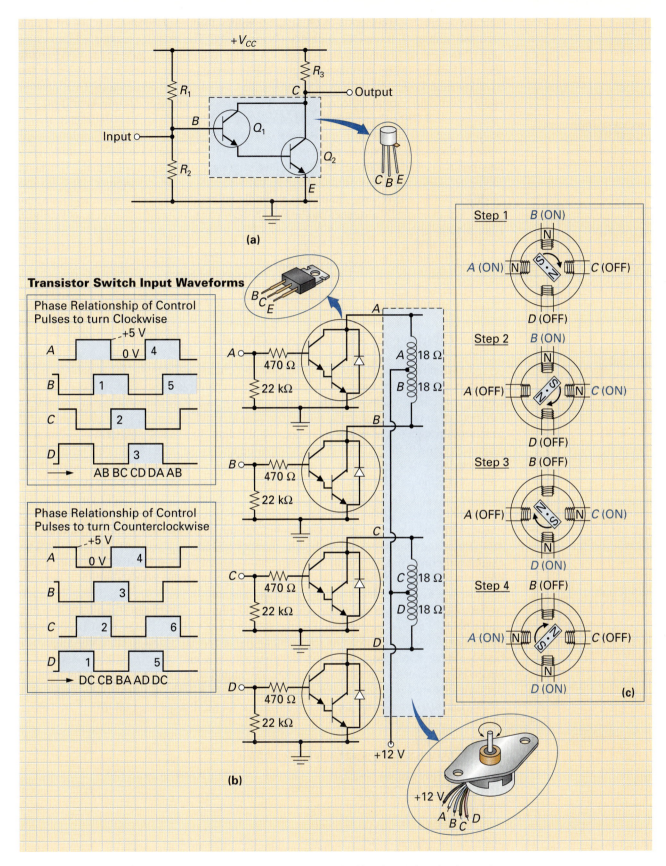

FIGURE 8-10 Darlington Pair DC Amplifier Circuits. (a) Basic Circuit. (b) Application: DC Stepper Motor Control. (c) Stepper Motor Clockwise Operation.

you can control the conduction of Q_2. In other words, turning Q_1 more ON will turn Q_2 more ON, and turning Q_1 more OFF will turn Q_2 more OFF. The overall common-emitter current gain of this circuit arrangement is effectively equal to the current gain of the first transistor (beta Q_1) multiplied by the current gain of the second transistor (beta Q_2). In practice, however, because the emitter current of Q_1 drives the base of Q_2, the base bias of Q_1 is generally made lower than normal to keep Q_1's emitter current from driving Q_2 into saturation.

Stepper Motor

A motor that rotates
in small angular steps.

Figure 8-10(b) shows how the high-current switching ability of a darlington pair can be made use of to control the operation of a **stepper motor.** The dc stepper motor is ideally suited for computer-controlled industrial applications because its rotor's position, speed, and direction can be controlled by a computer's two-state output control signals. The stepper motor derives its name from the fact that its rotor rotates in uniform angular steps. A variety of different types of stepper motors are available with a different number of steps per revolution. For example, if a stepper motor has 48 steps per revolution, each step will cause the rotor to turn 7.5°.

$$\text{Degrees per Step} = \frac{360°}{\text{Number of Steps}}$$

$$= \frac{360°}{48} = 7.5° \text{ per Step}$$

Table 8-1 lists some of the more frequently used steps-per-revolution sizes of stepper motors.

Let us now examine how the darlington pair circuit in Figure 8-10(b) operates. A pulse pattern is applied to the inputs of the four darlington pair transistors when the computer controlling this circuit wants to set the motor in motion. A HIGH to any of the inputs will turn ON both of the transistors in the darlington pair and switch ground from the emitter of the transistor through to the end of its connected winding. Explained another way, when a darlington is turned ON, its respective coil will be energized because the coil's center tap is connected to +12 V and its end is connected to ground via the darlington. The internal shunt-connected clipper diodes within the darlington pair transistors are used to protect the driver's circuit from the counter emf transient that is generated by the stator coils. The basic operation of the stepper motor is shown in Figure 8-10(c). For ease of explanation, only four stator poles and one rotor are shown, providing four steps. In reality, there would be many more stator and rotor poles, providing the number of steps required. The four-step clockwise excitation sequence is shown in the simplified diagram in Figure 8-10(c). In step 1, stator poles *A* and *B* are energized to produce two north poles so the rotor's south pole will align itself between stator poles *A* and *B*. In the second step, stator poles *B* and *C* are energized so the rotor's south pole will move clockwise and align itself between *B* and *C*. In step three, stator poles *C* and *D* are energized so the rotor's south pole will align itself between these two stator poles. In step four, stator poles *D* and *A* are energized so the rotor aligns itself between these two stators. The next stepping action would energize stator poles *A* and *B*, which puts us back to our original position. The signal sequence needed to control this stepper motor is shown in the timing diagrams in Figure 8-10(b). To drive the stepper motor clockwise, the stator coils are energized in the forward sequence *AB, BC, CD, DA, AB,*

TABLE 8-1   Typical Stepper Motor Sizes

STEPS PER REVOLUTION	ROTOR ANGLE MOVEMENT PER STEP
240	1.5°
180	2.0°
144	2.5°
72	5.0°
48	7.5°
24	15°
12	30°

and so on, which was the sequence shown in Figure 8-10(c). To drive the stepper motor counterclockwise, the stator coils are energized in the reverse sequence *DC, CB, BA, AD, DC,* and so on.

Differential DC Amplifier Circuits

Figure 8-11(a) illustrates a basic differential dc amplifier circuit. This circuit contains two transistors (Q_1 and Q_2), which are matched to have identical characteristics, each of which has its own collector resistor and voltage-divider bias. The distinguishing feature of this circuit is the sharing of an emitter resistor by both transistors. Most differential amplifiers are not constructed using the discrete or individual components used in Figure 8-11(a). Differential amplifiers are available in integrated circuit (IC) form, as shown in Figure 8-11(b). In this circuit, the voltage-divider bias resistors are not needed because the base–emitter junction of the transistor is forward biased by applying a negative emitter supply voltage ($-V_{EE}$). The symbol for this circuit is shown in Figure 8-11(c).

With two inputs and two outputs, the differential amplifier can have many different input/output combinations. Figures 8-11(d), (e), and (f) illustrate the more common differential amplifier circuit arrangements. Let us discuss each of these circuit arrangements in detail.

Single-Input, Single-Output Connection. In this arrangement, the input signal is applied to the base of Q_1 (V_{B_1}) and the output is taken from the collector of Q_2 (V_{C_2}), as shown in Figure 8-11(d). Transistor Q_1 acts as an emitter follower (common collector) and therefore amplifies the signal input, producing an output emitter voltage (V_E) that will follow the input signal applied to the base. This signal voltage developed across the emitter resistor will be applied to the emitter of Q_2, which will function as a common–base amplifier stage, developing an output voltage at its collector (V_{C_2}). The amount of voltage gain achieved by this circuit is equal to the voltage gain of the common base amplifier stage only (Q_2) because Q_1 will have no voltage gain, only current gain and impedance isolation.

Single-Input, Differential-Output Connection. In this arrangement, the input signal is applied to the base of Q_1 (V_{B_1}) and two outputs are available at each of the transistor's collectors (V_{C_1} and V_{C_2}), as shown in Figure 8-11(e). Transistor Q_1 amplifies the input signal and produces an output at its collector (V_{C_1}) that is out of phase with the input signal because Q_1 is operating as a common–emitter amplifier stage. The changes in Q_1's emitter current will be developed across the emitter resistor R_2 (V_E), and this voltage variation will be in phase with the input signal at the base of Q_1 (V_{B_1}). Transistor Q_2 amplifies the emitter voltage signal from Q_1, acting as a common–base amplifier stage, so there is no inversion between the emitter input of Q_2 and the collector output of Q_2 (V_{C_2}). The two separate outputs at the collector of Q_1 relative to ground, and at the collector of Q_2 relative to ground, are 180° out of phase. In most applications, the differential amplifier would not be used to provide two outputs with respect to ground in this way. Instead, the output signal voltage would be equal to the voltage difference between the collector of Q_1 and the collector of Q_2 (V_{out}), as shown in the single-input differential-output circuit in Figure 8-11(e). Because the voltages at the collectors of Q_1 and Q_2 are 180° out of phase, a very large voltage difference— and therefore voltage gain—will exist between these two points.

Differential-Input, Differential-Output Connection. In this arrangement, the differential amplifier is used to its full advantage, and its action demonstrates how the differential, or difference, amplifier derived its name. Referring to the differential-input, differential-output circuit in Figure 8-11(f), you can see that two input signals are applied to the base of Q_1 and Q_2. The output is equal to the voltage difference between the collector of Q_1 and the collector of Q_2.

When both input signals are in phase with one another, they are referred to as **common-mode input signals,** as seen in the waveforms in Figure 8-11(f). If both input signal voltages are of the same amplitude and phase ($V_{B_1} = V_{B_2}$), the difference between these two inputs is zero, and this is the voltage that appears at the output ($V_{out} = 0\,V$). This occurs because the two identical input signal voltages turn both Q_1 and Q_2 ON by the same amount. Therefore, their

Common-Mode Input Signals

Two input signals in phase with one another.

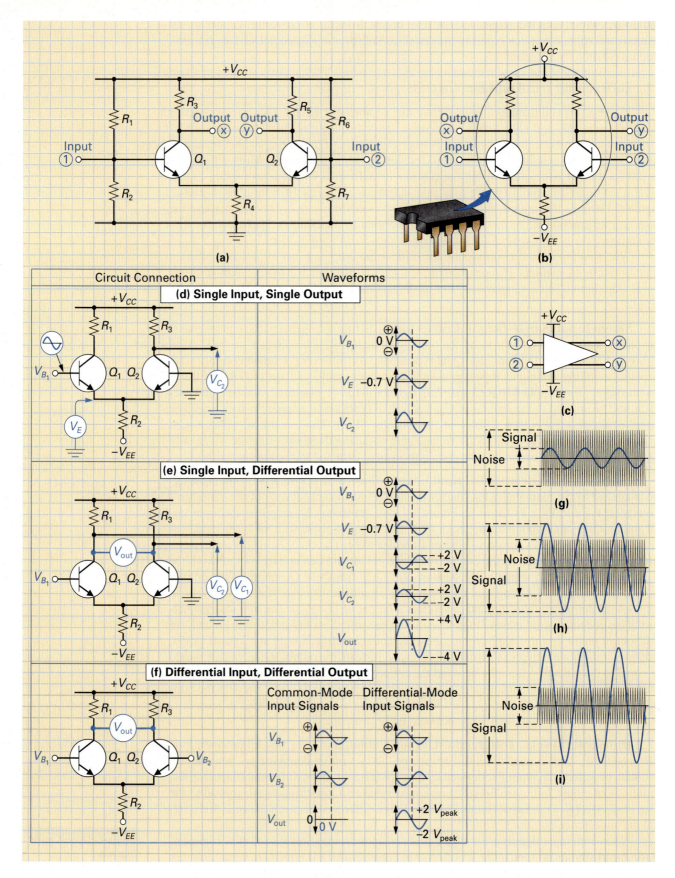

FIGURE 8-11 Differential Amplifier Circuits. **(a)** Basic Discrete Circuit. **(b)** Basic IC Circuit. **(c)** Symbol. **(d), (e), (f)** Modes of Operation. **(g), (h), (i)** Noise.

emitter currents, collector currents, and collector output voltages are the same ($V_{C_1} = V_{C_2}$). If the collector voltage at Q_1 is equal to the collector voltage of Q_2, then the output voltage, which is equal to the potential difference between V_{C_1} and V_{C_2}, will be 0 V.

If both of the input signals are out of phase with one another, they are referred to as **differential-mode input signals,** as seen in the waveforms in Figure 8-11(f). If both input signal voltages are out of phase with one another, the output voltage will be equal to the difference between the two input signal voltages. Let us see why this happens by examining each transistor's operation separately.

Differential-Mode Input Signals

Two input signals out of phase with one another.

When V_{B_1} swings positive, Q_1 will turn more ON. Due to Q_1's common–emitter action, its collector voltage will swing negative. The emitter of Q_1, on the other hand, will swing positive, and this will be applied to Q_2's emitter, which will turn Q_2 more OFF. At the same time, Q_2's base is being driven negative by its input signal and so its collector voltage will swing positive and its emitter voltage will swing negative. This negative voltage at the emitter of Q_2 will be applied to the emitter of Q_1 and turn Q_1 more ON. In summary, the positive input voltage at the base of Q_1 and the negative voltage at the emitter of Q_1 from Q_2 will work together to turn Q_1 more ON. On the other hand, the negative input voltage at the base of Q_2 and the positive voltage at the emitter of Q_2 from Q_1 will work together to turn Q_2 more OFF. The result is that V_{B_1} and V_{B_2} work together to produce two amplified output signals that are 180° out of phase. This difference between the two collector signal voltages means that there will be a large difference in potential across the output voltage terminals.

Common-Mode Rejection. This ability of the differential amplifier to reject common-mode input signals is its key advantage and accounts for why it is perhaps the most important of all the dc amplifiers. The question you may be asking at this stage is: What are common-mode input signals, and why should we want to reject them? The answer is that temperature changes and noise are common-mode input signals, and they are unwanted signals.

Removing Amplifier Temperature Instability. As you know, temperature variations within electronic equipment affect the operation of semiconductor materials and therefore semiconductor devices. These temperature variations can cause the dc output voltage of the first stage to drift away from its normal Q point, or no-input-signal bias level. The second-stage amplifier will amplify this voltage change in the same way as it would amplify any dc input signal and so would all of the following amplifier stages. The increase or decrease in the normal Q-point bias for all of the amplifier stages will get progressively worse due to this thermal instability. The final stage may have a Q point that is so far off its mid position that an input signal may drive it into saturation or cutoff, causing signal distortion. With the difference amplifier, the voltages between the collector of Q_1 and ground and the collector of Q_2 and ground will tend to increase and decrease by exactly the same amount as temperature changes. This common-mode change that occurs within every amplifier due to temperature changes will not appear at the output of the differential amplifier due to its **common-mode rejection.**

Common-Mode Rejection

Rejection, by a differential amplifier, of the common-mode change that occurs within every amplifier due to temperature changes.

Removing Amplifier Noise. The second common-mode input signal that the differential amplifier removes is noise. It is often necessary to amplify low-level signals from low-sensitivity sources such as microphones, light detectors, and other transducers. High-gain amplifiers are used to increase the amplitude of the input signal until it is large enough to drive or control a load such as a loudspeaker. The 60-Hz ac power line, or any other electrical variation, can induce a noise signal along with the input signal at the input of this high-gain multiple-stage amplifier. If this noise signal is larger than the desired signal, as shown in Figure 8-11(g), the signal will never be able to be extracted from the noise. Under normal conditions, the noise from a microphone should be at least 30 dB below the signal level for voice communication, as shown in Figure 8-11(h), and 60 dB below the signal level for quality music recording and broadcasting, as shown in Figure 8-11(i). These noise signals will be induced at all points in the circuit and be identical in amplitude and phase. The differential amplifier will block these unwanted signals because they will be present at both inputs of the differential amplifier (noise will be a common-mode input). A true input signal, on the other hand, will appear at the two inputs of the differential amplifier as a differential input signal and be amplified.

To summarize, the desired signal that appears on only one input (single-input mode) or both inputs (differential-input mode) of a differential amplifier will be amplified and applied to the output. On the other hand, unwanted signals caused by temperature variations or noise will appear as common-mode input signals and be rejected. A differential amplifier's ability to provide a high **differential gain (A_{VD})** and a low **common-mode gain (A_{CM})** is a measure of a differential amplifier's performance. This ratio is called the **common-mode rejection ratio (CMRR)** and is calculated with the following formula:

Differential Gain (A_{VD})

The amplification of differential-mode input.

Common-Mode Gain (A_{CM})

The amplification of common-mode input. This value is typically less than 1.

Common-Mode Rejection Ratio (CMRR)

The ratio of an operational amplifier's differential gain to the common-mode gain.

$$\text{CMRR} = \frac{A_{VD}}{A_{CM}}$$

Looking at this formula, you can see that the higher the A_{VD} (differential gain), or the smaller the A_{CM} (common-mode gain), the higher the CMRR value, and therefore the better the differential amplifier. This ratio can also be expressed in dBs by using the following formula:

$$\text{CMRR} = 20 \times \log \frac{A_{VD}}{A_{CM}}$$

■   **EXAMPLE:**

If a differential amplifier has a differential gain of 5,000 and a common-mode gain of 0.5, what is the amplifier's CMRR? Express the answer in standard gain and dBs.

■   *Solution:*

$$\text{CMRR} = \frac{A_{VD}}{A_{CM}} = \frac{5,000}{0.5} = 10,000$$

$$\text{CMRR} = 20 \times \log = \frac{A_{VD}}{A_{CM}} = 20 \times \log 10,000 = 20 \times 4 = 80 \text{ dB}$$

A CMRR of 10,000 or 80 dB means that the differential desired input signals will be amplified 10,000 times more than the unwanted common-mode input signals.

The versatility, temperature stability, and noise rejection of the differential amplifier make it an ideal candidate for a variety of applications. One such application is the "operational amplifier," an IC amplifier that has almost completely replaced the discrete, or individual-component, amplifier circuits. The operational amplifier, which is covered in Chapter 12, contains several differential-amplifier stages, and so we will be discussing the differential amplifier's characteristics in more detail along with this important application.

SELF-TEST EVALUATION POINT FOR SECTION 8-2

Use the following questions to test your understanding of Section 8–2.

1. Direct-current amplifiers are generally used to boost a weak signal from a transducer. (true/false)
2. How does the inclusion of an emitter resistor improve the amplifier's temperature stability?
3. What is the advantage of the complementary dc amplifier?
4. The gain of a multistage amplifier is always slightly less than the _____ of all the stage gains.
5. The darlington pair provides a very high _____ gain.
6. The advantage of the differential dc amplifier is that it provides a very _____ differential gain and a very _____ common-mode gain.

Audio is a Latin word meaning "I hear," which is why the term *audio-frequency range* includes all of the frequencies that the human voice can produce and the human ear can respond to. This audio-frequency range is from about 20 Hz to 20,000 Hz. An **audio-frequency (AF) amplifier** is a circuit containing one or more amplifier stages designed to amplify an audio-frequency signal. These audio amplifiers are used within any electronic system that processes, transmits, or receives audio signals, such as radio transmitters and receivers, television transmitters and receivers, telephone communication systems, music systems, multimedia computer systems, and so on. The two basic types of audio amplifiers are the AF voltage amplifier, which is generally class-A biased, and AF power amplifier, which is generally class-B or class-AB biased. Let us begin by discussing the appearance, operation, and characteristics of the AF voltage amplifier.

Audio-Frequency (AF) Amplifier
Amplifiers for ac signals between 20 Hz and 20,000 Hz.

8-3-1 *Audio-Frequency (AF) Voltage Amplifiers*

A voltage amplifier is used to boost the voltage level of a weak input signal so that it is large enough in amplitude to control a power amplifier. When used in this manner, voltage amplifiers are often referred to as **preamplifiers** or preamps because they precede the main power amplifier, which is used to boost signal current so that the final output can control a low-impedance load, such as a loudspeaker.

Preamplifiers
Amplifiers that precede the main power amplifier and boost signal current so that the final output can control a low-impedance load.

Single-Stage AF Voltage-Amplifier Circuits

Figure 8-12(a) shows a basic audio-frequency common-emitter voltage-amplifier circuit. The transistor Q_1 is class-A biased by the voltage-divider network R_1 and R_2. The input coupling capacitor (C_1) prevents Q_1's dc base bias voltage from being applied back to the source, while the output coupling capacitor (C_2) prevents Q_1's dc collector bias voltage from being applied to the load. Collector resistor R_3 operates in conjunction with the transistor to develop an output signal voltage, and resistor R_4 provides emitter feedback for temperature stability. Capacitor C_E has been added to remedy a problem that occurs when an amplifier is amplifying an ac signal. To explain this in detail, a signal variation at the base causes a variation in base current (I_B) and a corresponding change in output current (I_C). These variations in transistor current produce a varying voltage drop across R_3, but because R_4 is now included in the circuit for temperature stability, a similar voltage swing is also developed across R_4. Therefore, as the input signal swings positive, the transistor's V_{BE} bias voltage is increased, causing an increase in I_B and I_E and therefore an increase in the voltage drop across R_4. This increase in the voltage developed across R_4 will increase the positive potential at the transistor's emitter and therefore decrease or degenerate the original forward bias of the transistor's base–emitter (PN) junction. This **degenerative feedback** dramatically reduces the gain of any ac signal amplifier but can be remedied by including a capacitor in parallel with R_4, as shown in Figure 8-12(a). Because a capacitor basically operates as a closed switch to ac and an open switch to dc, the ac signal appearing at the emitter will now be shunted around R_4, preventing the degenerative feedback. The capacitor will not interfere with the dc feedback signal developed across R_4 to compensate for changes in temperature. Although the temperature stability feedback provided by R_4 varies up and down and is also degenerative, its changes occur at a much slower rate and can be considered dc and will suffer no interference from the capacitor.

Degenerative Feedback
Also called negative feedback, degenerative feedback is a method for coupling a portion of the output of a circuit back so that it is 180° out of phase with the input signal. This reduces amplification and stabilizes the circuit to reduce distortion and noise.

Like the dc amplifier, the voltage gain is equal to the ratio of the transistor's total collector resistance to emitter resistance. Because the audio amplifier is dealing with ac signals, the voltage gain of an audio-frequency amplifier will be equal to the ratio of total ac collector resistance to total ac emitter resistance.

$$A_{V(AC)} = \frac{R_{C(AC)}}{R_{E(AC)}}$$

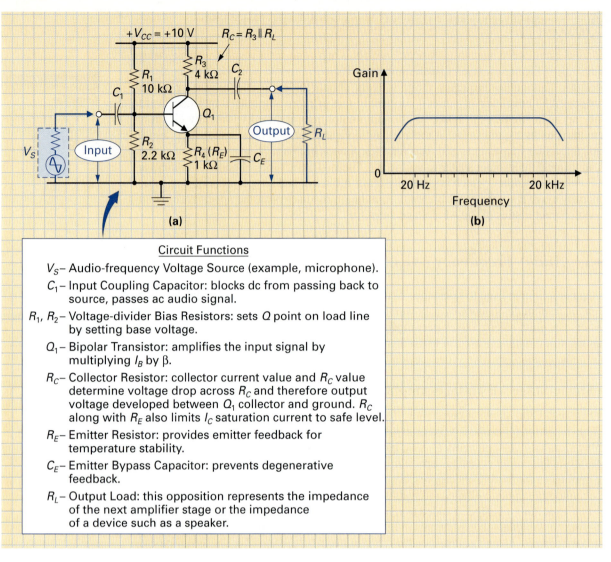

FIGURE 8-12 **Single-Stage Audio-Frequency (AF) Voltage Amplifier. (a) Basic Circuit. (b) Frequency Response.**

Because the collector current from the transistor has two paths—one through the collector resistor R_3 and the second through the load resistance R_L—R_3 and R_L are effectively in parallel with one another. Therefore, the total ac collector resistance is equal to

$$R_{C(AC)} = \frac{R_3 \times R_L}{R_3 + R_L}$$

The ac emitter resistance of a small-signal voltage amplifier will be a lot lower than the value of the emitter resistor due to the bypass capacitor C_E and can be calculated with the formula

$$R_{E(AC)} = \frac{25 \text{ mV}}{I_E}$$

Let us now see how we can use these formulas to calculate the voltage gain of a typical AF voltage amplifier.

EXAMPLE:

Referring to the class-A-biased AF voltage amplifier in Figure 8-12(a), calculate
1. its voltage gain
2. the signal voltage output if the input was 125 μV

Solution:

1. To begin with we will have to use our voltage-divider bias formulas to determine V_B, V_E, and I_E.

$$V_B = \frac{R_2}{R_1 + R_2} \times V_{CC} = \frac{2.2 \text{ k}\Omega}{10 \text{ k}\Omega + 2.2 \text{ k}\Omega} \times 10 \text{ V} = 1.8 \text{ V}$$

$$V_E = V_B - 0.7 \text{ V} = 1.8 \text{ V} - 0.7 \text{ V} = 1.1 \text{ V}$$

$$I_E = \frac{V_E}{R_E} = \frac{1.1 \text{ V}}{1 \text{ k}\Omega} = 1.1 \text{ mA}$$

Now that V_B, V_E, and I_E are known, we can calculate ac collector resistance, ac emitter resistance, and, finally, ac voltage gain.

$$R_{C(AC)} = \frac{R_3 \times R_L}{R_3 + R_L} = \frac{4 \text{ k}\Omega \times 15 \text{ k}\Omega}{4 \text{ k}\Omega + 15\text{k}\Omega} = \frac{60 \text{ M}\Omega}{19 \text{ k}\Omega} = 3.16 \text{ k}\Omega$$

$$R_{E(AC)} = \frac{25 \text{ mV}}{I_E} = \frac{25 \text{ mV}}{1.1 \text{ mA}} = 22.73 \text{ }\Omega$$

$$A_{V(AC)} = \frac{R_{C(AC)}}{R_{E(AC)}} = \frac{3.16 \text{ k}\Omega}{22.73 \text{ }\Omega} = 139$$

2. The ac signal voltage at the output is 139 times greater than the input voltage. For an input signal voltage that is 125 μV, the output voltage will be

$$V_{out(AC)} = V_{in(AC)} \times A_{V(AC)} = 125 \text{ μV} \times 139 = 17.4 \text{ mV}$$

EXAMPLE:

Calculate the efficiency of an AF voltage amplifier if it has the following circuit conditions:

$$V_{out} = 2 \text{ V}$$
$$R_L = 1 \text{ k}\Omega$$
$$V_{CC} = 12 \text{ V}$$
$$I_{CC} = 1 \text{ mA}$$

Solution:

As discussed previously, the efficiency of an amplifier is equal to

$$\eta = \frac{\text{Power Delivered to Load by Amplifier } (P_L)}{\text{DC Power Consumed by Amplifier } (P_S)} \times 100\%$$

Because the voltage supplied to the 1-kΩ load is 2 V, the power delivered to the load will be

$$P_L = \frac{V_L^2}{R_L} = \frac{2 \text{ V}^2}{1 \text{ k}\Omega} = 4 \text{ mW}$$

The dc power supplied to the amplifier is equal to

$$P_S = V_{CC} \times I_{CC} = 12 \text{ V} \times 1 \text{ mA} = 12 \text{ mW}$$

If the power delivered to the load is equal to 4 mW, and the power supplied to the amplifier by the dc power supply is equal to 12 mW, the class-A amplifier's efficiency in Figure 8-12(a) will be

$$\eta = \frac{P_L}{P_S} \times 100\% = \frac{4 \text{ mW}}{12 \text{ mW}} \times 100\% = 33.3\%$$

The single-stage class-A AF voltage amplifier shown in Figure 8-12(a) will provide a large voltage gain for a wide range of input signal frequencies. The frequency response of this audio frequency amplifier is shown in Figure 8-12(b). Looking at this curve, you can see that the amplifier cannot be used to amplify dc, or very low-frequency ac, signals because of the opposition or reactance of the input and output coupling capacitors. To amplify dc, or very low-frequency ac, signals, a dc amplifier would have to be used, which, as you know, does not include any dc blocking components in the signal path, such as capacitors and transformers. As the ac input signal frequency increases, its cycle time or period decreases, and eventually this cycle time is faster than the transit time needed for a charge carrier to pass through the transistor. At this time, the gain of the transistor will fall to almost zero, as shown at the upper end of the frequency response curve in Figure 8-12(b), because the transistor's internal currents do not have enough time to respond to the high-frequency input signal changes.

Multiple-Stage AF Voltage-Amplifier Circuits

If a single-stage AF amplifier cannot provide the amount of gain needed, a multiple-stage AF circuit can be used that includes two or more cascaded amplifier stages. Figure 8-13(a) shows a two-stage **RC-coupled AF voltage amplifier.** A capacitor is commonly used to connect one amplifier stage to another. When used in this way, the capacitor is called a **coupling capacitor,** which means that it connects two circuits together electronically so that the signal from one circuit can pass to the next circuit with no distortion. You may remember from your introductory dc/ac electronics that the reactance or opposition of a capacitor is inversely proportional to frequency.

$$X_C = \frac{1}{2\pi fC}$$

The coupling capacitor actually performs two functions:

1. Due to its low reactance to ac signals, it will pass the audio-frequency information signal from the first stage to the second stage with no distortion, and

2. Due to its high reactance to dc, the coupling capacitor will isolate the dc bias voltages of one stage from the dc bias voltages of another.

Referring to Figure 8-13(a), you can see that coupling capacitor C_2 performs these two functions of coupling the ac signal from the collector of Q_1 to the base of Q_2, but isolating the dc collector bias voltage of Q_1 (+5.6V) from the dc base bias of Q_2 (+1.8V).

The disadvantage of the RC-coupled amplifier is that maximum power is not transferred between one stage and the next. In the RC-coupled amplifier's first stage, the collector current has two paths: one through the collector resistor R_3 and the second through the low-input impedance of the second amplifier stage. The low-input impedance of the second stage and the first stage's collector resistor R_3 are therefore effectively in parallel with one another. This means that the overall load resistance of the first stage is reduced, and the voltage developed at the output and applied to the next stage is also reduced. Explained another way, the low-input impedance of the second stage (caused by the parallel connection of R_3 and the input impedance of Q_2) will "load" the first stage, causing a high-output current ($I\uparrow$) due to the low-equivalent resistance ($R\downarrow$). A small voltage will be developed across the output of the first stage ($V_{out}\downarrow$). With the output voltage of the first stage low, maximum power

RC-Coupled AF Voltage Amplifier

A multiple-stage AF voltage amplifier coupled by a capacitor.

Coupling Capacitor

A capacitor used to connect one amplifier stage to another.

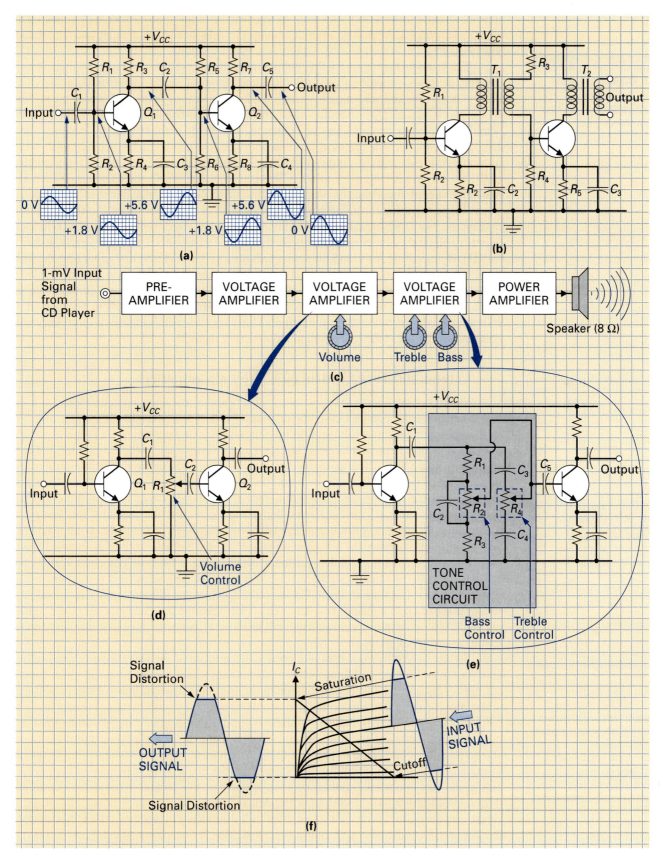

FIGURE 8-13 Multiple-Stage AF Voltage Amplifiers. (a) *RC* Coupled. (b) Transformer Coupled. (c) Application: CD Music Amplifier System. (d) Volume Control. (e) Tone Control. (f) Class-A Distortion.

is not being transferred from one stage to the next. To achieve an overall high voltage gain, several *RC*-coupled AF voltage-amplifier stages may be needed to compensate for the individual-stage gains, which will be slightly less than expected.

Transformer-Coupled AF Voltage Amplifier

A multiple-stage AF voltage amplifier coupled by a transformer.

The **transformer-coupled AF voltage amplifier** is shown in Figure 8-13(b). As you can see, the transformer T_1 is used to couple the output of the first stage to the input of the second stage. The ac audio signal is amplified by Q_1 and then developed across the primary of T_1, which acts as Q_1's collector load. Through electromagnetic induction, the ac audio signal is coupled to the secondary of T_1, where it is applied to the base of Q_2, causing a variation in the base bias and therefore base current. Like the first stage, the primary of T_2 serves as a collector load, and the output of this amplifier stage is developed across the secondary of T_2. As well as providing dc isolation between amplifier stages, the transformers can also match the output impedance of one amplifier stage to the input impedance of the following stage. Using the different inductive reactance values of the transformer's primary and secondary, the low-input impedance of the second amplifier stage (typically 4 kΩ to 6 kΩ) can appear to equal the higher output impedance of the first amplifier stage (typically 40 kΩ to 60 kΩ). By matching input and output impedances in this way, the low-input impedance of the second amplifier stage will not load down the first amplifier stage, and so a high value of signal current and signal voltage (and therefore signal power) will be applied to the second stage. Transformer coupling is often used in audio-frequency amplifiers because of its efficient transfer of power between stages. However, the transformer is a more expensive coupling method than the less efficient *RC* coupling method.

Volume and Tone Control of AF Voltage Amplifiers

Figure 8-13(c) shows the basic block diagram of a music system's amplifier. A 1-mV music signal from a compact disc player is applied first to a preamplifier, which would be similar to the single-stage voltage-amplifier circuit discussed previously in Figure 8-12. The output signal of this preamplifier is then applied to a two-stage voltage amplifier, which is shown in Figure 8-13(d). Unlike the previously discussed multiple-stage amplifiers, the overall gain of this amplifier can be controlled by the variable resistor R_1. This variable resistor will control the volume level or loudness of the music system by controlling the voltage developed across R_{C_1} and therefore the voltage applied to the second stage. For example, because C_1 has a very low capacitive reactance, nearly all of the signal voltage from Q_1 will be developed across the variable resistor R_1. If the volume control is turned "down" (wiper is moved down), a smaller resistance exists between the base and emitter of the following stage (Q_2). A smaller signal voltage will be developed across this resistance and applied to the base of Q_2 as an input. On the other hand, as the volume control is turned "up" (wiper is moved up), a greater resistance exists between the base and emitter of the following stage "Q_2." A greater signal voltage will be developed across this resistance and applied to the base of Q_2 as an input.

Treble

High audio frequencies normally handled by a tweeter in a sound system.

Bass

Low audio frequencies normally handled by a woofer in a sound system.

Tone

A term describing both the bass and treble of a sound signal.

Figure 8-13(e) shows how a circuit can be inserted between a two-stage voltage amplifier to control the **treble** and **bass** of the music, which combined is called the **tone** of the music. Because people hear the same music in different ways, music systems will generally provide these controls so that the amplifier's sensitivity to certain frequencies is either emphasized or deemphasized. If the bass and treble controls are placed in their mid position, the frequency response of an audio-frequency amplifier is generally flat, as was shown previously in Figure 8-12(b). If these controls are adjusted, however, the frequency response, or audio-frequency amplifier gain at certain frequencies, will be modified. For example, the bass control will either increase or decrease the gain (or response) of the audio-frequency amplifier to low-end bass frequencies, while the treble control will either increase or decrease the response of the audio-frequency amplifier to high-end treble frequencies. Referring to the circuit in Figure 8-13(e), let us now examine how this overall music signal tone control is achieved.

The large-value coupling capacitors C_1 and C_5 will have a very low reactance, and therefore the audio frequency signal will pass through these capacitors with very little opposition. Capacitor C_2 will shunt the high audio frequencies around the bass control resis-

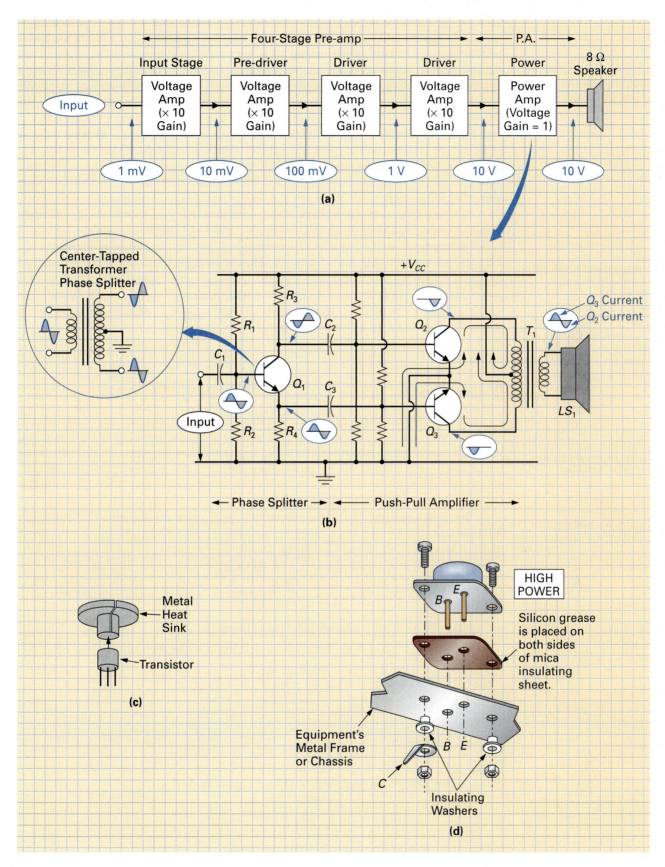

FIGURE 8-14   Basic Audio-Frequency (AF) Power Amplifiers. (a) Music Amplifier Block Diagram. (b) A Push-Pull Power Amplifier with a Transistor Phase-Splitter Circuit. (c) Low-Power Heat Sink. (d) High-Power Heat Sink.

tor R_2. Low audio frequencies (bass frequencies), however, will be developed across the voltage divider made up of R_1, R_2, and R_3. Moving the wiper of R_2 up will allow all of the bass frequencies to pass to the second stage with only little opposition, whereas moving the wiper of R_2 down will force the bass frequencies to pass through a greater resistance and therefore decrease their amplitude before they are passed to the following stage. With the treble control, capacitors C_3 and C_4 will have a very low reactance to high (treble) audio frequencies and a high reactance to the low (bass) audio frequencies. As a result, the high treble frequencies will be developed across the variable resistor R_4. Moving the wiper of R_4 up will allow all of the treble frequencies to pass to the second stage with only little opposition, whereas moving the wiper of R_4 down will force the treble frequencies to pass through a greater resistance and therefore decrease their amplitude before they are passed to the following stage.

A typical large amplifier system, like the one seen in Figure 8-13(c), may have several amplifier stages. The volume control and input signal level should be carefully adjusted so as not to heavily drive the amplifiers. For example, if the input signal level or volume is set too high, the signal will try to drive an amplifier beyond the limits of the power supply, as shown in Figure 8-13(f). This **class-A signal distortion** causes the top and bottom of the audio signal to be clipped when the signal amplitude reaches $+V_{CC}$ (transistor saturates) and 0 V (transistor is cut off). In practice, signal swings must be limited to about 80% of the power supply voltage for minimal distortion.

8-3-2 *Audio-Frequency (AF) Power Amplifiers*

Figure 8-14(a) repeats the simple block diagram of an audio-frequency music amplifier. Up until this point, we have only discussed the voltage-amplifier stages, which are used to increase the ac audio signal input of 1 mV to a large enough voltage to control the final-stage power amplifier. Although the **AF power amplifier** does not increase the signal voltage, it does supply the high output signal current (and therefore power) that is needed to operate a low-impedance load, such as a loudspeaker.

The power delivered to a load can be calculated using the standard V^2/R formula, as follows:

$$P_{RL} = \frac{V_{RL(\text{rms})}^2}{R_L}$$

P_{RL} = The ac power supplied to the load

V_{RL} = The rms voltage developed across the load

R_L = The load resistance

■ **EXAMPLE:**

Using the ac voltmeter readings given in Figure 8-14(a), calculate the power delivered by the audio-frequency amplifier to the 8-Ω loudspeaker load.

■ *Solution:*

Because a voltmeter is calibrated to display ac volts in rms, the power amplifier output voltage of 10 V can be entered directly into the formula.

$$P_{RL} = \frac{V_{RL(\text{rms})}^2}{R_L} = \frac{(10 \text{ V})^2}{8 \ \Omega} = 12.5 \text{ watts}$$

If a 4-Ω loudspeaker were connected to the output of the amplifier in Figure 8-14(a), instead of the 8-Ω loudspeaker, the output power delivered should be doubled.

$$P_{RL} = \frac{V_{RL(rms)}^2}{R_L} = \frac{(10 \text{ V})^2}{4 \ \Omega} = 25 \text{ watts}$$

However, the output power delivered could only be doubled if the amplifier circuit could supply the extra current without the lower-resistance load pulling down (or loading) the signal voltage. In reality, the power delivered would be less than double because the increase in circuit current will cause an increase in the heat power generated by the circuit and therefore a loss in the signal power delivered to the load. Most amplifiers are rated for a specific load resistance, and you should always check to see that the load is within specification. For example, most audio-frequency music system amplifiers are designed to drive loudspeaker loads of between 8 Ω to 15 Ω. Using a load resistance outside of the rated range can cause the amplifier to overdrive the speaker or the speaker to overload the amplifier.

Push-Pull AF Power-Amplifier Circuits

Figure 8-14(b) shows how two transistors (Q_2 and Q_3) can be connected to form a special type of power-amplifier circuit called a **push-pull power amplifier.** Before we discuss these two transistors, we will first need to examine the transistor that is driving the push-pull power amplifier, Q_1. The class-A biased transistor Q_1 operates as a **phase splitter,** providing two signal outputs: an inverted output from the collector that is 180° out of phase with the base input signal and an emitter signal that is in phase with the base input signal. The resistance values of R_3 and R_4 are chosen to compensate for the transistor's different emitter and collector characteristics to produce two **balanced output signal voltages** that are equal in amplitude but opposite in phase. As you can see in the inset in Figure 8-14(b), a center-tapped transformer could also be used to supply this two-phase output signal. However, a transistor phase splitter has a wider frequency response than a transformer. The complementary output signals from the phase splitter drive the inputs of the two class-B biased push-pull power transistors Q_2 and Q_3. Biased in class B means that each transistor will conduct for only half of the input signal cycle, as shown in the waveforms in Figure 8-14(b). To explain how and when each of the power transistors will conduct, let us examine each half-cycle of the input signal in more detail.

> **Push-Pull Power Amplifier**
>
> A balanced amplifier that uses two similar equivalent amplifying transistors working in phase opposition.
>
> **Phase Splitter**
>
> A circuit that takes a single input signal and produces two output signals that are 180° apart in time.
>
> **Balanced Output Signal Voltages**
>
> Signal voltages that are equal in amplitude but opposite in phase.

1. When the input signal swings positive, the emitter of Q_1 also swings positive, while its collector swings negative. Because both Q_2 and Q_3 are NPN transistors with their emitters at 0 V, a positive input voltage is needed to turn ON the transistor. Transistor Q_3 will conduct during this alternation of the input signal. Current will pass through Q_3's emitter-to-collector, through the lower half of the output transformer, and then to $+V_{CC}$, which is at the transformer's center tap. Transistor Q_2 remains cut OFF during this half of the input cycle.

2. When the input signal swings negative, the emitter of Q_1 also swings negative, while its collector swings positive. The positive signal voltage at the base of Q_2 will turn Q_2 ON, and current will pass through Q_2's emitter-to-collector, through the upper half of the output transformer, and then to $+V_{CC}$, which is at the transformer's center tap. Transistor Q_3 remains cut OFF during this half of the input cycle.

When Q_2 conducts, current passes through the upper half of the output transformer (which acts as Q_2's collector load) and a magnetic field is generated and coupled to the secondary. When Q_3 conducts, current passes through the lower half of the output transformer (which acts as Q_3's collector load) and a magnetic field of the opposite polarity will build up and be coupled to the secondary. Therefore, although only half of the cycle appears at the output of each transistor, the complete cycle is reconstructed at the secondary of the output transformer T_1, so a complete high-power ac audio-frequency signal drives the loudspeaker. The name "push-pull" is derived from this action because one transistor is responsible for driving current through the load in one direction (pushing), while the other transistor is responsible for driving current through the load in the opposite direction (pulling).

■ **EXAMPLE:**

If an oscilloscope were connected across the secondary of the output transformer T_1 and the ac audio-frequency signal applied to the 8-Ω speaker had a peak voltage of 25 V, what would be the power delivered to the speaker?

■ *Solution:*

To calculate the power delivered to the speaker, we will have to convert the peak voltage reading taken from the oscilloscope to an rms value.

$$\text{rms} = V_p \times 0.707 = 25 \text{ V} \times 0.707 = 17.7 \text{ V}$$

$$P_{RL} = \frac{V_{RL(\text{rms})}^2}{R_L} = \frac{(1.77 \text{ V})^2}{8 \text{ }\Omega} = 39.2 \text{ W}$$

Matched Transistors

Two output transistors that have the same part number and therefore the same operating characteristics.

To ensure that the two output half-cycles from the push-pull amplifier are identical, the two power transistors should be **matched transistors.** This means that the two output transistors must have the same part number and the same operating characteristics. For example, both Q_2 and Q_3 in Figure 8-14(b) should be 2N3902s, which are typical NPN silicon power transistors. If one of the transistors in a class-B amplifier fails, both transistors are normally replaced because a failure of one transistor generally damages or changes the characteristics of the other transistor.

Class-B Power-Amplifier Efficiency

A transistor's performance, or operating characteristics, is closely related to the temperature conditions. With power transistors, you have seen that the object is not to increase the voltage output, because most power amplifiers have a voltage gain of 1, which means that the output, voltage equals input voltage. The primary function of a power amplifier is to boost the output signal current so that the signal can operate a low-impedance (high-power) load. However, high current always goes hand-in-hand with high heat ($P = I^2R$). A class-B-operated transistor is much more efficient than a class-A power amplifier because it is not generating heat during the time intervals when it is cut off. This rest period between heavy work loads means that the class-B amplifier will work a lot more efficiently than a class-A power amplifier that is ON, or working, all of the time.

■ **EXAMPLE:**

Calculate the efficiency of an AF push-pull power amplifier under the following circuit conditions:

$$V_{\text{out}} = 2 \text{ V}$$
$$R_L = 16 \text{ }\Omega$$
$$V_{CC} = 20 \text{ V}$$
$$I_{CC} = 16 \text{ mA}$$

■ *Solution:*

As discussed previously, the efficiency of an amplifier is equal to

$$\eta = \frac{\text{Power Delivered to Load by Amplifier } (P_L)}{\text{DC Power Consumed by Amplifier } (P_S)} \times 100\%$$

Because the voltage supplied to the 16-Ω load is 2 V, the power delivered to the load will be

$$P_L = \frac{V_L^2}{R_L} = \frac{(2 \text{ V})^2}{16 \text{ }\Omega} = 250 \text{ mW}$$

The dc power supplied to the amplifier is equal to

$$P_S = V_{CC} \times I_{CC} = 20 \text{ V} \times 16 \text{ mA} = 320 \text{ mW}$$

If the power delivered to the load is equal to 250 mW, and the power supplied to the amplifier by the dc power supply is equal to 320 mW, the class-B push-pull amplifier's efficiency will be

$$\eta = \frac{P_L}{P_S} \times 100\% = \frac{250 \text{ mW}}{320 \text{ mW}} \times 100\% = 78.125\%$$

Power-Amplifier Heat Sinks

The high levels of current handled by power transistors are the reason they are usually connected to **heat sinks,** which are metal extensions of the transistor's heat radiating package. Figure 8-14(c) shows how a metal cap can be attached to the low-power transistor to increase the metal heat-radiating surface of the transistor. Figure 8-14(d) shows how high-power transistors can be mounted to a system's metal rear panel or chassis in order to increase the transistor's heat radiating metal surface area. Unless the transistor's collector is designed to be at ground potential, a mica insulating sheet is needed to isolate the transistor's collector (the case) from the grounded chassis. When an insulating sheet is used to isolate the collector case from the chassis in this way, both sides of the mica sheet are coated with silicon grease (a bad electrical conductor but good heat conductor), which will thermally connect the transistor's case to the chassis.

Complementary AF Power-Amplifier Circuits

Figure 8-15(a) shows another type of push-pull amplifier that uses the complementary nature of an NPN and a PNP transistor to perform the push-pull action. Because the NPN transistor (Q_1) will conduct when the input swings positive and the PNP transistor (Q_2) will conduct when the input swings negative, a phase splitter is not required and a power amplifier circuit can be constructed with fewer components. Transistors Q_1 and Q_2 are connected in series between $+V_{CC}$ and ground. Resistors R_1 and R_2 and diodes D_1 and D_2 form a voltage-divider circuit that biases Q_1 and Q_2 so that the $+V_{CC}$ supply voltage is divided equally across Q_1's collector–emitter and R_3 and Q_2's collector–emitter and R_4. Capacitor C_2 is included to isolate this dc bias voltage (appearing at the junction of R_3 and R_4) from the speaker. To be properly biased for class-B operation, each transistor's base should be 0.7 V with respect to its emitter and a total of 1.4 V (2 $\times$ 0.7 V) should exist between the base of Q_1 and the base of Q_2. This voltage difference is achieved by the diodes D_1 and D_2. Diodes D_1 and D_2 are also included to prevent thermal runaway, which occurs when a temperature increase in the transistor causes a corresponding current increase in the transistor, which causes a corresponding temperature increase, and so on. With D_1 and D_2 included in the circuit, a temperature increase causes an increasing current through Q_1 and Q_2 and D_1 and D_2 because all these devices are semiconductor components. The increase in current through D_1 and D_2 will decrease the voltage drop developed across the diodes, causing a decrease in the voltage difference between Q_1 and Q_2's base and therefore a decrease in the forward bias being applied to the transistors. As a result, an increase in current due to temperature will be counteracted by a decrease in forward bias and therefore a decrease in current. Emitter feedback resistors R_3 and R_4 also assist in this temperature stabilization by providing degenerative feedback.

Figure 8-15(b) shows how **crossover distortion** can occur if diodes D_1 and D_2 are not included to keep Q_1 and Q_2 biased correctly despite variations in temperature. The addition

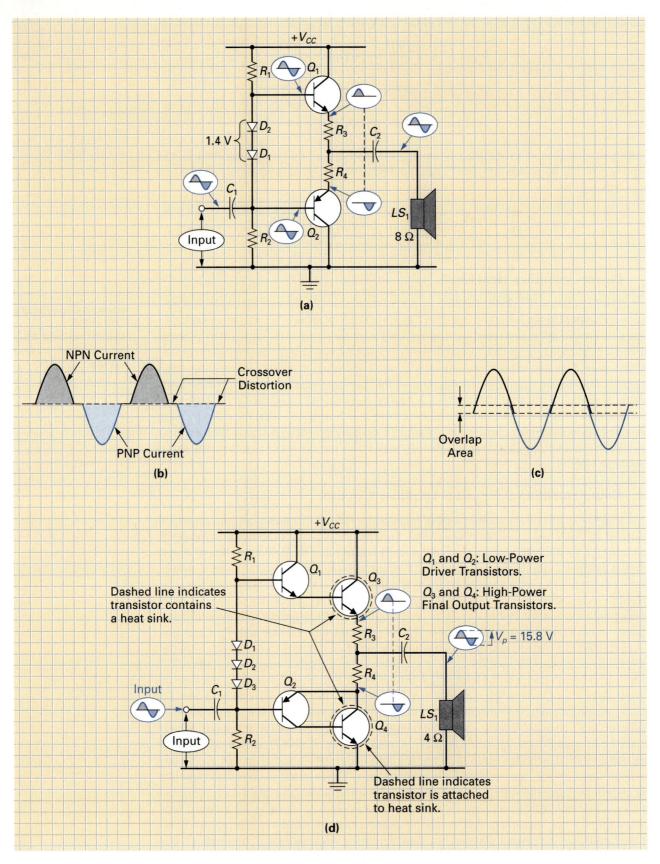

FIGURE 8-15 **Complementary Audio-Frequency (AF) Power Amplifiers. (a) Complementary Power-Amplifier Circuit. (b) Crossover Distortion. (c) Class-AB Biased. (d) Quasi-Complementary Power-Amplifier Circuit.**

of these two diodes in the input section ensures that each transistor is biased at exactly 0.7 V relative to its emitter (class B). Therefore, Q_1 is biased just OFF and ready to conduct the moment the input signal crosses zero and swings positive, and Q_2 is also biased just OFF and ready to conduct the moment the input signal swings negative.

In most practical push-pull power amplifiers, the two output transistors are class-AB biased to ensure that the output signal is not affected by crossover distortion. Unlike the class-B transistors that are biased OFF when no input signal is present, the class-AB amplifiers will have a small value of current flowing through both transistors when no input signal is present, as shown in Figure 8-15(c).

Once again, the two power transistors should be matched. In this case, however, one is an NPN and the other is a PNP, and you will have to check the device data sheet to find an NPN transistor-matched PNP counterpart. For example, the NPN 2N3904 transistor has the same operating characteristics and specifications as the PNP 2N3906. As before, if one transistor fails, you should replace both transistors just in case the other transistor has been damaged by the failure or its characteristics have been altered by the fault and will therefore damage the replacement.

Quasi-Complementary AF Power-Amplifier Circuits

Figure 8-15(d) shows how a higher-power output (10 watts or more) can be obtained by having a low-power push-pull transistor pair (Q_1 and Q_2) drive a high-power push-pull transistor pair (Q_3 and Q_4). This circuit is called **quasi-complementary** because the circuit action of two final output NPN transistors (Q_3 and Q_4) operate like (quasi) the NPN and PNP output transistors in a complementary-amplifier circuit. The advantage of this circuit is not only its higher output power due to the two stages of power amplification but also the fact that it uses two high-power final-stage NPN transistors, which are much cheaper than high-power PNP transistors.

When the input signal swings positive, NPN transistor Q_1 turns ON, while PNP transistor Q_2 turns OFF. With Q_2 OFF, Q_4 will receive no input base current, and so it will also cut OFF. Transistors Q_1 and Q_3 form an NPN darlington pair, and so this high-current arrangement provides an AF emitter-follower output signal to the load when the input signal is positive.

When the input signal swings negative, NPN transistor Q_1 turns OFF, while PNP transistor Q_2 turns ON. With Q_1 OFF, Q_3 will receive no input base current, and so it will also cut OFF. The negative signal alternation at the base of Q_2 will be inverted to a positive signal alternation, due to Q_2's base–collector inversion. This positive signal alternation at the collector of Q_2 will turn ON Q_4, which will again invert its positive input signal alternation to a negative signal alternation, which will then drive the loudspeaker. Like Q_1 and Q_3, Q_2 and Q_4 will operate together as a two-stage power amplifier.

The other difference with this circuit is that it has three diodes between the base of Q_1 and Q_2. This extra diode is needed because there are now three base–emitter junctions between Q_1 and Q_2's base (B–E of Q_1, B–E of Q_3, and B–E of Q_2).

Quasi-complementary
A circuit of two final output NPN transistors that operate like the NPN and PNP output transistors in a complementary amplifier circuit.

■ **EXAMPLE:**

Calculate the power delivered to the load by the power-amplifier circuit shown in Figure 8-15(d).

■ *Solution:*

$$V_{\text{rms}} = V_p \times 0.707 = 15.8 \text{ V} \times 0.707 = 11.17 \text{ V}$$

$$P_{RL} = \frac{V_{RL(\text{rms})}^2}{R_L} = \frac{(11.17 \text{ V})^2}{4 \text{ }\Omega} = 31.2 \text{ W}$$

Use the following questions to test your understanding of Section 8–3.

1. Audio-frequency amplifiers are designed to amplify all of the frequency that the human voice can produce and the human ear can respond to. This audio-frequency range is from about _____ Hz to _____ Hz.

2. What are the two basic types of audio-frequency amplifiers?

3. A _____ amplifier is used to boost the voltage level of a weak input signal so that it is large enough to control a _____ amplifier.

4. Why is an emitter capacitor generally included in parallel with an ac amplifier's emitter resistor?

5. A coupling capacitor is inserted between amplifier stages to provide _____ and _____ between amplifier stages.

6. A push-pull amplifier contains two transistors that are both class _____ biased.

7. What advantage(s) does the complementary power amplifier have over the basic push-pull power amplifier?

REVIEW QUESTIONS

Multiple-Choice Questions

1. A class _____ amplifier has its dc operating point set at cutoff.
   a. A   b. B   c. C   d. AB

2. A class _____ amplifier has its dc operating point set below cutoff.
   a. A   b. B   c. C   d. AB

3. A class _____ amplifier has its dc operating point set slightly above cutoff.
   a. A   b. B   c. C   d. AB

4. A class _____ amplifier has its operating point set near the center of the load line.
   a. A   b. B   c. C   d. AB

5. A darlington pair has
   a. three transistors
   b. a single base–emitter voltage drop
   c. a very low input impedance
   d. a very high current gain

6. The versatility, temperature stability, and noise rejection of the _____ dc amplifier make it an ideal candidate for many applications.
   a. darlington pair   c. complementary
   b. zener coupled   d. differential

7. A capacitor acts as a/an
   a. open switch to dc   c. closed switch to dc
   b. block to ac   d. open switch to ac

8. The efficiency of an amplifier is equal to the ratio of
   a. R_C to R_E c. P_L to P_S
   b. V_{out} to V_{in} d. all of the above

9. Which dc amplifier would be best suited for switching ON and OFF a high-current relay coil?
   a. C-B   c. Zener coupled
   b. Complementary   d. darlington pair

10. A stepper motor with 72 steps/revolution will cause the rotor to turn _____ for each step.
    a. 5°   c. 2°
    b. 7.5°   d. 15.5°

11. A differential amplifier should have a _____ differential gain and a _____ common-mode gain.
    a. high, high   c. low, low
    b. low, high   d. high, low

12. What are common-mode input signals?
    a. Voice signal   c. Transducer input signal
    b. Noise   d. None of the above

13. Audio-frequency voltage amplifiers are often referred to as _____ amplifiers.
    a. dc   b. pre-   c. power   d. IF

14. Audio-frequency power amplifiers are usually class _____ biased.
    a. A   b. B or AB   c. C   d. D

15. Audio-frequency voltage amplifiers are usually class _____ biased.
    a. A   b. B or AB   c. C   d. D

16. The capacitor that connects an ac ground to the emitter of an ac signal amplifier
    a. prevents degenerative feedback
    b. acts as an ac bypass for R_E
    c. performs the opposite of a coupling capacitor
    d. all of the above

17. Because power transistors are designed to boost _____, a large amount of _____ is generated. To ensure their operating characteristics are stable and their efficiency is good, they are operated with class _____ bias.
    a. voltage, heat, A   c. impedance, power, A
    b. voltage, current, C   d. current, heat, B

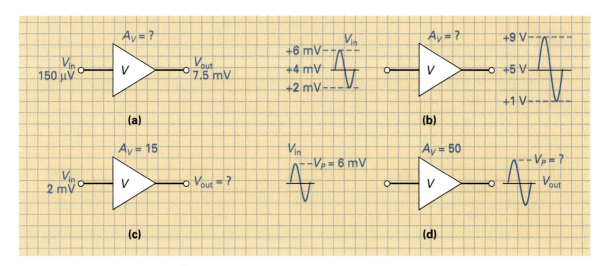

FIGURE 8-16 Voltage Amplifiers.

18. Audio-frequency power amplifiers are normally operated in class AB to prevent
   a. temperature stability
   b. degenerative feedback
   c. crossover distortion
   d. all of the above

19. If $R_C = 2.7$ kΩ and $R_L = 10$ kΩ, the equivalent collector load resistance will be
   a. 12.7 kΩ   **b.** 2.1 kΩ   **c.** 2.7 kΩ   **d.** 10 kΩ

20. The power gain of a class-B push-pull power amplifier is
   a. equal to the voltage gain
   b. equal to P_{out} divided by P_{in}
   c. less than the voltage gain
   d. equal to twice the amplifier's current gain

Practice Problems

21. Referring to the voltage amplifiers in Figure 8-16, calculate the unknown.

22. Referring to the current amplifiers in Figure 8-17, calculate the unknown.

23. Referring to the power amplifiers in Figure 8-18, calculate the unknown.

24. Referring to the actual system-amplifier circuits shown in Figure 8-19:
   a. Identify the bipolar transistor circuit configurations.
   b. Which of the circuits would provide a high voltage gain?
   c. Which of the circuits would provide a high current gain?
   d. Which of the circuits would provide a high power gain?
   e. Which of the circuits would be best suited as an impedance-matching circuit?
   f. Are these circuits designed for AF or dc signals?
   g. What type of biasing method is used for all three circuits?

25. Referring to Figure 8-20, calculate the
   a. amplifier's voltage gain
   b. amplifier's output voltage
   c. amplifier's efficiency

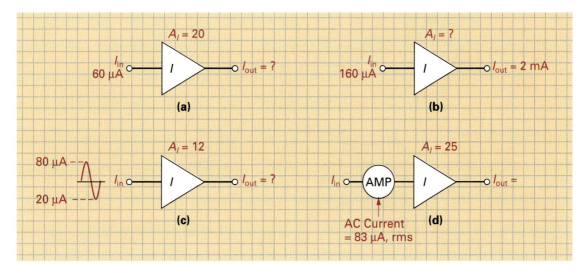

FIGURE 8-17 Current Amplifiers.

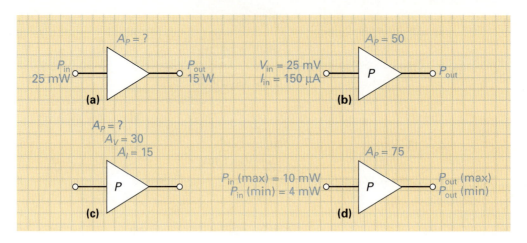

FIGURE 8-18 Power Amplifiers.

26. Identify the circuits shown in Figure 8-21, and briefly describe the circuits' function.

27. If a differential amplifier has a differential gain of 6,000 and a common-mode gain of 0.33, what would be the amplifier's common-mode rejection ratio?

28. Referring to the audio-frequency amplifier circuit in Figure 8-22, calculate the

   a. voltage gain
   b. output signal voltage for an input signal of 2 mV

29. Calculate the efficiency of the amplifier shown in Figure 8-22 if a 4.2-V signal is applied to a 1-kΩ load and $I_{CC} = 20$ mA.

30. What would be the power delivered to an 8-Ω speaker if a power amplifier outputs a 23-V rms signal?

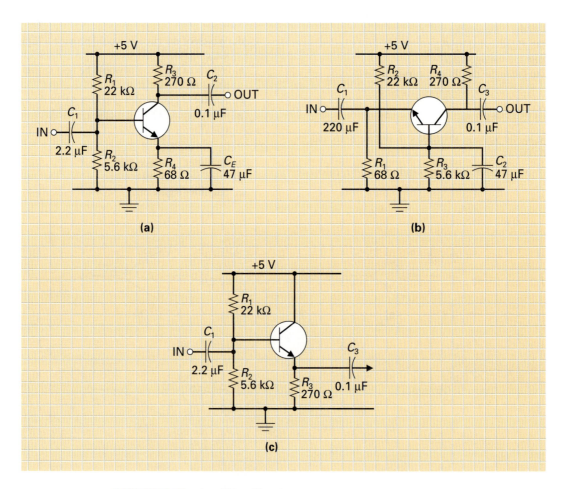

FIGURE 8-19 Amplifier Circuits.

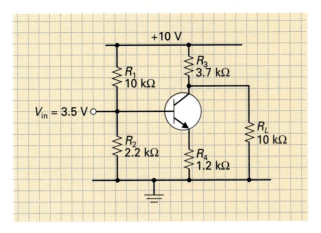

FIGURE 8-20 **A Direct-Current Amplifier.**

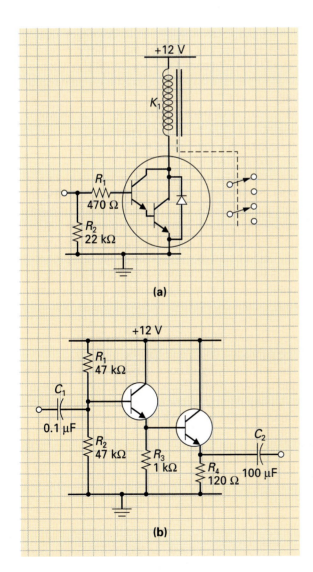

(a)

(b)

FIGURE 8-21 **Transistor Circuits.**

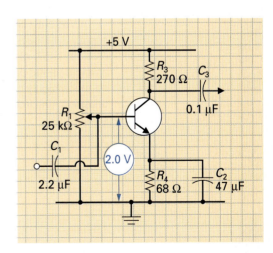

FIGURE 8-22 **An AF Voltage Amplifier.**

Oscillator Circuits

Making an Impact

John Von Neuman, a mathematics professor at the Institute of Advanced Studies, delighted in amazing his students by performing complex computations in his head faster than they could with pencil, paper, and reference books. He possessed a photographic memory and at parties in his home in Princeton, New Jersey, he gladly occupied center stage to recall from memory entire pages of books read years previously, the lineage of European royal families, and a store of controversial limericks. His memory, however, failed him in his search for basic items in a house he had lived in for 17 years. On many occasions when traveling, he would become so completely absorbed in mathematics that he would have to call his office to find out where he was going and why.

Born in Hungary, he was quick to demonstrate his genius. At the age of six, he would joke with his father in classical Greek. At the age of eight, he had mastered calculus, and in his midtwenties, he was teaching and making distinct contributions to the science of quantum mechanics, which is the cornerstone of nuclear physics.

Next to fine clothes, expensive restaurants, and automobiles that he had to replace annually due to smashups, he loved his work. His interest in computers began when he worked on the top secret Manhattan Project at Los Alamos, New Mexico, where he proved mathematically the implosive method of detonating an atom bomb. Working with the computers then available, he became aware that they could become much more than a high-speed calculator. He believed that they could be an all-purpose scientific research tool, and he published these ideas in a paper. This was the first document to outline the logical organization of the electronic digital computer, and it was widely circulated to all scientists throughout the world. In fact, even to this day, scientists still refer to computers as "Von Neuman machines."

Von Neuman collaborated on a number of computers of advanced design for military applications such as the development of the hydrogen bomb and ballistic missiles.

In 1957, at the age of 54, he lay in the hospital dying of bone cancer. Under the stress of excruciating pain, his brilliant mind began to break down. Because he had been privy to so much highly classified information, the Pentagon had him surrounded with only medical orderlies specially cleared for security for fear he might, in pain or sleep, give out military secrets.

Introduction

An oscillator is an electronic circuit that draws power from its dc supply to generate a continuously repeating ac output signal. Stated another way, the oscillator circuit is simply a signal generator converting its V_{CC} supply voltage input into a continuously repeating ac signal output.

Oscillators are used in a variety of applications such as radio and television receivers and transmitters. In other applications, the oscillator's ac signal is distributed throughout an electronic system to time or synchronize operations. When used in this way, the electronic oscillator can be compared to the pendulum of a clock. Just as the pendulum oscillates back and forth timing the minutes, the continuously repeating electronic oscillator's output signal is used to control the timing of circuit operations.

Like the amplifier circuit, some oscillator circuits are designed to generate low-frequency oscillations of only a few hertz, while other, high-frequency oscillators generate oscillations at several hundred gigahertz. Oscillators are generally classified into one of three basic groups based on the frequency-determining components used within the circuit. These three groups are

LC oscillators

Crystal oscillators

RC oscillators

In this chapter, we will study the operation and characteristics of typical bipolar transistor *LC*, *RC*, and crystal oscillator circuits.

9-1   *LC* OSCILLATOR CIRCUITS

As stated in the introduction, the oscillator circuit is simply a signal generator, converting its V_{CC} or dc supply voltage input into a continuously repeating ac signal output. In this section, we will discuss several different bipolar transistor *LC* oscillator circuits, which use an *LC* tuned circuit to set the frequency of oscillation and are all named after their inventors.

9-1-1   *Basic LC Oscillator Action*

To explain the basic action of an *LC* oscillator circuit, Figure 9-1(a) shows how an *LC* tank circuit can be "shock excited" into oscillation by a momentary surge of current from a dc power switch. Looking at the output of this circuit, you can see that the back-and-forth action of the circulating current within the tank generates a sine wave at the output. This sine-wave output, however, will slowly decrease in amplitude due to the energy lost in the coil's resistance. To prevent the oscillations from the tank from being "damped" in this way, we need to somehow replace the energy lost. Figure 9-1(b) shows how this can be achieved by closing the dc power switch once every cycle. The timing of this switch closure is crucial because the pulse of dc current must reinforce the tank's circulating current, and therefore be in phase with the tank's output sine wave, as seen in the waveform in Figure 9-1(b). Reinforcing energy that "regenerates" or "adds to" the sine-wave oscillations in this way is called **regenerative feedback** or **positive feedback.** Most oscillator circuits operate on this positive feedback principle, which means that an in-phase signal must be constantly injected into the oscillator circuit in order to replace the energy lost and maintain oscillations.

Although the oscillator circuit shown in Figure 9-1(b) would work, it is obvious that some sort of automatic electronic positive feedback system is needed instead of the very impractical manual mechanical positive feedback system. One of the simplest oscillator circuits

Regenerative (Positive) Feedback

Reinforcing energy that regenerates or adds to the sine-wave oscillations of an oscillator circuit.

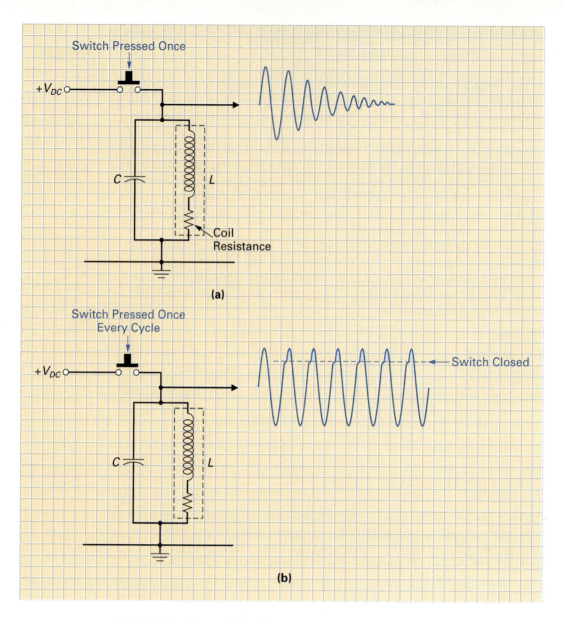

FIGURE 9-1 **Basic *LC* Oscillator Action.**

that makes use of a tuned *LC* tank and has an automatic electronic positive feedback loop is the Armstrong oscillator, which we will now examine.

9-1-2   *The Armstrong Oscillator*

Figure 9-2(a) shows the **Armstrong oscillator** circuit. Transistor Q_1 has the usual voltage-divider base bias (R_1 and R_2), emitter resistor and capacitor (R_3 and C_3), and input coupling capacitor (C_2). The frequency at which this circuit will oscillate is determined by the parallel resonant circuit C_1 and L_1, which is connected to the collector of transistor Q_1. Transformer T_1's primary coil (L_1) and secondary coil (L_2) are wound so that the voltage induced at the top of L_2 will be 180° out of phase with the voltage present at the bottom of L_1 (or collector of Q_1). In addition, this circuit includes a positive feedback loop to sustain oscillations between the circuit's output and the transistor's input.

In Figure 9-2(b), (c), (d), and (e), the bias components and coupling capacitors have been removed so that we can see more clearly the steps involved in this circuit's oscillating cycle.

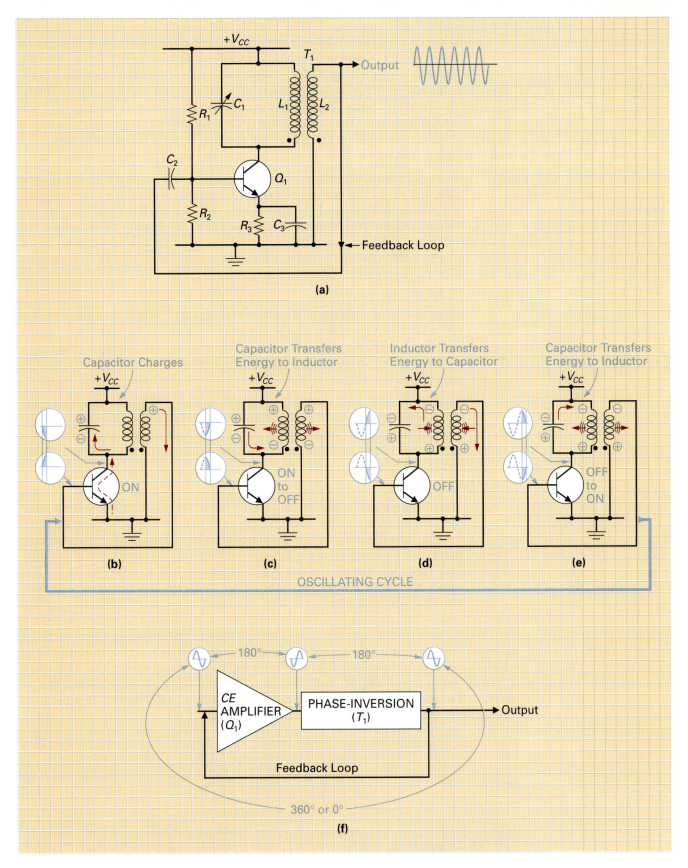

FIGURE 9-2   The Armstrong Oscillator.

Step 1: FIGURE 9-2(b)

When circuit power $(+V_{CC})$ is first applied, the voltage-divider bias resistors R_1 and R_2 cause Q_1 to turn ON and collector current to flow through Q_1 and the tank circuit to $+V_{CC}$. With Q_1 ON, its collector-voltage will be LOW, applying a LOW to the bottom of L_1. Due to the phase inversion of T_1, the top of L_2 will go HIGH. This HIGH positive voltage at the top of L_2 will be fed back to the base of Q_1 via the feedback loop. The conduction of Q_1 into saturation is increased, resulting in an increase in collector current and the charging of C_1 to the polarity shown.

Step 2: FIGURE 9-2(c)

When capacitor C_1 is fully charged, current through the tank circuit will drop to zero, and this lack of changing current will cause the transformer's induced positive voltage at the top of L_2 to drop to zero. Without this additional base–emitter forward bias from the feedback loop, Q_1 will turn OFF. With Q_1 OFF, capacitor C_1 will begin to discharge, transferring all of its energy to L_1 and initiating the tank's flywheel action. When C_1 is fully discharged, current through L_1 will cease and inductor L_1's magnetic field will begin to collapse.

Step 3: FIGURE 9-2(d)

As inductor L_1's magnetic field collapses, an opposite polarity is induced into the coil. Capacitor C_1 begins to charge as energy is transferred from the tank's coil to the capacitor. The positive polarity at the bottom of C_1 is applied to L_1, inducing a negative polarity at the top of L_2, driving Q_1 further into cutoff.

Step 4: FIGURE 9-2(e)

Continuing the flywheel action, capacitor C_1, which is now fully charged, begins to discharge through L_1 in the opposite direction. The coil's magnetic field once again begins to build. When capacitor C_1 is fully discharged, circuit current once again drops to zero and therefore the magnetic field around L_1 starts to collapse and C_1 begins to recharge.

Step 5: FIGURE 9-2(b)

The collapsing field of L_1 charges capacitor C_1 to the polarity shown. At the same time, the negative potential at the bottom of L_1 induces a positive potential at the top of L_2, and this potential forward biases Q_1, which turns ON. The transistor current assists the charging of capacitor C_1 and replaces the energy lost in the tank. Transistor Q_1 is therefore acting like an automatic switch, similar to the manual switch in Figure 9-1(b), replacing the energy lost in the tank circuit with synchronized positive feedback.

As you have seen in stepping through the oscillating cycle, capacitor C_1 and inductor L_1 set the frequency at which this circuit oscillates because their values control the rate at which energy is transferred back and forth within the tank. In many applications, the capacitor in the tuned circuit will be made variable so that the frequency of oscillation can be changed. The frequency at which this circuit will oscillate, and therefore the frequency of the sine-wave output, can be calculated with the formula

$$f_R = \frac{1}{2\pi\sqrt{L_1 \times C_1}}$$

■ **EXAMPLE:**

Calculate the frequency at which the circuit shown in Figure 9-2(a) will oscillate if $L_1 = 47$ mH and $C_1 = 0.033$ μF.

$$f_R = \frac{1}{2\pi \sqrt{L_1 \times C_1}} = \frac{1}{2\pi \sqrt{47 \text{ mH} \times 0.033 \text{ }\mu\text{F}}} = 4 \text{ kHz}$$

Another important point to stress is that the transistor Q_1 acts as a switch and is turned ON only when the output signal is positive and only when the transistor's collector current will assist the tank's circulating current in the charging of the capacitor. This "topping off" action is similar to someone giving a little push at the right time to a person on a playground swing.

To stress the idea of positive feedback in more detail, Figure 9-2(f) shows a block diagram of the Armstrong oscillator. Because a common–emitter amplifier has a 180° phase inversion between input and collector output, an additional 180° phase shift device is needed between the output and the input so that the feedback signal will be positive or reinforcing and the circuit will oscillate. The additional 180° phase shift is provided by the transformer T_1. When the base of Q_1 swings positive, the collector of Q_1 swings negative, the bottom of L_1 swings negative and the top of L_2 swings positive. This positive swing at the output is fed back to the input to reinforce the original positive input swing and sustain oscillations. As a result, the signal is shifted 360° as it is passed from input to output and then back to input. Because a phase shift of 360° is the same as not shifting the signal at all, the feedback signal is in phase with the input. If the phase-inverting transformer T_1 was not included and the collector output was fed directly back to the input, the feedback signal would oppose the input signal. This type of out-of-phase feedback is known as **degenerative feedback** or **negative feedback** and would result in a canceling of the input and therefore no output oscillations.

9-1-3   *The Hartley Oscillator*

Figure 9-3(a) shows the schematic of a **Hartley oscillator,** which, unlike the Armstrong oscillator, does not require a transformer. The 180° phase inversion needed for positive feedback is achieved by this circuit's identifying feature: the tapped inductor L_1. Transistor Q_1 has the usual bias and stability components; however, a radio-frequency choke (RFC) is connected in the collector to connect the dc power to the transistor while preventing the high-frequency oscillations at the output from passing to the $+V_{CC}$ line and back to the power supply. Coupling capacitor C_3 is included to block the dc collector bias from the output and couple ac collector variations to the tank. Coupling capacitor C_2 will block the dc base bias from the tank circuit, while coupling the resonant tank variations to the base of Q_1.

To describe this circuit's operation, we will once again break down the oscillating cycle to four steps, which are shown in Figures 9-3(b), (c), (d), and (e).

Step 1: Figure 9-3(b)

When circuit power $(+V_{CC})$ is first applied, the voltage-divider bias resistors R_1 and R_2 cause Q_1 to turn ON, and collector current will flow through Q_1 to the tank circuit. This current causes the magnetic field around L_1 to expand, and because the center tap of L_1 is grounded, the opposite ends of L_1 will be at different potentials. Making the top of L_1 LOW will make the bottom of L_1 HIGH, and this HIGH will be applied to the base of Q_1, causing it to quickly saturate.

Step 2: FIGURE 9-3(c)

When Q_1 is saturated, there will be no further increase in current, and the magnetic field produced by L_1 will collapse. The energy in the inductor will be transferred to the capacitor. As the bottom of C_1 is now charging to a negative potential—and this point is connected to the base of Q_1—the transistor will begin to feel a reverse bias voltage and start to cut OFF.

Degenerative (Negative) Feedback

Feedback in which a portion of the output signal is fed back 180° out of phase with the input signal. Also called degenerative feedback, inverse feedback, or stabilized feedback.

Hartley Oscillator

An *LC* tuned oscillator circuit in which the tank coil has an intermediate tap.

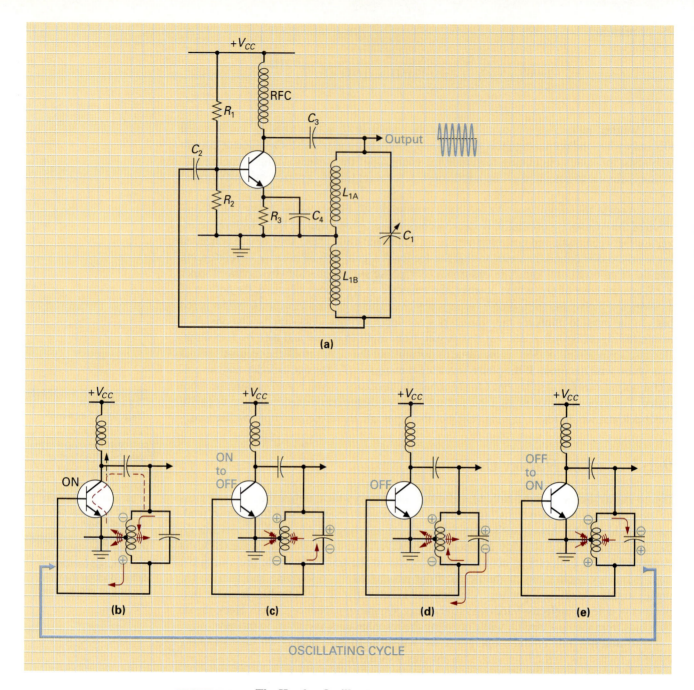

FIGURE 9-3 The Hartley Oscillator.

Step 3: FIGURE 9-3(d)

With Q_1 OFF, the tank's flywheel action takes over, and C_1 begins to discharge and transfer its energy from the capacitor to the inductor.

Step 4: FIGURE 9-3(e)

When capacitor C_1 is fully discharged, the magnetic field around L_1 starts to collapse and C_1 begins to recharge. Because the bottom of C_1 is charging to a positive potential—and this is connected to the base of Q_1—the transistor will begin to receive a forward bias and therefore start to turn ON. The transistor current assists the charging of capacitor C_1 and replaces the energy lost in the tank. Once C_1 is fully charged, flywheel action will take over, the capacitor's energy will be transferred to L_1 as shown in Figure 9-3(b), and the oscillating cycle will repeat.

The frequency at which this circuit oscillates is set by the tank components C_1 and L_T, and the frequency of the sine-wave output can be calculated with the formula

$$f_R = \frac{1}{2\pi\sqrt{L_T \times C_1}}$$

$$L_T = L_{1A} + L_{1B}$$

EXAMPLE:

Calculate the frequency at which the circuit shown in Figure 9-3(a) will oscillate if $L_{1A} = 47\ \mu H$, $L_{1B} = 22\ \mu H$, and $C_1 = 22\ nF$.

Solution:

$$L_T = L_{1A} + L_{1B} = 47\ \mu H + 22\ \mu H = 69\ \mu H$$

$$f_R = \frac{1}{2\pi\sqrt{L_T \times C_1}} = \frac{1}{2\pi\sqrt{69\ \mu H \times 22\ nF}} = 129.2\ kHz$$

9-1-4   *The Colpitts Oscillator*

The **Colpitts oscillator,** shown in Figure 9-4(a), is similar to the Hartley; however, in this circuit two capacitors are used instead of a center-tapped coil. These two capacitors, which are grounded at the center, are the identifying feature of this oscillator and will provide the needed 180° of phase shift for positive feedback.

> **Colpitts Oscillator**
> An *LC* tuned oscillator circuit in which two tank capacitors are used instead of a tapped coil.

To describe this circuit's operation, we will once again break down the oscillating cycle to four steps, which are shown in Figures 9-4(b), (c), (d), and (e).

Step 1: FIGURE 9-4(b)

When dc power is first applied to this circuit, the voltage-divider bias provided by R_1 and R_2 will cause Q_1 to conduct. With Q_1 ON, dc collector current will pass from Q_1's emitter to collector, through the radio-frequency choke, and to $+V_{CC}$. The drop in voltage at the collector of Q_1 will be applied to the top plate of C_1, causing C_1 to charge. Due to the ground at the center of C_1 and C_2, the bottom plate of C_2 will charge to a positive potential through L_1. This positive potential will be applied via the feedback loop to Q_1, sending it into saturation.

Step 2: FIGURE 9-4(c)

When Q_1 saturates, there is no further voltage change at the collector of Q_1, so C_1 and C_2 receive no further charge and the flywheel action of the tank circuit takes over. Capacitors C_1 and C_2 act together as they discharge through L_1, causing L_1's magnetic field to build up.

Step 3: FIGURE 9-4(d)

When capacitors C_1 and C_2 have fully discharged, current decreases to zero and the magnetic field around L_1 begins to collapse, inducing an opposite voltage into L_1, which then charges C_1 and C_2. Because the bottom plate of C_2 charges to a negative potential, and this is applied to the base of Q_1, the transistor will be driven into cutoff. The capacitors will then charge as L_1 transfers all of its energy into C_1 and C_2.

Step 4: FIGURE 9-4(e)

When capacitors C_1 and C_2 have received all of the energy from L_1, the flywheel action will again reverse causing C_1 and C_2 to discharge through L_1. When all en-

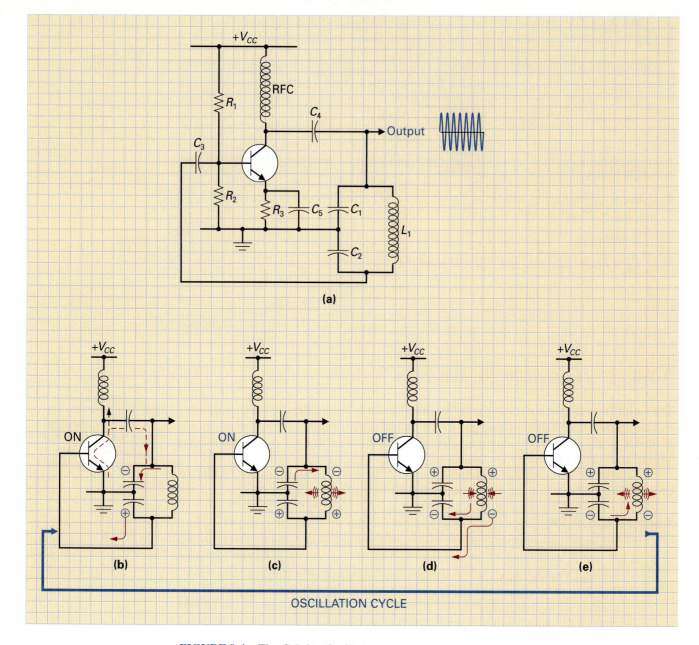

FIGURE 9-4   The Colpitts Oscillator.

ergy has been transferred to L_1, current will again drop to zero and the inductor's magnetic field will collapse. This induces a voltage into L_1 that will charge capacitors C_1 and C_2 to the polarity shown in Figure 9-4(b), and the cycle will repeat.

As with all of the previous LC oscillators, the frequency of oscillation is determined by the tank's inductive and capacitive values, which in this case are equal to L_1 and the series combination of C_1 and C_2.

$$f_R = \frac{1}{2\pi \sqrt{L_1 \times C_T}}$$

$$C_T = \frac{C_1 \times C_2}{C_1 + C_2}$$

EXAMPLE:

Calculate the frequency at which the circuit shown in Figure 9-4(a) will oscillate if $C_1 = 0.27\ \mu\text{H}$, $C_2 = 0.47\ \mu\text{H}$, and $L_1 = 0.6\ \text{mH}$.

Solution:

$$C_T = \frac{C_1 \times C_2}{C_1 + C_2} = \frac{0.27\ \mu\text{F} \times 0.47\ \mu\text{F}}{0.27\ \mu\text{F} + 0.47\ \mu\text{F}} = 0.17\ \mu\text{F}$$

$$f_R = \frac{1}{2\pi\sqrt{L_1 \times C_T}} = \frac{1}{2\pi\sqrt{0.6\ \text{mH} \times 0.17\ \mu\text{F}}} = 15.76\ \text{kHz}$$

SELF-TEST EVALUATION POINT FOR SECTION 9-1

Use the following questions to test your understanding of Section 9–1.

1. What are the three basic oscillator classifications?
2. What type of feedback is needed for an oscillator to continually oscillate?
3. Which component(s) provide the 180° phase shift in the following oscillator circuits?
   a. Armstrong
   b. Hartley
   c. Colpitts

9-2   CRYSTAL OSCILLATOR CIRCUITS

The *LC* oscillator circuits discussed in the previous section would typically have a 0.8% frequency drift. This frequency drift is due to circuit temperature changes, the aging of the components, and changes in the resistance of the load connected to the oscillator. If an *LC* oscillator were used to generate the timing signal for a wristwatch, a 0.8% frequency drift would mean that the watch may gain or lose approximately $11\frac{1}{2}$ minutes in a day.

$$60 \text{ minutes} \times 24 \text{ hours} = 1{,}440 \text{ minutes per day}$$
$$0.8\% \text{ of } 1{,}440 = 11.5 \text{ minutes per day}$$

If a high degree of oscillator stability is needed, crystal oscillators are used to generate the timing signal. Crystal-controlled oscillators will typically have a frequency drift of 0.0001%, and if used in a wristwatch, may gain or lose a maximum of half a minute a year.

$$60 \text{ minutes} \times 24 \text{ hours} \times 365 \text{ days} = 525{,}600 \text{ minutes per year}$$
$$0.0001\% \text{ of } 525{,}600 = 0.5 \text{ minutes per year}$$

Therefore, in applications where a high degree of frequency stability is needed, such as communication and computer systems, crystal-controlled oscillators are used. Crystal-controlled oscillators make use of a quartz crystal to control the circuit's frequency of operation. Before we discuss the operation of various crystal oscillator circuits, let us review the basic characteristics of a crystal.

9-2-1   *The Characteristics of Crystals*

The quartz crystal is made of silicon dioxide and is naturally a six-sided (hexagonal) compound with pyramids at either end, as seen in Figure 9-5(a). To construct an electronic com-

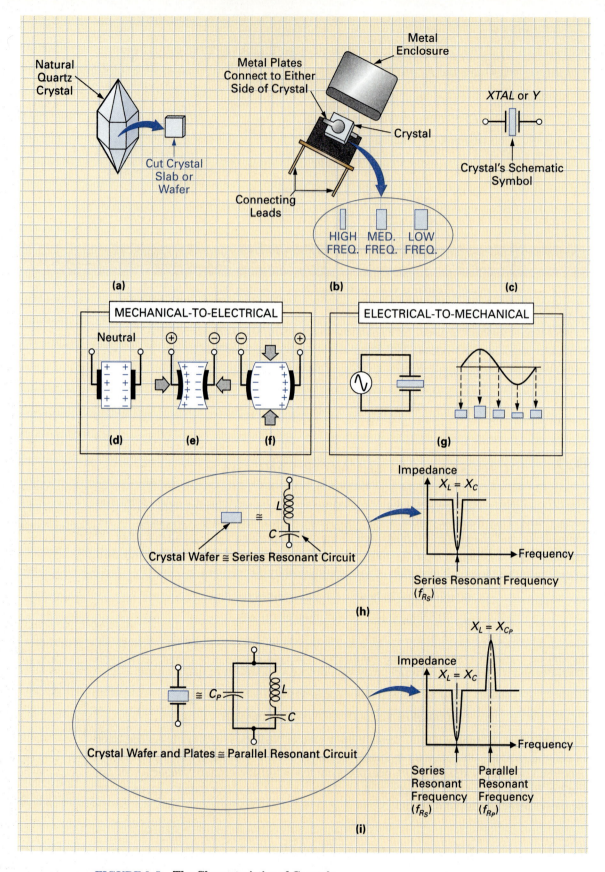

FIGURE 9-5   The Characteristics of Crystals.

ponent, a thin slice or slab of crystal is cut from the mother stone, mounted between two metal plates that make electrical contact, then placed in a protective holder, as seen in Figure 9-5(b). On a schematic diagram, a crystal is generally labeled either *XTAL* or *Y*, and has the symbol shown in Figure 9-5(c).

The crystal is basically operated as a transducer, or energy converter, transforming mechanical energy to electrical energy, as shown in Figure 9-5(d), (e), and (f), or electrical energy to mechanical energy, as shown in Figure 9-5(g).

Mechanical-to-Electrical Conversion

In Figure 9-5(d), you can see that a crystal normally has its internal charges evenly distributed throughout, and the potential difference between its two plates is zero. If the crystal is compressed by applying pressure to either side, as shown in Figure 9-5(e), opposite charges accumulate on either side of the crystal and a potential difference is generated. Similarly, if the crystal is expanded by applying pressure to the top and bottom, as shown in Figure 9-5(f), opposite charges accumulate on either side of the crystal, and a potential difference of the opposite polarity is generated. If a crystal were subjected to an alternating pressure that caused it to continually expand and compress, an alternating, or ac, voltage would be generated. Crystal microphones make use of this principle. Sound waves are applied to the microphone, and these mechanical waves continually compress and expand the crystal, generating an electrical wave that is equivalent to the original sound wave.

Electrical-to-Mechanical Conversion

Due to their composition, crystals have a "natural frequency of vibration." This means that if the ac voltage applied across a crystal matches the crystal's natural frequency of vibration, as shown in Figure 9-5(g), the crystal will physically expand and contract by a relatively large amount. If, on the other hand, the frequency of the applied ac voltage is either above or below the crystal's natural frequency, the vibration is only slight. Crystals are, therefore, very frequency selective—a characteristic we can make use of in filter circuits and oscillator circuits. This action is called the **piezoelectric effect,** which by definition is the tendency of a crystal to vibrate at a constant rate when it is subjected to a changing electric field produced by an applied ac voltage. The crystal's natural frequency of vibration is dependent on the thickness of the crystal between the two plates, as seen in the inset in Figure 9-5(b). By cutting a crystal to the right size, we can obtain a crystal that will naturally vibrate at an exact frequency. This rating is usually printed on the crystal's package.

When the frequency of the ac voltage applied to a crystal matches the crystal's natural frequency of vibration, the impedance or current opposition of the crystal drops to a minimum, as seen in Figure 9-5(h). This makes the crystal wafer or slab equivalent to a series resonant circuit, and, as you can see in the frequency response curve in Figure 9-5(h), the very sharp response at the crystal's series resonant frequency (f_{R_S}) means that the crystal is extremely frequency selective.

When a crystal is mounted between two connecting plates, the device is equivalent to the crystal's original series resonant circuit in parallel with a small value of capacitance due to the connecting plates (C_P), as seen in Figure 9-5(i). Referring to the frequency response curve in Figure 9-5(i), you can see that the crystal component will still respond in exactly the same way when the applied ac frequency is equal to the crystal's natural series resonant frequency. As frequency is increased beyond the series resonant frequency of the crystal, however, the reactance of C will decrease, and the crystal will become inductive. At a higher frequency, there will be a point at which the inductive reactance of the crystal is equal to the capacitive reactance of the connecting plates ($X_L = X_{C_P}$). Because the crystal's inductance is in parallel with the mounting plate's capacitance, the two form a parallel resonant circuit. Therefore, the device's impedance will be maximum at this parallel resonant frequency (f_{R_P}), as shown in Figure 9-5(i). Circuit applications can make use of either the crystal's series resonant selectivity or parallel resonant selectivity.

Most crystal oscillator circuits will make use of the crystal's series resonant frequency response to feed back only the desired frequency to the input and tightly maintain the fre-

Piezoelectric Effect
The tendency of a crystal to vibrate at a constant rate when subjected to a changing electric field.

quency stability of the oscillator. A crystal's series resonant tuned circuit will typically have a very high Q of 40,000. If you compare this to an LC tuned circuit that would typically have a Q of 200, it is easy to see why crystal-controlled oscillators are much more stable than LC-controlled oscillators.

9-2-2 *The Hartley Crystal Oscillator*

As you can see in Figure 9-6, the Hartley crystal oscillator is identical to the previously discussed LC Hartley oscillator, except for the crystal inserted in series with the feedback path. When the oscillator generates a frequency that is identical to the series resonant frequency of the crystal, the crystal will offer almost no opposition and the feedback signal will be maximum. If the oscillator's frequency begins to drift above or below the crystal's natural series resonant frequency, the impedance of the crystal will increase, causing the feedback to decrease. Because the frequency of the feedback signal controls the circuit's frequency of oscillation, the circuit is forced to return to the natural frequency of the crystal. Including a highly selective crystal in the feedback path means that this crystal-controlled Hartley oscillator will have a frequency drift rating that is almost 1,000 times better than an LC-controlled Hartley oscillator.

9-2-3 *The Colpitts Crystal Oscillator*

Referring to Figure 9-7, you can see that the Colpitts crystal oscillator is identical to the previously discussed LC Colpitts oscillator, except for the crystal inserted in series with the feedback path. This circuit will operate in exactly the same way as the previously discussed LC Colpitts oscillator; however, the circuit will have less frequency drift due to the inclusion of a crystal.

9-2-4 *The Pierce Crystal Oscillator*

Pierce Crystal Oscillator

An oscillator circuit in which a piezoelectric crystal is connected in a tank between output and input.

The **Pierce crystal oscillator** is shown in Figure 9-8, and, as you can see, it is almost identical to the Colpitts oscillator except for crystal Y_1, which replaces the tank's coil. Crystal Y_1

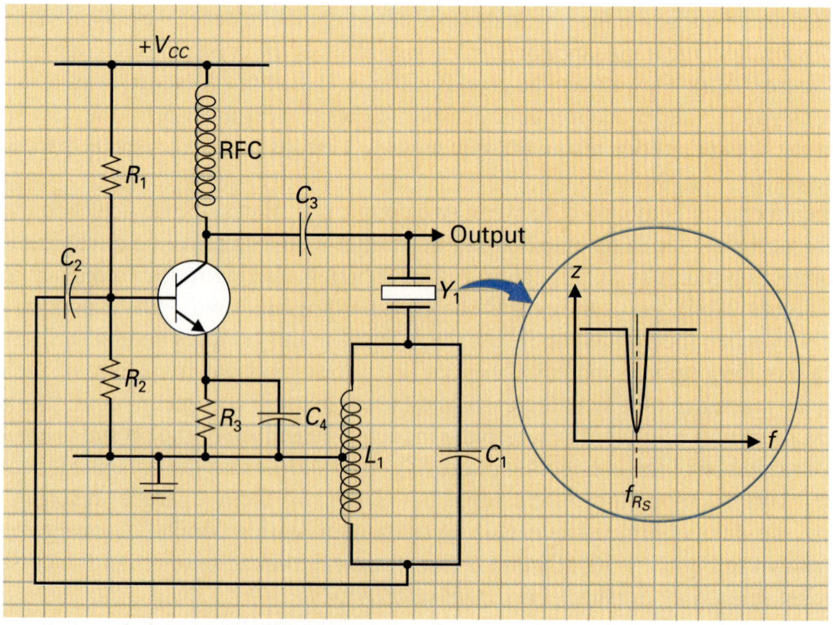

FIGURE 9-6 **The Hartley Crystal Oscillator.**

CHAPTER 9 / OSCILLATOR CIRCUITS

FIGURE 9-7   The Colpitts Crystal
             Oscillator.

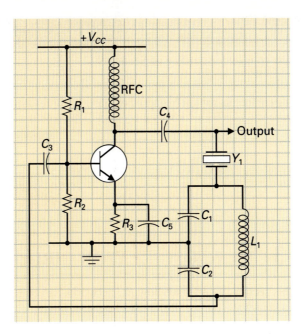

will be operating at its parallel resonant frequency, which means that it will offer a very high impedance to the oscillator's generated frequency. This action is not any different from the previous *LC* parallel resonant tank, which would also have a high impedance at resonance. However, now that a crystal controls the tank's impedance, the tuned circuit will be much more selective, and the oscillator's output will be more stable. At resonance, the tank circuit will have a very large impedance and therefore a large feedback voltage will be developed across C_2 and applied to the base of Q_1. If the oscillator's frequency drifts above or below resonance, the crystal's impedance will quickly decrease, decreasing the voltage developed across the tank and the amplitude of the feedback signal.

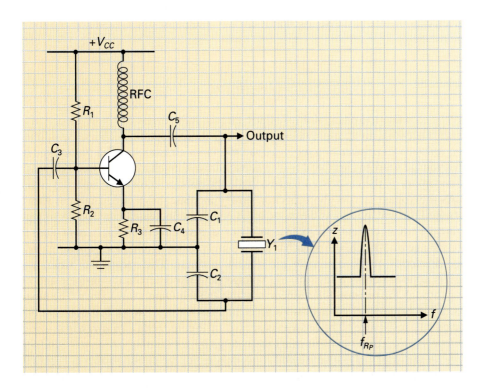

FIGURE 9-8   **The Pierce Crystal Oscillator.**

Use the following questions to test your understanding of Section 9–2.

1. What key advantage does a crystal oscillator circuit have over an *LC* oscillator circuit?

2. The physical size of the crystal determines its natural frequency of vibration, or resonant frequency. A thinly cut crystal will resonate at a _____ frequency, whereas a thickly cut crystal will resonate at a _____ frequency.

3. At a crystal's series resonant frequency, impedance will be _____, whereas at the parallel resonant frequency impedance will be _____.

4. Both the Hartley and Colpitts crystal oscillator circuits make use of the crystal's _____ resonant frequency, whereas the Pierce crystal oscillator makes use of the crystal's _____ resonant frequency.

9-3   *RC* OSCILLATOR CIRCUITS

In this section, we will see how *RC* networks can be used within oscillator circuits to generate a sine-wave output. Before we begin, however, let us examine why we need *RC* oscillators. Within electronic systems, you will see *LC* and crystal oscillators used to generate medium-frequency and high-frequency sine-wave oscillations. As previously stated, the choice between these two oscillator types is basically governed by two factors: frequency stability and price. If frequency stability is of prime importance, a crystal oscillator is used; if frequency stability is not crucial, the lower-priced *LC* oscillator can be used. The next question to answer is: Is it practical to use these oscillator circuits to generate low audio-frequency oscillations? The answer is no. At low frequencies, the size of the inductor needed for an AF tuned tank is very large, and therefore a low-frequency *LC* oscillator would be large in size and expensive. As far as the crystal oscillator is concerned, the physical size of the crystal makes it practical for crystals to be used only in applications requiring an output frequency of 50 kHz or greater. Although *RC* oscillators are not as stable as *LC* or crystal oscillators, they are inexpensive and are ideal in applications where we need to generate low-frequency oscillations of less than 10 kHz. In this section, we will discuss the operation and characteristics of two basic *RC* oscillator circuits: the phase-shift oscillator and the Wien-Bridge oscillator.

9-3-1   *The Phase-Shift Oscillator*

Before we examine the phase-shift oscillator circuit, let's review the basic theory of an *RC* phase-shift network. In your previous introductory dc/ac electronics course, you found that *when a voltage was applied to a purely capacitive circuit, the current led the voltage by 90°.* When a voltage source is applied across a capacitor with no charge, the circuit current is initially maximum as the capacitor charges, while the voltage across the capacitor is initially zero. As the voltage across the capacitor slowly increases, the circuit current slowly decreases until the voltage across the capacitor is equal but opposite to the source voltage, and therefore the circuit current is zero. In a purely capacitive circuit, circuit current leads the capacitor voltage by 90°. On the other hand, *in a purely resistive circuit the circuit current and voltage developed across a resistor are always in phase.* This means that when a source voltage is applied across a resistor, the voltage developed across the resistor and current passing through the resistor will immediately rise to their Ohm's law values. The next question is: What happens to voltage and current in a circuit or network that contains both capacitance and resistance? The answer is shown in Figure 9-9(a). *In a resistive-capacitive (RC) network, the circuit current leads the voltage by some value between 0° and 90°, depending on the values of resistance and capacitance.* For the circuit shown in Figure 9-9(a), the input voltage is applied across the series-connected capacitor and resistor, and the output voltage is taken across the resistor.

To review the formulas involved, the phase difference between the output voltage (which is the voltage developed across the resistor, V_R or V_{out}) and the input voltage (V_{in}) is dependent on the capacitive reactance (X_C) and resistance (R) of the RC network. For the values given in Figure 9-9(a), the capacitor's capacitive reactance will be

$$X_C = \frac{1}{2\pi fC} = \frac{1}{2 \times \pi \times 100\ \text{Hz} \times 0.6\ \mu\text{F}} = 2.65\ \text{k}\Omega$$

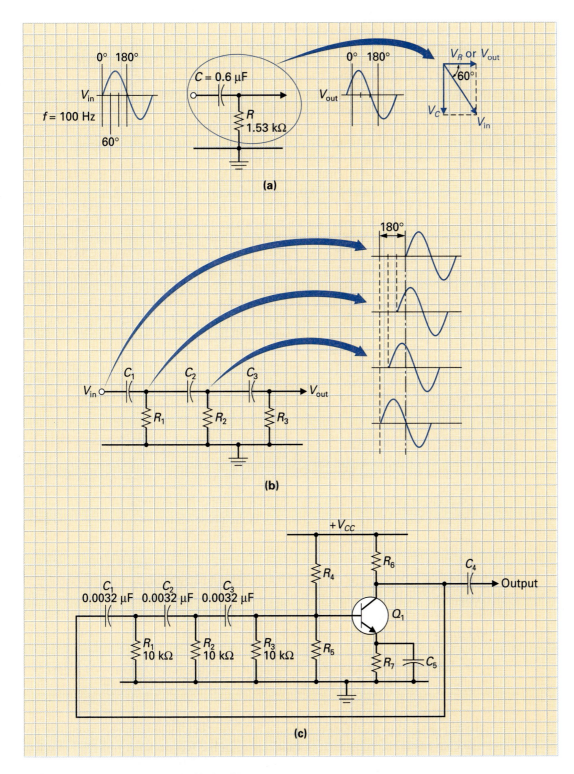

FIGURE 9-9 **The RC Phase-Shift Oscillator**

and therefore the degree of phase shift will be

$$\theta = \text{inv tan } \frac{X_C}{R} = \text{inv tan } \frac{2.65 \text{ k}\Omega}{1.53 \text{ k}\Omega} = 60°$$

Comparing the V_{in} waveform to the V_{out} waveform in Figure 9-9(a) or looking at the vector diagram, you can see that the output voltage leads the input voltage by 60°. As you have seen from the previous formulas, this phase shift is dependent on the values of R and C and the frequency of the applied ac input voltage. If the frequency of the input voltage were to change, the phase shift would also change.

Looking at Figure 9-9(b), you can see that if three 60° phase-shift networks were connected in series, the combined phase shift would be 180°. This 180° phase shift is exactly what is needed for positive feedback in an oscillator circuit. Therefore, if this network were placed between the collector output and base input of a common–emitter amplifier, the result would be a **phase-shift oscillator** circuit, as shown in Figure 9-9(c). Transistor Q_1 will typically have a voltage gain of between 40 and 50 so that it can compensate for the power loss in the RC phase-shift network. The circuit operates in much the same way as all of the previously discussed oscillator circuits. For instance, when power is first applied to the circuit, the bias resistors R_4 and R_5 ensure that Q_1 turns ON, resulting in a drop in voltage at the collector of Q_1. This decrease in voltage at the collector of Q_1 is inverted to an increase in voltage at the base of Q_1 by the 180° RC phase-shift network, which increases the forward bias applied to Q_1, sending it into saturation. When Q_1 saturates, there is no further change in the voltage at the collector of Q_1 and therefore the base of Q_1. Q_1 begins to cut off, causing its collector voltage to rise. This increase in voltage at the collector of Q_1 is inverted to a decrease in voltage at the base of Q_1 by the 180° RC phase-shift network, which decreases the forward bias applied to Q_1, sending it into cutoff. When the voltage at the collector of Q_1 reaches $+V_{CC}$, the voltage at the base of Q_1 will be the opposite, or 0 V, and the bias resistors R_4 and R_5 will provide sufficient base voltage to turn Q_1 ON. As Q_1 turns ON, there will be a drop in voltage at the collector of Q_1, which will be inverted to an increase in voltage at the base of Q_1 by the 180° RC phase-shift network. This increases the forward bias applied to Q_1, sending it into saturation. The oscillator action will then repeat with the continual rise and fall in voltage at the collector of Q_1, producing a sine wave at the output of the oscillator. The frequency of oscillation can be approximately calculated using the following formula:

$$f_R = \frac{1}{2 \times \pi \times \sqrt{6} \times R \times C}$$

R is the value of one of the resistors in the phase-shift network.

C is the value of one of the capacitors in the phase-shift network.

Phase-Shift Oscillator

An RC oscillator circuit in which the 180° phase shift is achieved with several RC networks.

■ **EXAMPLE:**

Calculate the frequency at which the phase-shift oscillator in Figure 9-9(c) will oscillate.

■ *Solution:*

$$f_R = \frac{1}{2 \times \pi \times \sqrt{6} \times R \times C} = \frac{1}{2 \times \pi \times 2.45 \times 10 \text{ k}\Omega \times 0.0032 \text{ } \mu\text{F}} = 2 \text{ kHz}$$

The RC phase-shift oscillator is typically only used as a fixed-frequency oscillator because any changes in the resistance or capacitance in the feedback network will interfere with the regenerative feedback. To improve oscillator frequency stability, many RC phase-shift oscillator circuits will have four 45°-RC networks, or six 30°-RC networks, which combined will still provide the 180° phase shift that is needed. Increasing the number of *RC* phase-shift networks will mean that each network will be responsible for less phase shift; therefore, frequency drift due to component aging and temperature will be reduced.

SELF-TEST EVALUATION POINT FOR SECTION 9-3

Use the following questions to test your understanding of Section 9–3.

1. *RC* oscillators are generally used in applications requiring a sine wave of less than _____.
2. For positive feedback, several *RC* phase-shift networks are needed to shift the feedback signal by _____ degrees.

REVIEW QUESTIONS

Multiple-Choice Questions

1. An oscillator's circuit basically converts
   a. ac to dc   **c.** ac to ac
   b. dc to ac   **d.** dc to dc

2. To oscillate, a tuned circuit will need an amplifier with
   a. regenerative feedback   **c.** negative feedback
   b. degenerative feedback   **d.** both (a) and (c) are true

3. Which of the *LC* oscillators makes use of a tuned transformer?
   a. Hartley oscillator   **c.** Armstrong oscillator
   b. Colpitts oscillator   **d.** Clapp oscillator

4. Which *LC* oscillator type makes use of a tapped inductor in the tuned circuit?
   a. Hartley oscillator   **c.** Armstrong oscillator
   b. Colpitts oscillator   **d.** Clapp oscillator

5. The _____ will produce an output sine-wave frequency that is determined by the values of an inductor in parallel with two series connected capacitors.
   a. Hartley oscillator   **c.** Armstrong oscillator
   b. Colpitts oscillator   **d.** Clapp oscillator

6. An oscillator circuit replaces the energy lost in the tank (due to its resistance) during the time the transistor is
   a. OFF   **b.** ON

7. If a high-frequency stable output is needed in a circuit application, the ideal choice is a(an) _____ oscillator.
   a. *LC*   **b.** crystal   **c.** *RC*

8. If a low-frequency oscillation of approximately 120 Hz is needed, the ideal choice would be a(an) _____ oscillator.
   a. *LC*   **b.** crystal   **c.** *RC*

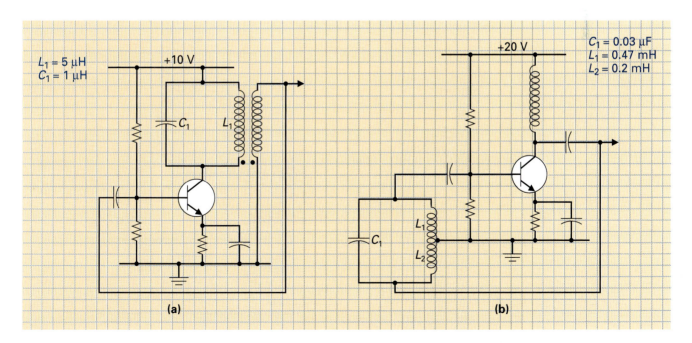

FIGURE 9-10 *LC* **Oscillator Circuits.**

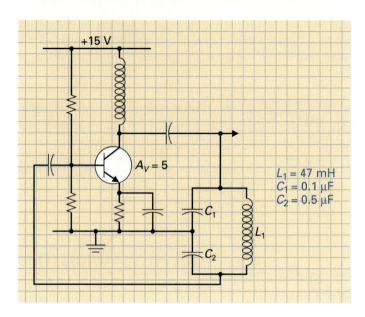

FIGURE 9-11 An *LC* Oscillator.

$L_1 = 47$ mH
$C_1 = 0.1$ µF
$C_2 = 0.5$ µF

9. In applications where a high-frequency sine-wave oscillation is needed, and the frequency stability is not important,
a(an) _____ oscillator is used.
a. *LC* **b.** crystal **c.** *RC*

10. If the ac signal frequency applied to a crystal matches the
crystal's natural series resonant frequency, the crystal's impedance will be
a. maximum **b.** minimum **c.** unaltered **d.** zero

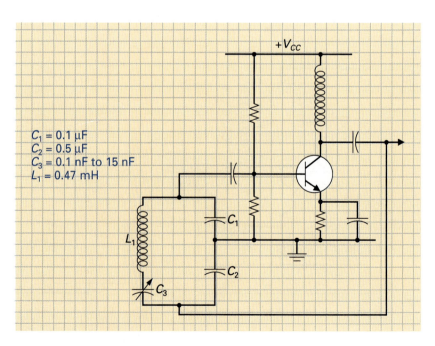

$C_1 = 0.1$ µF
$C_2 = 0.5$ µF
$C_3 = 0.1$ nF to 15 nF
$L_1 = 0.47$ mH

FIGURE 9-12 An *LC* Oscillator.

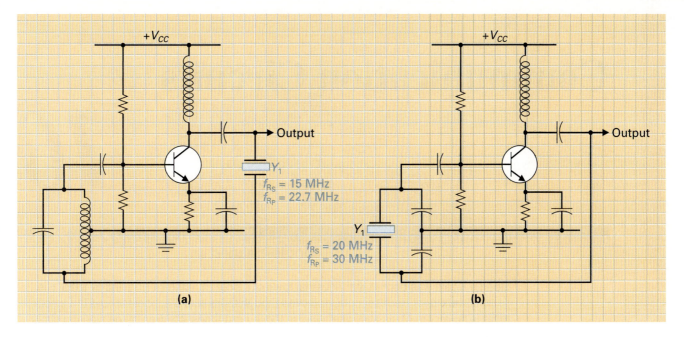

Y_1
$f_{RS} = 15$ MHz
$f_{RP} = 22.7$ MHz

Y_1
$f_{RS} = 20$ MHz
$f_{RP} = 30$ MHz

(a)   (b)

FIGURE 9-13   Crystal Oscillator Circuits.

Practice Problems

11. Identify the circuits shown in Figure 9-10, and then calculate their operating frequency.

12. Identify the circuit shown in Figure 9-11, and then calculate the circuit's operating frequency.

13. Identify the circuit shown in Figure 9-12, and then calculate its operating frequency range.

14. Identify the circuits shown in Figure 9-13, and then determine the operating frequency of these oscillator circuits.

15. Identify the circuit shown in Figure 9-14, and then calculate its operating frequency.

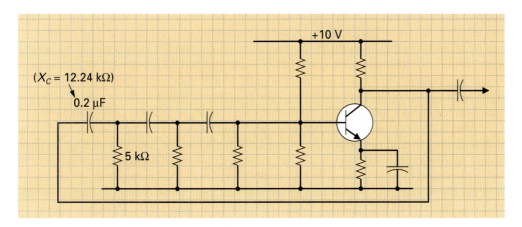

$(X_C = 12.24$ k$\Omega)$
0.2 µF
5 kΩ
+10 V

FIGURE 9-14   *RC* Oscillator Circuit.

Communication Circuits

Boole-Headed

George Boole was born in the industrial town of Lincoln in eastern England in 1815. His parents were poor tradespeople, and even though there was a school for boys in Lincoln, there is no record of his ever attending. In those hard times, children of the working class had no hope of receiving any form of education, and their lives generally followed the pattern of their parents. George Boole, however, broke the mold. He rose from these humble beginnings to become one of the most respected mathematicians of his day.

Boole's father had taught himself a small amount of mathematics, and because his son of six seemed to have a thirst for learning, he began to pass on his knowledge. At eight years old, Boole had surpassed his father's understanding and craved more. He quickly realized that his advancement was heavily dependent on understanding Latin. Luckily, a family friend who owned a local book shop knew enough about the basics of Latin to get Boole started. Once this friend had taught Boole all he knew, Boole continued with the books at his disposal. By the age of twelve he had mastered Latin and by fourteen he had added Greek, French, German, and Italian to his repertoire.

At the age of sixteen, however, poverty stood in his way. Because his parents could no longer support him, he was forced to take a job as a poorly paid teaching assistant. After studying all the material in the entire school system, he left four years later and opened his own school in which he taught all subjects. He discovered his command of mathematics was weak, so he began studying the mathematical journals at the local library in an attempt to stay ahead of his students. He quickly discovered that he had a talent for mathematics. As well as mastering all the present-day ideas, he began to develop some of his own, which were later accepted for publication. After a stream of articles, he became so highly regarded that he was asked to join the mathematics faculty at Queens College in 1849.

After accepting the position, Boole concentrated more on his ideas, one of which was to develop a system of symbolic logic. In this system, he created a form of algebra that had its own set of symbols and rules. Using this system, Boole could encode any statement that had to be proved (a proposition) into his symbolic language, then manipulate it to determine whether it was true or false. Boole's algebra has three basic operations that are simulated electronically by the logic gates AND, OR, and NOT. Using these three operations, Boole could add, subtract, multiply, divide, and compare. Boole's theory was that if all logical arguments could be reduced to one of two basic levels, the questionable middle ground would be removed, making it easier to arrive at a valid conclusion. Therefore, logic gates are binary in nature, dealing with only two entities: TRUE or FALSE, YES or NO, OPEN or CLOSED, ZERO or ONE, and so on.

At the time, Boole's system, later called *Boolean algebra*, was either ignored or criticized by colleagues who called it a folly with no practical purpose. Almost a century later, however, scientists would combine George Boole's Boolean algebra with binary numbers and make possible the digital electronic computer.

Introduction

Communication is defined as *a process by which information is exchanged*. The two basic methods of transferring information are the *spoken word* and the *written word*. Communication of the spoken word began face-to-face, then evolved into radio communications, telephone communications, and now video telecommunications. Communication of the written word began with hand-carried letters, then evolved into newspapers, the mail system, the telegraph, and now electronic mail.

At a very early stage, it was found that the distance we could communicate was important. Systems of long-distance communication developed, beginning with smoke signals and beating drums. However, it was not until the late 1800s that scientists discovered that a high-frequency electrical current passing through a conductor would radiate an energy wave that could be modified or altered to carry information. These radiated *electromagnetic waves,* or *radio waves,* could be used to communicate information of different types. For example (to name but a few applications), radio stations transmit audio information such as voice and music, television stations transmit moving-picture information, and cellular telephones transmit voice or facsimile (graphic) data. All of these radio communication transmitters and receivers need *high-frequency amplifiers* to boost the amplitude of the high-frequency information signals in both the transmitting circuits and receiving circuits.

In Chapters 8 and 9, we studied basic amplifier principles and the operation, characteristics, and testing of amplifier circuits and oscillator circuits. In this chapter, we will continue our coverage of amplifiers, concentrating on amplifier circuits that operate at progressively higher frequencies. These circuits are named after the particular frequency that they are designed to amplify: *radio-frequency (RF) amplifiers, intermediate-frequency (IF) amplifiers,* and *video-frequency (VF) amplifiers.*

10-1 RADIO-FREQUENCY (RF) AMPLIFIER CIRCUITS

Radio-frequency (RF) amplifiers are mainly used in communication systems, so we will first need to understand the basics of a simple communication system if we are to see why RF amps are so important. As an example, Figure 10-1 shows the block diagram of a basic AM (amplitude-modulated) transmitter, with each block representing a bipolar transistor circuit. The AF voltage amplifier and AF power amplifier circuits were discussed in Chapter 8 and oscillator circuits were covered in Chapter 9, so we already know that these circuits are used to generate a continually repeating ac signal. This leaves us with three remaining circuits, all of which contain an RF amplifier.

Before proceeding to the operation of these RF amplifier circuits, let us examine the basic function of each of the blocks or circuits in the AM transmitter shown in Figure 10-1. The high-frequency oscillator supplies a repeating, high-frequency, constant-amplitude, sine-wave signal to the RF amplifier/multiplier circuit that triples the input frequency and applies its output to the RF amplifier/modulator. Using the example values given in Figure 10-1, the oscillator applies a 300-kHz sine wave to the frequency multiplier, which converts this input to a 900-kHz sine wave (300 kHz × 3 = 900 kHz). The radio-frequency sine wave from the RF frequency multiplier circuit is applied to the RF amplifier/modulator along with an audio-frequency input signal from the microphone, which is amplified by an AF voltage and AF power amplifier. The RF amplifier/modulator combines the low-AF information signal and the high-RF wave, producing an amplitude-modulated (AM) output signal. With **amplitude modulation,** each electromagnetic transmitting station is assigned its own transmit frequency by the FCC, or Federal Communications Commission, so that its radiated signal will not interfere with other stations. The frequency at which a station transmits is called its **carrier frequency.** For example, your local AM radio stations each have their own as-

Radio-Frequency (RF) Amplifiers

An amplifier that has one or more transistor stages designed to amplify radio-frequency signals.

Amplitude Modulation

The process by which the amplitude of the high-frequency carrier is changed or varied to follow the amplitude of the low-frequency information signal.

Carrier Frequency

The frequency generated by a radio transmitter, or the average frequency of the emitted wave when modulated by an information signal.

Modulation

The process of converting a low-frequency information signal to a high-frequency carrier signal.

Electric or Voltage Field

A field or force that exists in the space between two different potentials or voltages.

Magnetic or Current Field

Magnetic lines of force traveling from the north to the south pole of a magnet.

Electromagnetic Field

Field having both an electric (voltage) and magnetic (current) field.

RF Amplifier/ Multiplier Circuit

A radio-frequency amplifier circuit in which the frequency of the output is an exact multiple of the input frequency.

signed carrier frequency in the AM radio band, and this high radio frequency acts as a carrier for the station's audio-frequency (AF) information signals (speech and music). Each station will have to convert their lower-frequency audio signals to their assigned higher carrier frequency before transmitting. This process of converting a low-frequency information signal to a high-frequency carrier signal is called **modulation.** The process by which the amplitude of the high-frequency carrier is changed or varied to follow the amplitude of the low-frequency information signal is called amplitude modulation. This action is best seen by looking at the change in the waveforms in Figure 10-1. By studying the RF amplifier/modulator's output waveform, you can see that the amplitude of the carrier, which is the high-frequency sine wave from the frequency multiplier, varies at the same rate as the audio signal, which originated from the microphone. In fact, the envelope of the carrier is an exact replica of the audio information signal. The final RF fixed-tuned amplifier, which in our example would be tuned to 900 kHz, is included to boost the amplitude-modulated or amplitude-varied carrier signal before it is applied to the antenna. It is, in fact, the voltage that is applied across the antenna by the final-stage RF amp that generates an **electric (electro) or voltage field.** The current that is generated by the final-stage RF amp and passes through the antenna generates a **magnetic or current field.** These combined fields make up an **electromagnetic field** that radiates away from the antenna in all directions at the speed of light (approximately 186,000 miles per second).

The radio frequency band includes any frequencies from about 10 kHZ to about 30,000 MHz. Let us now examine, in more detail, the three bipolar transistor RF amplifier circuits discussed in Figure 10-1.

10-1-1   *An RF Amplifier/Multiplier Circuit*

Figure 10-2 repeats the basic AM transmitter block diagram and focuses in on the operation of the **RF amplifier/multiplier circuit.** This block is included in the basic AM transmitter

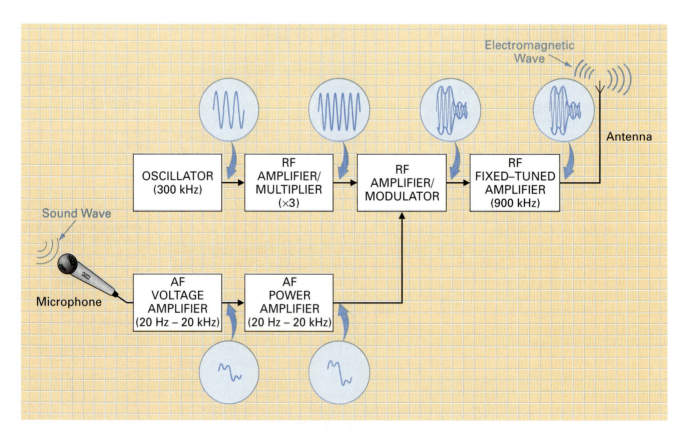

FIGURE 10-1   **Basic AM (Amplitude Modulated) Transmitter.**

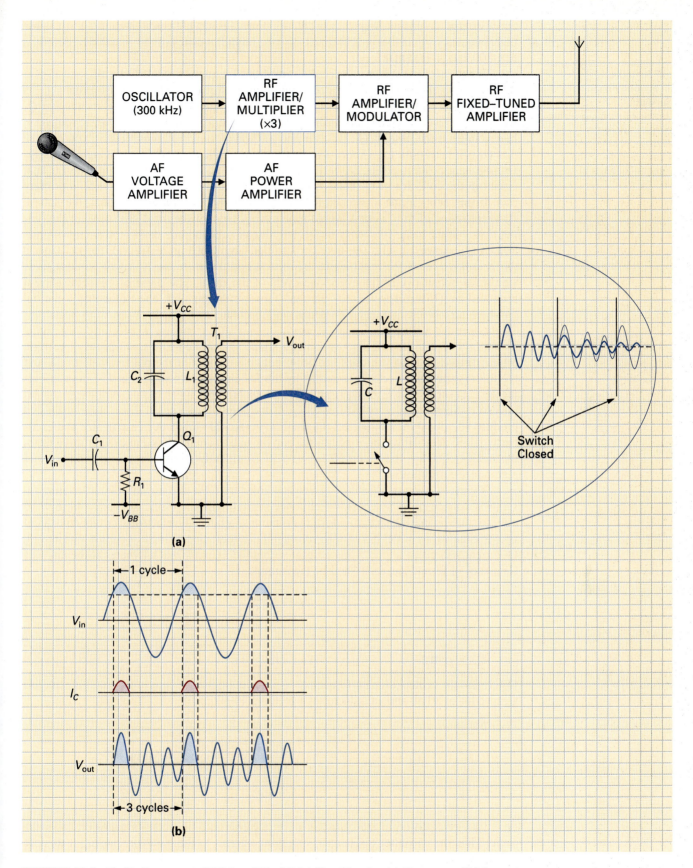

FIGURE 10-2 Radio Frequency (RF) Amplifier/Multiplier Circuits. (a) Frequency Tripler Circuit. (b) Frequency Tripler Input/Output Waveforms.

to convert the stable low-frequency oscillation from the 300-kHz oscillator circuit to a stable high-frequency signal that is three times (3×) the input frequency (input = 300 kHz, output = 3 × 300 kHz = 900 kHz).

Figure 10-2(a) shows a basic **RF amplifier/frequency tripler circuit,** and Figure 10-2(b) shows the circuit's associated input/output waveforms. The bipolar transistor Q_1 is class-C biased by the negative dc bias voltage ($-V_{BB}$) and resistor R_1. This means that Q_1 will only conduct when the input voltage is large enough to overcome the negative reverse bias voltage at the base of Q_1. Referring to the waveforms in Figure 10-2(b), you can see that Q_1 conducts only for a short time during the positive peak of the input signal voltage. During these positive peaks, Q_1 will turn ON and send a pulse of current into the LC tank circuit in the collector of Q_1 made up of capacitor C_2 and the inductance of T_1's primary (L_1). If this tank circuit was tuned to the same frequency as the input signal, the natural flywheel action of the LC tank would reproduce a copy of the input sine wave at the output that is of the same frequency but slightly larger in amplitude due to the amplifying action of Q_1. However, with the frequency tripler circuit in Figure 10-2(a), the LC tank is tuned to three times the input signal frequency, and therefore the tank will produce three sine-wave output cycles for every one cycle of the input. The pulse of collector current produced by Q_1 will *shock excite* the tank into oscillation, and, due to the tank's natural flywheel action, the parallel LC circuit will oscillate at its natural resonant frequency. When operated with class-C bias, it would seem that a transistor would introduce signal distortion because the full sine-wave input signal is distorted to a small pulse of current. Signal distortion, however, does not occur because the rest of the positive half-cycle, and the full negative half-cycle, is recreated by the LC tank's flywheel action. The inset in Figure 10-2(a) shows transistor Q_1 simplified to a switch that, when closed, will cause the LC tank to oscillate at its resonant frequency. If the transistor switch did not close again and produce additional current pulses, the oscillations would slowly decrease in size and eventually fall to zero. However, if additional current pulses are produced at exactly three times the input frequency, these current pulses will continually reenergize the tank circuit and maintain the oscillations.

To review, the name *flywheel action* is derived from the fact that this circuit's action resembles a mechanical flywheel, which, once started, will keep spinning back and forward until friction reduces the magnitude of the rotations to zero. The electronic circuit equivalent of a mechanical flywheel is a parallel resonant LC circuit. Supplying this circuit with a pulse of current will cause a circulating current within the LC tank that alternates at the circuit's frequency of resonance, calculated by the formula

$$f_R = \frac{1}{2\pi \sqrt{LC}}$$

■ **EXAMPLE:**

Determine the center resonant frequency of a tuned amplifier circuit if the parallel tank has a capacitor of 80 pF and an inductor of 1 mH.

■ *Solution:*

$$f_R = \frac{1}{2\pi \sqrt{LC}} = \frac{1}{2\pi \sqrt{(1 \text{ mH}) \times (80 \text{ pF})}} = 562.2 \text{ kHz}$$

Energy is stored in the capacitor in an electric field between its plates on one half-cycle. Then the capacitor discharges, supplying current to the inductor, causing it to store energy in a magnetic field during the other half-cycle. Once the inductor's magnetic field has built up to a maximum, the magnetic field will begin to collapse and supply a charge current to the capacitor. After the capacitor has charged to a maximum, it will discharge, supplying a current back to the inductor, and so on. The oscillations within a tank circuit will eventually fall to zero, due mainly to the coil resistance of the inductor. The energy "storing action" accounts for why this circuit is also called a *tank*.

The circuit in Figure 10-2(a) is called a *frequency tripler* because the output frequency is three times the input frequency, as seen in Figure 10-2(b). To ensure that the pulse of current occurs at the right point to sustain oscillations, the tank must be "tuned" to an exact multiple of the input frequency so that the current pulse from Q_1 will reinforce the oscillations. Frequency-doubler and frequency-quadrupler circuits are often used in RF circuits to convert a stable low-frequency oscillation to a stable high-frequency oscillation. Operating with class-C bias means that Q_1 is conducting (and consuming power) for only a small portion of time; therefore, the circuit is more efficient than a class-A circuit. Signal distortion does not occur with a class-C biased transistor if an *LC* tank is included in the circuit because the natural flywheel action of a parallel resonant circuit will recreate the entire sine-wave cycle at the output.

10-1-2   *An RF Amplifier/Modulator Circuit*

Figure 10-3 again repeats the basic AM transmitter block diagram and, in this case, focuses on the operation of a bipolar transistor RF amplifier/modulator circuit.

Referring to the circuit in Figure 10-3, you can see that the RF carrier from the frequency-multiplier circuit is applied through coupling capacitor C_1 to the base of the class-A voltage-divider biased transistor Q_1. The AF modulating signal from the AF amplifiers is applied to the emitter circuit of Q_1 via the transformer T_1. Remember that there is a very large frequency difference between the RF carrier input and the AF modulating signal input. For example, using the values chosen in our basic AM transmitter block diagram, the RF carrier input frequency is 900 kHz, while the AF modulating signal frequency could typically be about 900 Hz, making the RF carrier frequency 1,000 times higher than the AF modulating signal frequency.

$$\frac{900 \text{ kHz}}{900 \text{ Hz}} = 1,000$$

Let's see how this circuit combines these two input signals to produce a 900-kHz RF sine-wave output that varies in amplitude in accordance with the AF modulating signal input. Coupling capacitors C_1 and C_3 and emitter-bypass capacitor C_2 have values that will offer no opposition to the 900-kHz RF carrier signal. So in terms of the RF carrier input signal, this circuit is simply a common–emitter amplifier stage. With regard to the AF modulating signal, the emitter current of Q_1 is determined by the resistance of R_4 and the voltage developed across R_4. The voltage on the top of R_4 is determined by the voltage-divider bias resistors R_1 and R_2 and the 0.7-V drop between Q_1's base and emitter. When the AF modulating signal input swings positive, the voltage at the bottom of R_4 increases and the voltage difference (or drop) across R_4 decreases, causing a decrease in I_E. On the other hand, when the AF modulating signal input swings negative, the voltage at the bottom of R_4 decreases and the voltage difference or drop across R_4 increases, causing an increase in I_E. In Chapter 8, it was stated that the voltage gain of an amplifier stage is equal to the ratio of the transistor's ac collector resistance to ac emitter resistance.

$$A_V = \frac{R_{C(AC)}}{R_{E(AC)}}$$

As the AF modulating input signal increases and decreases, the ac emitter resistance of Q_1 is increased and decreased and the gain of Q_1 is increased and decreased. To be specific, as the AF modulating signal swings positive, the voltage across R_4 decreases, causing I_E to decrease, the ac emitter resistance to increase, and the transistor's gain to decrease. On the other hand, as the AF modulating signal swings negative, the voltage across R_4 increases, causing I_E to increase, the ac emitter resistance to decrease, and the transistor's gain to increase. The AF modulating signal increases and decreases the gain of Q_1 and therefore the

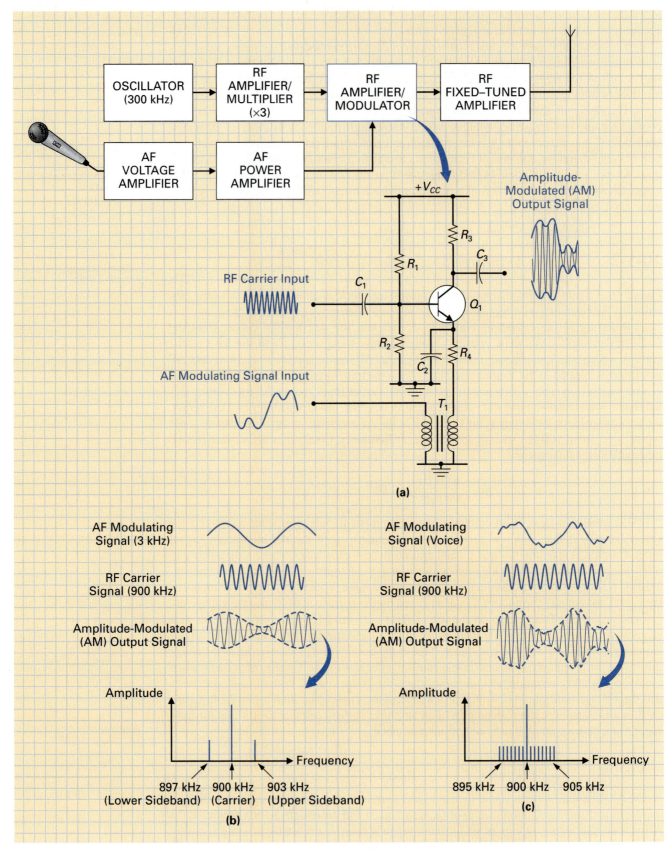

FIGURE 10-3 **Radio-Frequency (RF) Amplifier/Modulator Circuit.**

amplitude of the signal being amplified, which is the RF carrier signal. This **emitter modulator** circuit shows how an RF carrier signal can be amplitude modulated so that its envelope will follow the changes in the AF information signal.

Although communication techniques will be discussed in a later chapter, we will need to examine our amplitude-modulated output wave in a little more detail so that we can better understand RF amplifier circuits. By referring to the waveforms shown in Figure 10-3(b), you can see that if a 900-kHz RF carrier is modulated by a constant 3-kHz AF tone, a 900-kHz signal is produced that has a 3-kHz envelope. If we were to study the frequency components within this amplitude-modulated wave, we would find that it is made up of a 900-kHz center carrier frequency with two other frequency components at 897 kHz and 903 kHz, as seen in the frequency spectrum shown in Figure 10-3(b). These two additional frequencies are called **sidebands:** the 897-kHz signal is called the **lower sideband,** and the 903-kHz signal is called the **upper sideband.** In this example, the AM radio frequency signal would have a bandwidth of 6 kHz.

$$\text{BW} = f_{\text{HI}} - f_{\text{LO}} = 903 \text{ kHz} - 897 \text{ kHz} = 6 \text{ kHz}$$

In most applications, the RF carrier is modulated by a voice or music signal instead of a constant tone, as shown in Figure 10-3(c). In this case, the modulating signal contains many different frequencies which are constantly changing. Because one set of sidebands (upper and lower) is produced for each modulating frequency, the resulting AM waveform will have the frequency spectrum shown in Figure 10-3(c). In most applications, the bandwidth is always twice the highest modulating frequency.

> ■ **EXAMPLE:**
>
> If a 690-kHz carrier is voice modulated by a signal that is between 500 Hz to 5 kHz, what will be the transmission's bandwidth?
>
> ■ *Solution:*
>
> The bandwidth is always twice the highest modulating frequency.
>
> $$\text{BW} = 2 \times f_{\text{AF(HI)}} = 2 \times 5 \text{ kHz} = 10 \text{ kHz}$$
>
> Centered at 690 kHz, the lower sideband will begin at 685 kHz and the upper sideband will extend up to 695 kHz.

The carrier, therefore, remains at a constant amplitude and frequency at all times; however, the sidebands are constantly changing amplitude and frequency in accordance with the modulating signal.

10-1-3   *An RF Fixed-Tuned Amplifier Circuit*

Figure 10-4(a) continues our coverage of the basic AM transmitter by showing a final-stage **RF fixed-tuned amplifier circuit.** This circuit contains an *LC* tuned circuit in both the input and the output that will pass a select band of frequencies and reject all other frequencies outside the band. In this basic AM transmitter example, the tuned circuits will both be tuned to our transmit frequency of 900 kHz, and therefore our amplitude-modulated RF carrier will be amplified and passed on to the transmitting antenna.

A Tuned Circuit's Bandwidth and Selectivity

At resonance, a series *LC* tank has a minimum impedance and therefore a maximum current, whereas a parallel *LC* tank circuit has a maximum impedance and minimum current at resonance. Because the characteristics of the *LC* tuned circuit are so important to the op-

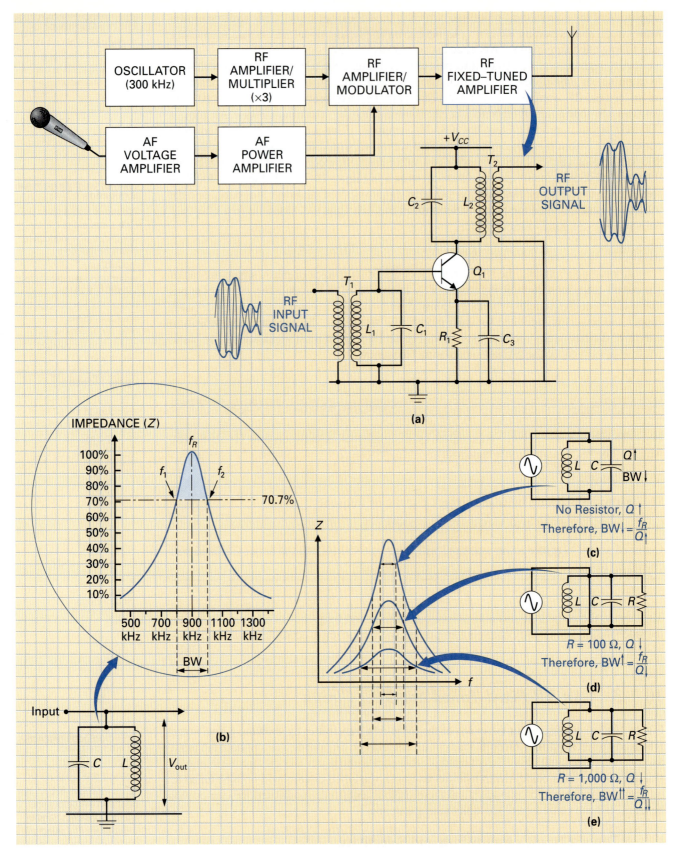

FIGURE 10-4 Fixed-Tuned Radio-Frequency (RF) Amplifier Circuit. (a) Basic Circuit. (b) Parallel Resonant Circuit's Response Curve. (c), (d), (e) Bandwidth Increase as Q Decreases.

eration of the RF amplifier, let us briefly review the tuned circuit material discussed previously in dc/ac electronics. Referring to Figure 10-4(b), you can see that a parallel resonant *LC* circuit has a maximum impedance at the resonant frequency. At frequencies above and below resonance, the impedance falls off to a low value. This high impedance at resonance is important in a tuned RF amplifier circuit because the voltage developed across a parallel resonant circuit is determined by the impedance of the resonant circuit. This impedance determines the amplitude of the signal voltage delivered to the output.

$$V_{out} = I \times Z$$

Figure 10-4(b) shows a parallel resonant *LC* circuit and its frequency response curve in the inset. At resonance, the parallel circuit has a maximum impedance ($Z\uparrow$) and therefore the voltage applied to the output will also be maximum at this frequency ($V_{out}\uparrow$). At frequencies above and below resonance, the impedance falls to a lower value ($Z\downarrow$) and therefore the voltage applied to the output also falls to a lower value ($V_{out}\downarrow$).

The bandwidth of a tuned circuit is the band of frequencies that will be passed through to the output and that have a usable power level of 50% or greater. These frequencies that are above the half-power points correspond to the 70.7% points on the impedance axis of the frequency response curve shown in Figure 10-4(b). Between frequencies f_1 and f_2, the amplitude of the output voltage signal will be greater than 70.7% of the maximum voltage. The bandwidth of a standard AM radio station is generally always twice the highest modulating frequency. Therefore, if the highest AF modulating frequency is 3 kHz, the bandwidth will be 6 kHz. To properly pass an AM carrier of 900 kHz and its sidebands within a 6-kHz bandwidth, we would have to make sure that frequencies between 897 kHz and 903 kHz are coupled to the output with a usable power level of 50% or greater.

The bandwidth of the parallel *LC* circuit is a very important characteristic because it determines the **selectivity** of the tuned circuit and the selectivity of the RF amplifier. By definition, selectivity is the ability of a circuit to select the wanted signal(s) and reject the unwanted signal(s). For example, a circuit with a wide bandwidth will pass a wide band of frequencies through to the output at a usable output voltage, and this circuit would be said to have a "poor selectivity." On the other hand, a circuit with a very narrow bandwidth will pass very few frequencies through to the output at a usable voltage level, and therefore this circuit is said to be "highly selective." Bandwidth and selectivity are inversely proportional: a circuit with a large bandwidth is not very selective, while a circuit with a small bandwidth is very selective.

Selectivity

The ability of a circuit to select the wanted signal(s) and reject the unwanted signal(s).

$$\text{Selectivity} \propto \frac{1}{\text{Bandwidth}}$$

The bandwidth of a tuned circuit, and therefore the tuned circuit's selectivity, is largely determined by the *Q* of the circuit.

$$\text{BW} = \frac{f_R}{Q}$$

This *Q* factor indicates the quality of a parallel resonance circuit and is a ratio of the tuned circuit's reactance to resistance.

$$Q = \frac{X}{R}$$

Stated another way, the *Q* of a tank circuit is equal to the ratio of energy stored in the tank to energy lost in the tank. Because the only resistance in the tank is that of the inductor's winding resistance, the *Q* of a tuned circuit is more specifically equal to the inductive reactance of the coil divided by the coil's winding resistance.

$$Q = \frac{X_L}{R_W}$$

X_L = Reactance of the coil at resonance
R_W = resistance of the winding

An amplifier's tuned circuit is tuned to 800 kHz and has a reactance-to-resistance ratio 6.2 to 1. Calculate

1. The tuned circuit's Q,
2. The bandwidth of the tuned circuit at the ratio of 6.2 to 1, and
3. The bandwidth of the tuned circuit if a resistor is added to the tuned circuit, doubling the tuned circuit's resistance.

■ *Solution:*

1. As stated in the question, the reactance is 6.2 times greater than the resistance, so the Q of the tuned circuit will be

$$Q = \frac{X}{R} = \frac{6.2}{1} = 6.2$$

2.
$$BW = \frac{f_R}{Q} = \frac{800 \text{ kHz}}{6.2} = 129 \text{ kHz}$$

3. If the resistance of the tuned circuit is doubled, the Q will be halved to 3.1. Therefore, the bandwidth will be

$$BW = \frac{f_R}{Q} = \frac{800 \text{ kHz}}{3.1} = 285 \text{ kHz}$$

The bandwidth, and therefore selectivity, of a tuned circuit can be controlled by simply increasing or decreasing the tuned circuit's resistance, as shown in Figure 10-4(c), (d), and (e). For example, no additional resistance other than that of the inductor's coil resistance (capacitors have almost no resistance) will result in a low value of tuned circuit resistance ($R\downarrow$), a high circuit Q ($Q\uparrow$), and small bandwidth (BW$\downarrow$), and the circuit will have a very high circuit selectivity. On the other hand, by increasing the tuned circuit's resistance ($R\uparrow$), we will decrease the circuit's Q ($Q\downarrow$), increasing the circuit's bandwidth (BW$\uparrow$), and the tuned circuit will have a very broad selectivity.

Comparing the three frequency response curves in Figure 10-4(c), (d), and (e), it would seem that a high Q is always the best situation because it yields a higher impedance at resonance and therefore a higher gain and selectivity. However, this is not always the case. For example, radio-frequency amplifiers in television receivers must have a broad bandwidth of 6 MHz, whereas radio-frequency amplifiers in a transmitter will have a narrow bandwidth to prevent unwanted signals from being transmitted.

Typically, the Q of an RF amplifier's tuned circuit is adjusted by adding in a parallel resistance, called a **swamping resistor,** to increase the bandwidth.

Swamping Resistor

A parallel resistance added to the RF amplifier tuned circuit to increase the bandwidth.

Transformer Coupling and Frequency Response

Figure 10-5 concentrates on the characteristics of the RF amplifier's tuned-input and tuned-output transformers. By including step-up voltage transformers in an RF amplifier, additional voltage gain can be achieved. These transformers can also determine the tuned circuit's frequency response by controlling the transformer's **coefficient of coupling.** To review, the coefficient of coupling for a transformer is the ratio of the number of magnetic lines of force that cut the secondary compared to the number of magnetic flux lines being produced by the primary. For example, Figures 10-5(a), (b), and (c) show how the coefficient of coupling between a transformer's primary and secondary can be changed to alter a tuned circuit's bandwidth. As an example, let us imagine that we wish to pass all of the upper- and lower-frequency components of a station between f_1 and f_2 and block all other frequencies.

Coefficient of Coupling

The degree of coupling that exists between two circuits.

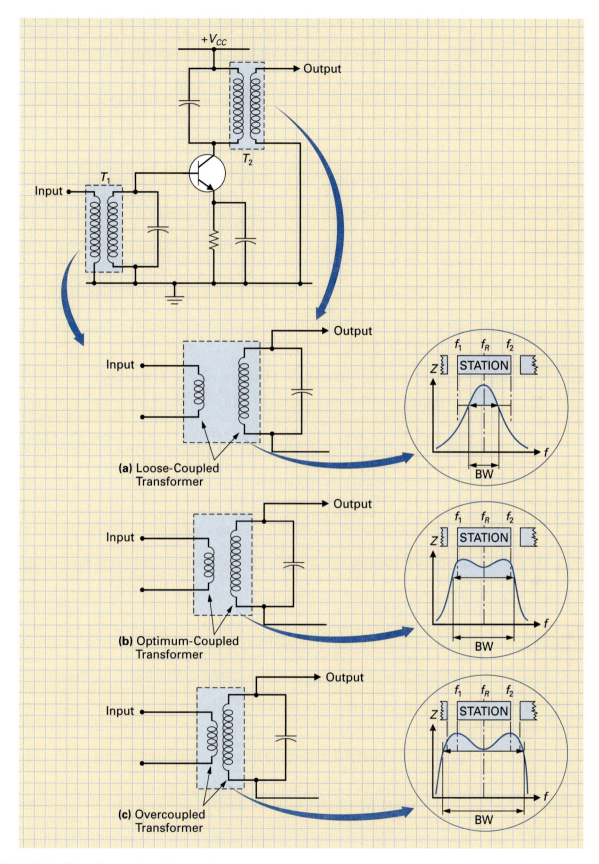

FIGURE 10-5   **Transformer Coupling and Frequency Response.**

Loose-Coupled Transformer

A transformer in which the coils are far apart and less signal energy is transferred between primary and secondary.

In Figure 10-5(a), the coefficient is low because the transformer coils are far apart and less signal energy is transferred between primary and secondary. Looking at the frequency response for this **loose-coupled transformer,** you can see that the transformer passes the higher-amplitude center frequency (or resonant frequency). However, frequencies either side of resonance (f_1 and f_2) are not coupled to the secondary with sufficient amplitude. This means that the lower-frequency (f_1) elements of our modulating signal and the higher-frequency (f_2) elements of our modulating signal will not be passed, causing signal frequency distortion.

In Figure 10-5(b), the coefficient of coupling has been increased so that more signal energy is transferred, causing an increase in the bandwidth and therefore a better response to a signal's resonant frequency and associated higher- and lower-frequency components.

Overcoupled

A transformer in which the coils are so close together that a dip appears in the center of the response curve, causing the resonant frequency component to be of a lower amplitude than the upper- and lower-frequency components of the signal.

In Figure 10-5(c), the transformer's primary and secondary have been **overcoupled** and, as a result, a dip appears in the center of the response curve-causing the resonant frequency component to be of a lower amplitude than the upper- and lower-frequency components of the signal. The other disadvantage with an overcoupled transformer is that its extremely wide bandwidth will permit unwanted outside signals to pass through to the next stage.

Ideally, transformers have optimum coupling, as shown in Figure 10-5(b), so that the circuit provides an equal response to all the desired frequencies between f_1 and f_2. Frequencies outside the f_1 to f_2 band will meet with a reduced impedance, and the voltage gain for these circuits will be low and outside the pass band of the amplifier.

Neutralizing Miller-Effect Capacitance in Transistors

Miller-Effect Capacitance (C_M)

An undesirable inherent capacitance that exists between the junctions of transistors.

Figure 10-6 shows our fixed-tuned RF amplifier circuit and details the effects of an undesirable inherent capacitance within transistors called **Miller-effect capacitance (C_M).** A small value of capacitance exists between the junctions of all transistors, as shown in Figure 10-6(a). These **junction capacitances** are determined by the physical size of the junction, the spacing between the transistors' leads, and the junction bias condition. At low frequencies these capacitances have very little effect because X_C is large. However, when frequency increases, X_C decreases, and the capacitance between a transistor's collector and base, which is known as the transistor's Miller capacitance [as shown in Figure 10-6(b)], becomes a problem: Its low-impedance path will couple a portion of the output signal back to the input. Because a phase inversion exists between the base and collector of a common emitter amplifier, this feedback voltage will be opposite in polarity to the input signal voltage, causing the input signal voltage to be decreased and the amplifier's gain to also be decreased. This type of out-of-phase feedback is called **negative feedback** because it has a degenerative effect.

Junction Capacitance

The small value of capacitance that exists between the junctions of all transistors.

Negative Feedback

An out-of-phase feedback in which the feedback voltage is opposite in polarity to the input signal voltage, causing the input signal to be decreased.

Figure 10-6(c) shows how the negative feedback caused by Miller capacitance can be neutralized. To counter the effect of Miller capacitance in high-frequency amplifiers, a **neutralizing capacitor (C_N)** is included to couple an in-phase signal back to the base of the RF transistor to compensate for the out-of-phase signal coupled back to the base of the transistor by the Miller capacitance C_M. Coupling an in-phase voltage back to the base of Q_1 in this way effectively neutralizes the out-of-phase feedback voltage also being fed back to the base of Q_1.

Neutralizing Capacitor (C_N)

A capacitor that couples an in-phase signal back to the base of the RF transistor to compensate for the out-of-phase signal coupled back to the base of the transistor by the Miller capacitance.

10-1-4   *An RF Variable-Tuned Amplifier Circuit*

Up until this point, we have seen how the RF amplifier can be used in a basic AM transmitter. In this section, we will see how the RF amplifier finds application in a basic AM receiver.

Tuned AM Receiver

An AM receiver circuit that can have its components varied so that the circuit responds to one selected frequency yet heavily attenuates all other frequencies.

Figure 10-7(a) shows the simplified block diagram of a **tuned AM receiver.** The electromagnetic waves transmitted by all of the different local AM radio stations will be present at the receiving antenna, which is simply a piece of wire. These electromagnetic signals will cut through the receiving antenna and induce an amplitude-modulated signal that corresponds in wave shape to the original AM transmitted signal; however, its amplitude will now be considerably lower, typically only a few hundred microvolts. To boost this very weak

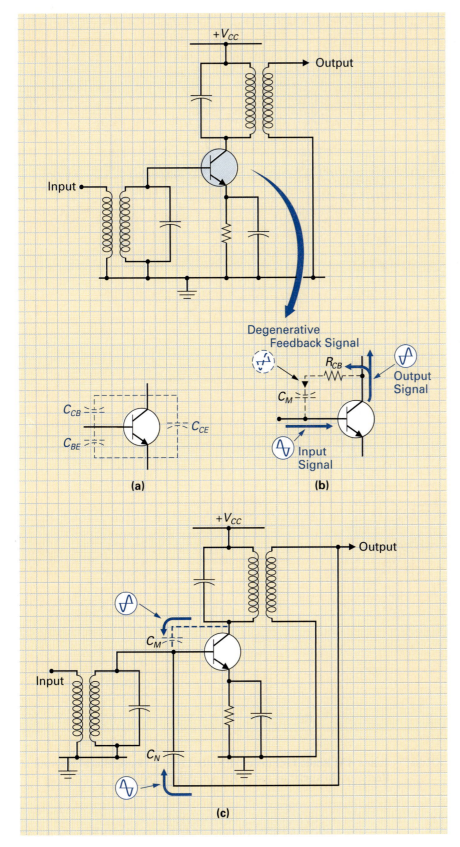

FIGURE 10-6 Neutralizing Miller-Effect Capacitance in Transistors.

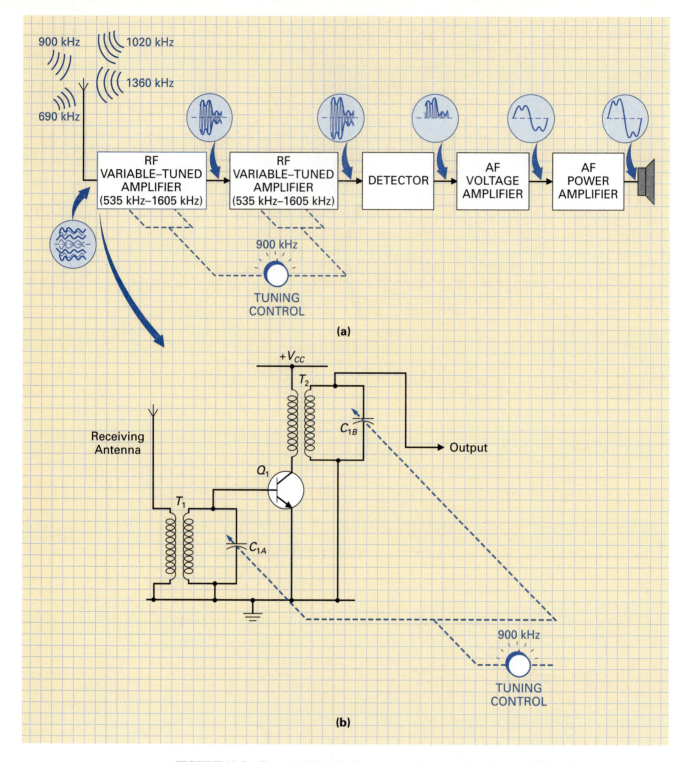

FIGURE 10-7 Tuned AM Radio Frequency (RF) Receiver. (a) Basic Block Diagram of RF Receiver. (b) Variable-Tuned RF Amplifier Circuit.

signal, receiver circuits will generally contain several front-end RF amplifier stages to boost the received signal to a more usable amplitude before it is applied to the detector.

Before discussing the detector circuit, let us examine these **variable-tuned RF amplifier** stages in more detail. With every local AM radio station signal at the input of the receiver, the question is: How do we tune in or select the desired station? The answer is to once again include *LC* tank circuits as band-pass filters in the RF amplifier circuit, as shown in Figure 10-7(b). Unlike the previously discussed fixed-tuned RF amplifier circuit, however, these tank circuits contain variable capacitors that can be adjusted to control the resonant frequency of the *LC* tuned circuit and therefore the band of frequencies or station that will be passed on to the next stage. The dashed line between the two capacitors C_{1A} and C_{1B} is used to indicate that these variable capacitors are **ganged,** or mechanically linked by a common shaft, so that the operator can adjust the "tuning control" to simultaneously set the band-pass frequency for both tuned circuits and therefore select which station is passed, or filtered out or blocked. These tuned circuits are tunable over a wide range of frequencies, which for the AM band would be between 535 kHz to 1605 kHz.

Ganged
Mechanically linked by a common shaft so that adjustments can be made simultaneously.

Let us now examine the operation of the variable-tuned RF amplifier circuit shown in Figure 10-7(b). The amplitude-modulated induced signals from the antenna are transformer-coupled to the first parallel *LC* tank circuit made up of capacitor C_{1A} and the secondary of T_1. In the example in Figure 10-7(b), the tuning control is set to 900 kHz, and so the input tuned circuit will pass this frequency to the base of Q_1 and reject all others. To be more specific, the 900-kHz signal will see the parallel tuned circuit as a very high impedance, and therefore nearly all of the signal voltage will be developed across the tank and applied to the base of Q_1. All other frequencies will see a very low tank impedance and be shunted to ground. The selected RF signal is then amplified by Q_1 and transformer-coupled to the output tuned circuit made up of T_2's secondary and capacitor C_{1B}. The output tank, which is tuned to the exact same frequency as the input tank circuit, is included to make the RF amplifier highly selective. Transformer input and output coupling is used to both isolate amplifier stages and match impedances for a high gain.

Let us now return to the basic block diagram of the tuned AM receiver shown in Figure 10-7(a) to discuss the function of the remaining blocks. As discussed previously, each information signal is placed on its own unique carrier frequency to prevent its transmission from interfering with other station transmissions. However, the human ear cannot hear high radio-frequency signals, and so a demodulator, or **detector circuit,** is included in every receiver to recover the AF information signal from the radio-frequency (RF) carrier. The recovered audio-frequency information signal is then amplified by an AF voltage and power amplifier so that the signal strength is large enough to drive the speaker.

Detector Circuit
The circuit that recovers the audio-frequency information signal from the radio-frequency carrier.

The tuned AM radio-frequency receiver shown in Figure 10-7(a) had two key disadvantages. The first was that several ganged RF variable-tuned amplifier stages were needed to increase the signal level to a usable voltage level before it could be applied to the demodulator. This large number of RF amplifiers was necessary because negative feedback is always present when amplifying high-frequency signals. The second disadvantage of this receiver was that, because all of the RF amplifier stages had to be ganged, the capacitor was both bulky and expensive. As a result, the tuned AM receiver circuit was replaced by the "superheterodyne receiver," which will be discussed in the following section.

SELF-TEST EVALUATION POINT FOR SECTION 10-1

Use the following questions to test your understanding of Section 10–1.

1. Radio-frequency amplifiers are used to amplify radio-frequency signals from about
   a. 20 Hz to 20 kHz   c. 10 kHz to 30 GHz
   b. 10 Hz to 35 MHz   d. 20 kHz to 20 MHz

2. Give the full names for the following abbreviations:
   a. RF   b. AM   c. FCC

3. Which RF amplifier circuit has a collector-tuned circuit that is tuned to an exact multiple of the input frequency?

4. What would be the bandwidth of an AM 1500 kHz carrier if it were modulated by a 3.5 kHz AF tone?

5. An RF amplifier with a narrow bandwidth has a _____ selectivity.

6. When a swamping resistor is added to an amplifier's tuned circuit, its bandwidth is _____ , and its gain is _____

10-2   INTERMEDIATE-FREQUENCY (IF) AMPLIFIER CIRCUITS

In the previous section, we discovered that the "tuned AM receiver" needed several high-frequency variable-tuned RF amplifier stages to boost the weak input signal to a usable voltage level and an expensive ganged capacitor for each tuned circuit in each RF amplifier stage. These two disadvantages were overcome by using a radio receiving technique that has been so successful it is used in nearly every AM radio, FM radio, and television receiver. This special type of receiver is called a **superheterodyne receiver,** with the word *heterodyning* meaning "frequency converting."

10-2-1   *The Superheterodyne Receiver*

To demonstrate the operation of a superheterodyne receiver, Figure 10-8(a) shows the block diagram of a basic superhet AM receiver. As before, the received AM signal is first amplified by a variable-tuned RF amplifier stage. The output of this stage is then applied to the distinct feature of this type of receiver, its **RF amplifier/mixer circuit.** This mixer circuit is a special circuit designed to combine, or mix, two input frequencies. The two input frequencies that are combined in the mixer circuit are the received AM signal from the RF variable-tuned amplifier and the RF sine-wave signal from the **RF local oscillator circuit.** Referring to the tuning control in Figure 10-8(a), you can see that it is ganged to control both the tuning of the front-end RF amplifier and the frequency of the RF local oscillator. The frequency of the RF local oscillator is controlled so that its frequency is 455 kHz greater than the received RF input frequency. For example, let us assume that the tuning control is set to 900 kHz. The mechanical linkage from the tuning control to the RF amplifier will vary its internal tuned circuit capacitors so that they will pass the band of frequencies centered at 900 kHz. At the same time, the mechanical linkage from the tuning control to the RF local oscillator will vary its internal tuned circuits so that it will produce an RF sine-wave frequency that is 455 kHz greater than the received RF signal.

$$f_{\text{RF Amp}} = 900 \text{ kHz}$$
$$f_{\text{LO}} = \text{RF} + 455 \text{ kHz} = 1355 \text{ kHz}$$

These two input signals are combined in the RF amplifier/mixer circuit, producing four signals at the output. These output signals are

1. The original RF input signal from the RF amplifier (900 kHz)
2. The original RF signal from the local oscillator (1355 kHz)
3. The sum of the received RF signal and the local oscillator signal (900 kHz + 1355 kHz = 2255 kHz),
4. And the difference between the received RF signal and the local oscillator signal (1355 kHz − 900 kHz = 455 kHz)

10-2-2   *An RF Amplifier/Mixer Circuit*

Figure 10-8(b) shows an example of a typical RF amplifier/mixer circuit. The received AM signal from the RF amplifier and the RF sine-wave signal from the RF local oscillator are both coupled to the base of Q_1, where they are amplified and combined to produce four signal combinations at the transistor's collector. The **double-tuned transformer** in the collector of Q_1 is fixed-tuned to the difference frequency of 455 kHz; therefore, only this signal is passed to the output.

Because the frequency of the RF local oscillator is controlled to always be 455 kHz above the tuned frequency of the RF amplifier, the difference output signal from the mixer

Superheterodyne Receiver

RF receiver that converts all RF inputs to a common intermediate frequency (IF) before demodulation.

RF Amplifier/Mixer Circuit

A circuit designed to combine or mix two input frequencies.

RF Local Oscillator Circuit

The radio-frequency oscillator in a superheterodyne receiver.

Double-Tuned Transformer

A transformer having both a tuned primary and tuned secondary.

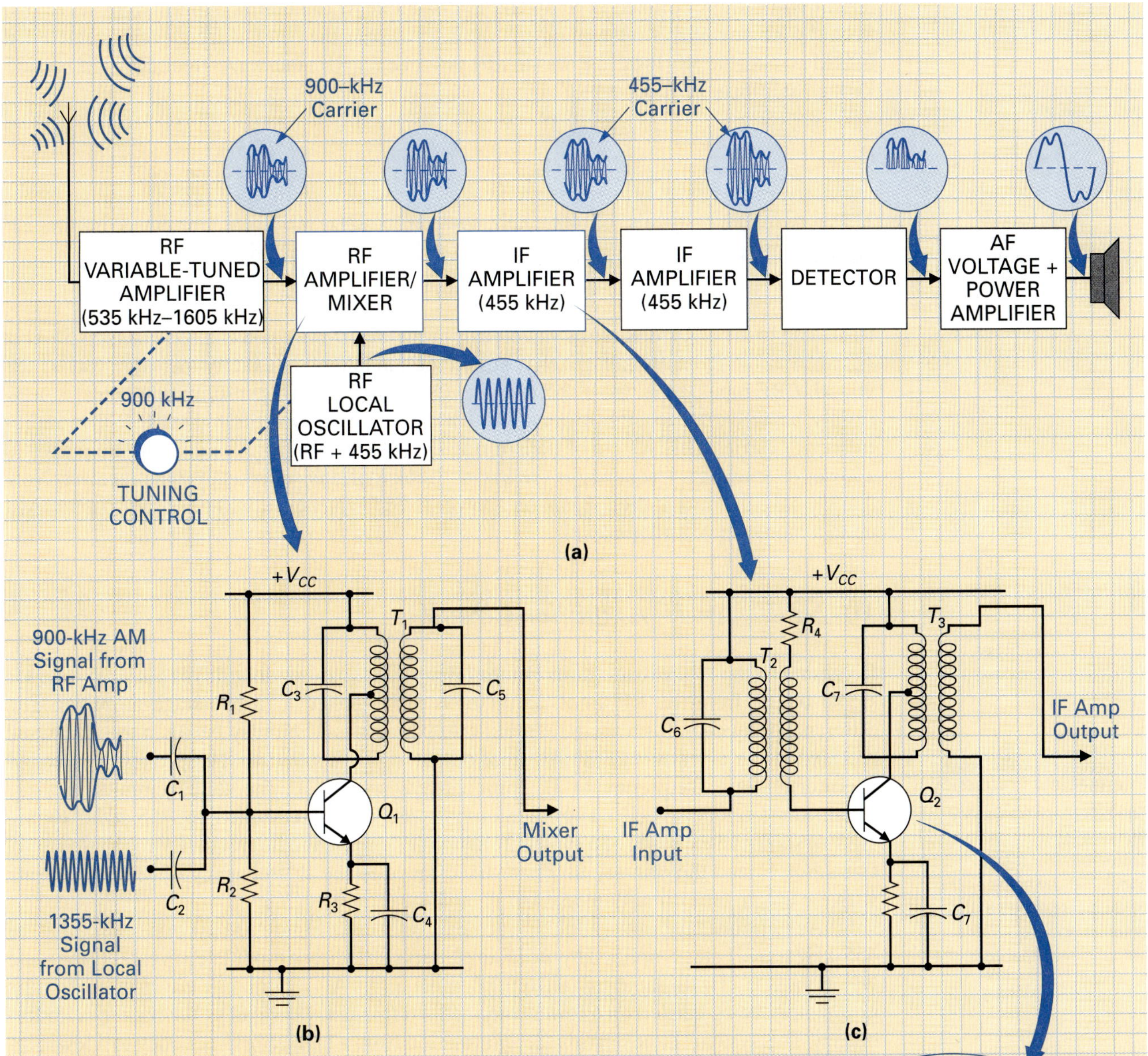

900–kHz Carrier
455–kHz Carrier
RF VARIABLE-TUNED AMPLIFIER (535 kHz–1605 kHz)
RF AMPLIFIER/ MIXER
IF AMPLIFIER (455 kHz)
IF AMPLIFIER (455 kHz)
DETECTOR
AF VOLTAGE + POWER AMPLIFIER
RF LOCAL OSCILLATOR (RF + 455 kHz)
900 kHz
TUNING CONTROL
(a)
+VCC
900-kHz AM Signal from RF Amp
1355-kHz Signal from Local Oscillator
R1
R2
R3
C1
C2
C3
C4
C5
T1
Q1
Mixer Output
(b)
+VCC
R4
T2
T3
C6
C7
Q2
IF Amp Input
IF Amp Output
(c)

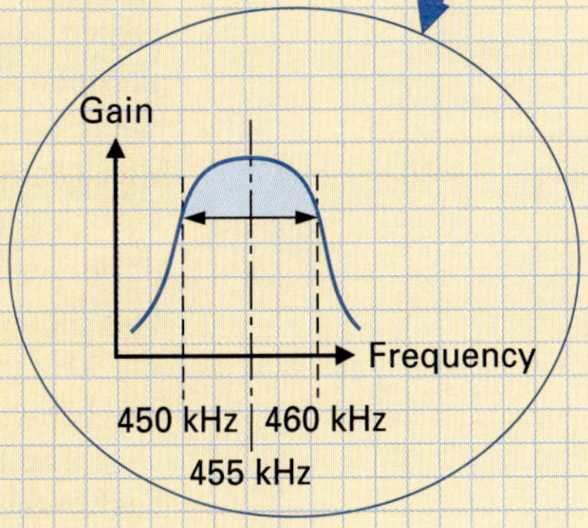

Gain
Frequency
450 kHz
460 kHz
455 kHz

COMMUNICATION TYPE	RF RECEIVED	TYPICAL IF	IF BANDWIDTH
AM RADIO	535 kHz–1605 kHz	455 kHz	10 kHz
FM RADIO	88 MHz–108 MHz	10.7 MHz	150 kHz
TV CHANNEL 2–6	54 MHz–88 MHz	41–47 MHz	6 MHz
TV CHANNEL 7–13	174 MHz–216 MHz	41–47 MHz	6 MHz
TV CHANNEL 14–83	470 MHz–890 MHz	41–47 MHz	6 MHz

(d)

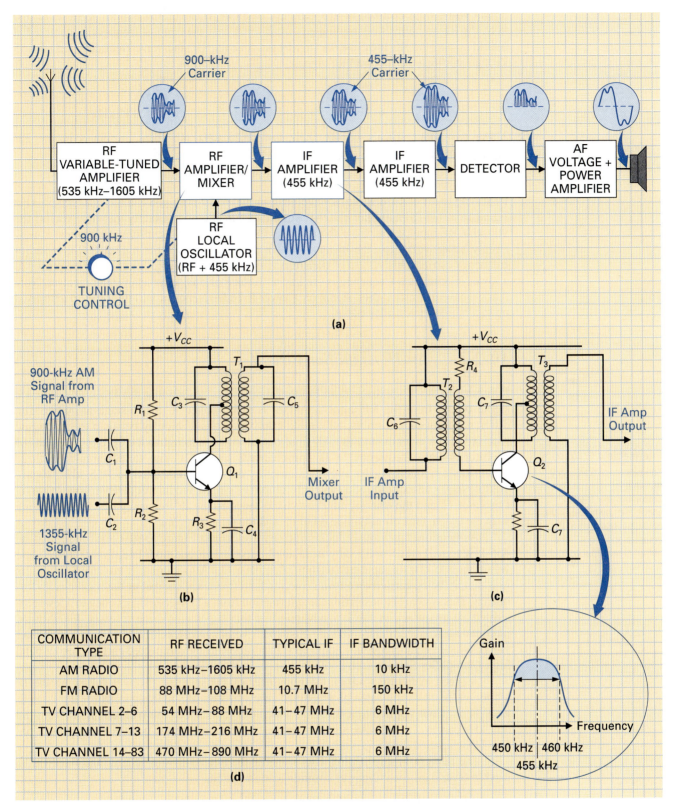

FIGURE 10-8 **Superheterodyne AM Radio-Frequency Receivers. (a) Basic Block Diagram of RF Receiver. (b) RF Amplifier/Mixer Circuit. (c) Intermediate-Frequency (IF) Amplifier Circuit. (d) Typical IF Frequencies.**

circuit will always be fixed at 455 kHz. For example, if the tuning control were set to 690 kHz, the RF amplifier's internal tuned circuits will pass the band of frequencies centered at 690 kHz, and the RF local oscillator will produce an RF sine-wave frequency that is 455 kHz greater than the received RF signal.

$$f_{\text{RF Amp}} = 690 \text{ kHz}$$

$$f_{\text{LO}} = \text{RF} + 455 \text{ kHz} = 690 \text{ kHz} + 455 \text{ kHz} = 1145 \text{ kHz}$$

These two input signals of 690 kHz and 1145 kHz are then combined in the RF amplifier/mixer circuit, and the difference of 455 kHz is extracted.

$$f_{\text{Mixer}} = f_{\text{LO}} - f_{\text{RF Amp}} = 1145 \text{ kHz} - 690 \text{ kHz} = 455 \text{ kHz}$$

Returning to the block diagram's waveforms in Figure 10-8(a), you can see that the 455-kHz signal from the RF amplifier mixer is amplitude modulated in exactly the same way as the 900-kHz RF input signal from the antenna. Our AF-modulating information signal is now simply being carried on the lower 455-kHz carrier instead of the higher 690-kHz radio-frequency carrier. This system of transferring the information from a high radio frequency to a lower intermediate frequency is the key advantage of this superheterodyne receiver. Now that we have the information signal on a lower fixed frequency, it will be easier to amplify it up to a more usable voltage level.

10-2-3   *An IF Amplifier Circuit*

Returning to the block diagram in Figure 10-8(a), you can see that the 455-kHz difference signal output from the RF amplifier/mixer is next applied to a two-stage amplifier. The difference frequency from the RF amplifier/mixer is about midway between the RF signal received at the antenna and the AF final output signal applied to the speaker. It is called an **intermediate frequency** or IF (pronounced "eye-eff"). The tuned circuits in these intermediate-frequency (IF) amplifier circuits are also fixed-tuned to 455 kHz. Figure 10-8(c) shows an example of a typical IF amplifier circuit and its frequency response curve. Capacitor C_6 and the primary of T_2 are fixed-tuned to the IF frequency of 455 kHz, along with the output tank circuit made up of C_7 and the primary of T_3.

As mentioned previously, looking at the waveforms in Figure 10-8(a), you can see that the 455-kHz signal being amplified by the IF amplifiers is amplitude modulated in exactly the same way as the 900-kHz RF input signal from the antenna. Our AF-modulating information signal is now simply being carried on a lower intermediate frequency instead of the higher radio frequency. This system of transferring the information from a high radio frequency to a lower intermediate frequency is the key advantage of this superheterodyne receiver. Fixed-tuned intermediate frequency amplifiers are less expensive, and have less high-frequency losses, than variable-tuned radio-frequency amplifiers. In addition, as we discovered earlier, narrow-band high-Q amplifier circuits, such as the IF amp, have higher gains.

As seen in Figure 10-8(a), the remainder of the superhet receiver following the IF amplifiers is much the same as the tuned receiver. The 455-kHz signal from the final IF amplifier stage is applied to a detector, which extracts the AF information signal from the 455-kHz carrier. Then the AF signal is applied to an amplifier and boosted in amplitude before being applied to the speaker.

The table in Figure 10-8(d) lists some of the typical IF frequencies and bandwidths used in several different types of communication circuits.

Intermediate Frequency

The difference frequency from the RF amplifier/mixer that is about midway between the RF signal received at the antenna and the AF final output signal applied to the speaker.

Use the following questions to test your understanding of Section 10–2.

1. In a superheterodyne receiver, the RF input signal is converted to a lower _____.
   a. RF   **b.** VF   **c.** AF   **d.** IF
2. The _____ circuit in a superhet receiver combines the RF input frequency and the RF signal from the local oscillator.
3. What are the two advantages of transferring an input signal onto an IF carrier instead of an RF carrier?
4. Are IF amplifiers fixed-tuned or variable-tuned?

10-3   VIDEO-FREQUENCY (VF) AMPLIFIER CIRCUITS

Video-frequency (VF) amplifiers are used in televisions, computer monitors, video games, virtual reality systems, and radar systems to amplify picture-information signals, which are also called **video signals.** These video signals include a very wide range of frequencies, spanning from about 10 Hz to 5 MHz; therefore, the video amplifier must provide a high gain over a very wide bandwidth. To begin with, let us discuss how we can modify a basic amplifier circuit in order to get it to respond to this wide band of frequencies.

Video-Frequency (VF) Amplifiers
Amplifiers used in televisions, computer monitors, video games, virtual reality systems, and radar systems to amplify picture-information signals (video signals).

10-3-1   *Video Amplifier Shunt Capacitance*

As discussed previously in the RF amplifier section, a small value of capacitance exists between all the junctions of a transistor. These junction capacitances are determined by the physical size of the junction, the spacing between the transistor's leads, and the junction bias condition. In addition, because many of the tracks on a printed circuit board are in close proximity to one another, they possess a certain value of capacitance that is inherent in every circuit, called *stray-wire capacitance*.

Video Signals
A signal that contains visual information for television or radar systems.

Referring to Figure 10-9(a), you can see a basic RC-coupled amplifier with all of the circuit's inherent shunt capacitances drawn in using dashed lines. The shunt output capacitance of transistor Q_1 ($C_{out(Q_1)}$) is made up of the transistor's collector-to-emitter capacitance (C_{CE}) and the **output stray-wire capacitance (C_{SO}).** At the other end, we have the input capacitance of Q_2, which is made up of Q_2's collector-to-base Miller capacitance (C_M), base-to-emitter capacitance (C_{BE}), and the input stray-wire capacitance (C_{SI}). Because all of these capacitance values are in parallel, we can add them together to obtain a total shunt capacitance value (C_T). This total shunt capacitance value is typically 160 pF and is shown in Figure 10-9(b). When the amplifier is amplifying low-frequency signals, this capacitance has very little effect because X_C is large. However, as the frequency of the signal being amplified increases, X_C decreases and the total shunt capacitance becomes a problem because it shunts some of the signal voltage to ground. The point at which the voltage of the signal frequency drops below 70.7% is called the **high-frequency cutoff** ($f_{H(\text{cutoff})}$) and can be calculated with the formula

Output Stray-Wire Capacitance (C_{SO})
The capacitance that is formed between any conductor and an adjacent conductor.

$$f_{H(\text{cutoff})} = \frac{1}{2\pi R_C C_T}$$

High-Frequency Cutoff ($F_{H(\text{cutoff})}$)
The point at which the voltage of the signal frequency drops below 70.7%.

■ EXAMPLE:

Calculate the high-frequency cutoff for the values given in Figure 10-9(b).

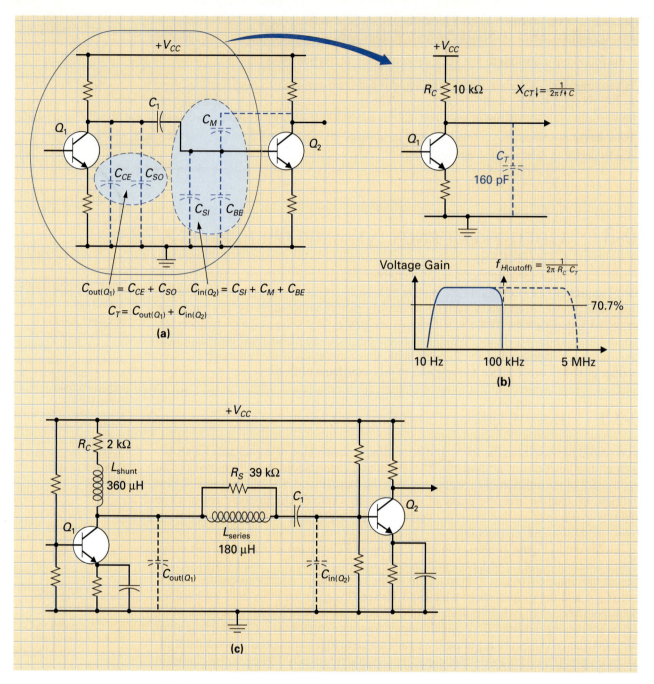

FIGURE 10-9 **Video Amplifier Circuits. (a) Shunt Capacitances. (b) High-Frequency Cutoff. (c) Frequency Compensation.**

■ *Solution:*

$$f_{H(\text{cutoff})} = \frac{1}{2\pi R_C C_T} = \frac{1}{2 \times \pi \times 10 \text{ k}\Omega \times 160 \text{ pF}}$$

$$= \frac{0.159}{(10 \times 10^3) \times (160 \times 10^{-12})} = 99.47 \text{ kHz}$$

This high-frequency cutoff at 100 kHz is shown in the frequency response curve in Figure 10-9(b). Unfortunately, the needed frequency response for video signals is up to 5

MHz, as shown by the dashed line. To increase the frequency response of a video amplifier we could decrease the value of R_C.

■ EXAMPLE:

What would be the high-frequency cutoff if R_C were reduced to 2 kΩ, as shown in Figure 10-9(c)?

■ Solution:

This would extend our frequency range to

$$f_{H(\text{cutoff})} = \frac{1}{2\pi R_C C_T} = \frac{1}{2 \times \pi \times 2 \text{ k}\Omega \times 160 \text{ pF}} = 497.4 \text{ kHz}$$

However, decreasing R_C will also reduce the voltage gain of the amplifier ($A_V = R_C/R_E$). Therefore, more video amplifier stages will be needed to achieve the same overall gain.

10-3-2   *Video Amplifier Peaking Coils*

Another way to compensate for this signal loss at high frequencies is to somehow cancel the effect of the circuit's inherent shunt capacitances. This is achieved by using the **peaking coils** shown in the video amplifier circuit in Figure 10-9(c). The *shunt peaking coil* (L_{shunt}) included in the collector of Q_1 will compensate for the output capacitance of Q_1 ($C_{\text{out}(Q_1)}$). By carefully choosing the value of the shunt peaking coil, we can cause L_{shunt} and $C_{\text{out}(Q_1)}$ to resonate at the higher frequencies. Because L_{shunt} and $C_{\text{out}(Q_1)}$ are in parallel with one another (collector current will split), this parallel resonance circuit will have a high impedance at resonance and therefore develop a larger output voltage. A *series peaking coil* (L_{series}) can also be used to increase the high-frequency response of a video amplifier by compensating for the input capacitance of Q_2 ($C_{\text{in}(Q_2)}$), as seen in Figure 10-9(c). Because the inductive reactance of this series peaking coil will increase with frequency ($X_L \propto f$), this coil will isolate the output capacitance of Q_1 from the input capacitance of Q_2, and therefore the large total shunt capacitance is now reduced to two smaller-value capacitances. Because $C_{\text{out}(Q_1)}$ is smaller than C_T, a larger value of R_C can be used, and therefore the amplifier will have a larger voltage gain. The swamping resistor R_S in parallel with the series peaking coil reduces the Q of the inductor to widen the response, or bandwidth, of the coil so that it will respond well to all video frequencies.

Peaking Coils
Coils included in a circuit to compensate for signal loss at high frequencies.

10-3-3   *Basic Television Receiver Circuits*

Video amplifiers find their largest application in television receivers. Figure 10-10 shows the basic block diagram of a television receiver and examples of some typical television receiver circuits.

Figure 10-10(a) shows a typical VHF fixed-tuned RF amplifier circuit. The tuned circuits, made up of L_1 with C_1 and C_3 with the primary of T_1, reject frequencies that are outside of the selected band. Transistor Q_1 operates as a class-A amplifier, with its gain controlled by an **automatic gain control (AGC)** input. This control voltage originates from a circuit that automatically senses the strength of the input signal, then increases or decreases the gain of the amplifier to ensure that the gain is almost constant over a wide range of input signal voltages. For example, when the input signal is too strong, the AGC circuit uses degenerative feedback to reduce the input signal strength, whereas if the input signal is weak, the degenerative feedback is reduced to increase the amplifier gain. By ground shielding the transistor Q_1,

Automatic Gain Control (AGC)
The control voltage from a circuit that automatically senses the strength of the input signal then increases or decreases the gain of the amplifier to ensure that the gain is almost constant over a wide range of input signal voltages.

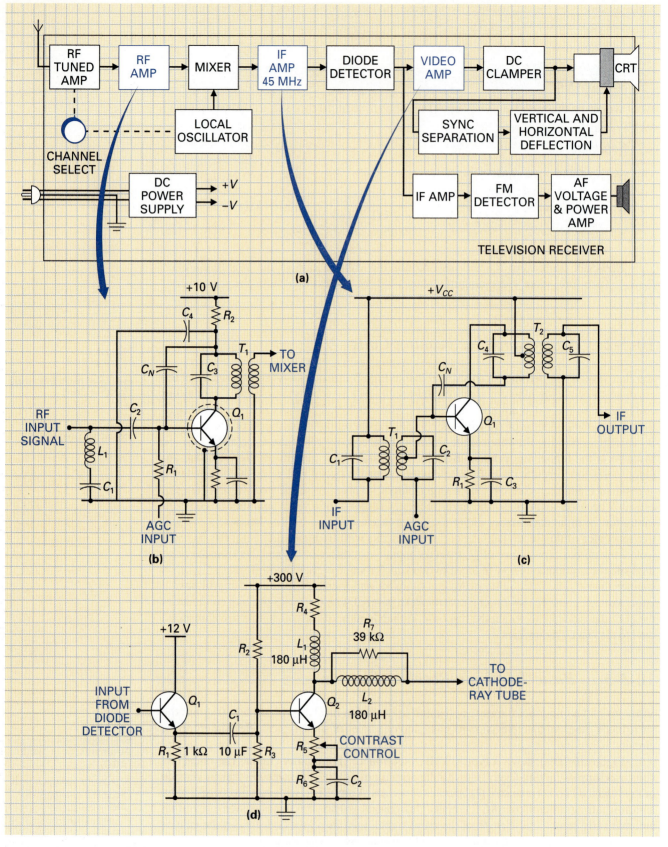

FIGURE 10-10 Television Receiver Circuits. (a) Basic Television Receiver Block Diagram. (b) RF Amplifier. (c) IF Amplifier. (d) VF Amplifier.

any stray RF signals will be shunted to ground instead of being introduced into the signal as noise. The noise level at this early amplifier stage should be kept to a minimum because any noise introduced will be amplified along with the desired signal by all of the following amplifier stages. Resistor R_2 and capacitor C_4 form a decoupling filter circuit that will prevent any of the RF signal from reaching the +10-V power supply line and then interfering with other circuits, while the neutralizing capacitor (C_N) compensates for Miller capacitance.

Figure 10-10(b) shows a typical IF amplifier for a TV receiver, which is double-transformer-tuned to the intermediate frequency of 45 MHz. Once again, the gain of Q_1 is controlled by the AGC control input, and capacitor C_N is included to neutralize the Miller effect.

Figure 10-10(c) shows a typical video-frequency amplifier for a television receiver. This black-and-white television amplifier will provide a voltage gain of about 30 for video signals from about 10 Hz to 3.5 MHz (for color, the bandwidth will have to be extended to 5 MHz). Transistor Q_1 is an emitter follower, included to act as a high-impedance load for the previous detector stage and to act as a low-impedance source for the following stage. The video output signal from Q_1 is developed across R_1 and then coupled to the base of Q_2 via the large-value capacitor C_1, which will have a low reactance to low-frequency signals. The common-emitter amplifier stage (Q_2) makes use of a very high V_{CC} supply voltage, which is needed to drive the television's cathode-ray tube (CRT). Resistor R_5 is not bypassed and will be degenerative, controlling the gain of this video amplifier and, as a result, controlling the **contrast** of the picture. For example, by adjusting the TV's contrast control to decrease R_5's resistance (moving R_5's wiper up), the gain of Q_2 will be increased, and the difference between the black and white dots on the TV screen will increase causing an increase in contrast. Looking at the collector of the video amplifier Q_2, you can see that a shunt peaking coil (L_1), a series peaking coil (L_2), and a swamping resistor (R_7) have been added to extend the amplifier's high-frequency cutoff point.

Contrast

The difference between the light and dark areas in a video picture. High-contrast pictures have dark blacks and brilliant whites. Low-contrast pictures have an overall gray appearance.

SELF-TEST EVALUATION POINT FOR SECTION 10-3

1. Where are video-frequency amplifiers used?
2. Video-frequency amplifiers are used to amplify video-frequency signals from about
   a. 20 Hz to 20 kHz
   b. 10 Hz to 5 MHz
   c. 10 kHz to 30 GHz
   d. 20 kHz to 20 MHz
3. Why are peaking coils included in most VF amplifier circuits?
4. Decreasing the value of R_C in a video amplifier will increase the amplifier's high-frequency cutoff and reduce the amplifier's gain. (True/False)

10-4 TUNING HIGH-FREQUENCY AMPLIFIERS

One of the biggest problems with tuned circuit amplifiers is that their tuned circuits are not tuned to the desired center frequency. The reasons for this problem can range from capacitor and inductor tolerances to the inherent stray and junction circuit capacitance values that are always present. To compensate for these problems, most tuned amplifier circuits either have a built-in variable inductor as seen in Figure 10-11(a) or a built-in variable capacitor as seen in Figure 10-11(b). A variable inductor is more frequently used because it is less expensive than a variable capacitor.

The next question is: How do we **tune** the amplifier? A simple and accurate method is

1. First connect an ac current meter between V_{CC} and the tuned circuit, as shown in Figure 10-11(c).

Tune

To adjust the resonance of a circuit so that it will select the desired frequency.

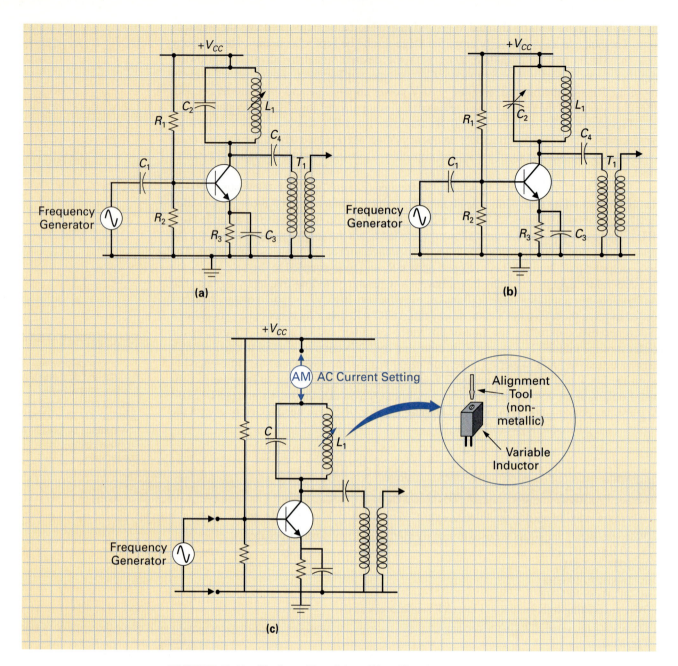

FIGURE 10-11   **Tuning a Tuned Amplifier Circuit.**

2. Next, connect a frequency generator to the input of the amplifier, and set it to the desired frequency.

3. Finally, adjust the variable inductor (L_1) until the meter shows a large dip or null in current.

Remember that because a parallel tuned circuit will have a maximum impedance at resonance, all we are doing is adjusting the value of inductance (and therefore the tuned circuit resonant frequency) until the current drops to a minimum. At frequencies above and below resonance, a small value of current will pass through the tank; however, when current drops to a minimum, we know that the tuned circuit's impedance is at a maximum, and therefore the tank is tuned to the same frequency as the frequency generator's input frequency.

Another important point to remember with this procedure is to always use a nonmetallic screwdriver because the proximity of any metal to the electric and magnetic fields of an adjustable capacitor or inductor will change its operation, and therefore change the frequency to which the tank is tuned.

Use the following questions to test your understanding of Section 10–4.

1. Why do amplifer circuits have to be tuned?
2. How is a high-frequency amplifier's tank circuit tuned?

REVIEW QUESTIONS

Multiple-Choice Questions

1. The radio frequency at which a station transmits is called its _____ frequency.
   a. intermediate   c. carrier
   b. superheterodyne   d. neutralizing

2. The voltage applied across a transmitting antenna generates a/an _____ field, while the current passing through the antenna generates a/an _____ field.
   a. electric, voltage   c. current, electric
   b. electric or voltage, current   d. electron, electric

3. Which of the following circuits could be used to vary the amplitude of an RF carrier in accordance with the AF signal?
   a. RF amplifier/multiplier circuit
   b. RF fixed-tuned amplifier circuit
   c. RF variable-tuned amplifier circuit
   d. RF amplifier/modulator circuit

4. Which of the following circuits could be used to double the frequency of an RF input sine-wave signal?
   a. RF amplifier/multiplier circuit
   b. RF fixed-tuned amplifier circuit
   c. RF variable-tuned amplifier circuit
   d. RF amplifier/modulator circuit

5. Which of the following circuits is used in a receiver to tune in or select a desired station?
   a. RF amplifier/multiplier circuit
   b. RF fixed-tuned amplifier circuit
   c. RF variable-tuned amplifier circuit
   d. RF amplifier/modulator circuit

6. Which of the following circuits is used to boost the amplitude of an RF signal of a specific frequency?
   a. RF amplifier/multiplier circuit
   b. RF fixed-tuned amplifier circuit
   c. RF variable-tuned amplifier circuit
   d. RF amplifier/modulator circuit

7. An amplifier with a high Q will have a _____ bandwidth and therefore a _____ selectivity.
   a. small, good   c. small, poor
   b. large, good   d. large, poor

8. To neutralize the _____ feedback caused by the Miller effect in high-frequency amplifiers, a _____ is usually included to couple an _____ signal back to the base.
   a. negative, resistor, out-of-phase
   b. positive, capacitor, in-phase
   c. negative, capacitor, in-phase
   d. positive, resistor, out-of-phase

9. With a superheterodyne receiver, the received RF signal is converted to a lower frequency called a/an _____ before the information signal is extracted.
   a. VF   c. RF
   b. IF   d. MF

10. Intermediate-frequency amplifiers are always
    a. fixed-tuned   b. variable-tuned

11. The standard IF frequency for an AM radio receiver is
    a. 41 MHz   c. 10 kHz
    b. 10.7 MHz   d. 455 kHz

12. Video amplifiers should ideally have a frequency response of _____ to _____.
    a. 10 kHz to 30,000 MHz   c. 10 Hz to 5 MHz
    b. 41 MHz to 47 MHz   d. 20 Hz to 20 kHz

13. To increase the high-frequency gain of a video amplifier, we could _____ the value of R_C, and/or use _____.
    a. decrease, peaking coils
    b. decrease, swamping resistors
    c. increase, shunt capacitances
    d. increase, peaking capacitors

14. An amplifier's low-frequency response will not be affected by the circuit's inherent shunt capacitance.
    a. True   b. False

15. A tuned amplifier has a tuned circuit load, therefore load _____ and amplifier _____ will vary with frequency.
    a. center frequency, power rating
    b. capacitance, frequency
    c. impedance, gain
    d. both (a) and (b) are true

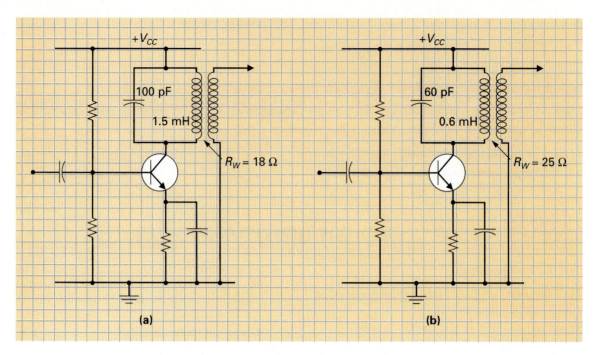

FIGURE 10-12   Tuned Amplifier Circuits.

Practice Problems

16. Referring to Figure 10-12, calculate the resonant frequency of the amplifier's tank circuits.

17. Calculate the Q of the tuned circuits shown in Figure 10-12.

18. Calculate the bandwidth of the tuned circuits shown in Figure 10-12.

19. Calculate the high-frequency cutoff of the video amplifier circuit shown in Figure 10-13.

20. If a 1360-kHz carrier is voice modulated by a signal that is between 200 Hz to 20 kHz, what will be the transmission's upper and lower sidebands and bandwidth?

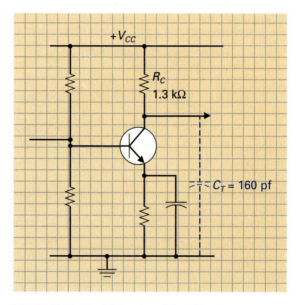

FIGURE 10-13   Video Amplifier Circuit.

Field Effect Transistors

11

Introduction

In the previous chapters, we have concentrated on all aspects of the bipolar junction transistor or BJT. We have examined its construction, operation, characteristics, testing, and basic circuit applications. In this chapter, we will examine another type of transistor, the *field effect transistor,* which is more commonly called an FET (pronounced "eff-ee-tee"). Like the BJT, the FET has three terminals, can operate as a switch, and can be used in digital circuit applications. It can also operate as a variable resistor and be used in analog or linear circuit applications. In fact, as we step through this chapter, you will see many similarities between the BJT and FET. You will also notice a few distinct differences between these two transistor types, and these differences are what make the BJT ideal in some applications and the FET ideal in other applications.

There are two types of field effect transistors or FETs. One type is the *junction field effect transistor,* which is more typically called a JFET (pronounced "jay-fet"). The other type is the *metal-oxide semiconductor field effect transistor,* which is more commonly called a MOSFET (pronounced "moss-fet"). In this chapter, we will examine the operation, characteristics, applications, and testing of these two types of field effect transistors.

11-1   JUNCTION FIELD EFFECT TRANSISTOR (JFET)

TIME LINE

In 1961, Steven Hofstein devised the field effect transistor used in MOS (metal-oxide semiconductor) integrated circuits.

Like the bipolar junction transistor, the **junction field effect transistor** or **JFET** is constructed from *n*-type and *p*-type semiconductor materials. However, the JFET's construction is very different from the BJT's construction, and therefore we will need to first see how the JFET device is built before we can understand how it operates.

11-1-1   *JFET Construction*

Junction Field Effect Transistor (JFET)

A field effect transistor made up of a gate region diffused into a channel region. When a control voltage is applied to the gate, the channel is depleted or enhanced and the current between source and drain is thereby controlled.

n-Channel JFET

A junction field effect transistor having an *n*-type channel between source and drain.

p-Channel JFET

A junction field effect transistor having a *p*-type channel between source and drain.

Gate

One of the field effect transistor's electrodes (also used for thyristor devices).

Source

One of the field effect transistor's electrodes.

Drain

One of the field effect transistor's electrodes.

Channel

A path for a signal.

Just as the bipolar junction transistor has two basic types (NPN BJT or PNP BJT), there are two types of junction field effect transistor, the **n-channel JFET** and **p-channel JFET.** The construction and schematic symbols for these two JFET types are shown in Figure 11-1.

To begin with, let us examine the construction of the more frequently used *n*-channel JFET, shown in Figure 11-1(a). This type of JFET basically consists of an *n*-type block of semiconductor material on top of a *p*-type substrate, with a U-shaped *p*-type section attached to the surface of a *p*-type substrate. Like the BJT, the JFET has three terminals—the **gate, source,** and **drain.** The gate lead is attached to the *p*-type substrate, and the source and drain leads are attached to either end of an *n*-type channel that runs through the middle of the U-shaped *p*-type section. In the simplified two-dimensional view in the inset in Figure 11-1(a), you can see the *n*-type **channel** that exists between the *n*-channel JFET's source and drain. The schematic symbol for the *n*-channel JFET is shown in Figure 11-1(b), and, as you can see, the gate lead's arrowhead points into the device. To aid your memory, you can imagine this arrowhead as a P-N junction diode, as shown in the inset in Figure 11-1(b). The gate lead is connected to the diode's anode, which is a *p*-type material, and the source and drain leads are connected to either end of the diode's cathode, which is an *n*-type material. Because the source-to-drain channel is made from an *n*-type material, this is an *n*-channel JFET. The *p*-type gate and *n*-type source and drain make this *n*-channel JFET equivalent to an NPN BJT, which has a *p*-type base and *n*-type emitter and collector, as shown in the inset in Figure 11-1(b).

Figure 11-1(c) shows the construction of the *p*-channel JFET. This JFET type is constructed in exactly the same way as the *n*-channel JFET except that the gate lead is attached to an *n*-type substrate, and the source and drain leads are attached to either end of a *p*-type channel. Looking at the schematic symbol for the *p*-channel JFET in Figure 11-1(d), you can see that the gate's arrowhead points out of the device. To help you distinguish the symbols used for the *n*-channel JFET from those used for the *p*-channel JFET, once again imagine the arrowhead as a P-N junction diode, as shown in the inset in Figure 11-1(d). Because the gate lead is connected to the diode's cathode, the gate must be an *n*-type material. Because the source and drain leads are connected to either end of the diode's anode, the source-to-drain channel is therefore made from a *p*-type material, and this is a *p*-channel JFET. The *n*-type gate and *p*-type source and drain make this *p*-channel JFET equivalent to a PNP BJT, which has an *n*-type base and *p*-type emitter and collector, as shown in the inset in Figure 11-1(d).

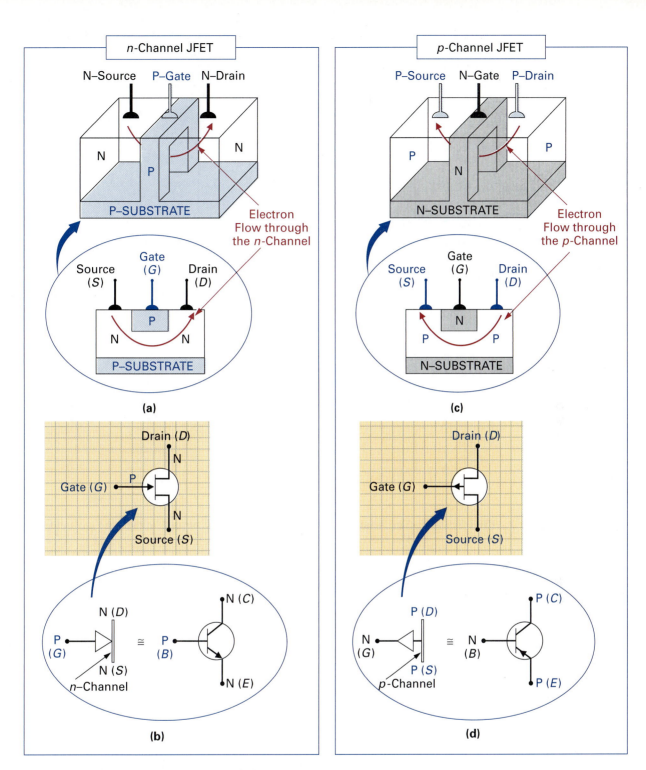

FIGURE 11-1 The Junction Field Effect Transistor (JFET) Types.

11-1-2 *JFET Operation*

As you know, an NPN bipolar transistor needs both a collector supply voltage ($+V_{CC}$) and a base–emitter bias voltage (V_{BE}) in order to operate correctly. The same is true for the JFET, which requires both a **drain supply voltage ($+V_{DD}$)** and a **gate–source bias voltage (V_{GS})**, as shown in Figure 11-2(a). The $+V_{DD}$ bias voltage is connected between the drain and source of the *n*-channel JFET and will cause a current to flow through the *n*-channel. This

Drain Supply Voltage ($+V_{DD}$)

The bias voltage connected between the drain and source of the JFET, which causes current to flow.

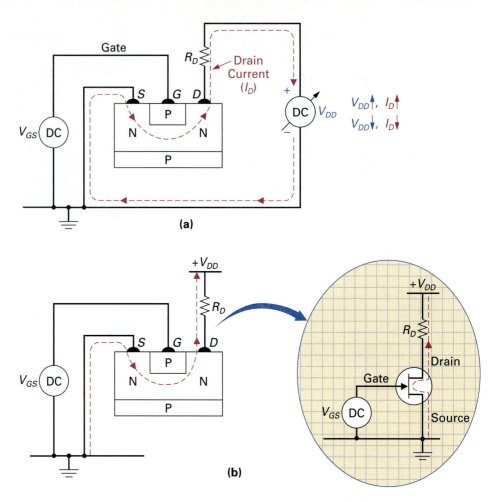

FIGURE 11-2 The Relationship Between a JFET's DC Supply Voltage ($+V_{DD}$) and Output Current (I_D).

source-to-drain current—which is made up of electrons because they are the majority carriers within an *n*-type material—is called the JFET's **drain current (I_D).** The value of drain current passing through a JFET's channel is dependent on two elements: the value of $+V_{DD}$ applied between the drain and source, and the value of V_{GS} applied between gate and source. Let us examine in more detail why these applied voltages control the value of drain current passing through the JFET's channel.

The Relationship Between $+V_{DD}$ and I_D

The value of $+V_{DD}$ controls the amount of drain current between source and drain because it is this supply voltage that controls the potential difference applied across the channel, as seen in Figure 11-2(a). Therefore, an increase in the voltage applied across the JFET's drain and source ($+V_{DD}\uparrow$) will increase the amount of drain current ($I_D\uparrow$) passing through the channel. Similarly, a decrease in the voltage applied across the JFET's drain and source ($+V_{DD}\downarrow$) will decrease the amount of drain current ($I_D\downarrow$) passing through the channel. In most cases, a schematic diagram will show the V_{DD} supply voltage connection to the JFET as a source connection to ground and a drain connection up to $+V_{DD}$, as shown in Figure 11-2(b).

The Relationship Between V_{GS} and I_D

The value of V_{GS} controls the amount of drain current between source and drain because it is this voltage that controls the resistance of the channel. Figure 11-3(a) shows that

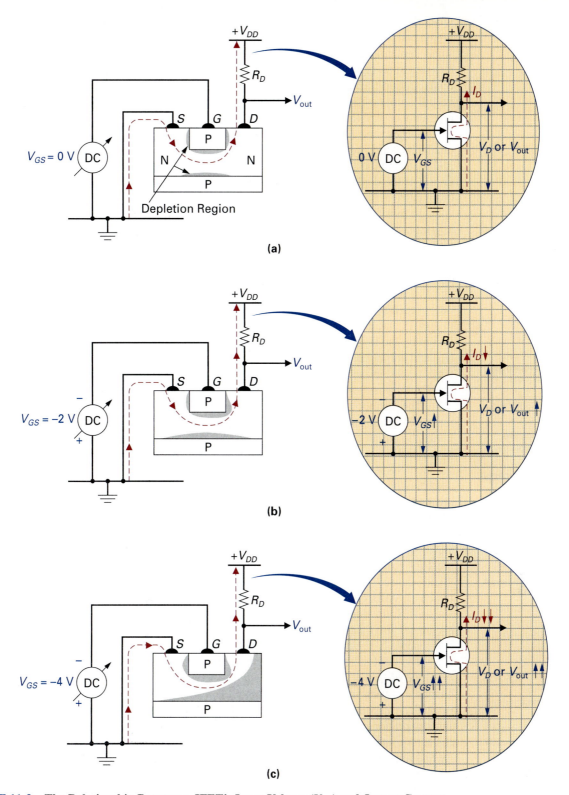

FIGURE 11-3 The Relationship Between a JFET's Input Voltage (V_{GS}) and Output Current (I_D).

when the V_{GS} bias voltage is 0 volts (which is the same as connecting the gate to ground), there is no potential difference between the gate and source. With no bias voltage applied, a small depletion layer will form and spread into the channel. Although it appears as though two depletion regions exist, in fact they are both part of the same depletion region that extends around the wall of the n-channel. This extremely small depletion region will offer very little opposition to I_D, and so drain current will be large. In Figure 11-3(b), V_{GS} is increased to -2 V (made more negative), and therefore the gate-to-source P-N junction will be further reverse biased. This causes an increase in the depletion region, a decrease in the channel's width, and a decrease in drain current. In Figure 11-3(c), V_{GS} is increased to -4 V, and therefore the gate-to-source P-N junction will be further reverse biased, causing an increase in the depletion region, a decrease in the channel's width, and a further decrease in drain current.

In most circuit applications, the $+V_{DD}$ supply voltage is maintained constant and the V_{GS} input voltage is used to control the resistance of the channel and the value of the output current, I_D. This can be seen more clearly in the insets in Figure 11-3. Because the output voltage (V_D or V_{out}) is dependent on the resistance between the JFET's source and drain, by controlling the value of I_D, we can control the output voltage. For instance, if the input voltage V_{GS} is made more negative, the resistance of the channel will be increased, causing the output current I_D to decrease, and therefore the voltage developed between drain and ground (V_D or V_{out}) to increase. The gate-to-source junction of an FET is normally always reverse biased by the input voltage (V_{GS}), and it is this input voltage that controls the output current (I_D) and output voltage (V_{out}). Because the gate-to-source junction of an FET is normally always reverse biased by the input voltage (V_{GS}), there will be no input current. This characteristic accounts for the FET's naturally high input impedance.

The operation of an FET is very different from that of the BJT, which normally uses an input voltage to forward bias the base-to-emitter junction and vary the input current (which varies the output current), and therefore the output voltage. This is the distinct difference between an FET and a BJT. An FET's input junction is normally reverse biased, and therefore the input voltage controls the output current. A BJT's input junction is normally forward biased and therefore the input current controls the output current. This difference is why BJTs are known as **current-controlled devices** and FETs are known as **voltage-controlled devices.** In fact, the name "field effect transistor" is derived from this voltage-control action because the applied input voltage will generate an electric field. It is this electric field that varies the size of the depletion region and therefore the resistance of the channel between the FET's drain and source output terminals. In other words, the "effect" of the electric "field" causes "transistance," which is the transferring of different values of resistance between the output terminals. The term *junction* is attached to this type of FET because of the single P-N junction formed between the gate and the source-to-drain channel. Therefore, an n-channel JFET has a single P-N junction between gate to channel, and a p-channel JFET has a single N-P junction between gate to channel.

The field effect transistor is also often referred to as a **unipolar device** because only one type of semiconductor material exists between the output terminals (n-type or p-type channel between source and drain), and therefore the charge carriers have only one polarity (unipolar). Compare this to a BJT, which is a **bipolar device** because there is a change in semiconductor material between the output terminals (NPN or PNP between emitter and collector), and the charge carriers can be one of two polarities (bipolar, because both majority and minority carriers are used).

11-1-3 JFET Characteristics

Like the BJT, the JFET's response to certain variables is best described by using a graph. Figure 11-4(a) shows a graph plotting drain current (I_D) against drain-to-source voltage (V_{DS}). As you have probably already observed, this **drain characteristic curve** is very similar to a bipolar transistor's collector characteristic curve. Starting at 0 V and moving right along the horizontal axis, you can see that an increase in the drain supply voltage ($+V_{DD}$),

---

Current-Controlled Device

A device in which the input junction is normally forward biased and the input current controls the output current.

Voltage-Controlled Device

A device in which the input junction is normally reverse biased and the input voltage controls the output current.

Unipolar Device

A device in which only one type of semiconductor material exists between the output terminals; therefore the charge carriers have only one polarity (unipolar).

Bipolar Device

A device in which there is a change in semiconductor material between the output terminals (NPN or PNP between emitter and collector), so the charge carriers can be one of two polarities (bipolar).

Drain Characteristic Curve

A plot of the drain current (I_D) versus the drain-to-source voltage (V_{DS}).

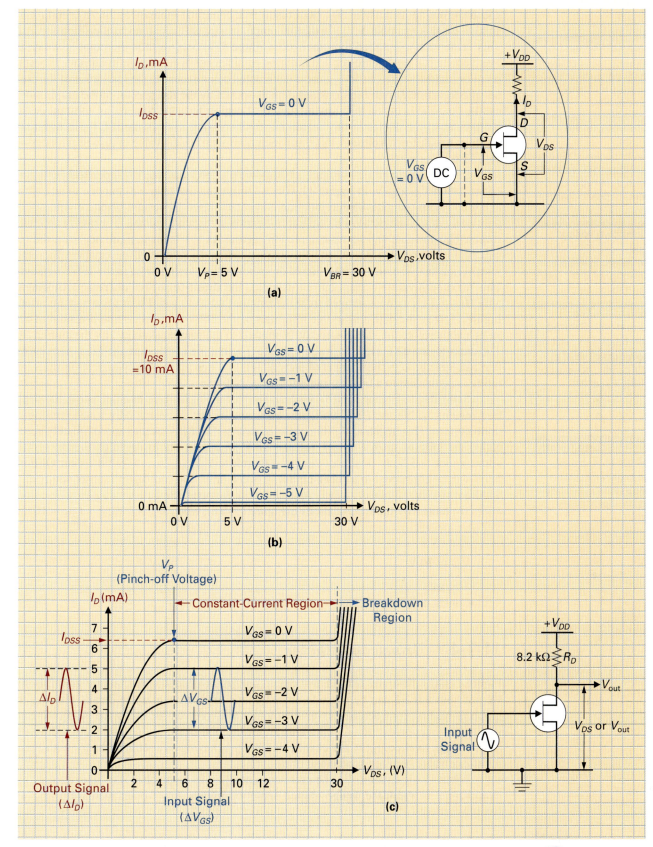

FIGURE 11-4 **JFET Characteristics.**

and therefore an increase in V_{DS}, will result in a continual increase in I_D. At a certain V_{DS} voltage (in this example 5 V), further increases in V_{DS} will cause no further increase in I_D. This value of V_{DS} is called the **pinch-off voltage (V_P)** because it is the point at which the bias voltage has caused the depletion region to pinch off or restrict drain current. From this point on, further increases in V_{DS} are counteracted by increases in the resistance of the channel and therefore I_D remains constant. This is shown by the flat portion of the graph in Figure 11-4(a) and is called the **constant-current region** because I_D remains constant despite changes in V_{DS}. If V_{DS} is further increased (by increasing $+V_{DD}$), the JFET will eventually reach its **breakdown voltage (V_{BR}),** at which time a damaging value of I_D will pass through the JFET.

In the example graph in Figure 11-4(a), we plotted what would happen to I_D as V_{DS} increased with V_{GS} at 0 V. In Figure 11-4(b) we will examine what will happen to an n-channel JFET when the gate–source junction is reverse biased by several negative voltages. As previously described in the JFET operation section, a negative voltage is normally applied to reverse bias the gate and set up a depletion region. As V_{GS} is made more negative, the gate will be further reverse biased and the corresponding I_D value will be smaller. Therefore, when V_{GS} is at 0 V, a maximum value of drain current is passing through the JFET's channel. This maximum value of drain current is called the **drain-to-source current with shorted gate (I_{DSS}).** This name is derived from the fact that when $V_{GS} = 0$ V, as shown in the inset in Figure 11-4(a), the gate and source terminals of the JFET are at the same potential of zero volts and therefore the gate is effectively shorted to the source, as shown by the dashed line. The drain-to-source current with shorted gate (I_{DSS}) rating is therefore the maximum current that can pass through the channel of a given JFET. When given on a specification sheet, this rating is equivalent to a bipolar transistor's $I_{C(Sat.)}$ rating.

Returning to Figure 11-4(b), you can see that if V_{GS} is made more negative, the depletion regions within the JFET will get closer and closer and eventually touch, cutting off drain current. This negative V_{GS} bias voltage that causes I_D to drop to approximately zero is called the **gate-to-source cutoff voltage or $V_{GS(OFF)}$.** In the example in Figure 11-4(b), when $V_{GS} = -5$ V, I_D is almost zero and therefore $V_{GS(OFF)} = -5$ V. When cut OFF, the JFET will be equivalent to an open circuit between drain and source, and subsequently all of the drain supply voltage (V_{DD}) will appear across the open JFET ($V_{DS} = V_{DD}$).

To summarize the specifications in Figure 11-4(b), when $V_{GS} = 0$ V, $I_D = I_{DSS} = 10$ mA,

$$V_P = 5 \text{ V}$$
$$V_{BR} = 30 \text{ V}$$
$$V_{GS(OFF)} = -5 \text{ V}$$
Constant-Current Region $= V_P$ to $V_{BR} = 5$ V to 30 V

11-1-4 *Transconductance*

Figure 11-4(c) illustrates a JFET test circuit and its associated characteristic graph. Before we see how a JFET can be made to amplify, let's first summarize the details given in this graph. If we first consider the curve when $V_{GS} = 0$ V, you can see that up to V_P, I_D increases in almost direct proportion to V_{DS}. This is because the depletion region is not sufficiently large enough to affect I_D, so the channel is simply behaving as a semiconductor with a fixed resistance value between source and drain.

When V_{DS} is equal to V_P, the drain current (I_D) will be pinched into an extremely narrow channel between the wedge-shaped depletion region. Any further increase in V_{DS} will have two effects:

1. It will increase the pinching effect on the channel, which will resist current flow, and
2. It will increase the potential between the drain and source, which will encourage current flow.

The net result is that channel resistance increases in direct proportion with V_{DS} and consequently I_D remains constant, as shown by the flat portion of the characteristic curve.

Assuming a fixed value of V_{DD}, any increase in the negative voltage of V_{GS} will cause a corresponding decrease in I_D. Therefore, beyond V_P, I_D is controlled by small-signal changes (such as the input signal) in V_{GS} and is independent of changes in V_{DS}. This section of the curve between V_P and V_{BR} is called the constant-current region.

Like the bipolar transistor, an FET can be used to amplify a signal, as shown in Figure 11-4(c). As before, the amount of amplification achieved is a ratio between output and input. For a bipolar transistor, the amount of gain is equal to the ratio of input current to output current (beta). For an FET, there is no input current, and therefore an FET's gain is equal to the ratio of output current change (ΔI_D) to input voltage change (ΔV_{GS}). This ratio is called the FET's **transconductance** (symbolized δ_m).

$$\delta_m = \frac{\Delta I_D}{\Delta V_{GS}}$$

δ_m = Transconductance in siemens (S)
ΔI_D = Change in drain current
ΔV_{GS} = Change in gate–source voltage

Transconductance
Also called mutual conductance, it is the ratio of a change in output current to the initiating change in input voltage.

EXAMPLE:

Calculate the transconductance of the FET for the example shown in Figure 11-4(c).

Solution:

$$\delta_m = \frac{\Delta I_D}{\Delta V_{GS}} = \frac{5 \text{ mA} - 2 \text{ mA}}{-1 \text{ V} - (-3 \text{ V})} = \frac{3 \text{ mA}}{2 \text{ V}} = 1.5 \text{ millisiemens (mS)}$$

A high-gain FET will produce a large change in I_D for a small change in V_{GS}, resulting in a high transconductance figure ($\delta_m\uparrow$).

11-1-5　*Voltage Gain*

Because transconductance is the ratio of output current change (ΔI_D) to input voltage change (ΔV_{GS}), it is no surprise that this ratio is used to determine a JFET's voltage gain. The voltage gain formula is as follows

$$A_V = \delta_m \times R_D$$

EXAMPLE:

Calculate the voltage gain for the circuit example shown in Figure 11-4(c).

Solution:

$$A_V = \delta_m \times R_D = 1.5 \text{ mS} \times 8.2 \text{ k}\Omega = 12.3$$

This means that the output voltage will be 12.3 times greater than the input voltage.

11-1-6   *JFET Data Sheet*

Throughout this section, we have used certain JFET specifications in our calculations, such as $V_{GS(OFF)}$ and I_{DSS}. As an example, Figure 11-5 shows the data sheet for a typical *n*-channel JFET. As before, notes have been inserted within the data sheet to describe any confusing ratings; however, most of these ratings are self-explanatory.

SELF-TEST EVALUATION POINT FOR SECTION 11-1

Use the following questions to test your understanding of Section 11–1.

1. Give the full names of the following abbreviations:
   a. BJT
   b. JFET
   c. MOSFET
2. Name the two different types of JFETs.
3. The BJT is a _____ operated device while the FET is a _____ operated device.
4. The gate-source junction of a JFET is always _____ biased.
5. When VGS = 0 V, ID = IDSS. (True/False)
6. When VGS = VGS(OFF), ID = max. (True/False)
7. _____ is a ratio of an FET's output current change to input voltage change.
8. A JFET has a _____ input impedance due to its _____ biased gate-source junction.

11-2   JFET BIASING

The biasing methods used in FET circuits are very similar to those employed in BJT circuits. In this section, we will examine the circuit calculations for the three most frequently used JFET biasing methods: gate biasing, self-biasing and voltage-divider biasing.

11-2-1   *Gate Biasing*

Figure 11-6(a) shows a gate-biased JFET circuit. The gate supply voltage ($-V_{GG}$) is used to reverse bias the gate–source junction of the JFET. With no gate current, there can be no voltage drop across R_G and therefore the voltage at the gate of the JFET will equal the dc gate supply voltage.

$$V_{GS} = V_{GG}$$

In the example in Figure 11-6,

$$V_{GS} = V_{GG} = -1.5 \text{ V}$$

Knowing V_{GS}, we can calculate I_D if the JFET's current (I_{DSS}) and voltage ($V_{GS(OFF)}$) specification limits are known by using the following formula:

$$I_D = I_{DSS} \left(1 - \frac{V_{GS}}{V_{GS(OFF)}} \right)^2$$

DEVICE: 2N5484 Through 2N5486—N-Channel JFET

Most of these specifications are self-explanatory. $V_{GS\,(OFF)}$ which has been given throughout is listed in the OFF characteristics, while I_{DSS} is listed in the ON characteristics. The only confusing maximum rating is "Forward Gate Current" because the gate is never normally forward biased (V_{GS} = a negative voltage or zero volts). This rating indicates that if the gate accidentally becomes forward biased, gate current must not exceed 10 mA dc or the JFET will be destroyed.

2N5484
2N5486

CASE 29-04, STYLE 5
TO-92 (TO-226AA)

1 Drain

3 Gate

2 Source

JFET
VHF/UHF AMPLIFIERS

N-CHANNEL — DEPLETION

MAXIMUM RATINGS

Rating	Symbol	Value	Unit
Drain-Gate Voltage	V_{DG}	25	Vdc
Reverse Gate-Source Voltage	V_{GSR}	25	Vdc
Drain Current	I_D	30	mAdc
Forward Gate Current	$I_{G(f)}$		mAdc
Total Device Dissipation @ T_C = 25°C Derate above 25°C	P_D	310 2.82	mW mW/°C
Operating and Storage Junction Temperature Range	T_J, T_{stg}	−65 to +150	°C

ELECTRICAL CHARACTERISTICS (T_A = 25°C unless otherwise noted.)

Characteristic		Symbol	Min	Typ	Max	Unit
OFF CHARACTERISTICS						
Gate-Source Breakdown Voltage (I_G = −1.0 μAdc, V_{DS} = 0)		$V_{(BR)GSS}$	−25	−	−	Vdc
Gate Reverse Current (V_{GS} = −20 Vdc, V_{DS} = 0) (V_{GS} = −20 Vdc, V_{DS} = 0, T_A = 100°C)		I_{GSS}	− −	− −	−1.0 −0.2	nAdc μAdc
Gate Source Cutoff Voltage (V_{DS} = 15 Vdc, I_D = 10 nAdc)	2N5484 2N5485 2N5486	$V_{GS(off)}$	−0.3 −0.5 −2.0	− − −	−3.0 −4.0 −6.0	Vdc
ON CHARACTERISTICS						
Zero-Gate-Voltage Drain Current (V_{DS} = 15 Vdc, V_{GS} = 0)	2N5484 2N5485 2N5486	I_{DSS}	1.0 4.0 8.0	− − −	5.0 10 20	mAdc

$V_{GS\,(OFF)}$

I_{DSS}

FIGURE 11-5 Data Sheet for an *n*-Channel JFET. (Copyright of Motorola. Used by permission.)

In the example in Figure 11-6,

$$I_D = 20 \text{ mA} \left(1 - \frac{-1.5 \text{ V}}{-4 \text{ V}} \right)^2$$
$$= 20 \text{ mA } (1 - 0.375)^2$$
$$= 20 \text{ mA} \times 0.625^2$$
$$= 20 \text{ mA} \times 0.39 = 7.8 \text{ mA}$$

Now that I_D is known, we can calculate the voltage drop across R_D using Ohm's law.

$$V_{R_D} = I_D \times R_D$$

In the example in Figure 11-6,

$$V_{R_D} = I_D \times R_D = 7.8 \text{ mA} \times 1 \text{ k}\Omega = 7.8 \text{ V}$$

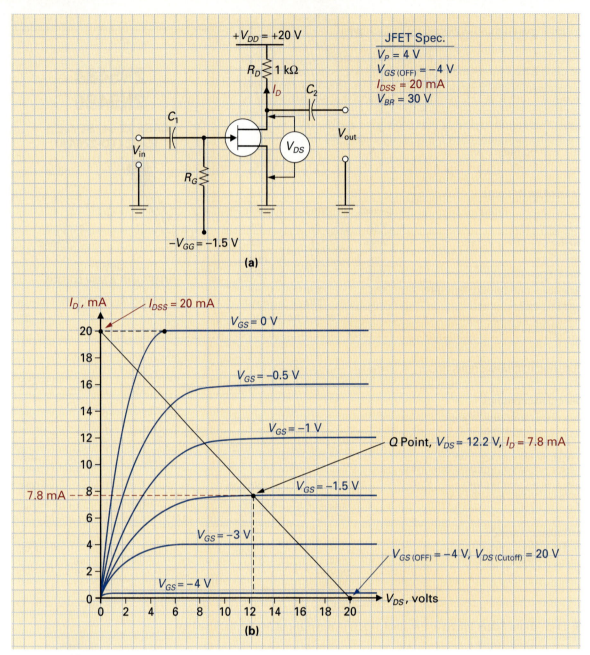

FIGURE 11-6 A Gate-Biased JFET Circuit. (a) Basic Circuit. (b) Drain Characteristic Curves and DC Load Line.

Because V_{R_D} plus V_{DS} will equal V_{DD}, we calculate V_{DS} once V_{R_D} is known with the following formula:

$$V_{DS} = V_{DD} - V_{R_D}$$

In the example in Figure 11-6,

$$V_{DS} = V_{DD} - V_{R_D} = 20\ \text{V} - 7.8\ \text{V} = 12.2\ \text{V}$$

Figure 11-6(b) shows the drain characteristic curves and dc load line for the example JFET circuit in Figure 11-6(a). Like the bipolar transistor, the JFET's dc load line extends between the maximum output current point, or saturation point (when the JFET is fully ON,

$I_{DSS} = 20$ mA), to the maximum output voltage point (when the JFET is cut OFF, $V_{DS} = 20$ V). The dc operating point, or Q point, which was determined with the previous calculations, is also plotted on the dc load line in Figure 11-6(b).

EXAMPLE:

Calculate the following for the circuit shown in Figure 11-7.

a. V_{GS}

b. I_D

c. V_{DS}

d. Maximum value of I_D

e. V_{DS} when $V_{GS} = V_{GS(OFF)}$

f. Q point

Solution:

a. $V_{GS} = V_{GG} = -3$ V

b. $I_D = I_{DSS}\left(1 - \dfrac{V_{GS}}{V_{GS(OFF)}}\right)^2 = 15 \text{ mA}\left(1 - \dfrac{-3 \text{ V}}{-6 \text{ V}}\right)^2 = 15 \text{ mA } (1 - 0.5)^2 = 3.75 \text{ mA}$

c. $V_{DS} = V_{DD} - V_{R_D}$ (since $V_{R_D} = I_D \times R_D$, we can substitute)

$V_{DS} = V_{DD} - (I_D \times R_D) = 15 \text{ V} - (3.75 \text{ mA} \times 1.2 \text{ k}\Omega) = 15 \text{ V} - 4.5 \text{ V} = 10.5 \text{ V}$

d. Maximum value of $I_D = I_{DSS} = 15$ mA

e. When $V_{GS} = V_{GS(OFF)}$, the JFET is cut off and equivalent to an open switch between drain and source. In this condition all of the drain supply voltage will appear across the open JFET.

$$V_{DS(cutoff)} = V_{DD} = 15 \text{ V}$$

f. The dc operating or Q point is

$$V_{GS} = -3 \text{ V}$$
$$I_D = 3.75 \text{ mA}$$
$$V_{DS} = 10.5 \text{ V}$$

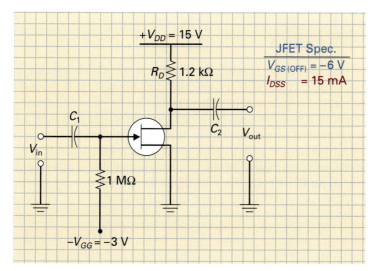

FIGURE 11-7 A Gate-Biased JFET Circuit Example.

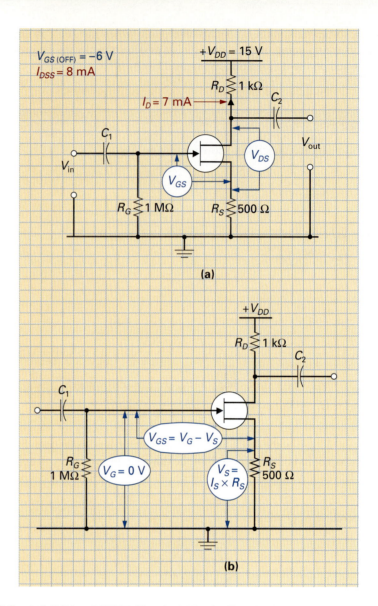

FIGURE 11-8   A Self-Biased JFET Circuit. (a) Basic Circuit. (b) How R_S Develops a $-V_{GS}$.

11-2-2   *Self Biasing*

Figure 11-8(a) shows how to self bias a JFET circuit. One advantage of this biasing method over gate biasing is that only a single drain supply voltage is needed (V_{DD}) instead of both V_{DD} and a negative gate supply voltage ($-V_{GG}$). The other difference you may have noticed is that a source resistor (R_S) has been included and R_G has been connected to ground. Although this arrangement seems completely different from the gate bias circuit, the inclusion of R_S and the grounding of R_G will achieve the same result, which is to reverse bias the JFET's gate–source junction. Figure 11-8(b) illustrates how this is achieved. Because there is no gate current in a JFET circuit ($I_G = 0$), all of the current flowing into the source will travel through the channel and flow out of the drain. Therefore,

$$I_S = I_D$$

For the example in Figure 11-8,

$$I_S = I_D = 7 \text{ mA}$$

Now that I_S is known, we can calculate the voltage drop across the source resistor (V_{R_S}) and therefore the voltage at the JFET's source (V_S).

$$V_{R_S} = V_S = I_S \times R_S$$

For the example in Figure 11-8,

$$V_{RS} = V_S = I_S \times R_S = 7 \text{ mA} \times 500 \text{ } \Omega = +3.5 \text{ V}$$

Because $I_G = 0$ A, there will be no voltage drop across R_G, and so the voltage at the gate of the JFET will be 0 V.

$$V_G = 0 \text{ V}$$

Now that we know that $V_S = +3.5$ V and $V_G = 0$ V, we can see how the JFET's gate–source junction is reverse biased. To reverse bias a gate-biased JFET, we simply made the gate voltage negative with respect to the source that is at 0 V. With a self-biased JFET, we achieve the same result by making the source voltage positive with respect to the gate that is at 0 V. This makes the gate of the JFET negative with respect to the source. This potential difference from gate-to-source (V_{GS}) is therefore equal to

$$V_{GS} = V_G - V_S$$

Because $V_S = I_S \times R_S$ and $V_G = 0$ V, we can substitute the previous formula to obtain

$$V_{GS} = 0 \text{ V} - (I_S \times R_S)$$

or

$$V_{GS} = -(I_S \times R_S)$$

or because $I_S = I_D$,

$$V_{GS} = -(I_D \times R_S)$$

In the example in Figure 11-8,

$$V_{GS} = -(I_S \text{ or } I_D \times R_S)$$
$$= -(7 \text{ mA} \times 500 \text{ } \Omega)$$
$$= -3.5 \text{ V}$$

If $-V_{GS}$ and R_S are known, we could transpose the preceding equation to calculate I_D.

$$I_D = \frac{V_{GS}}{R_S}$$

In the example in Figure 11-8,

$$I_D = \frac{V_{GS}}{R_S} = \frac{3.5 \text{ V}}{500 \text{ } \Omega} = 7 \text{ mA}$$

The final calculation is to determine the voltage at the JFET's drain with respect to ground (V_D) and the drain-to-source voltage drop across the JFET.

$$V_D = V_{DD} - V_{R_D}$$

Because $V_{R_D} = I_D \times R_D$,

$$V_D = V_{DD} - (I_D \times R_D)$$

In the example in Figure 11-8,

$$V_D = V_{DD} - (I_D \times R_D) = 15\text{ V} - (7\text{ mA} \times 1\text{ k}\Omega) = 15\text{ V} - 7\text{ V} = 8\text{ V}$$

Now that the voltage drops across R_D (V_{R_D}) and R_S (V_{R_S}) are known, we can calculate the voltage drop across the JFET's drain to source (V_{DS}).

$$V_{DS} = V_{D_D} - (V_{R_D} + V_{R_S})$$

In the example in Figure 11-8,

$$V_{DS} = V_{DD} - (V_{R_D} + V_{R_S}) = 15\text{ V} - (7\text{ V} + 3.5\text{ V}) = 15\text{ V} - 10.5\text{ V} = 4.5\text{ V}$$

EXAMPLE:

Calculate the following for the circuit shown in Figure 11-9.

   a. V_S

   b. V_{GS}

   c. V_{DS}

   d. I_D maximum

   e. V_{DS} when the JFET is OFF

   f. V_D

Solution:

   a. Because $I_S = I_D$, $V_S = I_S \times R_S = 4\text{ mA} \times 500\ \Omega = 2\text{ V}$

   b. $V_{GS} = V_G - V_S = 0\text{ V} - 2\text{ V} = -2\text{ V}$

   c. $V_{DS} = V_{DD} - (V_{R_D} + V_{R_S}) = 10\text{ V} - [(I_D R_D) + 2\text{ V}]$

$$= 10\text{ V} - [(4\text{ mA} \times 1.2\text{ k}\Omega) + 2\text{ V}] = 10\text{ V} - (4.8\text{ V} + 2\text{ V})$$

$$= 10\text{ V} - 6.8\text{ V} = 3.2\text{ V}$$

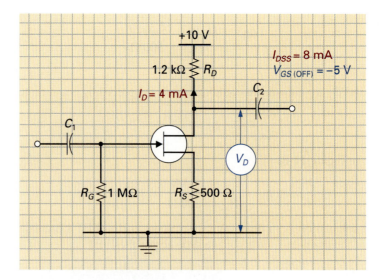

FIGURE 11-9 A Self-Biased JFET Circuit Example.

CHAPTER 11 / FIELD EFFECT TRANSISTORS

d. I_D maximum $= I_{DSS} = 8$ mA

e. $V_{DS(\text{cutoff})} = V_{DD} = 10$ V

f. $V_D = V_{DS} + V_{R_S} = 3.2$ V $+ 2$ V $= 5.2$ V

As previously mentioned, one advantage of this self-biased JFET method is that only a drain supply voltage is needed (V_{DD}). The gate supply voltage (V_{GG}) is not needed due to the inclusion of a source resistor that reverse biases the JFET's gate–source junction by applying a positive voltage to the source with respect to the 0 V on the gate. This method of effectively sending back a negative voltage from the source to the gate is known as "negative feedback." It not only enables us to bias a JFET with one supply voltage, it also provides temperature stability. Any change in the ambient temperature will cause a change in the semiconductor JFET's conduction, which would move the JFET's Q point away from its desired setting. The inclusion of R_S will prevent the Q point from shifting due to temperature in the same way as a BJT's emitter resistor. If temperature were to increase, for instance ($T\uparrow$), the resistance of the semiconductor would decrease ($R\downarrow$) because all semiconductor materials have a negative temperature coefficient of resistance ($T\uparrow, R\downarrow$), and this will cause the channel current to increase. If the drain current increases ($I_D\uparrow$), the voltage drop across R_S will increase (V_{R_S} or $V_S\uparrow = I_D\uparrow \times R_S$). This increase in V_S will increase the gate–source reverse voltage ($-V_{GS}\uparrow$), causing the JFET's channel to get narrower and the drain current to decrease ($I_D\downarrow$) and counteract the original increase. Similarly, a decrease in temperature will cause a decrease in I_D, which will decrease the gate–source reverse bias, resulting in an increase in I_D. The Q point will remain relatively stable despite changes in temperature when a JFET circuit has a source resistor included.

11-2-3   *Voltage-Divider Biasing*

Referring to the voltage-divider biased JFET circuit shown in Figure 11-10, you will probably notice that it is very similar to the voltage-divider-biased BJT circuit discussed previously. Like the self-biased circuit, the inclusion of a source resistor stabilizes the Q point despite ambient temperature changes. In addition, using a voltage divider to determine the gate–source bias voltage ensures that V_{GS}, and therefore the circuit, has increased stability.

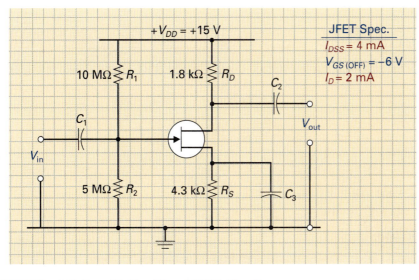

FIGURE 11-10   A Voltage-Divider-Biased JFET Circuit.

The gate voltage (V_G) is calculated using the following voltage-divider formula:

$$V_{R_2} \text{ or } V_G = \frac{R_2}{R_1 + R_2} \times V_{DD}$$

For the example in Figure 11-10,

$$V_{R_2} \text{ or } V_G = \frac{R_2}{R_1 + R_2} \times V_{DD} = \frac{5 \text{ M}\Omega}{10 \text{ M}\Omega + 5 \text{ M}\Omega} \times 15 \text{ V} = 5 \text{ V}$$

Because $I_D = I_S$ ($I_G = 0$) and the drain resistance and current are known, we can next calculate the voltage drop across the source resistor (V_{R_S}), drain resistor (V_{R_D}), and the JFET's source–drain junction (V_{DS}).

$$I_S = I_D$$

$$I_S = I_D = 2 \text{ mA}$$

$$V_{R_S} = I_S \times R_S$$

$$V_{R_S} = 2 \text{ mA} \times 4.3 \text{ k}\Omega = 8.6 \text{ V}$$

$$V_{R_D} = I_D \times R_D$$

$$V_{R_D} = 2 \text{ mA} \times 1.8 \text{ k}\Omega = 3.6 \text{ V}$$

$$V_{DS} = V_{DD} - (V_{R_S} + V_{R_D})$$

$$V_{DS} = 15 \text{ V} - (8.6 \text{ V} + 3.6 \text{ V}) = 2.8 \text{ V}$$

Now that the JFET's gate and source voltages are known (V_G and V_S), we can calculate the value of gate–source reverse bias $-V_{GS}$.

$$V_{GS} = V_G - V_S$$

$$V_G = V_{R_2}, \qquad V_S = V_{R_S}$$

For the example in Figure 11-10,

$$V_{GS} = 5 \text{ V} - 8.6 \text{ V} = -3.6 \text{ V}$$

■ EXAMPLE:

Calculate the following for the voltage-divider-biased JFET circuit shown in Figure 11-11:

 a. V_G
 b. I_S
 c. V_S
 d. V_{DS}
 e. V_{GS}
 f. V_D, when $V_{GS} = V_{GS(OFF)}$
 g. I_D, when $V_{GS} = 0 \text{ V}$

■ Solution:

 a. $V_G = \dfrac{R_2}{R_1 + R_2} \times V_{DD} = \dfrac{10 \text{ M}\Omega}{100 \text{ M}\Omega + 10 \text{ M}\Omega} \times 30 \text{ V} = 2.7 \text{ V}$

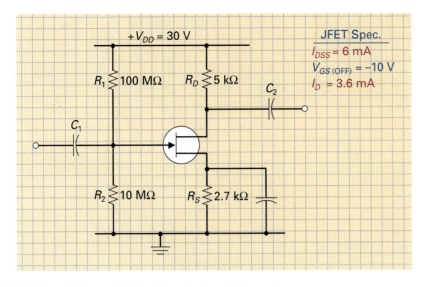

FIGURE 11-11 A Voltage-Divider-Biased Circuit Example.

b. $I_S = I_D = 3.6$ mA

c. $V_S = V_{R_S} = I_S \times R_S = 3.6$ mA $\times$ 2.7 kΩ = 9.7 V

d. $V_{DS} = V_{DD} - (V_{R_S} + V_{R_D}) = 30$ V $- [9.7$ V $+ (I_D \times R_D)]$
$= 30$ V $- [9.7$ V $+ (3.6$ mA $\times 5$ k$\Omega)] = 30$ V $- (9.7$ V $+ 18$ V$) = 2.3$ V

e. $V_{GS} = V_G - V_S = 2.7$ V $- 9.7$ V $= -7$ V

f. When $V_{GS} = V_{GS(OFF)}$, JFET is OFF and $V_D = V_{DD} = 30$ V

g. When $V_{GS} = 0$ V, $I_D =$ maximum $= I_{DSS} = 6$ mA

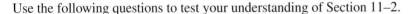

SELF-TEST EVALUATION POINT FOR SECTION 11-2

Use the following questions to test your understanding of Section 11–2.

1. What two advantages does self bias have over gate bias?

2. What component in a JFET circuit provides temperature stability?

3. With a self-biased JFET circuit the source voltage is _____ with respect to the gate voltage which is _____. This makes the gate of the JFET _____ with respect to the source.

4. Like self bias, _____ bias has negative feedback and therefore maintains the Q point stable.

11-3   JFET CIRCUIT CONFIGURATIONS

The three JFET circuit configurations are illustrated in Figure 11-12 along with their typical circuit characteristics. Like the bipolar transistor configurations, the term *common* is used to indicate which of the JFET's leads is common to both the input and output. In this section, we will examine the characteristics of these three configurations: common–source, common–gate, and common–drain.

11-3-1   *Common–Source (C–S) Circuits*

Similar to its bipolar counterpart, the common–emitter configuration, the **common–source configuration** is the most widely used JFET circuit; it is detailed in Figure 11-12(a). The

Common–Source Configuration

An FET configuration in which the source is grounded and common to the input and output signal.

	Voltage Gain	Input Impedance	Output Impedance	Circuit Appearance and Application	Waveforms
Common-Source **(a)**	5-10 (Voltage Amp)	Very High 1-15 MΩ	Low 2-10 kΩ	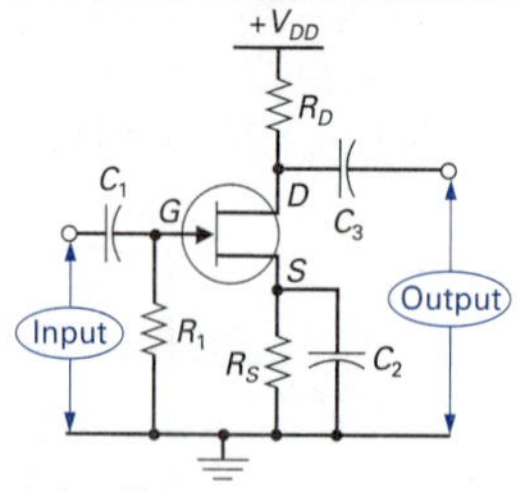 Most widely used FET configuration. It is mainly used as a voltage amplifier, however it is also used as an impedance matching device and can handle the high radio frequency signals.	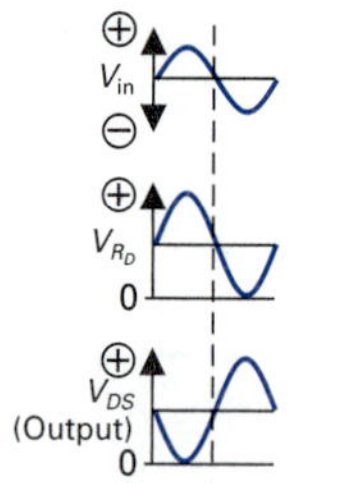 V_{in} and V_{out} are out of phase (180° phase shift)
Common-Gate **(b)**	2-5	Very Low 200-1500 Ω	Medium 5-15 kΩ	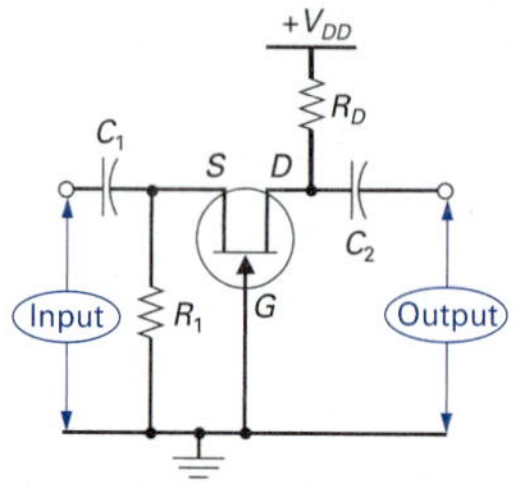 This configuration is used to amplify radio frequency signals due to its very stable nature at high frequencies. It is also used as a buffer to match a low impedance source to a high impedance load.	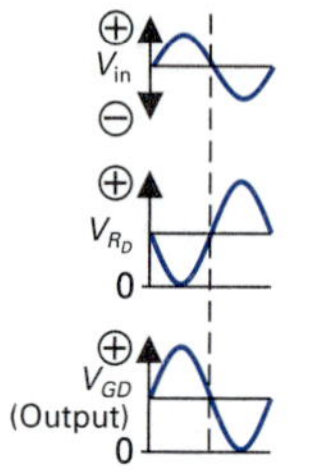 V_{in} and V_{out} are in phase (0° phase shift)
Common-Drain **(c)**	0.98	Very High 1500 MΩ	Low 10 kΩ	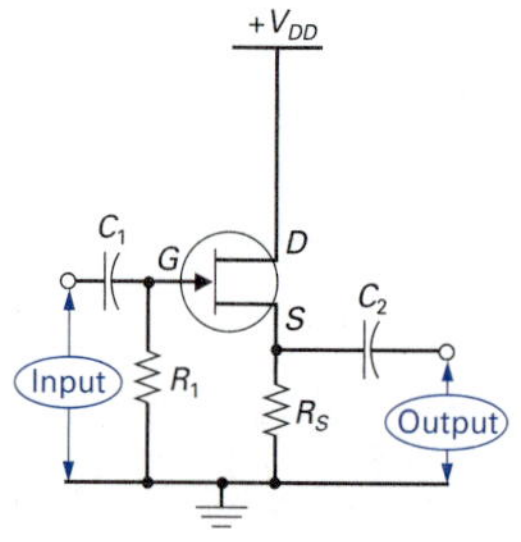 This amplifier is commonly called a source-follower as the source follows whatever is applied to the gate. Its very high input impedance will not load down (and therefore not distort) signals from high-impedance signal sources, such as a microphone, and its low output impedance is ideal to drive a low-impedance load such as an audio amplifier.	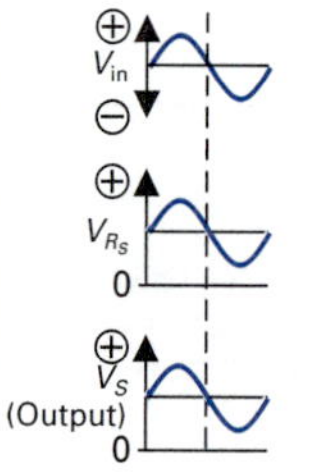 V_{in} and V_{out} are in phase (0° phase shift)

FIGURE 11-12   **JFET Circuit Configurations.**

input is applied between the gate and source and the output is taken between the drain and source, with the source being common to both input and output. The ac input will pass through the coupling capacitor C_1 and be superimposed on the dc gate–source bias voltage provided by resistor R_1, which sets up the dc operating or Q point. As the signal input changes, it will cause a change in gate voltage, which will cause a corresponding change in the output drain current. The output voltage developed between the FET's drain and ground is 180° out of phase with the input because an increase in $V_{in}\uparrow$, and therefore $V_{GS}\uparrow$, will cause an increase in $I_D\uparrow$, a decrease in the voltage drop across the FET ($V_{DS}\downarrow$), and a decrease in the output voltage $V_{out}\downarrow$. Resistor R_S is included to provide temperature stability and, as with the bipolar transistor, the source decoupling capacitor C_2 is included to prevent degenerative feedback.

When a small ac input signal is applied to the gate of a common–source amplifier, the variations in voltage at the gate control the JFET, which effectively acts as a variable resistor, varying the output drain current. These changes in drain current will vary the voltage drop across R_D and the drain-to-source voltage drop, which, with R_S, determines the output voltage. Referring to the characteristics listed in Figure 11-12(a), you can see that the output voltage (V_{out}) of the common–source JFET configuration can be five to ten times larger than the gate control input voltage (V_{in}). If a high amount of voltage gain is desired, R_D is made relatively large (typically, greater than 20 kΩ) and the JFET is biased so that its drain-to-source resistance is also high. A larger resistance will develop a larger voltage.

Also listed in the common–source characteristics in Figure 11-12(a) is the very high input resistance and the relatively low output resistance of this circuit. The high input resistance is due to the JFET's reverse-biased gate–source junction, which permits no gate input current and therefore has a very large resistance. This key characteristic means that the common–source JFET circuit is ideal in applications where we need to provide voltage amplification but do not want to load down a source that can only generate a small input signal. Such applications include the following:

1. Digital circuits in which the outputs of many circuits are connected to one another and therefore the output resistances of all the circuits load one another. As a result, the signals generated by these circuits are small, and a circuit is needed that will not load the signal source but will still provide voltage gain.

2. Analog circuits in which it can amplify both dc and low- and high-frequency ac input signal voltages. The C–S circuit's high input impedance makes it ideal at the front end of systems such as the first RF amplifier stage following the antenna and the first stage in a voltmeter, in which it will not load the source yet will amplify a wide range of input signal voltages.

11-3-2   *Common–Gate (C–G) Circuits*

The **common–gate circuit configuration** shown in Figure 11-12(b) is very similar to its bipolar counterpart, the common–base circuit. The input is applied between the source and gate, while the output appears across the drain and gate. Self-bias resistor R_1 sets up the static Q point, and the input is applied through the coupling capacitor C_1 and will cause a change in the JFET's source voltage. An increase in source voltage will cause a decrease in the V_{GS} forward bias (n-type source is driven positive), a decrease in I_D, a decrease in the voltage drop across R_D, and therefore an increase in the voltage dropped between the FET's drain and gate. Because the voltage developed across the JFET's drain and gate is applied to the output, an increase in the input produces an increase in the output, and so the input and output voltage are in phase with one another. Similarly, as the input voltage decreases, the gate–source forward bias will increase. Therefore, I_D will increase, and there will be more voltage developed across R_D and less voltage developed at the output.

Referring to the common–gate characteristics listed in Figure 11-12(b), you can see that this circuit can be used to provide a small voltage gain. Because the input is applied to the JFET's high-current source terminal, the input resistance is very low. This low input resistance and relatively high output resistance makes the circuit ideal in applications where we need to efficiently transfer power between a low-resistance source and a high-resistance load.

11-3-3   *Common–Drain (C–D) Circuits*

Comparable to the bipolar transistor's common–collector or emitter follower, the **common–drain configuration** shown in Figure 11-12(c) is sometimes called a **source follower** because the source output voltage follows in polarity and amplitude the input voltage at the gate. Once again, self-bias resistor R_1 sets up the quiescent operating point, and an ac gate input voltage will cause a variation in I_D. When the input voltage at the gate swings positive, the FET will conduct more current, less voltage will be developed across the FET drain-to-source, and therefore more voltage will be developed across R_S and the output. Similarly, a decrease in the input voltage will cause the resistance of the JFET's drain–source junction to increase. Therefore, V_{DS} will increase and V_{R_L}, or V_{out}, will decrease.

Referring to the common–drain characteristics listed in Figure 11-12(c), you can see that the output voltage is slightly less than the input voltage (circuit does not provide any voltage gain). The input resistance of the common–drain circuit configuration is extremely high due to the JFET's reverse-biased gate and R_S connection, and the output resistance is relatively very low. Inserting a common–drain circuit between a high-resistance source and a low-resistance load will ensure that the two opposite resistances are matched and power is efficiently transferred.

SELF-TEST EVALUATION POINT FOR SECTION 11-3

Use the following questions to test your understanding of Section 11–3.

1. Which FET circuit configuration is most widely used like its BJT common-emitter counterpart?
2. Which FET circuit configuration is also known as a source follower?
3. Which JFET circuit configuration could provide a high input impedance and a good value of voltage gain?
4. Which JFET circuit configuration is best suited for providing a very high input impedance and low output impedance?

11-4   JFET APPLICATIONS

It is the high input impedance of the JFET, and therefore its ability not to load a source, and the voltage amplification ability that are mainly made use of in circuit application. Like the bipolar junction transistor, the JFET can be made to function as a switch or as a variable resistor. Let us begin by examining how the JFET's switching ability can be made use of in digital or two-state circuits.

11-4-1   *Digital (Two-State) JFET Circuits*

As a switch, the JFET makes use of only two points on the load line: saturation (in which it is equivalent to a closed switch between source and drain) and cutoff (in which it is equivalent to an open switch between source and drain). Figure 11-13(a) shows an ON/OFF JFET switch circuit, and its associated load line is shown in Figure 11-13(b). Figure 11-13(c) shows the input/output voltages for each of the circuit's two operating states. When $V_{GS} = V_{GS(OFF)}$ (-4 V), the JFET is cut OFF (lower end of the load line) and is equivalent to an open switch between source and drain. With the JFET's drain–source open, $I_D = 0$ mA and the drain supply voltage will be applied to the output ($V_{DS} = V_{out} = +V_{DD}$). On the other hand, when $V_{GS} = 0$ V, the JFET is saturated (upper end of the load line) and is equivalent to a closed switch between source and drain. With the JFET's drain–source closed, $I_D =$ max $= I_{DSS}$ and the 0 V at the source will be applied to the output ($V_{DS} = V_{out} = 0$ V).

A typical FET application in digital circuits would be a buffer circuit, which is used to isolate one device from another. The high input impedance of the FET does not load the input circuit or circuits, while the low output impedance of the FET provides a high output current to the output circuit. The high output current and buffering or isolating characteristics of these circuits account for why they are also called *buffer drivers*. The schematic symbol of the buffer driver is shown in Figure 11-13(d).

The high input impedance ($Z_{in}\uparrow$) of the FET is also made use of in other FET integrated circuits (ICs). When Z_{in} is high, circuit current is low ($I\downarrow$) and therefore power dissipation is low ($P_D\downarrow$). This condition is ideal for digital integrated circuits (ICs), which contain thousands of transistors all formed onto one small piece of silicon. The low power

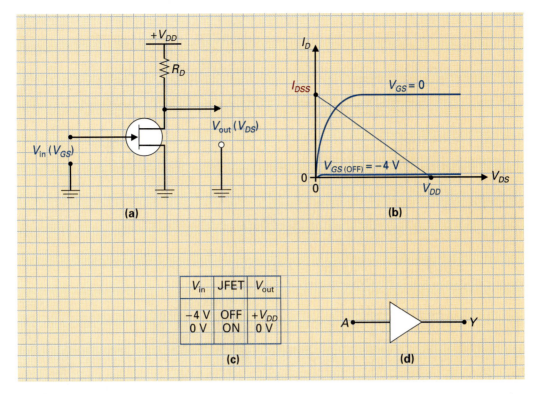

FIGURE 11-13 The JFET's Switching Action—Digital Circuit Applications. (a) Basic Switching Circuit. (b) Load Line. (c) Input/Output Voltages. (d) Digital Buffer-Driver Schematic Symbol.

dissipation of the JFET enables us to densely pack many more components into a very small area.

11-4-2 *Analog (Linear) JFET Circuits*

In two-state applications, the JFET is made to operate between the two extreme points of saturation (0 Ω) and cutoff (max Ω). By controlling the gate–source bias voltage (V_{GS}), the resistance between the JFET's drain and source (R_{DS}) can be changed to be any value between 0 Ω and maximum Ω. The JFET can therefore be made to act as a variable resistor, with an increase in negative V_{GS} causing a larger R_{DS}. In contrast, a decrease in negative V_{GS} causes a smaller R_{DS}. This is illustrated in Figure 11-14(a).

It is the reverse-biased gate–source junction of a JFET that gives the JFET its key advantage: *an extremely high input impedance* (typically in the high-MΩ range). In addition, the JFET can provide a small voltage gain and has been found to be a very low noise component. All these characteristics make it an ideal choice as an amplifier.

In previous chapters, we have seen how a light load (large resistance $R_L\uparrow$ and small $I\downarrow$) does not pull down the source voltage by any large amount, whereas a heavy load (small resistance $R_L\downarrow$ and large $I\uparrow$) will pull down the source voltage. A heavy load results in less output voltage ($V_{R_L}\downarrow$) and an increased current and heat loss at the source. The overall effect is that a small load resistance or impedance causes less power to be delivered to the load.

The circuit in Figure 11-14(b) shows how a common–source JFET has been connected to function as a preamplifier, which is a circuit that provides gain for a very weak input signal. In this example, the JFET preamplifier matches the high-impedance (small-signal) crystal microphone to a low-impedance power amplifier. The reverse-biased gate–source junction of a JFET preamplifier will offer a large load impedance to the source or microphone. This light load ($R_L\uparrow$) input resistance of the JFET will therefore permit most of the signal

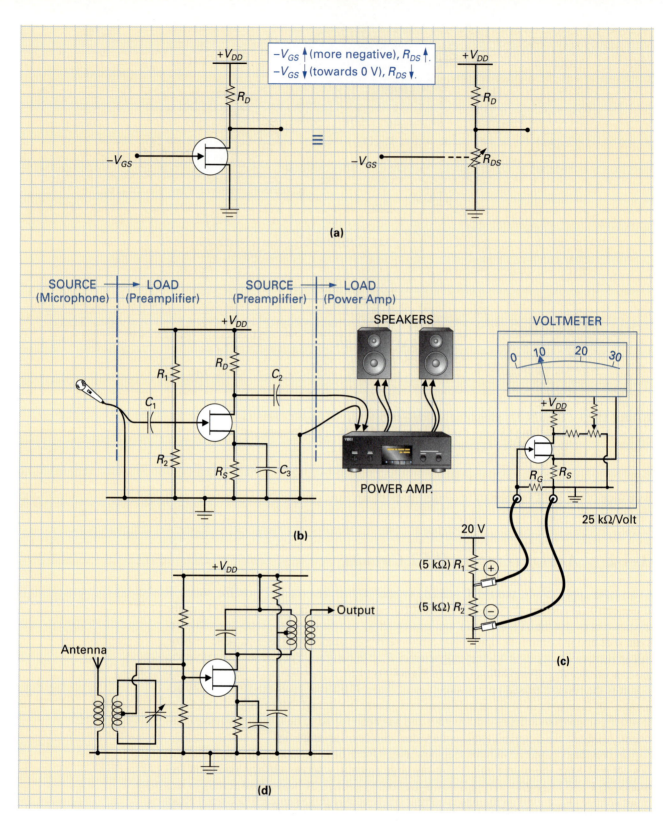

FIGURE 11-14 The JFET's Variable-Resistor Action—Analog Circuit Applications.
(a) Equivalent Circuit. (b) Application 1: An Audio Preamplifier Circuit.
(c) Application 2: A Voltmeter High Input Impedance Circuit. (d) Application
3: An RF Amplifier Circuit.

voltage being generated by the microphone to be applied to the JFET's gate and then be amplified. In other words, the high input impedance of the JFET amplifier circuit will not pull down the voltage signal being generated by the microphone. Therefore, maximum power will be transferred from source to load.

Figure 11-14(c) shows how a JFET at the front end of a voltmeter or oscilloscope will provide a very high input impedance and therefore not load the circuit under test. In this example, the meter will measure 10 V across R_2, and because the ohms per volt (Ω/V) rating of the voltmeter is 125 kΩ/V, the meter input impedance is

$$Z_{in} = \Omega/V \times V_{measured}$$
$$= 125 \text{ k}\Omega/V \times 10 \text{ V} = 1.25 \text{ M}\Omega$$

A 1.25-MΩ meter resistance in parallel with the 5-kΩ resistance of R_2 will have very little effect (1.25 MΩ in parallel with 5 kΩ = about 5 kΩ), so an accurate reading will be obtained.

Figure 11-14(d) shows how the JFET can be used as a radio-frequency (RF) amplifier. By studying this circuit, you can see that both the gate and drain contain tuned circuits in the same way as the previously discussed BJT RF amplifier circuits. However, there are two advantages that the JFET has over the bipolar transistor as a front-end RF amp.

1. The very weak signals injected into the antenna will have a very small value of current. Because the JFET is a voltage-controlled device, it requires no input current and it will respond well to the small-voltage signal variations picked up by the antenna.

2. The JFET is a very low noise component. Because any noise generated at the front end will be amplified along with the signal at each of the following amplifier stages, this JFET characteristic is ideal in this application.

SELF-TEST EVALUATION POINT FOR SECTION 11-4

Use the following questions to test your understanding of Section 11–4.

1. Which JFET characteristic is made use of in most circuit applications?
2. Why is the JFET ideal as an RF preamplifier?

11-5   TESTING JFETS

The transistor tester shown in Figure 11-15(a) can be used to test both BJTs and FETs. This tester can be used to determine

1. Whether an open or short exists between any of the terminals
2. The FET's transconductance/gain, and
3. The FET's value of I_{DSS} and leakage current

If a transistor tester is not available, the ohmmeter can be used to detect the most common failures: opens and shorts. Figure 11-15(b) indicates what resistance values should be obtained between the terminals of a good n-channel and p-channel JFET. Looking at these ohmmeter readings, you can see that because the JFET has only one P-N junction (gate-to-channel), it is relatively simple to test with an ohmmeter for an open or shorted junction.

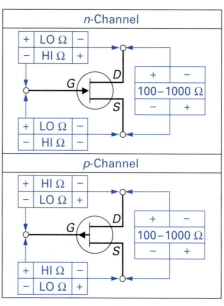

(a)   **(b)**

FIGURE 11-15   Testing JFETs.

SELF-TEST EVALUATION POINT FOR SECTION 11-5

Use the following question to test your understanding of Section 11–5.

1. Using a transistor tester to test a 2N5484 JFET, what typical readings should we obtain for the following:

a. $V_{GS(OFF)}$

b. $I_{DSS(MAXIMUM)}$

REVIEW QUESTIONS

Multiple-Choice Questions

1. A (an) _____ has three terminals called the gate, source, and drain.
   a. BJT   **c.** bipolar transistor
   b. zener   **d.** JFET

2. The _____ channel JFET schematic symbol has the arrow pointing out while the _____ channel JFET schematic symbol has the arrow pointing in.
   a. D–S, p   **b.** p, G–S   **c.** n, p   **d.** p, n

3. The BJT is a _____ controlled device, whereas the FET is a _____ controlled device.
   a. voltage, current   **b.** current, voltage

4. The gate–source junction of a JFET is always _____ biased since the input voltage is normally _____ or some _____ voltage.
   a. reverse, 0 V, negative   **c.** reverse, –4 V, positive
   b. forward, 0 V, negative   **d.** forward, –4 V, negative

5. When $V_{GS} = V_{GS(OFF)}$, $I_D = ?$
   a. I_{DSS} **c.** maximum
   b. zero   **d.** V_P

6. Which JFET circuit configuration is also known as a source follower?
   a. common–drain   **c.** common–gate
   b. common–source   **d.** Both (a) and (b) are true.

7. Which JFET circuit configuration provides a high input impedance and a good voltage gain?
   a. common–drain   **c.** common–gate
   b. common–source   **d.** Both (a) and (c) are true.

8. Which biasing method makes use of a $-V_{GG}$ supply voltage?
   a. Self biasing   **c.** Base biasing
   b. Gate biasing   **d.** Voltage-divider biasing

9. Which biasing method uses a source resistor?
   a. Self biasing   **c.** Voltage-divider biasing
   b. Gate biasing   **d.** Both (a) and (c)

10. Transconductance is a ratio of
    a. ΔI_D to ΔV_{DS} **c.** ΔI_D to ΔV_{GS}
    b. ΔV_{GD} to ΔI_D **d.** ΔV_{GS} to ΔV_{DS}

11. The input resistance of an FET is much higher than the input resistance of a BJT.
    a. True **b.** False

12. With a junction FET, as V_{GS} is made more negative the depletion region will _____ , the channel size will get _____ , and therefore I_D will _____ .
    a. decrease, larger, increase
    b. increase, smaller, decrease
    c. decrease, smaller, increase
    d. increase, larger, decrease

13. Which JFET circuit configuration has phase inversion between input and output?
    a. common–drain   **c.** common–gate
    b. common–source   **d.** both (b) and (c)

14. With a p-channel JFET, the gate should be _____ with respect to the source, and the drain should be _____ with respect to the source.
    a. negative, negative **c.** positive, positive
    b. negative, positive **d.** positive, negative

15. The current between _____ leads of a JFET is controlled by varying the reverse bias voltage applied to the _____ leads.
    a. gate and source, source and drain
    b. source and drain, drain and gate
    c. source and drain, gate and source
    d. drain and gate, gate and source

Practice Problems

16. When the gate of a 2N5484 JFET is reverse biased (normal operation), the gate reverse current (I_{GSS}) = 1.0 nA when

$V_{GS} = -20$ V ($V_{DS} = 0$ V). Calculate the input or gate–source impedance of the JFET.

17. Calculate the following for the amplifier circuit in Figure 11-16:
    a. Transconductance **b.** Voltage gain

18. Calculate the following for the circuit in Figure 11-17:
    a. V_{GS} **d.** $I_{D(maximum)}$
    b. I_D **e.** Q point
    c. V_{DS}

19. Calculate the following for the circuit in Figure 11-18:
    a. V_S **b.** V_{GS} **c.** V_{DS}

20. Calculate the following for the circuit in Figure 11-19:
    a. V_G **b.** V_S **c.** V_{GS} **d.** V_{DS}

21. Referring to Figure 11-20,
    a. What is the circuit configuration?
    b. What is the circuit's voltage gain?
    c. What is a circuit like this typically used for?

22. Referring to the circuit in Figure 11-21,
    a. What is the circuit configuration?
    b. What will be the output when the input is 0 V?
    c. What will be the output when the input is −6 V?
    d. Is there phase inversion between input and output?

Troubleshooting Questions

23. Briefly list what characteristics of a JFET can be tested with a transistor tester.

24. How can an ohmmeter test JFETs?

25. Which of the JFETs in Figure 11-22 are good, and which are bad? If bad, state the suspected problem.

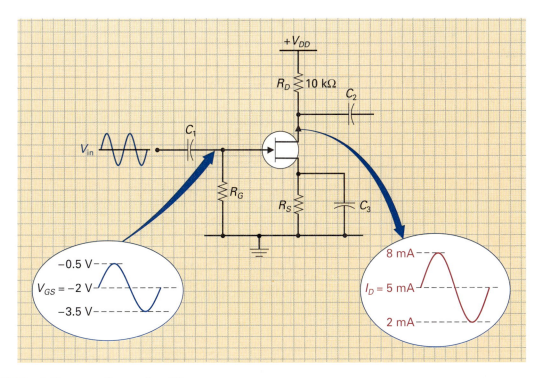

FIGURE 11-16 A Common–Source Amplifier.

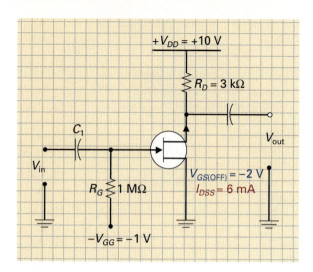

FIGURE 11-17   A Gate-Biased JFET Circuit.

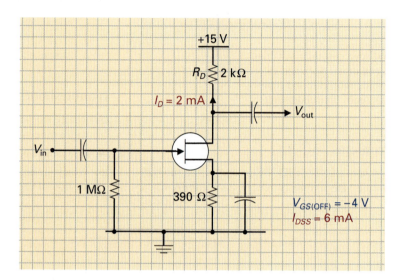

FIGURE 11-18   A Self-Biased JFET Circuit.

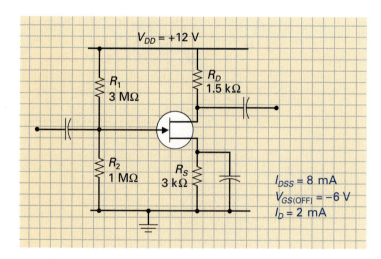

FIGURE 11-19   A Voltage-Divider-Biased JFET Circuit.

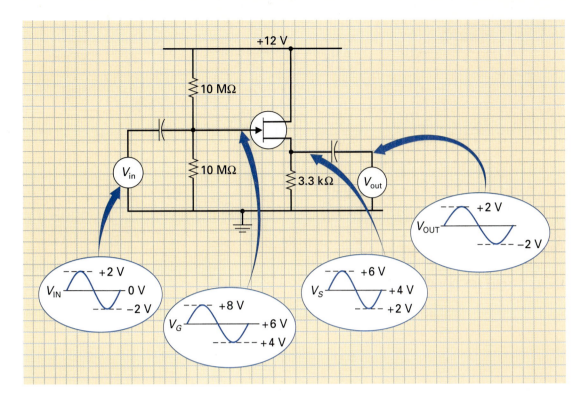

FIGURE 11-20   **An Analog JFET Circuit.**

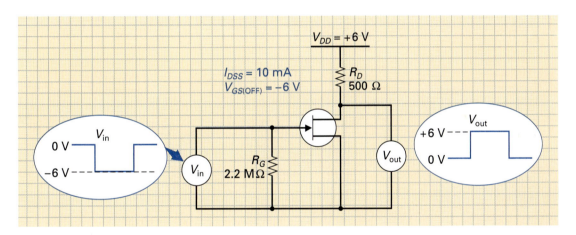

FIGURE 11-21   **A Digital JFET Circuit.**

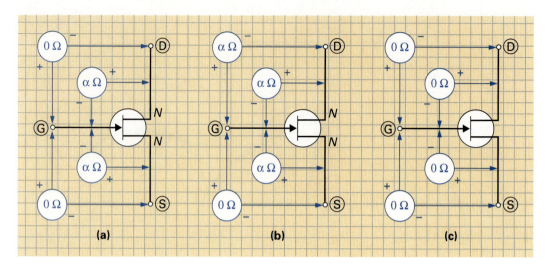

FIGURE 11-22 Testing JFETs with the Ohmmeter.

Field Effect Transistors (MOS)

12

Copy Master

Born in 1906 to two invalid Swedish immigrants, Chester Carlson worked part-time day and night to support his parents and still achieved excellent grades in high school. Sadly, when Carlson was seventeen both of his parents died, so he left New York and went to the California Institute of Technology where he graduated in 1930 with a degree in physics.

Carlson's first job was as a researcher for Bell Telephone Laboratories in New York. It was here that he saw the need for a machine that could copy documents—the method at that time was to have someone in a typing pool retype the original. Carlson realized that available photographic methods were too messy and time-consuming. He left Bell Labs in 1935 and began developing a clean and quick copying machine in the small rented bedroom that he had converted into a laboratory.

On the morning of October 23, 1938, Carlson statically charged a metal sulfur-coated plate by rubbing it with his handkerchief. He then exposed the plate to a glass slide that had on it the date and place "10-23-38 ASTORIA." When dry black powder on paper was pressed against the metal plate, the world's first photocopy was created.

Refining the design took Carlson several years because he had to develop more sensitive plates and a powder that would stick to paper. In fact, even after he had a good working prototype, it took him two years to find a company that was interested in manufacturing and selling his machine. However, in January, 1947, Haloid Company of Rochester, New York, which was a small photography firm, signed an agreement with Carlson for what he called his "dry printing machine." They were very dubious about whether it would really catch on and told Carlson that they would not pay him for his invention, but only give him a percentage of the profits. The company called the process Xerography, which meant "dry printing." After an almost overnight success, the company gave up all of its other products and renamed itself after its product, "Xerox Corporation."

The early Xerox machines in 1950 needed an operator to actuate the mechanism. However, by 1960 a fully automatic machine produced perfect copies by pressing one button.

Carlson's invention that nobody wanted transformed his lifestyle from an income that could support only a rented bedroom to royalties that paid him several million dollars a year.

Introduction

As discussed in the previous chapter, there are two basic types of field effect transistor, or FET. In the previous chapter, we concentrated on all aspects of the junction field effect transistor, or JFET. In this chapter, we will be examining the second type of FET, the metal-oxide semiconductor FET, or MOSFET (pronounced "moss-fet").

With the JFET, an input voltage of zero volts reverse biases the P-N junction, resulting in a maximum channel size and a maximum value of source-to-drain current. To decrease the size of the channel, the input voltage is made negative to further reverse bias the gate–source junction. This action depletes the channel of free carriers, reducing its size and therefore the source-to-drain current. This type of action is actually called *depletion-mode operation* because an input voltage is used to deplete the channel and therefore reduce the channel's size and current. The MOSFET does not have a P-N gate–channel junction like the JFET. It has a metal gate that is insulated from the semiconductor channel by a layer of silicon dioxide, hence the name *metal-oxide semiconductor*. Like all field effect transistors (FETs), the input voltage will generate an electric field that will have the effect of changing the channel's size.

The key difference between the JFET and MOSFET is that the JFET's input voltage would always have to be zero or negative in order to reverse bias the gate–source junction. With the MOSFET, the input voltage can be either a positive or negative voltage since gate current will always be zero because the gate is insulated from the channel. Let us examine each of these input voltage possibilities.

1. If the input voltage is negative, the resulting electric field depletes the channel, reducing its size, and the MOSFET is said to be operating in the depletion mode.

2. If the input voltage is positive, the resulting electric field enhances the channel, increasing its size, and the MOSFET is said to be operating in the enhancement mode.

The MOSFET can therefore be operated in either the depletion or enhancement mode due to its insulated gate. The two different types of MOSFET are given names based on their normal mode of operation. For instance, the *depletion-type MOSFET (D-type MOSFET or D-MOSFET)* should actually be called a *DE-MOSFET* because it can be operated in both the depletion mode and the enhancement mode, whereas the *enhancement-type MOSFET (E-type MOSFET or E-MOSFET)* is correctly named because it can be operated only in the enhancement mode. In this chapter, we will examine the construction, operation, characteristics, circuit biasing, applications, and testing of these two MOSFET types.

12-1 THE DEPLETION-TYPE (D-TYPE) MOSFET

The **depletion-type MOSFET** construction is slightly different from the JFET, and therefore we will need to first see how the D-MOSFET device is built before we can understand how it operates.

12-1-1 *D-MOSFET Construction*

Like the JFET and BJT, the D-MOSFET has two basic transistor types called the **n-channel D-type MOSFET** and **p-channel D-type MOSFET.** The construction and schematic symbols for these two D-MOSFET types are shown in Figure 12-1.

To begin with, let us examine the construction of the more frequently used *n*-channel D-type MOSFET, shown in Figure 12-1(a). This type of MOSFET basically consists of an *n*-type channel formed on a *p*-type substrate. A source and drain lead are connected to either end of the *n*-channel, and an additional lead is attached to the substrate. In addition, a thin insulating (silicon dioxide) layer is placed on top of the *n*-channel, and a metal plated area with a gate lead attached is formed on top of this insulating layer. Figure 12-1(b) shows the schematic symbol for an *n*-channel D-MOSFET; as you can see, the arrow on the substrate (*SS*) or base (*B*) lead points into the device. As a memory aid, imagine this arrowhead as a P-N junction diode, as shown in the inset. The source and drain leads are connected to either end of the diode's *n*-type cathode. Therefore, this device must be an *n*-channel D-MOSFET.

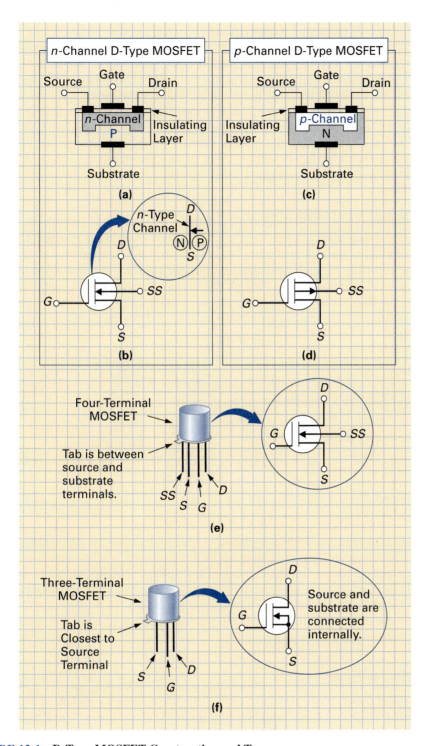

FIGURE 12-1   D-Type MOSFET Construction and Types.

The basic difference in the construction and schematic symbol of the *p*-channel D-MOSFET can be seen in Figures 12-1(c) and (d).

Figures 12-1(e) and (f) show how the MOSFET is available as a four-terminal or three-terminal device. In some applications, a separate bias voltage will be applied to the substrate terminal for added control of drain current and the four-terminal device will be used. In most circuit applications, however, the three-terminal device, which has its source and substrate lead internally connected, is all that is needed.

12-1-2   *D-MOSFET Operation*

Figure 12-2(a) shows the typical drain characteristic curves for an *n*-channel depletion-type MOSFET. As you can see, this set of curves has the same general shape as the JFET's set of drain curves and the BJT's set of collector curves. The key difference is that V_{GS} is plotted for both positive and negative values. This is because the D-MOSFET, which should actually be called a DE-MOSFET, can be operated in both the depletion mode (in which V_{GS} is a negative value) and the enhancement mode (in which V_{GS} is a positive value). To best understand the operation of the D-MOSFET, let us examine the three operation diagrams shown in Figures 12-2(b), (c), and (d).

Zero-Volt Operation

The center operation diagram, Figure 12-2(b), shows how the *n*-channel D-MOSFET will respond to a V_{GS} input of zero volts. When $V_{GS} = 0$ V, the gate and source terminals are at the same zero-volt potential, and therefore the gate is effectively shorted to the source. The value of drain current passing through the channel is called the I_{DSS} value (I_{DSS} is the drain-to-source current passing through the channel when the gate is shorted to the source). Therefore, when

$$V_{GS} = 0 \text{ V}, I_D = I_{DSS}$$

When zero volts is applied to the input of a D-type MOSFET, therefore, it will conduct a value of drain current. With no input, this device is ON, which is why the D-type MOSFET is known as a "normally ON" device.

Enhancement Mode

The upper operation diagram, Figure 12-2(c), shows how the *n*-channel D-MOSFET will respond when V_{GS} is made positive. In this condition, the channel is enhanced or widened, and the value of I_D is increased above I_{DSS}. Therefore, when

$$V_{GS} = +V, I_D > I_{DSS}$$

Let us examine in more detail why the channel is widened by a positive gate voltage. Because the valence-band holes in the *p*-type material (majority carriers) will be repelled by a positive gate voltage, and the conduction-band electrons in the *p*-type material (minority carriers) will be attracted to the channel by the positive gate voltage, there will be a buildup of electrons in the *p*-type material near the channel. This buildup of electrons in the *p*-type material below the channel will effectively widen the size of the channel, reducing its resistance, and therefore increasing I_D to a value greater than I_{DSS}.

Depletion Mode

The lower operation diagram, Figure 12-2(d), shows how the *n*-channel D-MOSFET will respond when V_{GS} is made negative. In this condition, the channel is depleted of free carriers, and therefore the value of I_D is decreased below I_{DSS}. Therefore, when

$$V_{GS} = -V, I_D < I_{DSS}.$$

To summarize the *n*-channel MOSFET's operation, when V_{GS} is either zero volts or a negative voltage, the *n*-channel D-MOSFET acts in almost exactly the same way as an *n*-channel JFET. However, unlike the JFET, the D-MOSFET can have a forward-biased gate-to-source P-N junction because the silicon-dioxide insulating layer prevents any current from passing through the gate and will still maintain a high input resistance. This dual operating ability is why the depletion-type MOSFET or D-MOSFET should actually be called a depletion-enhancement or DE-MOSFET.

The drain characteristic curves of the D-MOSFET can be used to plot the device's dc load line, as shown in Figure 12-2(a), with

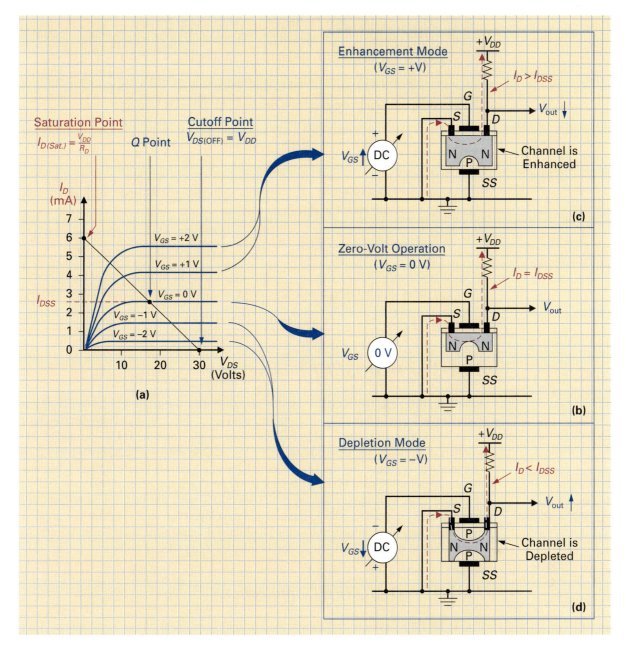

FIGURE 12-2 D-Type MOSFET Operation and Characteristics.

$$I_{D(\text{sat})} = \frac{V_{DD}}{R_D} \quad \text{at saturation}$$

and

$$V_{DS(\text{OFF})} = V_{DD} \text{ at cutoff}$$

As with the JFET, the D-MOSFET's transconductance is equal to the ratio of output current change (ΔI_D) to input voltage change (ΔV_{GS}),

$$\delta_m = \frac{\Delta I_D}{\Delta V_{GS}}$$

and the D-MOSFET's voltage gain is equal to

$$A_V = \delta_m \times R_D$$

EXAMPLE:

A D-MOSFET circuit has the following specifications:

$$I_{DSS} = 2 \text{ mA}, \qquad V_{GS(OFF)} = -6 \text{ V}, \qquad R_D = 3 \text{ k}\Omega, \qquad V_{DD} = 12 \text{ V}$$

Calculate the following two extremes on the D-MOSFET's load line:

    a. $I_{D(\text{sat})}$

    b. $V_{DS(OFF)}$

■ *Solution:*

    a. When the D-MOSFET is saturated, it is equivalent to a closed switch and therefore the only resistance is that of R_D.

$$I_{D(\text{sat})} = \frac{V_{DD}}{R_D} = \frac{12 \text{ V}}{3 \text{ k}\Omega} = 4 \text{ mA}$$

    b. When the D-MOSFET is cut off, it is equivalent to an open switch, and therefore the full drain supply voltage will appear across the open between drain and source.

$$V_{DS(OFF)} = V_{DD} = 12 \text{ V}$$

12-1-3   *D-MOSFET Biasing*

Like the JFET, the D-MOSFET can be configured in the same way as a common–drain, common–gate, or common–source circuit, with all of the dc and ac configuration character-istics being the same. The D-MOSFET is easier to bias than the JFET because of its ability to operate in either the depletion mode $(-V_{GS})$ or the enhancement mode $(+V_{GS})$. In fact, one of the most frequently used D-MOSFET biasing methods is to simply have no biasing at all, a method called **zero biasing** because the Q point is set at zero volts $(V_{GS} = 0 \text{ V})$, as seen in Figure 12-3. This makes biasing the D-MOSFET very simple because no gate or source bias voltages are needed. The ac input signal developed across R_G is therefore applied to the extremely high input impedance of the D-MOSFET, causing an increase and decrease in the conduction of the MOSFET above and below the $V_{GS} = 0 \text{ V}$ Q point.

Zero Biasing

A configuration in which no bias voltage is applied at all.

12-1-4   *D-MOSFET Applications*

The D-MOSFET is most frequently used in analog or linear circuit applications. This is be-cause the D-MOSFET can be very simply biased at a midpoint in the load line and then have

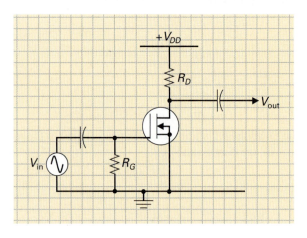

FIGURE 12-3   Zero Biasing a D-MOSFET.

its output current varied above and below this natural Q point in a linear fashion. This, coupled with the D-MOSFET's almost infinite input impedance and low noise properties, makes it ideal as a preamplifier at the front end of a system. Figure 12-4 shows how two D-MOSFETs can be used to construct a typical front-end **cascode amplifier circuit.** This cascode amplifier circuit consists of a self-biased common–source amplifier (Q_1) in series with a voltage-divider-biased common–gate amplifier (Q_2). The input signal (V_{in}) is applied to Q_1's gate, and the amplified output at Q_1's drain is then passed to Q_2's source, where it is further amplified by Q_2 before appearing at Q_2's drain, and therefore at the output (V_{out}).

The FET's only limiting factor is that its high input impedance starts to decrease as the input signal's frequency increases. Refer to the inset in Figure 12-4, which shows how the gate, insulator, and channel of a D-MOSFET form a capacitor. This input capacitance of typically 5 pF has very little effect at low input signal frequencies ($X_C\uparrow = 1/2\pi f\downarrow C$) because the input impedance is high ($X_C\uparrow$ therefore $Z_{in}\uparrow$) and the loading effect is negligible. At higher radio frequency ($X_C\downarrow = 1/2\pi f\uparrow C$), however, the input impedance is lowered ($X_C\downarrow$ therefore $Z_{in}\downarrow$) and the D-MOSFET loses its high input impedance advantage. To compensate for this disadvantage, FETs are often connected in series, as in Figure 12-4, so that their input capacitances are also in series. Recall that series-connected capacitors have a lower total capacitance than any of the individual capacitance values. Therefore, the overall input capacitance of two series-connected D-MOSFETs will be less than that of a single D-MOSFET, making this cascode amplifier ideal as a high radio-frequency (RF) amplifier: a low input capacitance ($C_{in}\downarrow$) means a high input reactance ($X_{C(in)}\uparrow$), and therefore a high input impedance ($Z_{in}\uparrow$) at high frequencies.

Cascode Amplifier Circuit

An amplifier circuit consisting of a self-biased common–source amplifier in series with a voltage-divider-biased common–gate amplifier.

12-1-5 *Dual-Gate D-MOSFET*

To compensate for the D-MOSFET's input capacitance problem, the **dual-gate D-MOSFET** was developed. The construction and schematic symbol for the dual-gate D-MOSFET is shown in Figure 12-5(a). In most applications, the dual-gate D-MOSFET is connected so that it acts as two series-connected D-MOSFETs, as shown in the cascode

Dual-Gate D-MOSFET

A metal-oxide semiconductor FET having two separate gate electrodes.

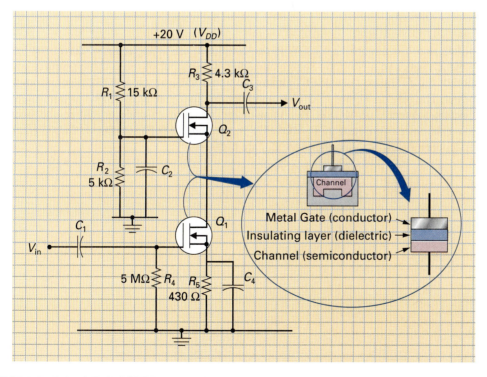

FIGURE 12-4   A D-MOSFET Analog Circuit Application—Cascode Amplifier.

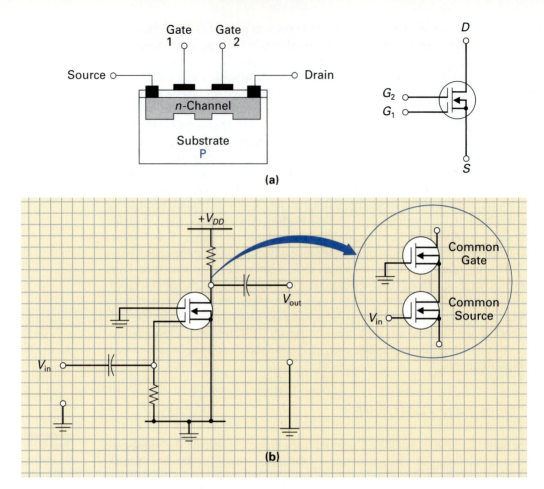

FIGURE 12-5 The Dual-Gate D-MOSFET. (a) Dual-Gate D-MOSFET Construction and Schematic Symbol. (b) Dual-Gate D-MOSFET Application—Cascode

amplifier circuit in Figure 12-5(b). With this amplifier, the ac input signal drives the lower gate, which acts like a common-source amplifier. The output of the common–source lower section of the dual-gate D-MOSFET drives the upper half, which acts like a common-gate amplifier. The inset in Figure 12-5(b) shows how the dual-gate D-MOSFET is equivalent to two series-connected D-MOSFETs. As with the previous cascode amplifier, the overall input capacitance of a dual-gate D-MOSFET is less than that of a standard D-MOSFET, and if capacitance is low, X_C, and therefore Z_{in}, are high.

Use the following questions to test your understanding of Section 12-1.

1. The two different types of MOSFETs are called the _____ type MOSFET and _____ type MOSFET.

2. True or False: The D-type MOSFET can be operated in both the depletion and enhancement mode.

3. The D-type MOSFET is a normally _____ (ON/OFF) device.

4. Which FET has a higher input impedance: JFET or MOSFET?

5. Why is the D-MOSFET ideal as a preamplifier?

   a. It can be mid-load-line biased when 0 V is applied.

   b. It has a high input impedance.

   c. It has low noise properties.

   d. All of the above.

6. The _____ MOSFET was developed to lower input capacitance so that it can handle high-frequency signals.

12-2   THE ENHANCEMENT-TYPE (E-TYPE) MOSFET

With an input of zero volts, a D-MOSFET will be ON and a certain value of current will pass through the channel between source and drain. If the input to the D-MOSFET is made positive, the channel is enhanced, causing the source-to-drain current to increase. If the input is made negative, the channel is depleted, causing the source-to-drain current to decrease. The D-MOSFET can therefore operate in either the enhancement or depletion mode and is called a "normally ON" device because it is ON when nothing (0 V) is applied.

The **enhancement-type MOSFET** or **E-MOSFET** can only operate in the enhancement mode. In other words, when the input is either zero volts or a negative voltage, the transistor is OFF and there is no source-to-drain current. However, when the input is made positive, the E-MOSFET will turn ON, resulting in a source-to-drain channel current. The E-MOSFET is therefore a "normally OFF" device because it is OFF when nothing (0 V) is applied.

> **Enhancement-Type MOSFET or E-MOSFET**
>
> A field effect transistor with an insulated gate (MOSFET) that can only be turned on if the channel is enhanced.

12-2-1   *E-MOSFET Construction*

As with all of the other transistor types, it is easier to understand the operation and characteristics of a device once we have seen how the component is constructed. The E-MOSFET has two basic transistor types, called the **n-channel E-type MOSFET** and **p-channel E-type MOSFET.** The construction and schematic symbols for these two E-MOSFET types are shown in Figure 12-6.

To begin with, let us examine the construction of the more frequently used n-channel E-type MOSFET, shown in Figure 12-6(a). Notice that no channel exists between the source and drain. Consequently, with no gate bias voltage, the device will be OFF. When the gate is made positive, however, electrons will be attracted from the substrate, causing a channel to be induced between the source and drain. This enhanced channel will permit drain current to flow, and any further increase in gate voltage will cause a corresponding increase in the size of the channel and therefore the value of I_D.

The schematic symbol for the n-channel E-type MOSFET can be seen in Figure 12-6(b). The construction and schematic symbol for the p-channel E-MOSFET can be seen in Figure 12-6(c) and (d). As with the D-MOSFET, both four-terminal and three-terminal

> **n-Channel E-Type MOSFET**
>
> An enhancement type MOSFET having an n-type channel between its source and drain terminals.

> **p-Channel E-Type MOSFET**
>
> An enhancement type MOSFET having a p-type channel between its source and drain terminals.

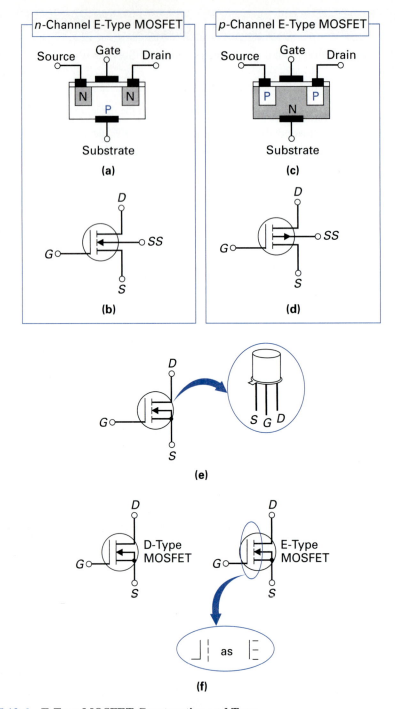

FIGURE 12-6 **E-Type MOSFET Construction and Type.**

devices are available, with the three-terminal device having a common connection between source and substrate, as shown in Figure 12-6(e).

The only difference between the E-type and D-type MOSFET symbols is the three dashed lines representing the drain, substrate, and source regions. The dashed line is used instead of the solid line to indicate that an E-MOSFET has a normally broken path, or channel, between drain, substrate, and source. To aid memory, the three dashed lines used in the E-MOSFET symbol could be thought of as the three horizontal prongs in the capital letter E, as shown in Figure 12-6(f).

12-2-2   *E-MOSFET Operation*

Figure 12-7(a) shows the typical drain characteristic curves for an *n*-channel enhancement-type MOSFET. As you can see, this set of drain curves is very similar to the D-MOSFET's set of drain curves; however, in this case only positive values of V_{GS} are plotted. Looking at the relationship between V_{GS} and I_D, you may have noticed that any increase in V_{GS} will cause a corresponding increase in I_D.

To best understand the operation of the E-MOSFET, let us examine the three operation diagrams shown in Figures 12-7(b), (c), and (d).

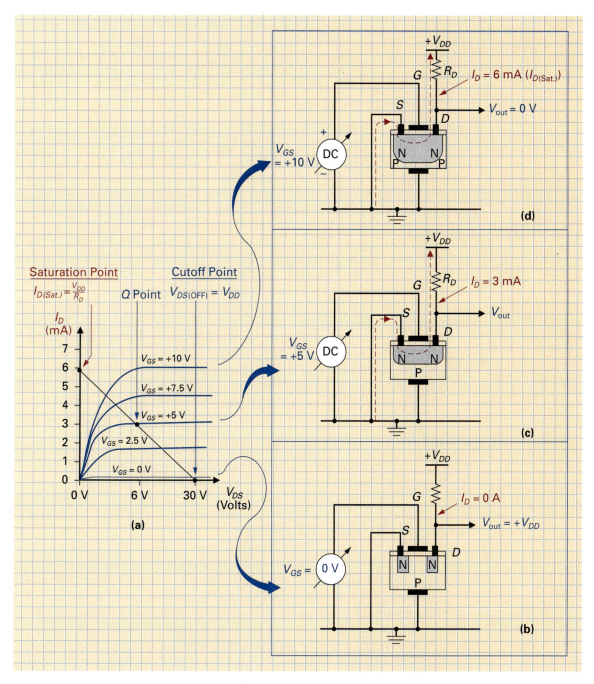

FIGURE 12-7   E-Type MOSFET Operation and Characteristics.

$V_{GS} = 0$ V Curve

The lower operation diagram, Figure 12-7(b), shows how the n-channel E-MOSFET will respond to a V_{GS} input of zero volts. When $V_{GS} = 0$ V, there is no channel connecting the source and drain, and therefore the drain current will be zero. As a result, $+V_{DD}$ will be present at the output because the E-MOSFET is equivalent to an open switch between drain and source.

$V_{GS} = +5$ V Curve

The center operation diagram, Figure 12-7(c), shows how the n-channel E-MOSFET will respond to a V_{GS} input of +5 volts. When $V_{GS} = +5$ V, the E-MOSFET will act in almost exactly the same way as an enhanced D-MOSFET. The positive gate voltage will repel the p-type material's majority carriers (holes) away from the gate, while attracting the p-type material's minority carriers (electrons) toward the gate. This action will form an n-type bridge between the source and drain, and therefore a value of I_D will flow between source and drain, as shown.

$V_{GS} = +10$ V Curve

The upper operation diagram, Figure 12-7(d), shows how the n-channel E-MOSFET will respond if the V_{GS} input is further increased to +10 volts. When $V_{GS} = +10$ V, the attraction of electrons and repulsion of holes within the p-type material is increased, causing the channel's width to increase and I_D to also increase. As a result, 0 V will be present at the output because the E-MOSFET is equivalent to a closed switch between drain and source.

In summary, when V_{GS} is zero volts, drain current is also zero. As the value of V_{GS} is increased (made more positive), the channel becomes wider, causing I_D to increase. On the other hand, as the value of V_{GS} is decreased (made less positive), the channel becomes narrower, causing I_D to decrease. In other words, when the input is either zero volts or a negative voltage, the transistor is OFF, and there is no source-to-drain current. However, when the input is made positive, the E-MOSFET will turn ON, resulting in a source-to-drain channel current. The E-MOSFET is therefore a "normally OFF" device because it is OFF when nothing (0 V) is applied.

Although it cannot be seen in the set of drain curves in Figure 12-7(a), V_{GS} will have to increase to a positive threshold voltage of about +1 V before a channel will be induced and a small value of drain current will flow. This "threshold level" is a highly desirable characteristic because it prevents noise or any low-level input signal voltage from turning ON the device. This advantage makes the E-type MOSFET ideally suited as a switch because it can be turned ON by an input voltage and turned OFF once the input voltage falls below the threshold level.

The p-channel E-type MOSFET operates in much the same way as the n-channel device except that holes are attracted from the substrate to form a p-channel and the V_{GS} and V_{DS} bias voltages are reversed.

12-2-3   *E-MOSFET Biasing*

Drain-Feedback Biased

A configuration in which the gate receives a bias voltage fed back from the drain.

Like the D-MOSFET, the E-MOSFET can be configured as a common–drain, common–gate, or common–source circuit. Unlike the D-MOSFET and JFET, the E-MOSFET cannot be biased using self bias or zero bias because V_{GS} must be a positive voltage. As a result, gate bias and voltage-divider bias can be used; however, more frequently E-MOSFETs are **drain-feedback biased,** as shown in Figure 12-8(a). In this example, R_D equals 8 kΩ, and R_G (which feeds back a positive voltage from the drain, hence the name *drain-feedback bias*) equals 100 MΩ. Because an E-MOSFET has an extremely high input impedance (due to the insulated gate), no current will flow in the gate circuit. With no gate current, there will be no voltage drop across the gate resistor ($V_{R_G} = 0$ V), and therefore the voltage at the gate will be at the same potential as the voltage at the drain.

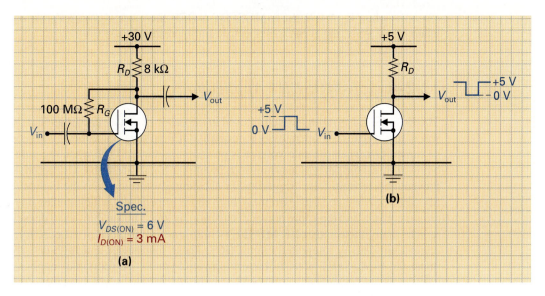

FIGURE 12-8 Biasing E-Type MOSFETs. (a) Drain-Feedback Biasing. (b) Voltage-Controlled Switching.

$$V_{GS} = V_{DS}$$

To help set up the Q point, most manufacturer's data sheets specify a load-line-midpoint drain current $I_{D(ON)}$ and drain voltage $V_{DS(ON)}$. In the example in Figure 12-8(a), when the E-MOSFET is ON, or conducting, and I_D = 3 mA, the E-MOSFET's drain-to-source voltage drop ($V_{DS(ON)}$) is 6 V. The value of R_D in Figure 12-8(a) has been chosen so that this E-MOSFET circuit will be biased at its specified Q point. To check, we can use the formula

$$V_{DS} = V_{DD} - (I_{D(ON)} \times R_D)$$

In the example in Figure 12-8(a),

$$V_{DS} = V_{DD} - (I_{D(ON)} \times R_D) = 30\,V - (3\,mA \times 8\,k\Omega)$$
$$= 30\,V - 24\,V = 6\,V$$

■ **EXAMPLE:**

A drain-feedback-biased E-MOSFET circuit has the following specifications: R_D = 1 kΩ, $I_{D(ON)}$ = 10 mA, V_{DD} = 20 V. Calculate V_{DS} and V_{R_D}.

■ *Solution:*

$$V_{DS} = V_{DD} - (I_{D(ON)} \times R_D)$$
$$= 20\,V - (10\,mA \times 1\,k\Omega)$$
$$= 20\,V - 10\,V = 10\,V$$

The constant drain supply voltage (V_{DD}) will be evenly divided across R_D and the E-MOSFET's drain-to-source (V_{DS} = 10 V, V_{R_D} = 10 V).

In most instances, the E-MOSFET will be used in digital two-state switching circuit applications. In this case, there will be no need for a gate resistor (R_G) because the input voltage (V_{in}) will either turn ON or OFF the E-MOSFET, as shown in Figure 12-8(b). For example, when $V_{in} = 0$ V, the E-MOSFET is OFF, therefore $V_{out} = +V_{DD} = +5$ V, whereas when $V_{in} = +5$ V, the E-MOSFET is ON and therefore $V_{out} = 0$ V.

12-2-4   *E-MOSFET Applications*

The E-MOSFET is more frequently used in digital or two-state circuit applications. One reason is that it naturally operates as a normally OFF voltage-controlled switch because it can be turned ON when the gate voltage is positive and turned OFF when the gate voltage falls below a threshold level. This threshold level is a highly desirable characteristic because it prevents noise from false triggering, or accidentally turning ON, the device. The other E-MOSFET advantage is its extremely high input impedance, which means that the device's circuit current, and therefore power dissipation, is low. This enables us to densely pack or integrate many thousands of E-MOSFETs onto one small piece of silicon, forming a high-component-density integrated circuit (IC). These low-power and high-density advantages make the E-MOSFET ideal in battery-powered small-size (portable) applications such as calculators, wristwatches, notebook computers, handheld video games, digital cellular phones, and so on.

12-2-5   *Vertical-Channel E-MOSFET (VMOS FET)*

As just mentioned, the E-MOSFET is generally used in two-state or digital circuit applications, where it acts as a normally OFF switch. In most digital circuit applications, the channel current is small and therefore a standard E-MOSFET can be used. However, if a larger current-carrying capability is needed, the **vertical-channel E-MOSFET** or VMOS FET can be used. Figure 12-9(a) shows the construction of a VMOS FET. The gate at the top of the device is insulated from the source (which is also at the top), and, as with all E-MOSFETs, no channel exists between the source terminal and the drain terminal with no bias voltage applied. The VMOS FET's semiconductor materials are labeled P, N+, and N− and indicate different levels of doping.

When this *n*-channel VMOS FET is biased ON (gate is made positive with respect to source), as seen in Figure 12-9(b), a vertical *n*-type channel is formed between source and drain. This channel is much wider than a standard E-MOSFET's horizontal channel, which is why VMOS FETs can handle a much higher drain current.

Figure 12-9(c) shows how the high-current capability of a VMOS FET can be made use of in an interfacing circuit application. In this circuit, a VMOS FET is used to interface a low-power source input signal to a high-power load. Referring to the current specifications of the VMOS FET, relay, and motor, you can see that each device is used to step up current. The standard E-MOSFET supplies a low-power input signal to the VMOS FET, which can handle enough current to actuate a relay whose contacts can handle enough current to switch power to the dc motor. To be more specific, if V_{in} is LOW, the standard E-MOSFET will turn OFF, producing a HIGH output to the gate of the VMOS FET, turning it ON. When the VMOS FET turns ON, it effectively switches ground through to the lower end of the relay coil, energizing the relay and closing its normally open contacts. The closed relay contacts switch ground through to the lower end of the motor, which turns ON because it now has the full +12 V supply across its terminals. The motor will stay ON as long as V_{in} stays LOW. If V_{in} were to go HIGH, the standard E-MOSFET inverter would produce a LOW output, which would turn OFF the VMOS FET, relay, and motor.

Other than its high-current capability, the VMOS FET also has a positive temperature coefficient of resistance, which means an increase in temperature will cause a decrease in drain current ($T\uparrow$, $R\uparrow$, $I\downarrow$), and this will prevent thermal runaway. This gives the VMOS FET a distinct advantage over the BJT power amplifier, which has a negative temperature coefficient of resistance ($T\uparrow$, $R\downarrow$, $I\uparrow$). This means that if maximum ratings are exceeded, an

Vertical-Channel E-MOSFET

An enhancement-type MOSFET that, when turned on, forms a vertical channel between source and drain.

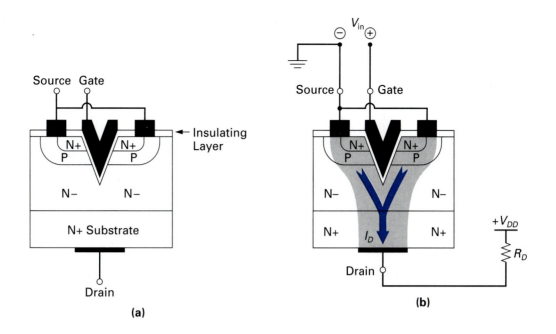

(a)

(b)

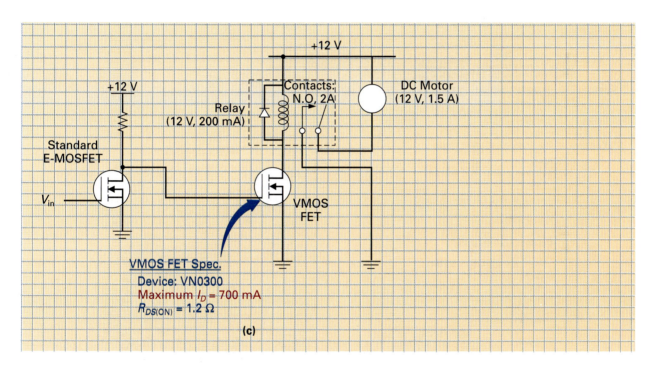

(c)

FIGURE 12-9 The Vertical Channel E-MOSFET (VMOS FET). (a) VMOS FET OFF. (b) VMOS FET ON. (c) VMOS FET Application Circuit—Low-Power Signal to High-Power Load Circuit.

increase in temperature will cause an increase in current, which will generate a further increase in heat (temperature) and current, and so on.

12-2-6 *MOSFET Data Sheets*

Throughout this chapter, we have used certain MOSFET specifications in our calculations, such as I_{DSS}, $V_{DS(ON)}$, $I_{D(ON)}$, and so on. As an example, Figure 12-10 shows the data sheet for a typical E-type MOSFET; as you can see, these specifications are listed.

DEVICE: 2N7002LT1—N-Channel Silicon E-MOSFET

MAXIMUM RATINGS

Rating	Symbol	Value	Unit
Drain-Source Voltage	V_{DSS}	60	Vdc
Drain-Gate Voltage (R_{GS} = 1 MΩ)	V_{DGR}	60	Vdc
Drain Current — Continuous TC = 25°C(1) TC = 100°C(1) — Pulsed (2)	I_D I_D I_{DM}	±115 ±75 ±800	mA
Gate-Source Voltage — Continuous — Non-repetitive (tp ≤50 μs)	V_{GS} V_{GSM}	±20 ±40	Vdc Vpk
Total Power Dissipation TC = 25°C TC = 100°C Derate above 25°C ambient	P_D	200 80 1.6	mW mW/°C

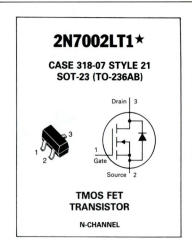

2N7002LT1★

CASE 318-07 STYLE 21
SOT-23 (TO-236AB)

TMOS FET TRANSISTOR

N-CHANNEL

ELECTRICAL CHARACTERISTICS (T_A = 25°C unless otherwise noted.)

Characteristic	Symbol	Min	Typ	Max	Unit
OFF CHARACTERISTICS					
Drain-Source Breakdown Voltage (V_{GS} = 0, I_D = 10 μA)	$V_{(BR)DSS}$	60	—	—	Vdc
Zero Gate Voltage Drain Current (V_{GS} = 0, V_{DS} = 60 V) TJ = 25°C TJ = 125°C	I_{DSS}	— —	— —	1.0 500	μAdc
Gate-Body Leakage Current Forward (V_{GS} = 20 Vdc)	I_{GSSF}	—	—	100	nAdc
Gate-Body Leakage Current Reverse (V_{GS} = −20 Vdc)	I_{GSSR}	—	—	−100	nAdc

(1) The Power Dissipation of the package may result in a lower continuous drain current.
(2) Pulse Width ≤ 300 μs, Duty Cycle ≤ 2.0%.

Characteristic	Symbol	Min	Typ	Max	Unit
ON CHARACTERISTICS*					
Gate Threshold Voltage (V_{DS} = V_{GS}, I_D = 250 μA)	$V_{GS(th)}$	1.0	—	2.5	Vdc
On-State Drain Current (V_{DS} ≥ 2.0 $V_{DS(on)}$, V_{GS} = 10 V)	$I_{D(on)}$	500	—	—	mA
Static Drain-Source On-State Voltage (V_{GS} = 10 V, I_D = 500 mA) (V_{GS} = 5.0 V, I_D = 50 mA)	$V_{DS(on)}$	 — —	 — —	 3.75 .375	Vdc
Static Drain-Source On-State Resistance (V_{GS} = 10 V, I_D = 500 mA) TC = 25°C TC = 125°C (V_{GS} = 5.0 V, I_D = 50 mA) TC = 25°C TC = 125°C	$r_{DS(on)}$	 — — — —	 — — — —	 7.5 13.5 7.5 13.5	Ohms
Forward Transconductance (V_{DS} ≥ 2.0 $V_{DS(on)}$, I_D = 200 mA)	g_{FS}	80	—	—	mmhos
DYNAMIC CHARACTERISTICS					
Input Capacitance (V_{DS} = 25 V, V_{GS} = 0, f = 1.0 MHz)	C_{iss}	—	—	50	pF
Output Capacitance (V_{DS} = 25 V, V_{GS} = 0, f = 1.0 MHz)	C_{oss}	—	—	25	pF
Reverse Transfer Capacitance (V_{DS} = 25 V, V_{GS} = 0, f = 1.0 MHz)	C_{rss}	—	—	5.0	pF
SWITCHING CHARACTERISTICS*					
Turn-On Delay Time (V_{DD} = 25 V, I_D ≅ 500 mA,	$t_{d(on)}$	—	—	30	ns
Turn-Off Delay Time R_G = 25 Ω, R_L = 50 Ω)	$t_{d(off)}$	—	—	40	ns
BODY-DRAIN DIODE RATINGS					
Diode Forward On-Voltage (I_S = 11.5 mA, V_{GS} = 0 V)	V_{SD}	—	—	−1.5	V
Source Current Continuous (Body Diode)	I_S	—	—	−115	mA
Source Current Pulsed	I_{SM}	—	—	−800	mA

*Pulse Test: Pulse Width ≤ 300 μs, Duty Cycle ≤ 2.0%.

$I_{D(ON)}$

$V_{DS(ON)}$

FIGURE 12-10   An E-Type MOSFET Data Sheet. (Courtesy of Motorola. Used by permission.)

12-2-7 *MOSFET Handling Precautions*

Certain precautions must be taken when handling any MOSFET device. The very thin insulating layer between the gate and the substrate of a MOSFET can easily be punctured if an excessive voltage is applied. Your body can build up extremely large electrostatic charges due to friction. If this charge came in contact with the pins of a MOSFET device, an electrostatic discharge (ESD) would occur, resulting in a possible arc across the thin insulating layer causing permanent damage. Most MOSFETs presently manufactured have zeners internally connected between gate and source to bypass high-voltage static or in-circuit potentials and protect the MOSFET. However, it is important to remember the following:

1. All MOS devices are shipped and stored in a "conductive foam" or "protective foil" so that all of the IC pins are kept at the same potential and therefore electrostatic voltages cannot build up between terminals.

2. When MOS devices are removed from the conductive foam, be sure not to touch the pins because your body may have built up an electrostatic charge.

3. When MOS devices are removed from the conductive foam, always place them on a grounded surface such as a metal tray.

4. When continually working with MOS devices, use a "wrist grounding strap," which is a length of cable with a 1 MΩ resistor in series. This prevents electrical shock if you come in contact with a voltage source.

5. All test equipment, soldering irons, and workbenches should be properly grounded.

6. All power in equipment should be off before MOS devices are removed or inserted into printed circuit boards.

7. Any unused MOSFET terminals must be connected because an unused input left open can build up an electrostatic charge and float to high voltage levels.

8. Any boards containing MOS devices should be shipped or stored with the connection side of the board in conductive foam.

12-2-8 *MOSFET Testing*

Like the BJT and JFET, MOSFETs can be tested with the transistor tester to determine (1) whether an open or short exists between any of the terminals, (2) the transistor's transconductance/gain, and (3) the value of I_{DSS} and leakage current.

If a transistor tester is not available, the ohmmeter can be used to determine the most common failures: opens and shorts. Figure 12-11 shows what resistance values should be obtained between the terminals of a good *n*-channel or *p*-channel D-MOSFET or E-MOSFET when testing with an ohmmeter. Remember that the gate-to-channel resistance of a MOSFET is always an open due to the insulated gate.

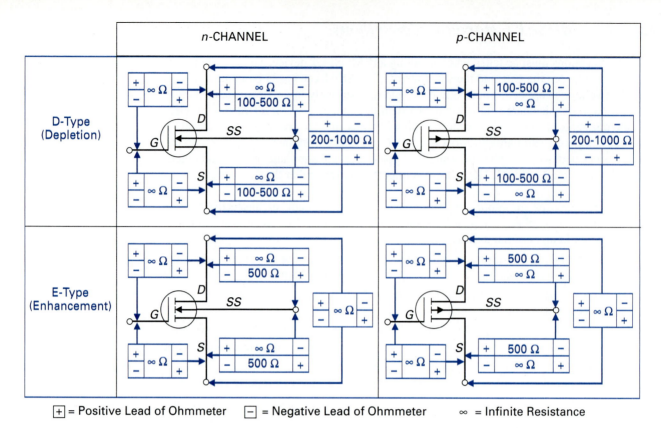

FIGURE 12-11 Testing MOSFETs with the Ohmmeter.

SELF-TEST EVALUATION POINT FOR SECTION 12-2

Use the following questions to test your understanding of Section 12-2.

1. True or false: The E-type MOSFET can be operated in both the depletion and enhancement mode.

2. The E-type MOSFET is a normally _____ (ON/OFF) device.

3. The E-type MOSFET is generally
   a. zero biased   c. drain-feedback biased
   b. base biased   d. none of the above

4. True or false: The E-MOSFET naturally operates as a voltage-controlled switch.

Multiple-Choice Questions

1. The _____ has an insulated gate that allows us to use either a positive or negative gate input voltage.
   a. JFET
   c. BJT
   b. MOSFET
   d. VJT

2. Which type of FET operates in the depletion mode?
   a. JFET
   c. D-MOSFET
   b. E-MOSFET
   d. Both (a) and (c)

3. Which type of FET makes use of an electric field to change the channel's size?
   a. JFET
   c. D-MOSFET
   b. E-MOSFET
   d. All of the above

4. The _____ has drain current when the gate voltage is zero and is therefore known as a normally _____ device.
   a. D-MOSFET, OFF
   c. E-MOSFET, OFF
   b. D-MOSFET, ON
   d. E-MOSFET, ON

5. The _____ has no drain current when gate voltage is zero and is therefore known as a normally _____ device.
   a. D-MOSFET, OFF
   c. E-MOSFET, OFF
   b. D-MOSFET, ON
   d. E-MOSFET, ON

6. Which type of FET operates in both the depletion and enhancement mode?
   a. JFET
   c. E-MOSFET
   b. D-MOSFET
   d. Both (b) and (c)

7. The dual-gate MOSFET is ideal for interfacing low-voltage devices to high-power loads.
   a. True   b. False

8. The vertical channel MOSFET is ideal for interfacing low-voltage devices to high-power loads.
   a. True   b. False

9. Which of the following transistors has the highest input impedance?
   a. BJT
   c. MOSFET
   b. JFET
   d. Both (a) and (c) are true

10. D-MOSFETs are more frequently used in _____ circuit applications whereas E-MOSFETs are more often used in _____ circuit applications.
    a. analog, digital
    c. digital, linear
    b. linear, analog
    d. two-state, digital

Practice Problems

11. Calculate the following for the circuit shown in Figure 12-12:
    a. $I_{D(sat)}$ b. $V_{DS(OFF)}$

12. Calculate the following for the circuit shown in Figure 12-13:
    a. I_D b. V_{DS}

13. What biasing method was used in Figures 12-12 and 12-13?

14. Identify the circuit in Figure 12-14. What will the approximate output voltage be if the input equals
    a. $V_{in} = 0$ V   b. $V_{in} = +5$ V

15. Referring to the circuit in Figure 12-15, answer the following questions:
    a. Is Q_1 ON or OFF, and why?
    b. Calculate V_{out} when $V_{in} = +5$ V.
    c. Calculate V_{out} when $V_{in} = 0$ V.

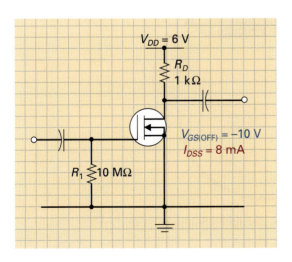

FIGURE 12-12   A D-MOSFET Amplifier Circuit.

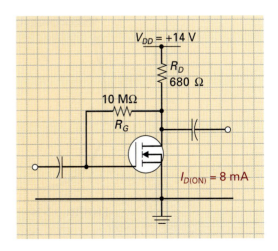

FIGURE 12-13   An E-MOSFET Amplifier Circuit.

Troubleshooting Questions

16. If you had a choice of placing the pins of a good MOSFET transistor on a plastic or metal tray, which would you choose?

17. What resistance would you expect the ohmmeter to show in Figure 12-16?

18. Are the MOSFETs in Figure 12-17 good or bad?

19. Is there a problem with the circuit in Figure 12-18? If so, what do you think it could be?

20. Is there a problem with the circuit in Figure 12-19? If so, what do you think it could be?

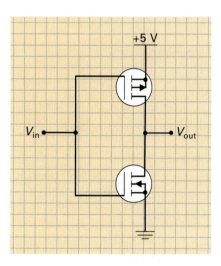

FIGURE 12-14   **MOSFET Circuit.**

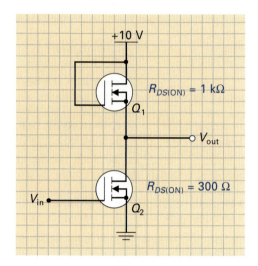

FIGURE 12-15   **A MOSFET Acting as a Load Resistor.**

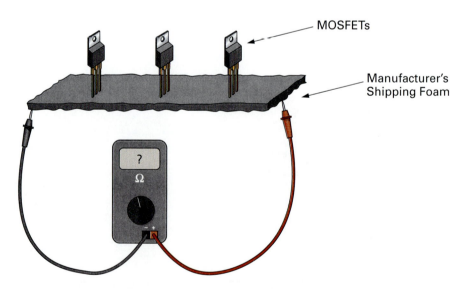

FIGURE 12-16   **MOSFETs in Manufacturer's Shipping Foam.**

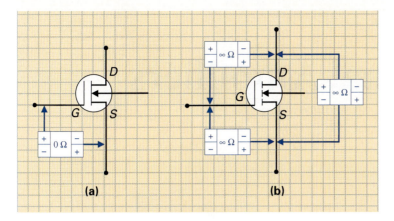

FIGURE 12-17 Testing MOSFETs with the Ohmmeter.

FIGURE 12-18 D-MOSFET Circuit Troubleshooting.

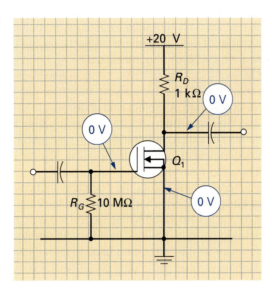

FIGURE 12-19 E-MOSFET Circuit Troubleshooting.

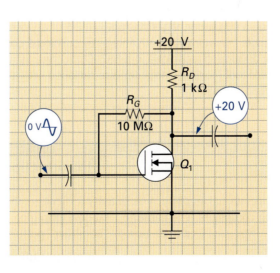

Operational Amplifiers

13

I See

Standing six feet six inches tall, Jack St. Clair Kilby was a quiet, introverted man from Kansas. Excited about the prospect of joining the very well-respected Massachusetts Institute of Technology (MIT) to further his education, he was thoroughly disappointed when he failed the mathematics entrance exam by three points. For the next 10 years he worked for a manufacturer of radio and television parts, paying particularly close attention to the new component on the block, the transistor.

In May of 1958, Texas Instruments, a new, fast-growing company who had developed the first commercially available silicon transistor just four years earlier, offered him a job in their development lab. A project was under way to print electronic components on ceramic wafers and then wire them and stack them together to make a circuit. The more Kilby became involved in the project, the more he realized how complicated and ridiculous the method was. The idea to miniaturize was a good one, but a different solution was needed.

Two months later in July, the company shut down for summer vacation, but Kilby was forced to work because he had not accrued vacation time. This proved to be a blessing in disguise for Kilby, who found himself in the lab with a lot of time and resources available to him to develop his idea. His idea was to build resistors and capacitors from the same semiconductor material that was being used to manufacture transistors. This would mean that all of the components that make up a circuit could be manufactured simultaneously on a single slice of semiconductor material.

A few months later Kilby presented his prototype to a very skeptical boss. It contained five components all connected by tiny wires with the complete assembly held together by large blobs of wax. Kilby suddenly found himself the owner of a patent and the richly deserved acclaim that always goes along with being first. The first integrated circuit, or IC, was born on a thin wafer of germanium just 2/5 of an inch long. Texas Instruments demonstrated its miniaturization advantage by building a computer for the Air Force using their newly developed technique in a 587 IC. Its rewards were immediately apparent when a 78 cubic foot monster computer was replaced with a more powerful unit measuring only 6.5 cubic inches.

Introduction

The operational amplifier was initially a vacuum tube circuit used in the early 1940s in analog computers. The name "operational amplifier" or "op-amp" was chosen because the circuit was used as a high-gain dc "amplifier" performing mathematical "operations."

These early circuits were expensive and bulky, and they found very little application until the semiconductor integrated circuit was developed in 1958 by Jack Kilby at Texas Instruments. Circuits that once needed hundreds of discrete or individual components can now be integrated into a single IC, making equipment smaller, more energy efficient, cheaper, and easier to design and troubleshoot.

Today's IC op-amp is a very high-gain dc amplifier that can have its operating characteristics changed by connecting different external components. This makes the op-amp very versatile, and it is this versatility that has made the op-amp the most widely used linear IC.

In this chapter, we will be examining the op-amp's operation, characteristics, typical circuit applications, and troubleshooting.

13-1   OPERATIONAL AMPLIFIER BASICS

To begin with, Figure 13-1 introduces the **operational amplifier,** or **op-amp,** by showing its schematic symbol in Figure 13-1(a) and internal circuit in Figure 13-1(b). It would be safe to say that you will not really be learning anything that has not already been covered because the op-amp's internal circuit is simply a combination of three previously covered amplifier circuits. These three circuits are all interconnected and contained within a single IC, and together they function as a "high-gain, high input impedance, low output impedance amplifier."

Operational Amplifier (Op-Amp)
Special type of high-gain amplifier.

13-1-1   *Op-Amp Symbol*

Referring again to Figure 13-1(a), you can see that the triangle-shaped amplifier symbol is used to represent the op-amp in an electronic schematic diagram. Comparing the two symbols, you may have noticed that in some cases the two power supply connections are not shown, even though power is obviously applied.

Let us now examine the op-amp's input and output terminals shown in Figure 13-1. The two op-amp inputs are labeled "−" and "+." The "−" or negative input is called the **inverting input** because any signal applied to this input will be amplified and inverted between input and output (output is 180° out of phase with input). On the other hand, the "+" or positive input is called the **noninverting input** because any signal applied to this input will be amplified but not inverted between input and output (output is in phase with input). An input signal will normally be applied to only one of these inputs; the other input is used to control the op-amp's operating characteristics.

Inverting Input
The inverting or negative input of an op-amp.

Noninverting Input
The noninverting or positive input of an op-amp.

The two power supply connections to the op-amp are labeled "+V" and "−V." Figure 13-1(c) shows how power to the op-amp can be supplied by dual supply voltages or by a single supply voltage. When two supply voltages are used (dual supply voltages), the voltage values are of the same value but of opposite polarity (for example, +12 V and −12 V). On the other hand, when only one supply voltage is used (single supply voltage), a positive or negative voltage is applied to its respective terminal while the other terminal is grounded (for example, +5 V and ground or −5 V and ground). Having both a positive and negative power supply voltage will allow the output signal to swing positive and negative, above and below zero. As with all high-gain amplifiers, however, the output voltage can never exceed the value of the +V and −V supply voltages.

13-1-2   *Op-Amp Packages*

The entire op-amp circuit is placed within one of two basic packages, shown in Figures 13-2(a) and (b). The TO-5 metal can package is available with 8, 10, or 12 leads, while the dual in-line through-hole and surface-mount packages typically have 8 or 14 pins.

As with all ICs, an identification code is used to indicate the device manufacturer, device type, and key characteristics. Figure 13-2(c) lists some of the more common

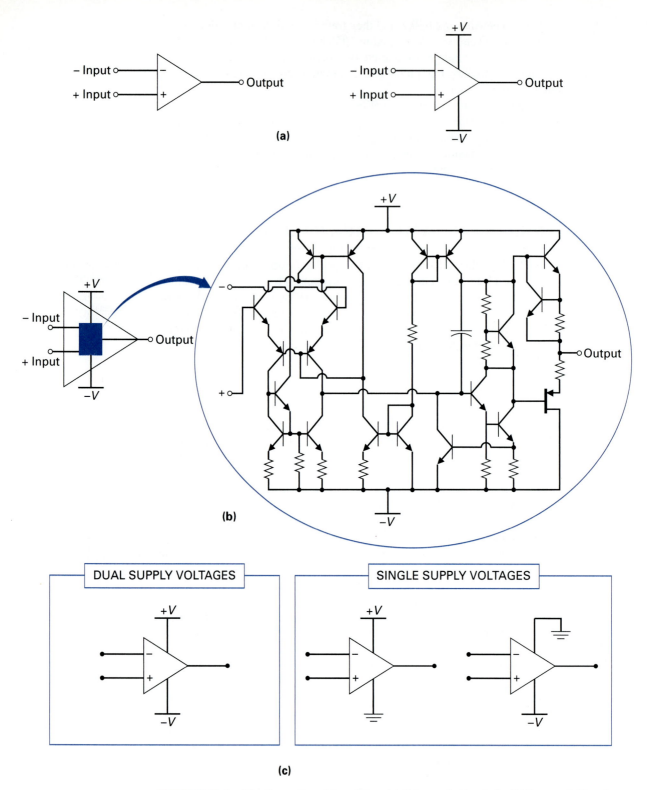

FIGURE 13-1   **The Operational Amplifier. (a) Schematic Symbols. (b) Internal Circuit. (c) Power Supply Connections.**

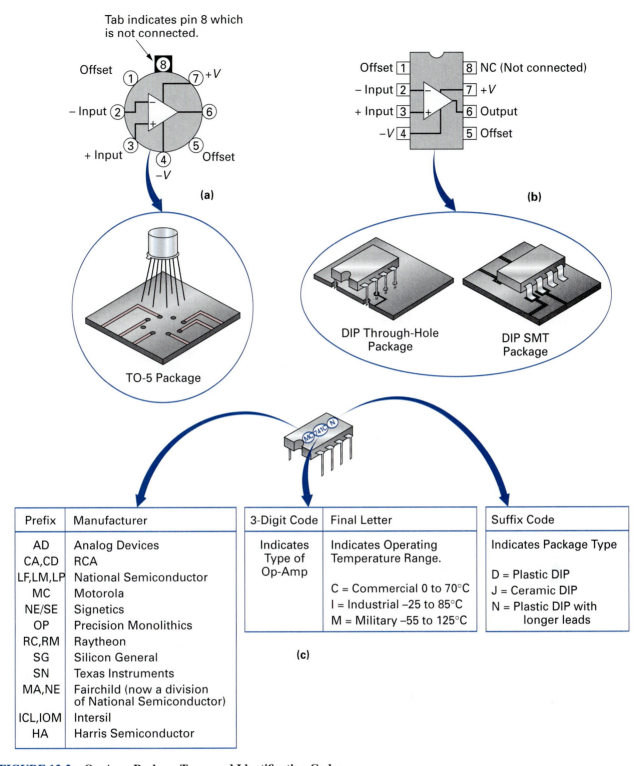

FIGURE 13-2 Op-Amp Package Types and Identification Codes.

manufacturer prefix codes, operating temperature codes, and package codes. In this example, the "MC 741C N" code indicates that the 741 op-amp is made by Motorola, it is designed for commercial application in which the temperature range is between 0 to 70°C, and the package is a through-hole DIP with longer leads.

Referring back to the IC packages in Figures 13-2(a) and (b), you can see that in addition to the two inputs, single output, and two power supply terminals, there are two

Balancing

Setting the output of an op-amp to zero volts when both inverting and noninverting inputs are at zero volts.

additional leads labeled "offset". These two inputs will normally be connected to a potentiometer that can be adjusted to set the output at zero volts when both inverting and noninverting inputs are at zero volts. **Balancing** the op-amp in this way is generally required due to imbalances within the op-amp's internal circuit.

SELF-TEST EVALUATION POINT FOR SECTION 13-1

Use the following questions to test your understanding of Section 13-1.

1. Is the operational amplifier a discrete component or an integrated circuit?

2. List the names of the op-amp's five terminals.

3. Name the two basic op-amp package types.

4. Briefly describe the meaning of each of the three parts in an op-amp's identification code.

13-2 OP-AMP OPERATION AND CHARACTERISTICS

As mentioned previously, the operational amplifier contains three amplifier circuits, and these three circuits are all interconnected and contained within a single IC. Referring to the block diagram of the op-amp in Figure 13-3(a), you can see that these three circuits are *a differential amplifier, a voltage amplifier, and an output amplifier.* Combined, these three circuits give the op-amp its key characteristics, which are *high gain, high input impedance,* and *low output impedance.* We will briefly review their characteristics because, combined, they determine the characteristics of the op-amp.

13-2-1 *The Differential Amplifier Within the Op-Amp*

The differential amplifier within the op-amp is connected to operate in its "differential-input, single-output mode." The operation of the differential amplifier in this mode is reviewed in Figure 13-3(b). When both input signals are equal in amplitude and in phase with one another, they are referred to as **common-mode input signals,** as seen in the waveforms in Figure 13-3(b). On the other hand, if the input signals are out of phase with one another, they are referred to as **differential-mode input signals,** as seen in the other set of waveforms in Figure 13-3(b). The differential amplifier will amplify differential input signals while rejecting common-mode input signals. The questions you may be asking at this stage are what are common-mode input signals and why should we want to reject them? The answers are as follows: Temperature changes and noise are common-mode input signals, and they are unwanted signals. Let us examine these common-mode signals in more detail.

Common-Mode Input Signals

Input signals to an op-amp that are in phase with one another.

Differential-Mode Input Signals

Input signals to an op-amp that are out of phase with one another.

1. Temperature variations within electronic equipment affect the operation of semiconductor materials and, therefore, the operation of semiconductor devices. These temperature variations can cause the dc output voltage of the first stage to drift away from its normal Q point. The second-stage amplifier will amplify this voltage change in the same way as it would amplify any dc input signal and so will all of the following amplifier stages. The increase or decrease in the normal Q-point bias for all of the amplifier stages will get progressively worse due to this thermal instability, and the final stage may have a Q point that is so far off of its midposition that an input signal may drive it into saturation or cutoff, causing signal distortion. With the differential amplifier, any change that occurs due to temperature changes will affect both stages and so will not appear at the output of the differential amplifier due to its **common-mode rejection.**

Common-Mode Rejection

The ability of a device to reject a voltage signal applied simultaneously to both input terminals.

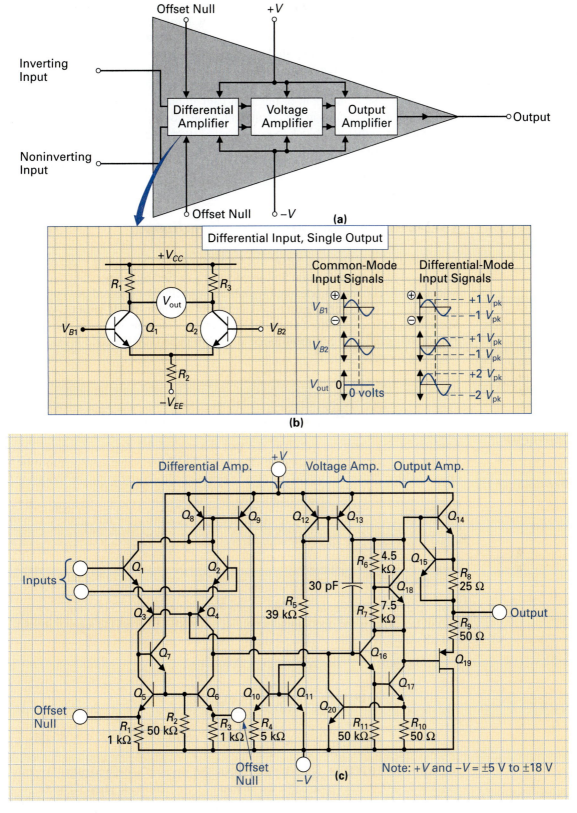

FIGURE 13-3 The Operational Amplifier's Internal Circuit. (a) Block Diagram.
(b) The Differential Amplifier's Operation. (c) Circuit Diagram.

2. The second common-mode input signal that the differential amplifier removes is noise. It is often necessary to amplify low-level signals from low-sensitivity sources such as microphones, light detectors, and other transducers. High-gain amplifiers are used to increase the amplitude of these small input signals up to a more usable level that is large enough to drive or control a load, such as a loudspeaker. The 60-Hz ac power line, or any other electrical variation, can induce a noise signal along with the input signal at the input of this high-gain amplifier. Since these noise signals will be induced at all points in the circuit, and be identical in amplitude and phase, the differential amplifier will block these unwanted signals because they will be present at both inputs of the differential amplifier (noise will be a common-mode input). A true input signal, on the other hand, will appear at the two inputs of the differential amplifier as a differential input signal and therefore be amplified.

13-2-2   *Common-Mode Rejection Ratio*

To summarize, a differential input will be amplified by the op-amp's differential amplifier and passed to the output, while unwanted signals caused by temperature variations or noise will appear as common-mode input signals and therefore be rejected. An op-amp's ability to provide a high **differential gain (A_{VD})** and a low **common-mode gain (A_{CM})** is directly dependent on its internal differential amplifier and is a measure of an op-amp's performance. This ratio is called the **common-mode rejection ratio (CMRR)** and is calculated with the following formula:

$$CMMR = \frac{A_{VD}}{A_{CM}}$$

Looking at this formula, you can see that the higher the A_{VD} (differential gain), or the smaller the A_{CM} (common-mode gain), the higher the CMRR value, and therefore the better the operational amplifier. This ratio can also be expressed in dBs by using the following formula:

$$CMMR = 20 \times \log \frac{A_{VD}}{A_{CM}}$$

EXAMPLE:

If an op-amp's differential amplifier has a differential gain of 5,000 and a common-mode gain of 0.5, what is the operational amplifier's CMRR? Express the answer in standard gain and dBs.

Solution:

$$CMMR = \frac{A_{VD}}{A_{CM}} = \frac{5,000}{0.5} = 10,000$$

$$CMMR = 20 \times \log \frac{A_{VD}}{A_{CM}} = 20 \times \log 10,000 = 20 \times 4 = 80 \text{ dB}$$

A CMRR of 10,000 or 80 dB means that the op-amp's desired input signals will be amplified 10,000 times more than the unwanted common-mode input signals.

13-2-3   *The Op-Amp Block Diagram*

Now that we have reviewed the differential amplifier's circuit characteristics, let us return to the op-amp block diagram in Figure 13-3(a). It is the op-amp's differential-amplifier stage that provides the good common-mode rejection and high differential gain. Because the op-amp's "−" and "+" inputs are applied to either base of the diff-amp, we know that input current will be very small. It is this circuit characteristic that provides the op-amp with another key feature, which is a high input impedance ($I\downarrow$, $Z\uparrow$). The voltage-amplifier stage following the diff-amp usually consists of several darlington-pair stages that provide an overall op-amp voltage gain of typically 50,000 to 200,000. The final-output stage consists of a complementary emitter-follower stage to provide a low output impedance and high current gain, so the op-amp can deliver up to several milliamps, depending on the value of the load.

13-2-4   *The Op-Amp Circuit Diagram*

The complete internal circuit of a typical op-amp can be seen in Figure 13-3(c). With integrated circuits, it is better to have transistors function as resistors wherever possible because they occupy less chip space than actual resistors. This accounts for why the circuit seems to contain many transistors that have their base and collector leads connected. You may also have noticed that no coupling capacitors have been used, so the op-amp can amplify both ac and dc input signals. As with most schematics, the inputs are shown on the left, output on the right, and power is above and below. As discussed previously, the two balancing, or **offset null, inputs** will normally be connected to an external potentiometer that can be adjusted to set the output at zero volts when both the inverting and noninverting inputs are at zero volts. Balancing the op-amp to find the zero-volt output point, or null, in this way is generally required due to slight imbalances within the op-amp's internal circuit.

> **Offset Null Inputs**
> The two balancing inputs used to balance an op-amp.

An important point to realize at this time is that the op-amp is a single component, and up until this time we have concentrated on an understanding of the op-amp's internal circuitry because this helps us to better understand the circuit's normal input/output relationships and characteristics. These operational characteristics are important if we are going to be able to isolate whether a circuit malfunction is internal or external to the op-amp. However, because it is impossible to repair any internal op-amp failures, we will not concentrate on every detail of the op-amp's internal circuit.

13-2-5   *Basic Op-Amp Circuit Applications*

Now that we have an understanding of the op-amp's characteristics, let us now put it to use in some basic circuit applications. To begin with, we will examine the comparator circuit.

The Open-Loop Comparator Circuit

Figure 13-4 shows how the op-amp can be used to function as a **comparator,** which is a circuit that is used to detect changes in voltage level. In Figure 13-4(a), the inverting input (−) of the op-amp is grounded and the input signal is applied to the op-amp's noninverting input (+). Referring to the associated waveforms, you can see that when the input swings positive relative to the negative input (which is 0), the output of the amplifier goes into immediate saturation due to the very large gain of the op-amp. For example, if the op-amp had a voltage gain of 25,000 ($A_V = 25{,}000$), even a small input of +25 mV would cause the op-amp to try and drive its output to 625 V ($V_{out} = V_{in} \times A_V = 25 \text{ mV} \times 25{,}000 = 625 \text{ V}$). Because the maximum possible positive output voltage cannot exceed the positive supply voltage ($+V$), the output goes to its maximum positive limit, which is equal to the $+V$ supply voltage. When the input swings negative, the amplifier is driven immediately into its opposite state (cutoff), and the output goes to its maximum negative limit, which is equal to the $-V$ supply voltage.

Figure 13-4(b) shows how a voltage divider made up of R_1 and R_2 can be used to supply the inverting input (−) of the op-amp with a reference voltage (V_{ref}) that can be determined by using the voltage-divider formula.

> **Comparator**
> An op-amp used without feedback to detect changes in voltage level.

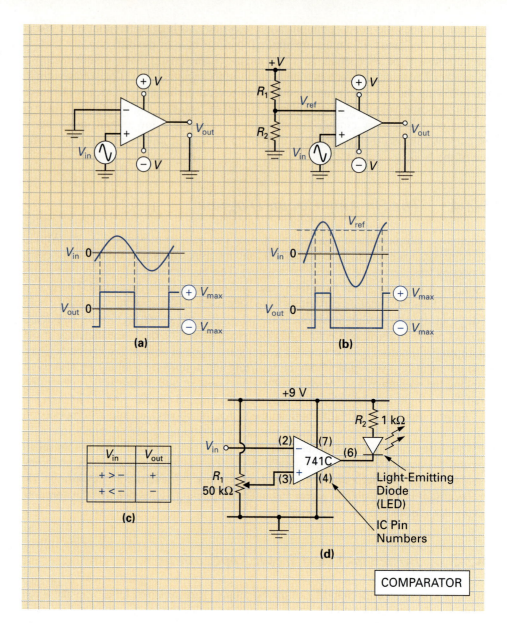

FIGURE 13-4   The Open-Loop Comparator Circuit.

$$V_{\text{ref}} = \frac{R_2}{R_1 + R_2} \times (+V)$$

Referring to the associated waveforms in Figure 13-4(b), you can see that whenever the ac input signal is more positive than the reference voltage, the output is positive. On the other hand, whenever the ac input signal is less than the reference voltage, the output is negative.

The table in Figure 13-4(c) summarizes the operation of the comparator circuit. In the first line of the table you can see that when the negative input is negative relative to the positive input, the output will go to its positive limit ($V_{\text{out}} = +V$). In the second line of the table you can see that when the opposite occurs (the negative input is positive, or the positive input is negative), the output will go to its negative limit ($V_{\text{out}} = -V$).

■ **EXAMPLE:**

Briefly describe the operation of the 741C op-amp comparator circuit shown in Figure 13-4(d).

■ **Solution:**

Potentiometer R_1 sets up the reference voltage (V_{ref}), which can be anywhere between 0 V and $+\times$ V. When the input voltage (V_{in}) to the negative input of the op-amp is greater than the positive reference voltage, the op-amp's output will be equal to the $-V$ voltage supply (which is ground), and so the LED will turn ON.

The Closed-Loop Inverting Amplifier Circuit

The op-amp is usually operated in either the **open-loop mode** or **closed-loop mode.** With the previously discussed comparator circuit, the op-amp was operating in its open-loop mode because there was no signal feedback from output to input. In most instances, the op-amp is operated in the closed-loop mode, in which there is signal feedback from output back to input. This feedback signal will always be out of phase with the input signal and therefore oppose the original signal, which is why it is called "degenerative or negative feedback." Negative feedback, however, is necessary in nearly all op-amp circuits for the following reasons:

1. Because the op-amp has such an extremely high gain, even a very small input signal will be amplified to a very large signal, which will drive the op-amp out of its linear region and into saturation and cutoff. Negative feedback will lower the op-amp's gain and therefore control the op-amp to prevent output waveform distortion.

2. Having such a high gain can cause the amplifier to go into oscillation due to positive feedback. Negative feedback prevents an amplifier from going into oscillation by reducing the op-amp's gain.

3. The open-loop gain of an op-amp can have a very large range of value for the same device. For example, the 741's open-loop gain can be anywhere from a minimum of 25,000 to 200,000. Including negative feedback in the op-amp circuit will reduce the gain to a consistent value so that the same part can be relied on to provide the same response.

Figure 13-5(a) shows how an op-amp can be connected as an **inverting amplifier circuit,** which produces an amplified output signal that is 180° out of phase with the input signal. Looking at the output voltage label ($-V_{out}$), notice that the negative symbol preceding V_{out} is being used to indicate the 180° phase inversion between input and output. In this circuit arrangement, the input signal (V_{in}) is applied through an input resistor (R_{in}) to the inverting input ($-$) of the op-amp, while the noninverting input ($+$) is connected to ground. A feedback loop is connected from the output back to the inverting input via the feedback resistor R_F.

Let us now take a closer look at the closed-loop feedback system that occurs within this amplifier circuit. If the applied input voltage was zero volts ($V_{in} = 0$ V), the differential input signal (which is the difference between the op-amp's "+" and "−" inputs) will be 0 V, because both the inverting and noninverting inputs will now be at 0 V. A differential input of zero volts therefore will generate an output of zero volts. If the input signal were now to swing positive toward $+5$ V, the output (V_{out}) would swing negative due to the internal op-amp circuit phase inversions. This negative output voltage swing would be applied back to the inverting input via R_F to counteract the original positive input change. The feedback path is designed so that it cannot completely cancel the input signal, for if it did, there would be no input and therefore no output or feedback. In most instances, the feedback voltage (V_F) will greatly restrain the input voltage change to the point that a $+5$ V input change at V_{in} will only be felt as a $+5$ microvolt change at the op-amp's inverting input. Therefore, even though V_{in} seems to change in values measured in volts, the inverting input of the op-amp will only change in values measured in microvolts. In fact, if the voltage at the inverting input of the op-amp is measured with a voltmeter, as shown in Figure 13-5(b), the "−" input appears to remain at 0 V due to the very minute change at the "−" input.

Open-Loop Mode

A control system that has no means of comparing the output with the input for control purposes.

Closed-Loop Mode

A control system containing one or more feedback control loops in which functions of the controlled signals are combined with functions of the commands that tend to maintain prescribed relationships between the commands and the controlled signals.

Inverting Amplifier Circuit

An op-amp circuit that produces an amplified output signal that is 180° out of phase with the input signal.

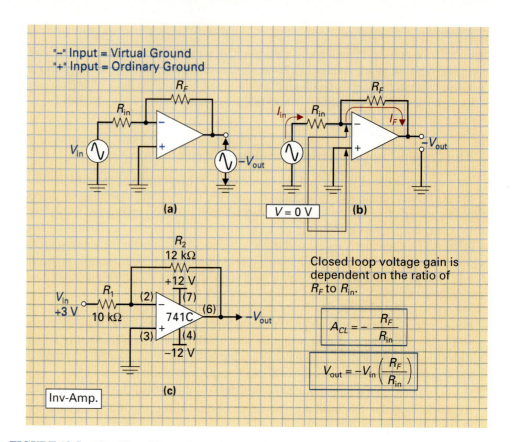

(a)

"−" Input = Virtual Ground
"+" Input = Ordinary Ground

R_F
R_{in}
V_{in}
$-V_{out}$

(b)

$V = 0\,V$

R_F
I_{in} R_{in} I_F
V_{out}

Closed loop voltage gain is dependent on the ratio of R_F to R_{in}.

$$A_{CL} = -\frac{R_F}{R_{in}}$$

$$V_{out} = -V_{in}\left(\frac{R_F}{R_{in}}\right)$$

(c)

R_2
12 kΩ
+12 V
V_{in} R_1 (2) (7)
+3 V 10 kΩ 741C (6) $-V_{out}$
(3) (4)
−12 V

Inv-Amp.

FIGURE 13-5   The Closed-Loop Inverting Amplifier Circuit.

The inverting input of the op-amp in this case would be defined as a **virtual ground,** which is different from an **ordinary ground.** A virtual ground is a voltage ground because this point is at zero volts; however, it is not a current ground because it cannot sink or conduct away any current. An ordinary ground, on the other hand, is at zero volts and can sink any amount of current.

Returning to the inverting amplifier circuit in Figure 13-5(b), we can now analyze this circuit in a little more detail now that we know its basic operation. To begin with, if the "−" input of the op-amp is at 0 V, then all of the input voltage (V_{in}) will be dropped across the input resistor (R_{in}). Therefore, the input current can be calculated, if we know the value of V_{in} and R_{in}, with the following formula:

$$I_{in} = \frac{V_{in}}{R_{in}}$$

Knowing that the left side of R_F is at 0 V means that all of the output voltage will be developed across R_F, and therefore the value of feedback current (I_F) can be calculated with the following formula:

$$I_F = \frac{-V_{out}}{R_F}$$

The extremely high input impedance of the op-amp means that only a very small fraction of the input current will enter the inverting input. In fact, nearly all of the current flowing through R_{in} and reaching the "−" op-amp input will leave this virtual ground point via the easiest path, which is through R_F. Therefore, it can be said that the feedback current is equal to the input current, or

$$I_{in} = I_F$$

If I_{in} (which equals V_{in}/R_{in}) and I_F (which equals $-V_{out}/R_F$) are equal, then

$$\frac{V_{in}}{R_{in}} = \frac{-V_{out}}{R_F}$$

If this is rearranged, we arrive at the following:

$$\frac{V_{out}}{V_{in}} = -\frac{R_F}{R_{in}}$$

Because the voltage gain of an amplifier is equal to

$$A_V = \frac{V_{out}}{V_{in}}$$

Closed-Loop Voltage Gain (A_{CL})

The voltage gain of an amplifier when it is operated in the closed-loop mode.

and because $V_{out}/V_{in} = -R_F/R_{in}$, the **closed-loop voltage gain (A_{CL})** of the inverting operational amplifier is equal to the ratio of R_F to R_{in}.

$$A_{CL} = -\frac{R_F}{R_{in}}$$

(Negative symbol preceding R_F/R_{in} indicates signal inversion)

By rearranging the previous equation $V_{out}/V_{in} = -R_F/R_{in}$, we can arrive at the following formula for calculating the output voltage of this circuit. Since

$$\frac{V_{out}}{V_{in}} = -\frac{R_F}{R_{in}}$$

$$V_{out} = -V_{in}\left(\frac{R_F}{R_{in}}\right)$$

The input impedance of this inverting op-amp is equal to the value of R_{in} because the input voltage (V_{in}) is developed across the input resistor (R_{in}).

■ **EXAMPLE:**

Calculate the output voltage of the inverting amplifier shown in Figure 13-5(c).

■ *Solution:*

$$V_{out} = -V_{in}\left(\frac{R_F}{R_{in}}\right) = -3\text{ V} \times \frac{12\text{ k}\Omega}{10\text{ k}\Omega} = -3\text{ V} \times 1.2 = -3.6\text{ V}$$

The Closed-Loop Noninverting Amplifier Circuit

Figure 13-6(a) shows how an op-amp can be configured to operate as a **noninverting amplifier circuit.** The input voltage (V_{in}) is applied to the op-amp's noninverting input ($+$), and therefore the output voltage (V_{out}) will be in phase with the input. To achieve negative feedback, the output is applied back to the inverting input ($-$) of the op-amp via the feedback network formed by R_F and R_1.

Let us now take a closer look at the closed-loop negative feedback system that occurs within this amplifier circuit. The output voltage (V_{out}) is proportionally divided across R_F and R_1, with the feedback voltage (V_F) developed across R_1 being applied to the inverting

Noninverting Amplifier Circuit

An op-amp circuit that produces an amplified output signal that is in phase with the input signal.

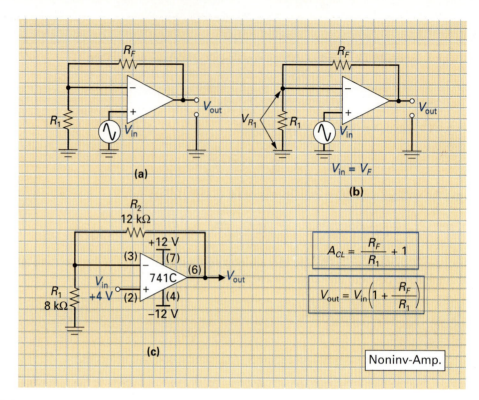

(a)

(b)

$$A_{CL} = \frac{R_F}{R_1} + 1$$

$$V_{out} = V_{in}\left(1 + \frac{R_F}{R_1}\right)$$

(c)

Noninv-Amp.

FIGURE 13-6 The Closed-Loop Noninverting Amplifier Circuit.

input ($-$) of the op-amp, as shown in Figure 13-6(b). Because V_{out} is in phase with V_{in}, the feedback voltage (V_F) will also be in phase with V_{in}, and therefore these two in-phase inputs to the op-amp will be common-mode input signals. As a result, feedback will be degenerative. However, because V_{out} is slightly larger than V_{in}, there will be a small difference between V_{in} and V_F, and this differential input will be amplified. To summarize, the noninverting op-amp provides negative feedback by feeding back an in-phase common-mode signal, and this degenerative feedback will lower the op-amp's gain to

1. Prevent output waveform distortion
2. Prevent the amplifier from going into oscillation
3. Reduce the gain of the op-amp to a consistent value

The very small microvolt difference between V_{in} and V_F will be amplified; however, because there is such a very small difference between these two input signals, it can be said that

$$V_{in} = V_F$$

Because the closed-loop voltage gain of any op-amp is equal to

$$A_{CL} = \frac{V_{out}}{V_{in}}$$

the gain of the noninverting amplifier could also be calculated with

$$A_{CL} = \frac{V_{out}}{V_F}$$

By using the voltage-divider formula, we can develop a formula for calculating V_F.

$$V_F = \frac{R_1}{R_1 + R_2} \times V_{out}$$

By rearranging this formula as follows:

$$\frac{V_{out}}{V_F} = \frac{R_1 + R_F}{R_1} \quad \text{or} \quad \frac{V_{out}}{V_F} = \frac{R_1}{R_1} + \frac{R_F}{R_1} = \frac{R_F}{R_1} + 1$$

and because $V_{out}/V_F = A_{CL}$, the noninverting op-amp's gain can also be calculated with the formula

$$A_{CL} = \frac{R_F}{R_1} + 1$$

Because the output voltage of an amplifier is equal to the product of input voltage and gain ($V_{out} = V_{in} \times A_{CL}$), we can add V_{in} to the previous closed-loop gain formula in order to calculate output voltage.

$$V_{out} = V_{in}\left(1 + \frac{R_F}{R_1}\right)$$

With the inverting op-amp, the input impedance is determined by the input resistor (R_{in}). With the noninverting amplifier, there is no resistor connected because V_{in} is applied directly into the very high input impedance of the op-amp. As a result, the noninverting amplifier circuit has an extremely high input impedance.

■ **EXAMPLE:**

Calculate the output voltage from the noninverting amplifier shown in Figure 13-6(c).

■ *Solution:*

$$V_{out} = V_{in}\left(1 + \frac{R_F}{R_1}\right) = 4\text{ V}\left(1 + \frac{12\text{ k}\Omega}{8\text{ k}\Omega}\right) = 4\text{ V} \times (1 + 1.5) = 4\text{ V} \times 2.5 = +10\text{ V}$$

The open-loop comparator and closed-loop inverting and noninverting amplifier circuits are just three of many op-amp application circuits. In the final section of this chapter, we will examine many other typical op-amp circuit applications, but for now let us review our understanding of the op-amp by examining a typical manufacturer's data sheet.

13-2-6 *An Op-Amp Data Sheet*

To better understand the characteristics of the op-amp, Figure 13-7 shows the data sheet for a 747, which is a 14-pin IC containing two 741 op-amps. Like most data sheets, this one contains a general description of the device, a pin-configuration diagram, a listing of maximum ratings, an internal-circuit diagram, and a listing of input/output characteristics. Notes have been included so that you will be able to understand the meaning of most key terms.

Figure 13-8 compares the characteristics and cost of several different op-amp types to the 741, which is very popular due to its good performance and low price. Referring to the cost-factor column, you can see, for example, that the LF351 is twice the price of the 741.

DEVICE: μA 747—Dual Linear Op-Amp IC

DESCRIPTION

The 747 is a pair of high-performance monolithic operational amplifiers constructed on a single silicon chip. High common-mode voltage range and absence of "latch-up" make the 747 ideal for use as a voltage-follower. The high gain and wide range of operating voltage provides superior performance in integrator, summing amplifier, and general feedback applications. The 747 is short-circuit protected and requires no external components for frequency compensation. The internal 6dB/octave roll-off insures stability in closed-loop applications. For single amplifier performance, see μA741 data sheet.

FEATURES

- No frequency compensation required
- Short-circuit protection
- Offset voltage null capability
- Large common-mode and differential voltage ranges
- Low power consumption
- No latch-up

PIN CONFIGURATION

N Package

	Pin		Pin	
INV. INPUT A	1		14	OFFSET NULL A
NON-INVERTING INPUT A	2	A	13	V + A
OFFSET NULL A	3		12	OUTPUT A
V–	4		11	NO CONNECT
OFFSET NULL B	5		10	OUTPUT B
NON-INVERTING INPUT B	6	B	9	V + B
INVERTING INPUT B	7		8	OFFSET NULL B

TOP VIEW   SL00100

ABSOLUTE MAXIMUM RATINGS

SYMBOL	PARAMETER	RATING	UNIT
V_S	Supply voltage	±18	V
$P_{D\ MAX}$	Maximum power dissipation T_A=25°C (still air)[1]	1500	mW
V_{IN}	Differential input voltage	±30	V
V_{IN}	Input voltage[2]	±15	V
	Voltage between offset null and V-	±0.5	V
T_{STG}	Storage temperature range	-65 to +150	°C
T_A	Operating temperature range	0 to +70	°C
T_{SOLD}	Lead temperature (soldering, 10sec)	300	°C
I_{SC}	Output short-circuit duration	Indefinite	

Explanation of Key Maximum Ratings

Supply Voltage: This is the maximum voltage that can be used to power the op-amp.

Maximum Power Dissipation: This is the maximum power the op-amp can dissipate.

Differential Input Voltage: This is the maximum voltage that can be applied across the + and – inputs.

Input Voltage: This is the maximum voltage that can be applied between an input and ground.

Operating Temperature Range: The temperature range in which the op-amp will operate within the manufacturer's specifications.

Output Short Circuit Duration: This is the amount of time that the op-amp's output can be short circuited to ground or to a supply voltage. This op-amp has an "indefinite" rating since it has an internal circuit that will turn OFF the op-amp's output and protect the internal circuitry if an output short occurs.

FIGURE 13-7   An Op-Amp Data Sheet. (Courtesy of Philips Semiconductors.)

Studying the key characteristics in this figure, you can see that the 741, for example, will typically have an input impedance of 2 MΩ, an output impedance of 75 Ω, an open-loop gain of 25,000, and a common-mode rejection ratio of 70 dB for any input signal from 0 Hz up to 1 MHz. The gain bandwidth column lists only the upper frequency limit because the op-amp has no internal coupling capacitors, and therefore the lower frequency limit extends down to dc signals, or 0 Hz.

DEVICE: μA 747—Dual Linear Op-Amp IC

EQUIVALENT SCHEMATIC

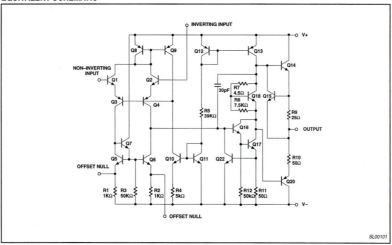

SL00101

DC ELECTRICAL CHARACTERISTICS

T_A=25°C, V_{CC} = ±15V unless otherwise specified.

SYMBOL	PARAMETER	TEST CONDITIONS	μA747C Min	μA747C Typ	μA747C Max	UNIT
V_{OS}	Offset voltage	R_S≤10kΩ		2.0	6.0	mV
		R_S≤10kΩ, over temp.		3.0	7.5	mV
$\Delta V_{OS}/\Delta T$				10		μV/°C
I_{OS}	Offset current			20	200	nA
		Over temperature		7.0	300	nA
$\Delta I_{OS}/\Delta T$				200		pA/°C
I_{BIAS}	Input current			80	500	nA
		Over temperature		30	800	nA
$\Delta I_B/\Delta T$				1		nA/°C
V_{OUT}	Output voltage swing	R_L≥2kΩ, over temp.	±10	±13		V
		R_L≥10kΩ, over temp.	±12	±14		V
I_{CC}	Supply current each side			1.7	2.8	mA
		Over temperature		2.0	3.3	mA
P_d	Power consumption			50	85	mW
		Over temperature		60	100	mW
C_{IN}	Input capacitance			1.4		pF
	Offset voltage adjustment range			±15		mV
R_{OUT}	Output resistance			75		Ω
	Channel separation			120		dB
PSRR	Supply voltage rejection ratio	R_S≤10kΩ, over temp.		30	150	μV/V
A_{VOL}	Large-signal voltage gain (DC)	R_L≥2kΩ, V_{OUT}=±10V	25,000			V/V
		Over temperature	15,000			V/V
CMRR	Common-mode rejection ratio	R_S≤10kΩ, V_{CM}=±12V Over temperature	70			dB

Explanation of Key Ratings

Input Offset Voltage: The voltage that must be applied to one input for the output voltage to be zero.
Input Offset Current: The difference of the two input bias currents when the output voltage is zero.
Input Bias Current: The average of the currents flowing into both inputs (ideally input bias currents are equal).
Input Resistance: This is the resistance of either input when the other input is grounded.

Output Resistance: This is the resistance of the op-amp's output.
Output Short-Circuit Current: Maximum output current that the op-amp can deliver to a load.

Supply Current: The current that the op-amp circuit will draw from the power supply.
Slew rate: The maximum rate of change of output voltage under large signal conditions.
Channel Separation: When two op-amps are within one package, there will be a certain amount of interference or "crosstalk" between op-amps.

FIGURE 13-7 (continued)

OP-AMP TYPE	INPUT IMPEDANCE	OUTPUT IMPEDANCE	OPEN-LOOP GAIN (Min.)	CMRR (dB)	COST FACTOR	SLEW RATE (V/µs)	GAIN BANDWIDTH	FEATURES
741 C	2 MΩ	75Ω	25,000	70	1	0.5	1 MHz	Low cost
101	800 kΩ	Low	25,000	70	1	0.5	1 MHz	Low cost
108 A	70 MΩ	Low	80,000	96	—	0.3	1 MHz	Precision low drift
351	High	Low	25,000	70	2	13	4 MHz	Low bias current
318	High	Low	25,000	70	5	70	15 MHz	High slew rate
357	High	Low	50,000	80	4	30	20 MHz	High CMRR
363	High	Low	1,000,000	94	45	—	2 MHz	Low noise; high rejection
356	High	Low	25,000	80	3	10	5 MHz	Improved 741

FIGURE 13-8   **Comparing the Characteristics of Several Op-Amps.**

SELF-TEST EVALUATION POINT FOR SECTION 13-2

Use the following questions to test your understanding of Section 13-2.

1. Can the op-amp be used to amplify dc as well as ac signal inputs?
2. What is the difference between an open-loop and closed-loop op-amp circuit?
3. List the key characteristics of an op-amp.
4. What are the three basic amplifier types within an op-amp?
5. Which of the following circuits is connected in an open-loop mode?
   a. Comparator
   b. Inverting amplifier
   c. Noninverting amplifier
6. Why is it important for an op-amp circuit to have negative feedback?

13-3   ADDITIONAL OP-AMP CIRCUIT APPLICATIONS

The operational amplifier's flexibility and characteristics make it the ideal choice for a wide variety of circuit applications. In fact, because the op-amp is the most frequently used linear IC, it is safe to say that you will find several op-amps in almost every electronic system. Although it is impossible to cover all of these applications, many circuits predominate, and others are merely variations on the same basic theme. In this section, we will concentrate on the operation and characteristics of all of the most frequently used op-amp circuit applications.

13-3-1   *The Voltage-Follower Circuit*

Voltage-Follower Circuit

An op-amp circuit that has a direct feedback to give unity gain so the output voltage follows the input voltage. It is used in applications where a very high input impedance and very low output impedance are desired.

Figure 13-9 shows how an op-amp can be connected to form a noninverting **voltage-follower circuit.** Using the noninverting closed-loop gain formula discussed previously, we can calculate the voltage gain of this circuit.

$$A_{CL} = \frac{R_F}{R_1} + 1 = \frac{0\ \Omega}{0\ \Omega} + 1 = 0 + 1 = 1$$

With a gain of 1, the output voltage will be equal to the input voltage—so what is the advantage of this circuit? The answer is the op-amp characteristics of a high input impedance

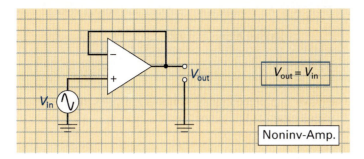

FIGURE 13-9 Voltage-Follower Circuit.

and a low output impedance. Similar to the BJT's emitter-follower and the FET's source-follower, the op-amp voltage-follower circuit derives its name from the fact that the output voltage follows the input voltage in both polarity and amplitude. This circuit is therefore ideal as a buffer, interfacing a high-impedance source to a low-impedance load.

13-3-2   *The Summing Amplifier Circuit*

The **summing amplifier circuit,** or adder amplifier, consists of two or more input resistors connected to the inverting input of an op-amp, as shown in Figure 13-10(a). This circuit will sum or add all of the input voltages, and therefore the output voltage will be

Summing Amplifier Circuit (or Adder Circuit)

An op-amp circuit that will sum or add all of the input voltages.

$$V_{out} = -(V_{in1} + V_{in2} + V_{in3})$$

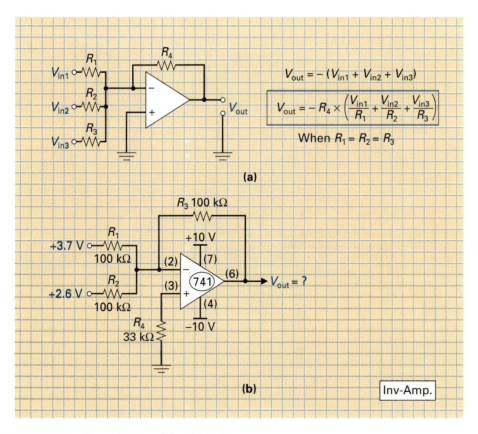

FIGURE 13-10 Summing Amplifier Circuit.

Using Ohm's law ($V = R \times I$), we can also calculate the output voltage with the formula

$$V_{out} = -R_4 \times \left(\frac{V_{in1}}{R_1} + \frac{V_{in2}}{R_2} + \frac{V_{in3}}{R_3} \right)$$

Once again, the negative sign preceding the formula indicates that the output signal will be opposite in polarity to the two or three input signals.

■ **EXAMPLE:**

Calculate the output voltage of the summing amplifier circuit shown in Figure 13-10(b).

■ *Solution:*

$$V_{out} = -(V_{in1} + V_{in2}) = -(3.7\,V + 2.6\,V) = -6.3\,V$$

13-3-3   *The Difference Amplifier Circuit*

Differential Amplifier Circuit

An op-amp circuit in which the output voltage is equal to the difference between the two input voltages.

Figure 13-11(a) shows how the op-amp can be connected to operate as a difference or **differential amplifier circuit.** In this circuit application, the op-amp will simply be making use of its first internal amplifier stage, which (as mentioned earlier) is a diff-amp. In this cir-

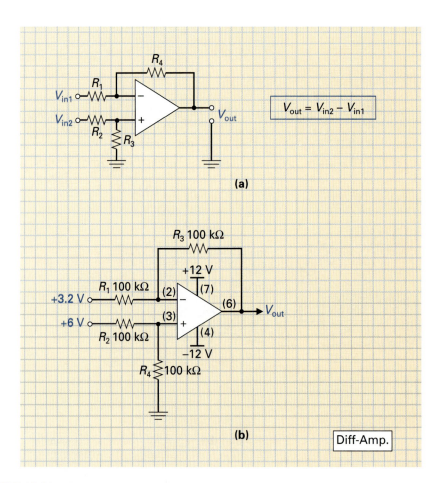

FIGURE 13-11   The Difference Amplifier Circuit.

cuit, all four resistors are normally of the same value, and the output voltage is equal to the difference between the two input voltages.

$$V_{out} = V_{in2} - V_{in1}$$

■ **EXAMPLE:**

Calculate the output voltage of the differential amplifier circuit shown in Figure 13-11(b).

■ *Solution:*

$$V_{out} = V_{in2} - V_{in1} = 6\,V - 3.2\,V = 2.8\,V$$

13-3-4 *The Differentiator Circuit*

The op-amp **differentiator circuit** (not to be confused with the previous differential circuit) is similar to the basic inverting amplifier except that R_{in} is replaced by a capacitor, as shown in Figure 13-12(a). Including a capacitor in any circuit means that we will develop problems as the frequency of the input signal increases, because capacitive reactance is inversely proportional to frequency. This means that the reactance of the input capacitor will decrease for input signals that are higher in frequency, and therefore the input voltage applied to the op-amp and output voltage from the op-amp will increase with frequency. Including an

Differentiator Circuit

A circuit whose output voltage is proportional to the rate of change of the input voltage. The output waveform is the time derivative of the input waveform.

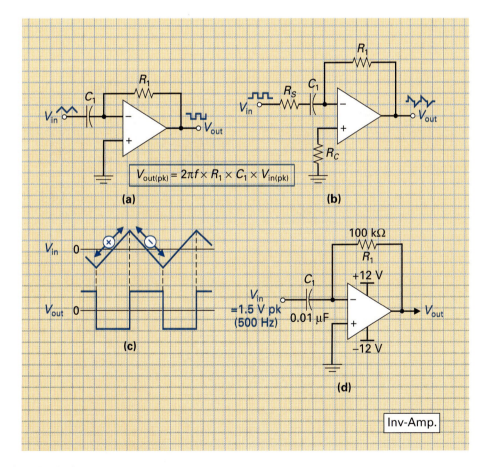

FIGURE 13-12   Differentiator Circuit.

additional resistor (R_S) in series with the input capacitor, as shown in Figure 13-12(b), will decrease the high-frequency gain because gain will now be a ratio of R_F/R_S. Figure 13-12(c) shows the differentiator's input/output waveforms, with the peak of the output square wave being equal to

$$V_{out(p)} = 2\,\pi f \times R_1 \times C_1 \times V_{in(p)}$$

EXAMPLE:

Calculate the peak output voltage of the differentiator circuit shown in Figure 13-12(d).

Solution:

$$V_{out(p)} = 2\,\pi f \times R_1 \times C_1 \times V_{in(p)}$$
$$= (2 \times \pi \times 500\ \text{Hz}) \times 100\ \text{k}\Omega \times 0.01\ \mu\text{F} \times (1.5\ \text{V}_p) = 4.7\ \text{V}$$

13-3-5 *The Integrator Circuit*

Figure 13-13(a) shows how the position of the resistor and capacitor in the differentiator circuit can be reversed to construct an op-amp **integrator circuit.** As in the differentiator circuit, the capacitor will alter the gain of the op-amp because its capacitive reactance changes with frequency. To compensate for this effect, a parallel resistor (R_P) is included in shunt with the capacitor, as shown in Figure 13-13(b), to decrease the low-frequency gain: gain will now be a ratio of R_P/R_1. Figure 13-13(c) shows the integrator's input/output waveforms, with the peak of the output triangular wave being equal to

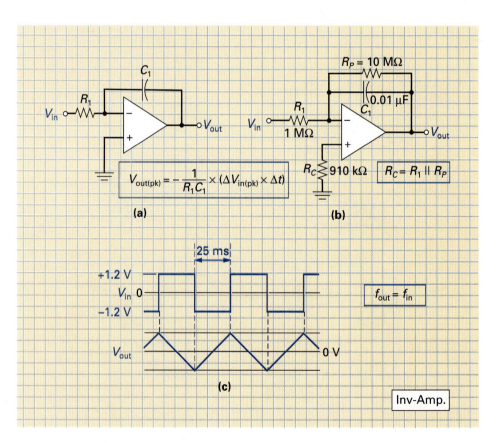

FIGURE 13-13   Integrator Circuit.

$$V_{out(p)} = \frac{1}{R_1 C_1} \times (\Delta V_{in(p)} \times \Delta t)$$

EXAMPLE:

Calculate the peak output voltage of the integrator circuit shown in Figure 13-13(b) considering the input/output waveforms given in Figure 13-13(c).

Solution:

$$V_{out(p)} = \frac{1}{R_1 C_1} \times (\Delta V_{in(p)} \times \Delta t) = \frac{1}{1\ M\Omega \times 0.01\ \mu F} \times (0\ V\ to\ 1.2\ V_p \times 25\ msec)$$
$$= 100 \times (1.2\ V_p \times 25\ msec) = 100 \times 0.03 = 3\ V_p$$

13-3-6 Signal Generator Circuits

Figure 13-14 shows how the op-amp can be connected to act as a signal generator, which is a circuit that will convert a dc supply voltage into a repeating output signal.

In Figure 13-14(a), the **twin-T sine-wave oscillator,** which has two T-shaped feedback networks, will generate a repeating sine-wave output at a frequency equal to

$$f_0 = \frac{1}{2\pi RC}$$

EXAMPLE:

Calculate the frequency of the oscillator shown in Figure 13-14(a).

Solution:

$$f_0 = \frac{1}{2\pi RC} = \frac{1}{2 \times \pi \times 6.8\ k\Omega \times 0.033\ \mu F} = 709\ Hz$$

In Figure 13-14(b), the op-amp has been connected to act as a **square-wave generator,** or, as it is more frequently called, a **relaxation oscillator.** The output signal is fed back to both the inverting and noninverting inputs of the op-amp. The capacitor will charge and discharge through R, controlling the frequency of the output square wave, which is equal to

$$f_0 = \frac{1}{2\ RC \log\left(\frac{2R_1}{R_2} + 1\right)}$$

Connecting the output of the square-wave generator in Figure 13-14(b) into the inputs of the circuits in Figures 13-14(c) and (d), we can generate a triangular or staircase output waveform. The **triangular-wave generator** in Figure 13-14(c) is simply the integrator circuit discussed previously, with its output frequency equal to the square-wave input frequency. When the switch is closed in the **staircase-wave generator** in Figure 13-14(d) the capacitor is bypassed and will therefore not charge. On the other hand, when the switch is open, C_2 will be charged by each input cycle, producing equal output steps that have the following voltage change:

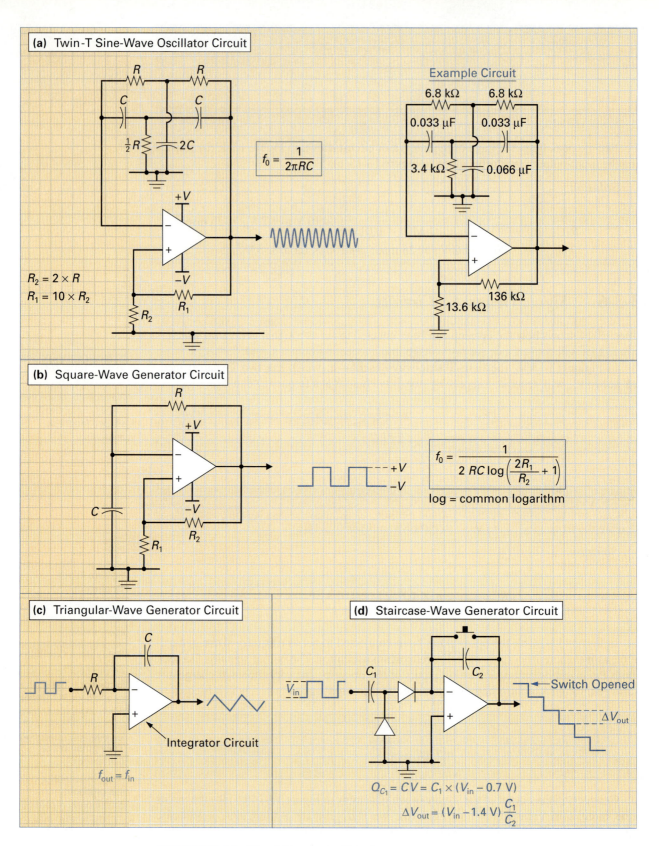

(a) Twin-T Sine-Wave Oscillator Circuit

$$f_0 = \frac{1}{2\pi RC}$$

$R_2 = 2 \times R$
$R_1 = 10 \times R_2$

Example Circuit

6.8 kΩ   6.8 kΩ
0.033 µF 0.033 µF
3.4 kΩ   0.066 µF

136 kΩ
13.6 kΩ

(b) Square-Wave Generator Circuit

$$f_0 = \frac{1}{2\,RC\,\log\!\left(\dfrac{2R_1}{R_2} + 1\right)}$$

log = common logarithm

(c) Triangular-Wave Generator Circuit

Integrator Circuit

$f_{out} = f_{in}$

(d) Staircase-Wave Generator Circuit

V_{in} C_1 C_2

Switch Opened

ΔV_{out}

$$Q_{C_1} = CV = C_1 \times (V_{in} - 0.7\ \text{V})$$

$$\Delta V_{out} = (V_{in} - 1.4\ \text{V})\,\frac{C_1}{C_2}$$

FIGURE 13-14   Signal Generator Circuits.

$$\Delta V_{\text{out}} = (V_{\text{in}} - 1.4 \text{ V}) \frac{C_1}{C_2}$$

13-3-7 *Active Filter Circuits*

Passive filters are circuits that contain passive or nonamplifying components (resistors, capacitors, and inductors) connected in such a way that they will pass certain frequencies while rejecting others. An **active filter,** on the other hand, is a circuit that uses an amplifier with passive filter elements to provide frequency paths with rejection characteristics. Active filters, like the op-amp circuits seen in Figure 13-15, have several advantages over passive filters.

1. Because the op-amp provides gain, the input signal passed to the output will not be attenuated, and therefore better response curves can be obtained.

2. The high input impedance and low output impedance of the op-amp means that the filter circuit does not interfere with the signal source or load.

3. Because active filters provide gain, resistors can be used instead of inductors, and therefore active filters are generally less expensive.

Figure 13-15 illustrates how the op-amp can be connected to form the four basic active filter types.

Active High-Pass Filter

Figure 13-15(a) illustrates the simple op-amp circuit, frequency response, and relevant formulas for an **active high-pass filter.** As before, the gain of this inverting amplifier is dependent on the ratio of R_F to R_{in}. When capacitors are included in any circuit, impedance (Z) must be considered instead of simply resistance, and gain is now equal to the ratio of feedback impedance to input impedance.

$$A_{CL} = -\frac{Z_F}{Z_{\text{in}}}$$

The input RC network will offer a high impedance to low frequencies, resulting in a low voltage gain. At high frequencies, the RC network will have a low impedance, causing a high voltage gain. The cutoff frequency for this circuit can be calculated with the following formula when $C_1 = C_2$:

$$f_C = \frac{1}{2\pi RC}$$

Active Low-Pass Filter

Figure 13-15(b) illustrates the op-amp circuit, frequency response curve, and relevant formulas for an **active low-pass filter.** At low frequencies, the capacitor's reactance is high and low-frequency signals will be passed to the op-amp's input to be amplified and passed to the output. As frequency increases, the capacitive reactance of C_1 will decrease; more of the signal will be shunted away from the op-amp and will not appear at the output. The cutoff frequency for this circuit can be calculated with the following formula when $R_1 = R_2$:

$$f_C = \frac{1}{2\pi RC}$$

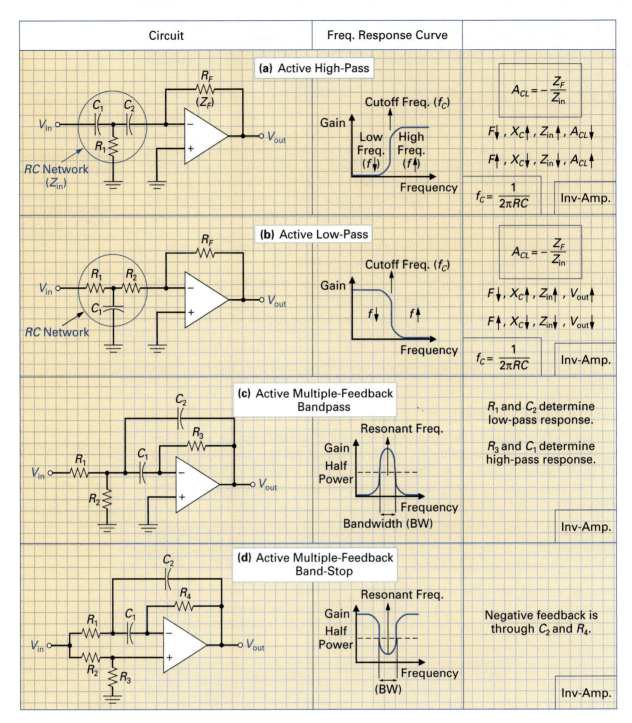

Circuit	Freq. Response Curve	
(a) Active High-Pass		$A_{CL} = -\dfrac{Z_F}{Z_{in}}$
		$F\downarrow,\, X_C\uparrow,\, Z_{in}\uparrow,\, A_{CL}\downarrow$
		$F\uparrow,\, X_C\downarrow,\, Z_{in}\downarrow,\, A_{CL}\uparrow$
		$f_C = \dfrac{1}{2\pi RC}$ Inv-Amp.
(b) Active Low-Pass		$A_{CL} = -\dfrac{Z_F}{Z_{in}}$
		$F\downarrow,\, X_C\uparrow,\, Z_{in}\uparrow,\, V_{out}\uparrow$
		$F\uparrow,\, X_C\downarrow,\, Z_{in}\downarrow,\, V_{out}\downarrow$
		$f_C = \dfrac{1}{2\pi RC}$ Inv-Amp.
(c) Active Multiple-Feedback Bandpass		R_1 and C_2 determine low-pass response.
		R_3 and C_1 determine high-pass response.
		Inv-Amp.
(d) Active Multiple-Feedback Band-Stop		Negative feedback is through C_2 and R_4.
		Inv-Amp.

FIGURE 13-15   **Active Filter Circuits.**

Active Band-Pass Filter

Active Band-Pass Filter

A circuit that uses an amplifier with passive filter elements to pass only a band of input frequencies.

Figure 13-15(c) illustrates how the op-amp can be connected to form an **active band-pass filter.** At frequencies outside of the band, V_{out} is fed back to the input without being at-tenuated, and therefore the input signal amplitude is almost equal to the feedback signal am-plitude. This results in almost complete cancellation of the signal and therefore a very small output voltage. On the other hand, for the narrow band of frequencies within the band, the feedback network will increase its amount of attenuation. This increase of attenuation means that a very small feedback signal will appear back at the negative input of the op-amp and will have a very small degenerative effect. As a result, the change at the input of the op-amp will be larger when the input signal frequencies are within this band, and the voltage out will also be larger.

Active Band-Stop Filter

Figure 13-15(d) illustrates how the op-amp can be connected to form an **active band-stop filter,** also known as a band-reject or notch filter. The basic operation of this circuit is opposite to that of the previously discussed band-pass filter. At frequencies outside of the band, the feedback signal will be heavily attenuated, and therefore the degenerative effect will be small and the output voltage large. On the other hand, at frequencies within the band, the feedback signal will not be heavily attenuated, and therefore the degenerative effect will be large and the output voltage small.

Active Band-Stop Filter
A circuit that uses an amplifier with passive filter elements to block a band of input frequencies.

SELF-TEST EVALUATION POINT FOR SECTION 13-3

Use the following questions to test your understanding of Section 13-3.

1. Which op-amp circuit provides a voltage gain of 1 and is used as a buffer?
2. Which op-amp circuit will sum all of the input voltages?
3. What is the basic circuit difference and input/output waveform difference between the integrator and differentiator circuit?
4. Which op-amp circuit will generate an output that is equal to the difference between the two inputs?
5. Sketch a circuit showing how the op-amp can be connected to generate a repeating square-wave output.
6. What is the difference between an active filter and a passive filter?

REVIEW QUESTIONS

Multiple-Choice Questions

1. When a differential amplifier is used in the differential-input, single-output mode, it has a _____ differential gain and a _____ common-mode gain.
   a. high, high   c. low, low
   b. high, low   d. low, high

2. The op-amp's internal circuit contains a _____ , _____ , and _____ amplifier stage.
   a. differentiator, current, power
   b. integrator, voltage, output
   c. darlington-pair, emitter-follower, summing
   d. a differential, darlington-pair, emitter-follower

3. The op-amp's differential-amplifier stage provides the op-amp with a
   a. low common mode gain
   b. high differential gain
   c. high input impedance
   d. all of the above

4. Which transistor circuit is used in the op-amp's final output stage to provide a low output impedance and high current gain?

   a. Common–emitter   c. Common–collector
   b. Common–base   d. Both (a) and (c)

5. Could an op-amp circuit be constructed using discrete components?
   a. Yes   b. No

6. The comparator is considered a/an _____ loop op-amp circuit.
   a. common   c. differential
   b. open   d. closed

7. What is the lower frequency limit of an op-amp?
   a. 20 Hz   b. 6 Hz   c. dc   d. 7.34 Hz

8. A virtual ground is a ground to _____ but not to _____ .
   a. current, voltage   b. voltage, current

9. The feedback loop in a closed-loop op-amp circuit provides
   a. positive feedback   d. both (a) and (b)
   b. negative feedback   e. both (b) and (c)
   c. degenerative feedback

10. The _____ input(s) of an op-amp is used to compensate for slight differences in the transistors in the differential-amplifier stage.
    a. inverting   b. dc offset   c. +V   d. V_{out}

Practice Problems

11. Explain the meaning of the following op-amp manufacturer codes:
   a. LM 318C N   **b.** NE 101C D

12. Calculate the common-mode rejection ratio of an op-amp if it has a common-mode gain of 0.8 and a differential gain of 27,000. Also give the answer in dBs.

13. Identify the circuit shown in Figure 13-16. Why must the input to this circuit always be a negative voltage?

14. What would be the output voltage from the circuit in Figure 13-16 if the input voltage were −1.6 V?

15. Identify the circuits shown in Figures 13-17(a) and (b), and then sketch the shape of the output waveform if a square wave were applied to the inputs.

16. Referring to the circuit shown in Figure 13-18, what would be the voltage out if a +7.3-V input were applied?

17. Identify the circuits shown in Figures 13-19(a) and (b), and then calculate the output voltages for the given input voltages.

18. Which of the circuits shown in Figure 13-20 is an active high-pass filter and which is an active low-pass filter?

19. Calculate the cutoff frequency of the circuit in Figure 13-20(a) if $C_1 = C_2 = 0.1$ μF, $R_1 = 10$ kΩ.

20. Calculate the cutoff frequency of the circuit in Figure 13-20(b), if $R_1 = R_2 = 33$ kΩ, $C_1 = 0.33$ μF

FIGURE 13-16   **A 741C Op-Amp Circuit.**

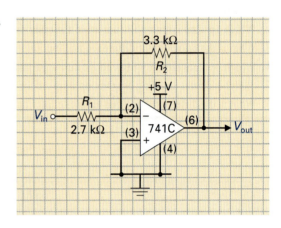

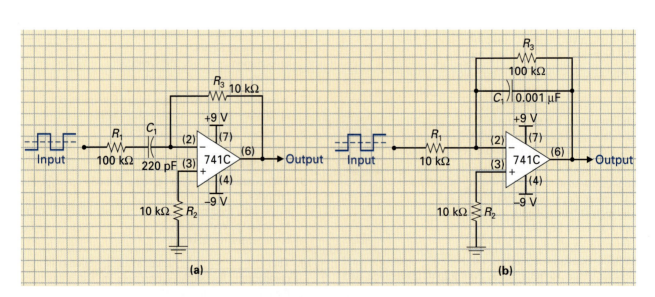

FIGURE 13-17   **Two Applications for the 741C Op-Amp.**

FIGURE 13-18 An Op-Amp Circuit.

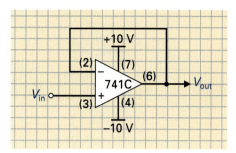

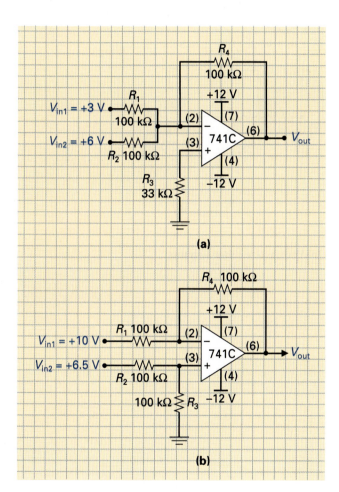

FIGURE 13-19 Op-Amp Circuit Examples.

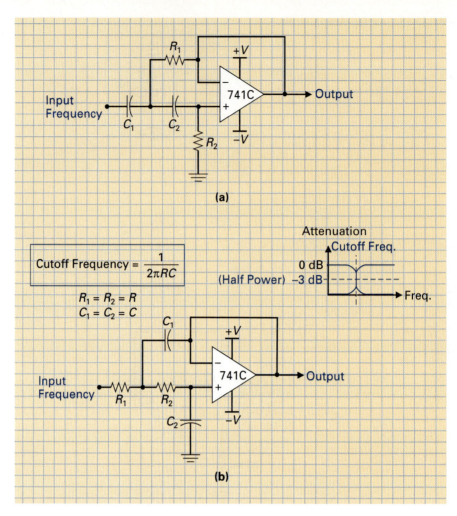

FIGURE 13-20 **Op-Amp Active Filter Circuits.**

Timers, Thyristors, and Transducers

14

Leibniz's Language of Logic

Gottfried Wilhelm Leibniz was born in Leipzig, Germany, in 1646. His father was a professor of moral philosophy and spent much of his time discussing thoughts and ideas with his son. Tragically, his father died when he was only six, and from that time on Leibniz spent hour upon hour in his late father's library reading through all of his books.

By the age of twelve he had taught himself history, Latin, and Greek. At fifteen he entered the University of Leipzig, and it was here that he came across the works of scholars such as Johannes Kepler and Galileo. The new frontiers of science fascinated him, so he added mathematics to his curriculum.

In 1666, while finishing his university studies, the twenty-year-old Leibniz wrote what he called modestly a schoolboy essay, "De Arte Combinatoria," which means "On the Art of Combination." In this work, he described how all thinking of any sort on any subject could be reduced to exact mathematical statements. Logic or, as he called it, the laws of thought, could be converted from the verbal realm—which is full of ambiguities—into precise mathematical statements. In order to achieve this, however, Leibniz stated that a "universal language" would be needed. Most of his professors found the paper either baffling or outrageous and this caused Leibniz not to pursue the idea any further.

After graduating from university, Leibniz was offered a professorship at the University of Nuremberg, which he turned down for a position as an international diplomat. This career proved not to be as glamorous as he imagined because most of his time was spent in uncomfortable horse-drawn coaches traveling between the European capitals.

In 1672, his duties took him to Paris where he met Dutch mathematician and astronomer Christian Huygens. After seeing the hours that Huygens spent on endless computations, Leibniz set out to develop a mechanical calculator. A year later Leibniz unveiled the first machine that could add, subtract, multiply, and divide decimal numbers.

In 1676 Leibniz began to concentrate more on mathematics. It was at this time that he invented calculus, which was also independently discovered by Isaac Newton in England. Leibniz's focus, however, was on the binary number system, which occupied him for years. He worked tirelessly to document the long combinations of ones and zeros that make up the modern binary number system and to perfect binary arithmetic. What is ironic is that for all his genius, Leibniz failed to make the connection between his 1666 essay and binary, which was the universal language of logic that he was seeking. It would be a century and a quarter after Leibniz's death (in 1716) when another self-taught mathematician named George Boole would discover it.

In this chapter, we will examine timers, thyristors, and transducers. As their name states, timer circuits generate timing signals that synchronize operations within electronic systems. Thyristors are generally used as dc and ac power control devices, and transducers are generally used as sensing, displaying, and actuating devices.

In the first section of this chapter, you will be introduced to the astable, monostable, and bistable multivibrator circuits and the 555 timer integrated circuit.

The semiconductor thyristor acts as an electronically controlled switch, switching power ON and OFF to adjust the average amount of power delivered to a load or connecting or disconnecting power from a load. Some thyristors are "unidirectional," which means that they will only conduct current in one direction (dc), while others are "bi-directional," which means that they can conduct current in either direction (ac). Thyristors are generally used as electronically controlled switches instead of the transistors because they have better power-handling capabilities and are more efficient.

A semiconductor transducer is an electronic device that converts one form of energy to another. Electronic transducers can be classified as either input transducers (such as the photodiode) that generate input control signals or output transducers (such as the LED) that convert output electrical signals to some other energy form. Input transducers are sensors that convert thermal, optical, mechanical, and magnetic energy variations into equivalent voltage and current variations. Output transducers, on the other hand, perform the exact opposite, converting voltage and current variations into optical or mechanical energy variations.

14-1   DIGITIAL TIMER AND CONTROL CIRCUITS

Timing is everything in digital logic circuits. To control the timing of digital circuits, a clock signal is distributed throughout the digital system. This square-wave **clock signal** is generated by a **clock oscillator,** and its sharp positive (leading) and negative (trailing) edges are used to control the sequence of operations in a digital circuit. In this section, we will discuss the **astable multivibrator,** which is commonly used as a clock oscillator, and the **monostable multivibrator,** which when triggered will generate a rectangular pulse of a fixed duration. To complete our discussion on multivibrator circuits, we will also discuss the **bistable multivibrator,** which is a digital control device that can be either set or reset.

14-1-1   *The Astable Multivibrator Circuit*

The astable multivibrator circuit, seen in Figure 14-1(a), is used to produce an alternating two-state square or rectangular output waveform. This circuit is often called a **free-running multivibrator** because the circuit requires no input signal to start its operation. It will simply begin oscillating the moment the dc supply voltage is applied.

The circuit consists of two *cross-coupled bipolar transistors,* which means that there is a cross connection between the base and the collector of the two transistors Q_1 and Q_2. This circuit also contains two *RC* timing networks: R_1/C_1 and R_2/C_2.

Let us now examine the operation of this astable multivibrator circuit. When no dc supply voltage is present ($V_{CC} = 0$ V), both transistors are OFF, and therefore there is no output. When a V_{CC} supply voltage is applied to the circuit (for example, $+5$ V), both transistors will receive a positive bias base voltage via R_1 and R_2. Although both Q_1 and Q_2 are matched bipolar transistors, which means that their manufacturer ratings are identical, no two transistors are ever the same. This difference, and the differences in R_1 and R_2 due to re-

Clock Signal

Generally a square wave used for the synchronization and timing of several circuits.

Clock Oscillator

A device for generating a clock signal.

Astable Multivibrator

A device commonly used as a clock oscillator.

Monostable Multivibrator

A device that when triggered will generate a rectangular pulse of fixed duration.

Bistable Multivibrator

A digital control device that can be either set or reset.

Free-Running Multivibrator

A circuit that requires no input signal to start its operation, but simply begins to oscillate the moment the dc supply voltage is applied.

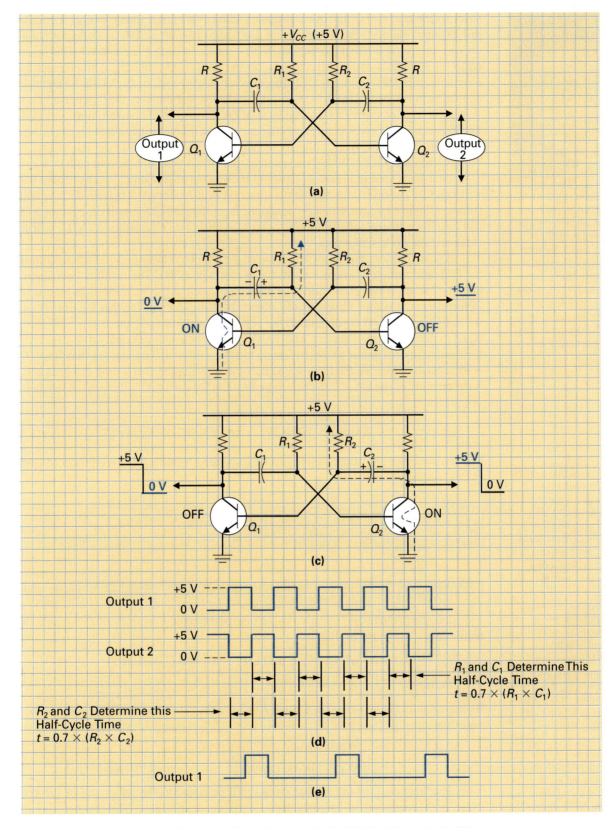

FIGURE 14-1 The Astable Multivibrator. (a) Basic Circuit. (b) Q_1 ON Condition. (c) Q_2 ON Condition. (d) Square-Wave Mode. (e) Rectangular-Wave Mode.

sistor tolerances, means that one transistor will turn ON faster than the other. Let us assume that Q_1 turns on first, as seen in Figure 14-1(b). As Q_1 conducts, its collector voltage decreases because it is like a closed switch between the collector and emitter. This decrease in collector voltage is coupled through C_1 to the base of Q_2, causing it to conduct less and eventually turn OFF. With Q_2 OFF, its collector voltage will be high (+5 V) because Q_2 is equivalent to an open switch between the collector and emitter. This increase in collector voltage is coupled through C_2 to the base of Q_1, causing it to conduct more and eventually turn fully ON. The cross coupling between these two bipolar transistors will reinforce this condition with the LOW Q_1 collector voltage keeping Q_2 OFF and the HIGH Q_2 collector voltage keeping Q_1 ON. With Q_1 equivalent to a closed switch, a current path now exists for C_1 to charge, as seen in Figure 14-1(b). As soon as the charge on C_1 reaches about 0.7 V, Q_2 will conduct because its base–emitter junction will be forward biased. This condition is shown in Figure 14-1(c). When Q_2 conducts, its collector voltage will drop, cutting OFF Q_1 and creating a charge path for C_2. As soon as the charge on C_2 reaches 0.7 V, Q_1 will conduct again and the cycle will repeat.

The output waveforms switch between the supply voltage ($+V_{CC}$) when a transistor is cut OFF and zero volts when a transistor is saturated (ON). The result is two square-wave outputs that are out of phase with one another, as seen in the waveforms in Figure 14-1(d). Referring to the output waveforms in Figure 14-1(d), you can see that the time constant of R_1 and C_1 and R_2 and C_2 determine the complete cycle time. If the R_1/C_1 time constant is equal to the R_2/C_2 time constant, both halves of the cycle will be equal (50% duty cycle) and the result will be a square wave. Referring to Figure 14-1(d), you can see that the R_1/C_1 time constant will determine the time of one half-cycle, while the R_2/C_2 time constant will determine the time of the other half-cycle. The formula for calculating the time of one half-cycle is equal to

$$t = 0.7 \times (R_1 \times C_1) \quad \text{or} \quad t = 0.7 \times (R_2 \times C_2)$$

The frequency of this square wave can be calculated by taking the reciprocal of both half-cycles, which will be

$$f = \frac{1}{1.4 \times RC}$$

■ **EXAMPLE:**

Calculate the positive and negative cycle time and circuit frequency of the astable multivibrator circuit in Figure 14-1, if

$$R_1 \text{ and } R_2 = 100 \text{ k}\Omega \qquad C_1 \text{ and } C_2 = 1 \text{ μF}$$

■ *Solution:*

Because the time constant of R_1/C_1 and R_2/C_2 are the same, each half-cycle time will be the same and equal to

$$t = 0.7 \times (R \times C)$$
$$= 0.7 \times (100 \text{ k}\Omega \times 1 \text{ μF}) = 0.07 \text{ sec or } 70 \text{ msec}$$

The frequency of the astable circuit will be equal to the reciprocal of the complete cycle, or the reciprocal of twice the half-alternation time.

$$f = \frac{1}{1.4 \times RC} = \frac{1}{1.4 \times (100 \text{ k}\Omega \times 1 \text{ μF})} = 7.14 \text{ Hz}$$

or
$$f = \frac{1}{2 \times t} = \frac{1}{2 \times 70 \text{ msec}} = \frac{1}{0.14} = 7.14 \text{ Hz}$$

If the time constants of the two *RC* timing networks in the astable circuit are different, however, the result will be a rectangular or pulse waveform, as seen in Figure 14-1(e). In this instance, the same formula can be used to calculate the time for each alternation, and the frequency will be equal to the reciprocal of the time for both alternations.

14-1-2 *The Monostable Multivibrator Circuit*

The astable multivibrator is often referred to as an "unstable multivibrator" because it is continually alternating or switching back and forth and therefore has no stable condition or state. The monostable multivibrator has, as its name implies, one (mono) stable state. The circuit will remain in this stable state indefinitely until a trigger is applied and forces the monostable multivibrator into its unstable state. It will remain in its unstable state for a small period of time and then switch back to its stable state and await another trigger. The monostable multivibrator is often compared to a gun and is called a **one-shot multivibrator** because it will produce one output pulse or shot for each input trigger.

Referring to the monostable multivibrator circuit in Figure 14-2(a), you can see that the monostable is similar to the astable except for the trigger input circuit and for the fact that it has only one *RC* timing network. To begin with, let us consider the stable state of the monostable. Components R_2, D_1, and R_5 form a voltage divider, the values of which are chosen to produce a large positive Q_2 base voltage. This large positive base bias voltage will cause Q_2 to saturate (turn heavily ON), which in turn will produce a LOW Q_2 collector voltage, which will be coupled via R_4 to the base of Q_1, cutting it OFF. The circuit remains in this stable state (Q_2 ON, Q_1 OFF) until a **trigger input** is received.

Referring to the timing waveforms in Figure 14-2(b), you can see how the circuit reacts when a positive input trigger is applied. The pulse is first applied to a differentiator circuit (C_2 and R_5) that converts the pulse into a positive and a negative spike. These spikes are then applied to the positive clipper diode D_1, which only allows the negative spike to pass to the base of Q_2. This negative spike will reverse bias Q_2's base–emitter junction, turning Q_2 OFF and causing its collector voltage to rise to $+V_{CC}$, as seen in the waveforms. This increased Q_2 collector voltage will be coupled to the base of Q_1, turning it ON. The monostable multivibrator is now in its unstable state, which is indicated in the second color in Figure 14-2(a). In this condition, C_1 will charge as shown by the dashed current line. However, as soon as the voltage across C_1 reaches 0.7 V (which is dependent on the R_2/C_1 time constant), it will force Q_2 to conduct, which in turn will cause Q_1 to cut OFF and the monostable to return to its stable state. The output pulse width or pulse time (*t*) seen in Figure 14-2(b), can be calculated with the same formula used for the astable multivibrator:

$$t = 0.7 \times (R_2 \times C_1)$$

The one-shot multivibrator is sometimes used in *pulse-stretching* applications. For example, referring to the waveforms in Figure 14-2(b), imagine the input positive trigger pulse is 1 μsec in width and the *RC* time constant of R_2 and C_1 is such that the output pulse width (*t*) is 500 μsec. In this example, the input pulse would be effectively stretched from 1 μsec to 500 μsec. The monostable multivibrator, or one-shot timer circuit, is also used to introduce a *time delay*. Referring again to the waveforms in Figure 14-2(b), imagine a differentiator circuit connected to the output of the monostable circuit. If the output pulse width was again set to 500 μsec, there would be a 500-μsec delay between the differentiated negative edge of the input pulse and the differentiated negative edge of the output pulse.

Figure 14-2(c) shows the logic symbol for a monostable (one-shot) multivibrator. Nearly all one-shot circuits in use today are in integrated form. These IC one-shots operate in exactly the same way as their discrete component counterparts. For example, the 74LS123 IC contains two fully independent monostable multivibrators. Like the symbol in Figure 14-2(c), the 74LS123 has pins for connecting external timing resistors and capacitors.

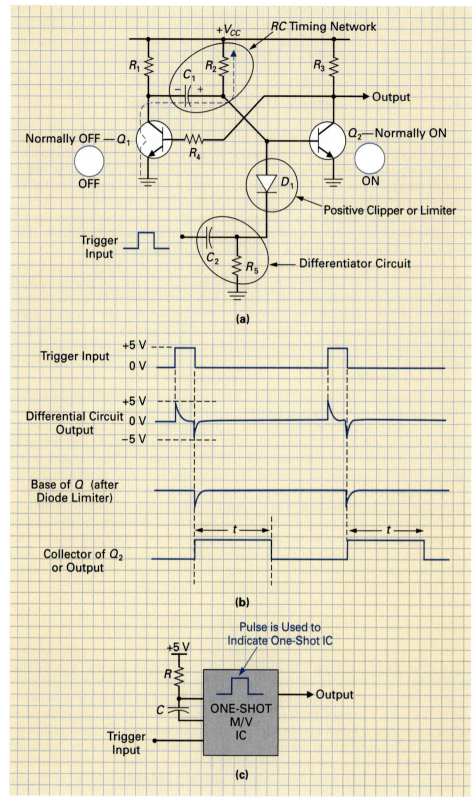

FIGURE 14-2 The Monostable Multivibrator Circuit. **(a)** Bipolar Transistor Circuit. **(b)** Input/Output Timing Waveforms. **(c)** One-Shot IC.

CHAPTER 14 / TIMERS, THYRISTORS, AND TRANSDUCERS

14-1-3 *The Bistable Multivibrator Circuit*

The bistable multivibrator has two (bi) stable states; its bipolar transistor circuit is illustrated in Figure 14-3(a). The circuit has two inputs, called the "SET" and "RESET" inputs, and these inputs drive the base of Q_1 and Q_2. The two outputs from this circuit are taken from the collectors of Q_1 and Q_2 and are called "Q" and "$\overline{Q}$" (pronounced "Q not"). The $\overline{Q}$ output derives its name from the fact that its voltage level is always the opposite of the Q output. For example, if Q is HIGH, $\overline{Q}$ will be LOW, and if Q is LOW, $\overline{Q}$ will be HIGH. The bistable multivibrator circuit is often called an ***S-R* (set-rest) flip-flop** because

> *A pulse on the SET input will "flip" the circuit into the set state (Q output is set HIGH), while*
>
> *A pulse on the RESET input will "flop" the circuit into its reset state (Q output is reset LOW).*

Figure 14-3(b) shows the logic symbol for a *S-R* or *R-S* flip-flop, or bistable multivibrator circuit.

To fully understand the operation of the circuit, refer to the waveforms in Figure 14-3(c). When power is first applied, one of the transistors will turn ON first and, because of the cross coupling, turn the other transistor OFF. Let us assume that the circuit in Figure 14-3(a) starts with Q_1 ON and Q_2 OFF. The low voltage (approximately 0.3 V) on Q_1's collector will be coupled to Q_2's base, thus keeping it OFF. The high voltage on Q_2's collector (approximately +5 V) will be coupled to the base of Q_1, keeping it ON. This condition is called the **reset state** because the primary output (output 1, or Q) has been reset to binary 0, or 0 V. The cross-coupling action between the transistors will keep the transistors in the reset state (output 1 or $Q = 0$ V and output 2 or $\overline{Q} = 5$ V) until an input appears.

By following the waveforms in Figure 14-3(c), you can see that the first input to go active is the SET input at time t_1. This positive pulse will be applied to the base of Q_2 and forward bias its base–emitter junction. As a result, Q_2 will go ON and its LOW collector voltage will be applied to output 2 ($\overline{Q}$). This LOW on Q_2's collector will also be cross-coupled to the base of Q_1, turning it OFF and therefore making output 1 (Q) go HIGH. When a pulse appears on the SET input, the circuit will be put in its **set state,** which means that the primary output (output 1 or Q) will be set HIGH. Studying the waveforms in Figure 14-3(c) once again, you will notice that after the positive SET input pulse has ended, the bistable will still remain **latched** or held in its last state, due to the cross coupling between the transistors. This ability of the bistable multivibrator to remain in its last condition or state explains why the *S-R* flip-flop is also called an ***S-R* latch.**

By following the waveforms, you can see that a RESET pulse occurs at time t_2 and resets the primary output (output 1 or Q) LOW. The flip-flop then remains latched in its reset state, until a SET pulse is applied to the set input at time t_3, setting Q HIGH. Finally, a positive RESET pulse is applied to the reset input at time t_4, and Q is reset LOW.

The operation of the *S-R* flip-flop is summarized in the truth table or function table shown in Figure 14-3(d). When only the R input is pulsed HIGH (reset condition), the Q output is reset to a binary 0 or reset LOW ($\overline{Q}$ will be the opposite, or HIGH). On the other hand, when only the S input is pulsed HIGH (set condition), the Q output is set to a binary 1 or set HIGH ($\overline{Q}$ will be the opposite, or LOW). When both the S and R inputs are LOW, the *S-R* flip-flop is said to be in the **no-change** or latch condition because there will be no change in the output Q. For example, if the output Q is SET, and then the S and R inputs are made LOW, the Q output will remain SET, or HIGH. On the other hand, if the output Q is RESET, and then the S and R inputs are made LOW, the Q output will remain RESET, or LOW.

The external circuits driving the S and R inputs will be designed so that these inputs are never both HIGH, as shown in the last condition in the table in Figure 14-3(d). This is called the *race condition* because both bipolar transistors will have their bases made positive and therefore they will race to turn ON and then shut the other transistor OFF via the cross coupling. This input condition is not normally applied because the output condition is unpredictable.

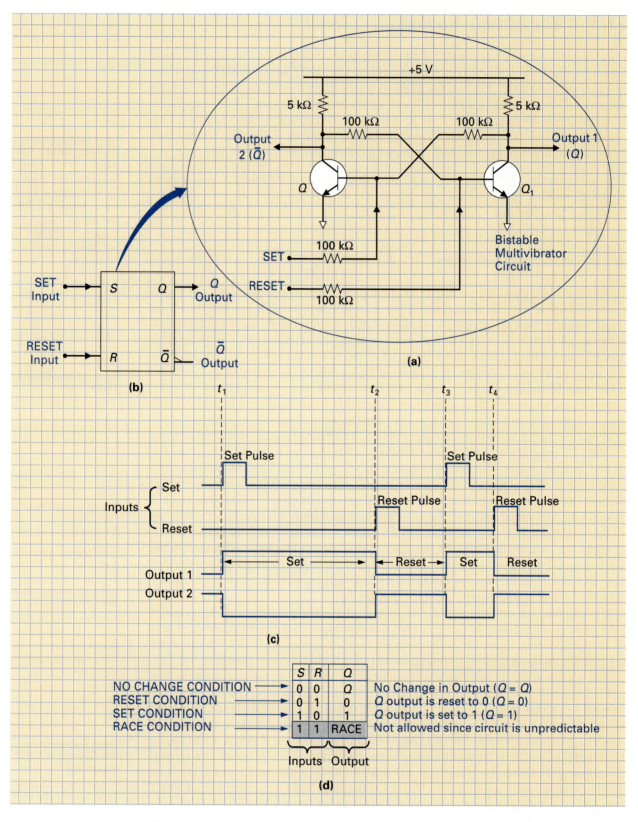

FIGURE 14-3 The Bistable Multivibrator or Set-Rest ($S = R$ or $R = S$) Flip-Flop.
(a) Bipolar Transistor Circuit. (b) *S-R* Symbol. (c) Timing Waveforms.
(d) Truth Table.

The Schmitt trigger circuit discussed previously is a bistable multivibrator circuit because its output voltage can be either one of two states. That is, its output can be either SET HIGH or RESET LOW, based on the voltage level of the input control voltage.

The bistable multivibrator *S-R* flip-flop or *S-R* latch has become one of the most important circuits in digital electronics. It is used in a variety of applications ranging from data storage to counting and frequency division.

14-1-4   *The 555 Timer Circuit*

One of the most frequently used low-cost integrated circuit timers is the *555 timer*. Its IC package consists of 8 pins, as seen in Figure 14-4(a), and derives its number identification from the distinctive voltage-divider circuit, seen in Figure 14-4(b), consisting of three 5-kΩ

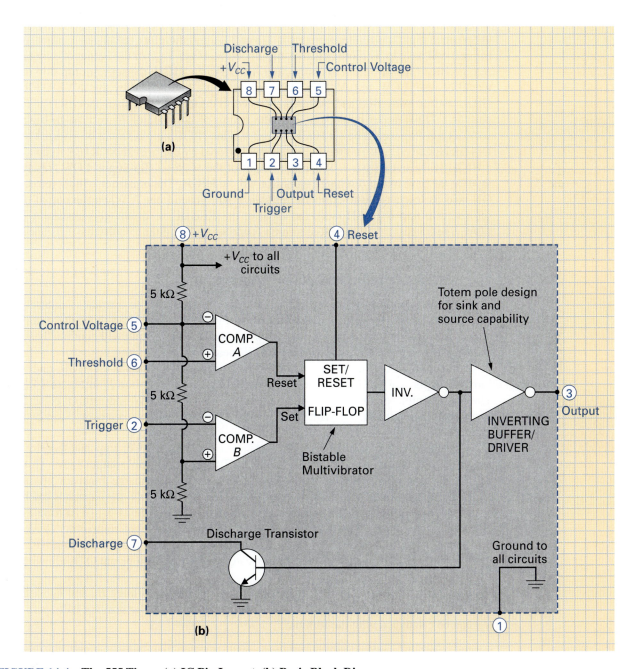

FIGURE 14-4   The 555 Timer. (a) IC Pin Layout. (b) Basic Block Diagram.

resistors. It is a highly versatile timer that can be made to function as an astable multivibrator, monostable multivibrator, frequency divider, or modulator, depending on the connection of external components.

Nearly all the IC manufacturers produce a version of the 555 timer, which can be labeled in different ways, for example SN72 555, MC14 555, SE 555, and so on. Two 555 timers are also available in a 16-pin dual-IC package that is labeled with the numbers 556.

Basic 555 Timer Circuit Action

Referring to the basic block diagram in Figure 14-4(b), let us examine the basic action of all the devices in a 555 timer circuit. The three 5-kΩ resistors develop reference voltages at the inputs of the two comparators A and B. As previously mentioned, a comparator is a circuit that compares an input signal voltage to an input reference voltage and then produces a YES/NO or HIGH/LOW decision output. The negative input of comparator A will have a reference that is 2/3 of V_{CC}, and therefore the positive input (pin 6, threshold) will have to be more positive than 2/3 of V_{CC} for the output of comparator A to go HIGH. With comparator B, the positive input has a reference that is 1/3 of V_{CC}, and therefore the negative input (pin 2, trigger) will have to be more negative, or fall below, 1/3 of V_{CC} for the output of comparator B to go HIGH. If the output of comparator A were to go HIGH, the set/reset flip-flop output would be reset LOW, to 0 V. This LOW output would be inverted by the INVERTER to a HIGH and then inverted and buffered (boosted in current) by the final INVERTER to appear as a LOW at the output pin 3. If the output of comparator B were to go high, the set/reset flip-flop would be set HIGH, to +5 V. This HIGH output would be inverted by the INVERTER to a LOW and then inverted and buffered by the final INVERTER to appear as a HIGH at the output pin 3. When the output of the set/reset flip-flop is LOW (reset), the input at the base of the discharge transistor will be HIGH. The transistor will therefore turn ON, and its low emitter-to-collector resistance will ground pin 7. When the output of the set/reset flip-flop is HIGH (set), the input at the base of the discharge transistor will be LOW. The transistor will therefore turn OFF, and its high emitter-to-collector resistance will cause pin 7 to ground.

The 555 Timer as an Astable Multivibrator

Figure 14-5(a) shows how the 555 timer can be connected to operate as an astable or free-running multivibrator. The waveforms in Figure 14-5(b) show how the externally connected capacitor C will charge and discharge and how the output will continually switch between its positive ($+V_{CC}$) and negative (0 V) peaks.

To begin with, let us assume that the output of the S-R flip-flop is set HIGH, and therefore the output will be HIGH (blue condition in Figure 14-5(a), time T_1 in Figure 14-5(b)). The HIGH output of the S-R flip-flop will be inverted to a LOW and turn OFF the 555's internal discharge transistor. With this transistor OFF, the external capacitor C can begin to charge towards $+V_{CC}$ via R_A and R_B.

At time T_2, the capacitor's charge has increased beyond 2/3 of V_{CC}, and therefore the output of comparator A will go HIGH and RESET the S-R flip-flop's output LOW. This will cause the output (pin 3) of the 555 to go LOW. However, the discharge transistor's base will be HIGH, and so it will turn ON. With the discharge transistor ON, the capacitor can begin to discharge (black condition in Figure 14-5(a), time T_2 in Figure 14-5(b)). At time T_3, the capacitor's charge has fallen below 1/3 of V_{CC}, or the trigger level of comparator B. As a result, comparator B's output will go HIGH and SET the output of the S-R flip-flop HIGH, or back to its original state. The discharge transistor will once again be cut OFF, allowing the capacitor to charge and the cycle to repeat.

As you can see in Figure 14-5(a), the capacitor charges through R_A and R_B to 2/3 of V_{CC} and then discharges through R_B to 1/3 of V_{CC}. As a result, the positive half-cycle time (t_p) can be calculated with the formula

$$t_p = 0.7 \times C \times (R_A + R_B)$$

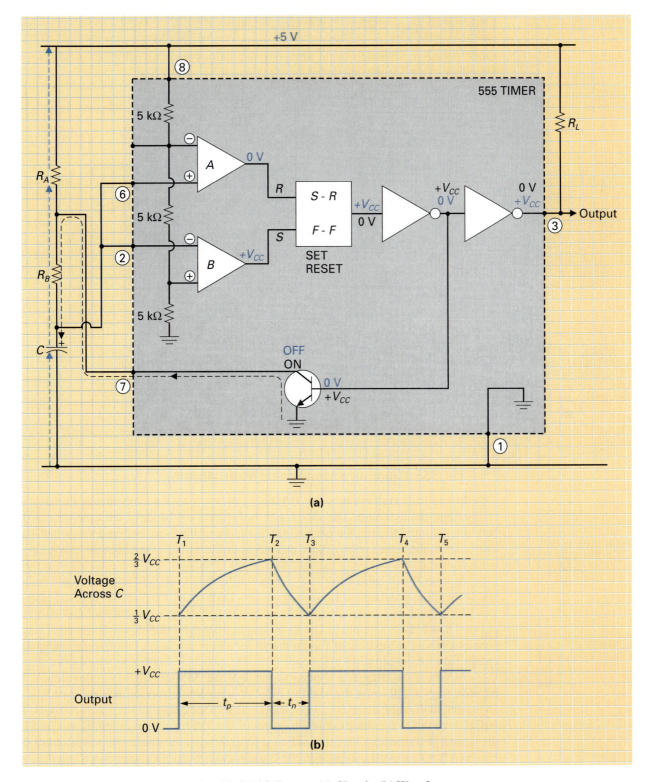

FIGURE 14-5 **The 555 Timer as an Astable Multivibrator. (a) Circuit. (b) Waveforms.**

The negative half-cycle time (t_n) can be calculated with the formula

$$t_n = 0.7 \times C \times R_B$$

The total cycle time will equal the sum of both half-cycles ($t = t_p + t_n$), and the frequency will equal the reciprocal of time ($f = 1/t$).

Calculate the positive half-cycle time, negative half-cycle time, complete cycle time, and frequency of the 555 astable multivibrator circuit in Figure 14-5, if

$$R_A = 1 \text{ k}\Omega \qquad R_B = 2 \text{ k}\Omega \qquad C = 1 \text{ }\mu\text{F}$$

■ *Solution:*

The positive half-cycle will last for

$$t_p = 0.7 \times C \times (R_A + R_B)$$
$$= 0.7 \times 1 \text{ }\mu\text{F} \times (1 \text{ k}\Omega + 2 \text{ k}\Omega) = 2.1 \text{ msec}$$

The negative half-cycle will last for

$$t_n = 0.7 \times C \times R_B$$
$$= 0.7 \times 1 \text{ }\mu\text{F} \times 2 \text{ k}\Omega = 1.4 \text{ msec}$$

The complete cycle time will be

$$t = t_p + t_n$$
$$= 2.1 \text{ msec} + 1.4 \text{ msec} = 3.5 \text{ msec}$$

The frequency of this 555 astable multivibrator will be

$$f = \frac{1}{t}$$

$$= \frac{1}{3.5 \text{ msec}} = 285.7 \text{ Hz}$$

The 555 Timer as a Monostable Multivibrator

Figure 14-6(a) shows how the 555 timer can be connected to operate as a monostable, or one-shot, multivibrator. The waveforms in Figure 14-6(b) show the time relationships between the input trigger, the charge and discharge of the capacitor, and the output pulse. The width of the output pulse (P_W) is dependent on the values of the external timing components R_A and C.

At time T_1 in Figure 14-6(b), the set-reset flip-flop (*S-R F-F*) is in the reset condition and is therefore producing a LOW output. This LOW from the *S-R F-F* is inverted by the INVERTER and then inverted and buffered by the final stage to produce a LOW (0 V) output from the 555 timer at pin 3. The LOW output from the *S-R F-F* will be inverted and appear as a HIGH at the base of the discharge transistor, turning it ON and providing a discharge path for the capacitor to ground.

At time T_2, a trigger is applied to pin 2 of the 555 monostable multivibrator. This negative trigger will cause the negative input of comparator *B* to fall below 1/3 of V_{CC}, and so the output of comparator *B* will go HIGH and SET the output of the *S-R F-F* HIGH. This HIGH from the *S-R F-F* will send the output of the 555 timer (pin 3) HIGH and turn OFF the discharge transistor. Once the path to ground through the discharge transistor has been removed from across the capacitor, the capacitor can begin to charge via R_A to $+V_{CC}$, as seen in the waveforms in Figure 14-6(b). The output of the 555 timer remains HIGH until the charge on the capacitor exceeds 2/3 of V_{CC}. At this time (T_3), the output of comparator *A* will go HIGH, resetting the *S-R F-F* and causing the output of the 555 timer to go LOW and also turning ON the discharge transistor to discharge the capacitor. The circuit will then remain in this stable condition until a new trigger arrives to initiate the cycle once again.

The leading edge of the positive output pulse is initiated by the input trigger, and the trailing edge of the output pulse is determined by the R_A and C charge time, which is dependent on their values. Because the capacitor can charge to 2/3 of V_{CC} in a little more than

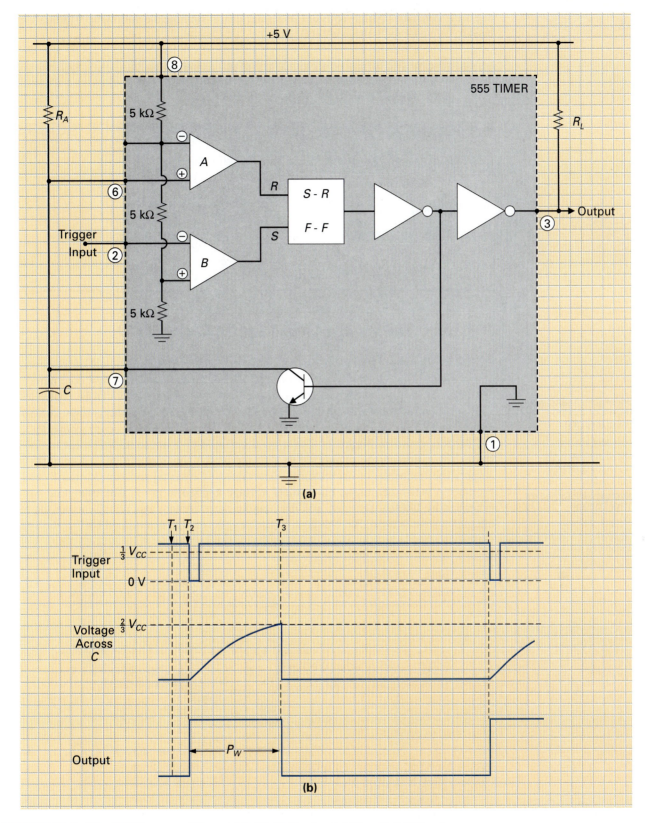

FIGURE 14-6 The 555 Timer as a Monostable Multivibrator. (a) Circuit. (b) Waveforms.

one time constant (1 time constant = 0.632, 2/3 = 0.633), the following formula can be used to calculate the pulse width (P_W):

$$P_W = 1.1 \times (R_A \times C)$$

EXAMPLE:

Calculate the pulse width of a 555 monostable multivibrator, if

$$R_A = 2 \text{ M}\Omega, \quad \text{and} \quad C = 1 \text{ }\mu\text{F}$$

Solution:

The width of the output pulse will be

$$P_W = 1.1 \times (R_A \times C)$$
$$= 1.1 \times (2 \text{ M}\Omega \times 1 \text{ }\mu\text{F}) = 2.2$$

14-1-5 *The Voltage Controlled Oscillator (VCO)*

Previously, in the oscillator circuits discussed in Chapter 9, the frequency of the repeating output signal was determined by an *LC, RC,* or crystal network. In many communication circuit applications, however, we need an oscillator whose frequency can be controlled by an input "control voltage." The circuit that achieves this function is the **voltage-controlled oscillator (VCO)** or, as it is sometimes called, a *voltage-to-frequency converter.* The NE/SE 566 shown in Figure 14-7 is a good example of a linear IC VCO. Figure 14-7(a) shows how a 566 can be connected to generate both a square-wave and a triangular-wave output. The frequency of oscillation is determined by external resistors R_1 and R_2 and capacitor C_1, which determine the **control voltage** applied to pin 5. The triangular wave is generated by linearly charging and discharging the external capacitor C_1, using the 566's internal current source. The charge and discharge levels are determined by the 566's inter-

Voltage-Controlled Oscillator (VCO)

An oscillator whose frequency can be controlled by an input control voltage.

Control Voltage

A voltage signal that starts, stops, or adjusts the operation of a device, circuit, or system.

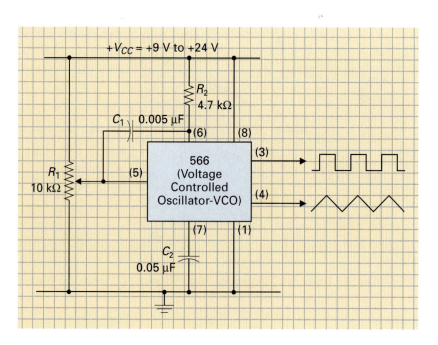

FIGURE 14-7 A Voltage-Controlled Oscillator (VCO).

nal Schmitt trigger, which is also used to generate the square-wave output. The triangular wave will typically have an amplitude of 2.4 V_{p-p}, and the square wave will typically have an amplitude of 5.4 V_{p-p}.

14-1-6 *Function Generator Circuit*

Function generators are test instruments whose output waveforms are used to test circuits by injecting an input into that circuit and then monitoring its response to the input. For example, sine-wave signals are normally used to test linear circuits such as amplifiers and filters, whereas square-wave signals are often used to test digital circuits such as logic gates. In the past, a function generator circuit was constructed using discrete components; today, however, a complete function generator circuit is available in one linear IC. Figure 14-8 shows how this IC could be connected to form a function generator circuit.

The 8038 function generator circuit in Figure 14-8 produces a 20-Hz to 100-kHz sine-wave, square-wave, or triangular-wave output. The frequency select switch (SW_1) is used to select the desired frequency range, and then resistor R_4 is used to adjust the frequency output to the desired frequency within that range. Switch SW_2 is used to select either the square-wave, triangular-wave, or sine-wave output. The position of this output selector switch will determine which waveform is applied to the output stage made up of an op-amp and power

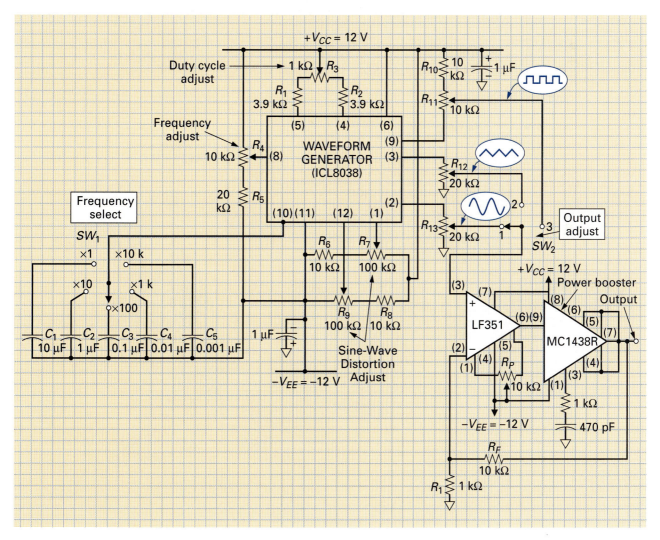

FIGURE 14-8 Function Generator Circuit.

amp. The output amplitude of the square wave can be varied by R_{11}, and the duty cycle of the square wave can be adjusted by R_3. Similarly, the triangular wave's output amplitude can be adjusted by R_{12} and the sine wave's output amplitude can be adjusted by R_{13}. Variable resistors R_7 and R_9 are included to fine tune the sine-wave output for minimum distortion.

14-1-7   *Phase-Locked Loop (PLL) Circuits*

Integrated circuit technology has made the **phase-locked loop (PLL) circuit** reasonably priced and therefore widely used in a variety of applications.

Phase-Locked Loop Operation

The operation of a phase-locked loop is best described by referring to the block diagram of the LM 565 PLL linear IC shown in Figure 14-9. As can be seen in this diagram, the PLL contains a phase comparator, amplifier and low-pass filter, and a voltage-controlled oscillator. The input frequency (f_{in}) is applied to one input of the phase comparator and compared with the output frequency (f_{out}) from the VCO. If a frequency difference exists be-

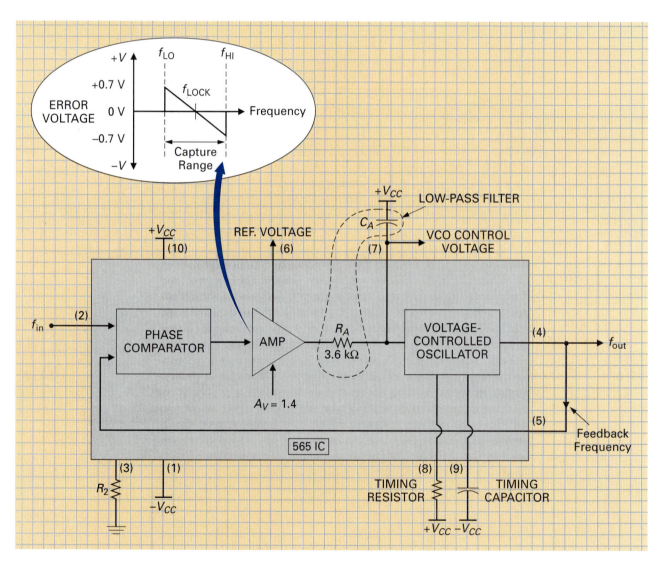

FIGURE 14-9   The Phase-Locked Loop (PLL).

tween the two comparator inputs, the comparator will detect the difference in phase and generate an **error voltage,** which will be amplified and filtered before it is applied as the control input voltage to the VCO. If the input frequency (f_{in}) and VCO output frequency (f_{out}) differ, the error voltage generated by the phase comparator will shift the VCO frequency until it matches the input frequency, and therefore $f_{in} = f_{out}$. When f_{in} matches the VCO frequency or f_{out}, the PLL is said to be in the **locked** condition. If the input frequency is varied, the VCO output frequency will follow or track f_{in}, provided that the input frequency is within the capture range of the PLL. Referring to the inset in Figure 14-9, you can see how the error voltage from the phase comparator varies while the PLL is trying to **capture** the input frequency. When the input frequency is outside the capture range of the PLL, the VCO will generate a **free-running** or idle frequency.

To summarize: a phase-locked loop circuit consists of a phase comparator that compares the output frequency of a voltage-controlled oscillator with an input frequency. The error voltage out of the phase comparator is then coupled via an amplifier and low-pass filter to the control input of the voltage-controlled oscillator to keep it in phase—and therefore at exactly the same frequency—as the input frequency. These phase-locked loop circuits will operate in the free-running state, capture state, or locked state.

Phase-Locked Loop Applications

The phase-locked loop circuit can be used in a variety of circuit applications, including the demodulation of FM radio signals, frequency multiplication and division, and signal regeneration. Most of these circuits will be covered in your electronic communication course. However, at this time let us examine how the PLL can be used with a crystal-controlled oscillator as a "frequency multiplier circuit."

PLL Frequency Multiplier Circuit   To generate several stable output frequencies, we would either have to have several crystal-controlled oscillators, as shown in Figure 14-10(a), or use frequency multiplier circuits to step up an input frequency, as shown in Figure 14-10(b). The disadvantage with the circuit in Figure 14-10(a) is that we will need several complete crystal oscillator circuits, and the disadvantage of the circuit in Figure 14-10(b) is that its frequency characteristics depend on an LC tuned circuit that will need frequent alignment. The PLL block diagram shown in Figure 14-10(c) and its associated circuit diagram shown in Figure 14-10(d) will generate several output frequencies that have the same frequency stability as a crystal oscillator's output without the need for several crystal oscillators or frequent LC tuning. As can be seen in Figure 14-10(c), a crystal oscillator applies a very stable 1-kHz square-wave input to the 565 PLL. When switch 1 is in the lower position, the output of the VCO is fed directly back to the phase comparator, and the circuit will operate in exactly the same way as the circuit in Figure 14-9. It will generate an output frequency that tracks the input frequency and is a very stable 1 kHz due to the locking action of the PLL. When switch 1 is in the mid position, a divide-by-5 circuit is switched into circuit between the VCO output and the phase comparator input. Referring to the waveforms in Figure 14-10(c), you can see that the divide-by-5 circuit will generate one output cycle for every 5 input cycles. This means that the VCO feedback signal applied to the phase comparator will be one-fifth of the output frequency. Because this signal will not be in phase with the 1-kHz signal from the crystal oscillator, the phase comparator will generate an error voltage instructing the VCO to increase its output frequency by 5 times. When the VCO generates an output frequency of 5 kHz, the divide-by-5 circuit will produce a 1-kHz output (5 kHz ÷ 5 = 1 kHz) to the phase comparator, which will match the phase comparator's other 1-kHz input from the crystal oscillator. As a result, the error voltage from the phase comparator will be zero and the VCO—and therefore the output—will remain locked at 5 kHz. The circuit diagram for this switch position is shown in Figure 14-10(d). If switch SW_1 of the circuit in Figure 14-10(c) is put in the upper position, a divide-by-10 circuit will be included between the VCO output and the phase comparator input. As a result, the PLL will generate an output frequency that is 10 times the input frequency ($f_{out} = 10 \times f_{in} = 10 \times 1$ kHz = 10 kHz).

Error Voltage

A voltage that is proportional to the error that exists between input and output.

Locked

To automatically follow a signal.

Capture

The act of gaining control of a signal.

Free-Running

Operating without any external control.

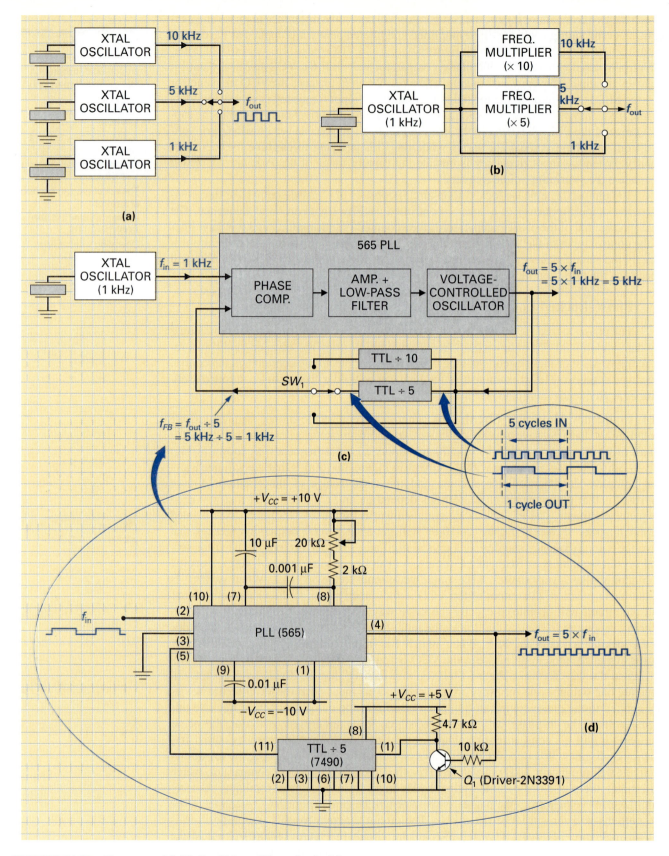

FIGURE 14-10 Frequency Multiplier Using a Phase-Locked Loop.

Use the following questions to test your understanding of Section 14-1.

1. The astable multivibrator is also called the _____ multivibrator.
2. Which of the multivibrator circuits has only one stable state?
3. Which multivibrator is also called a set-reset flip-flop?
4. Why is the set-reset flip-flop also called a latch?
5. The 555 timer derived its number identification from _____ .
6. List two applications of the 555 timer.
7. A VCO IC generates a repeating output waveform based on a _____ at its input.
8. Which three waveforms are generated by the 8038 function generator IC?
9. What are the three basic blocks in a phase-locked loop circuit?
10. What is included in the feedback path between the VCO and the phase comparator of a PLL when the PLL is used as a frequency multiplier?

14-2 THYRISTORS

The most frequently used thyristors are the silicon-controlled rectifier (SCR), triode ac semiconductor switch (TRIAC), diode ac semiconductor switch (DIAC), unijunction transistor (UJT), and programmable unijunction transistor (PUT). In this section, we will examine the operation, characteristics, applications, and testing of all these electronic devices.

14-2-1 *The Silicon-Controlled Rectifier (SCR)*

The **silicon-controlled rectifier** or **SCR** is the most frequently used of the thyristor family. Figure 14-11(a) shows how this device is a four-layered, alternately doped component with three terminals, labeled anode (*A*), cathode (*K*), and gate (*G*).

SCR Operation

To simplify the operation of the SCR, Figure 14-11(b) shows how the four layers of the SCR can be thought of, when split, as being a PNP and NPN transistor, and Figure 14-11(c) shows how this interconnection forms a **complementary latch circuit.** To operate as an ON/OFF switch, the SCR must be biased like a diode, with the anode of the SCR made positive relative to the cathode. The gate of the SCR is an active-HIGH input and must be triggered by a voltage that is positive relative to the cathode. In the example in Figure 14-11(c), you can see that the anode of the SCR is connected to +50 V via the load (R_L), the cathode is connected to 0 V, and the gate has a control input that is either 0 V or +5 V. When the gate input is at 0 V, the NPN and PNP transistors are OFF and the SCR is equivalent to an open switch, as shown in the inset in Figure 14-11(c). On the other hand, when the gate input is +5 V, the NPN transistor's base is made positive with respect to the emitter and so the NPN transistor will turn ON. Turning the NPN transistor ON will connect the 0 V at the emitter of the NPN transistor through to the PNP's base, causing it to also turn ON. Turning the PNP transistor ON will connect the large positive voltage on the PNP's emitter through to the collector and the base of the NPN transistor, keeping it ON even after the +5 V input trigger is removed. As a result, both transistors will be latched ON and held ON by one another, allowing a continuous flow of current from cathode to anode. A momentary positive input trigger, therefore, will cause the SCR to be latched ON and be equivalent to a closed switch, as seen in the inset in Figure 14-11(c).

Silicon-Controlled Rectifier or SCR

A three-junction, three-terminal, unidirectional P-N-P-N thyristor that is normally an open circuit. When triggered with the proper gate signal it switches to a conducting state and allows current to flow in one direction.

Complementary Latch Circuit

A circuit containing an NPN and PNP transistor that once triggered ON will remain latched ON.

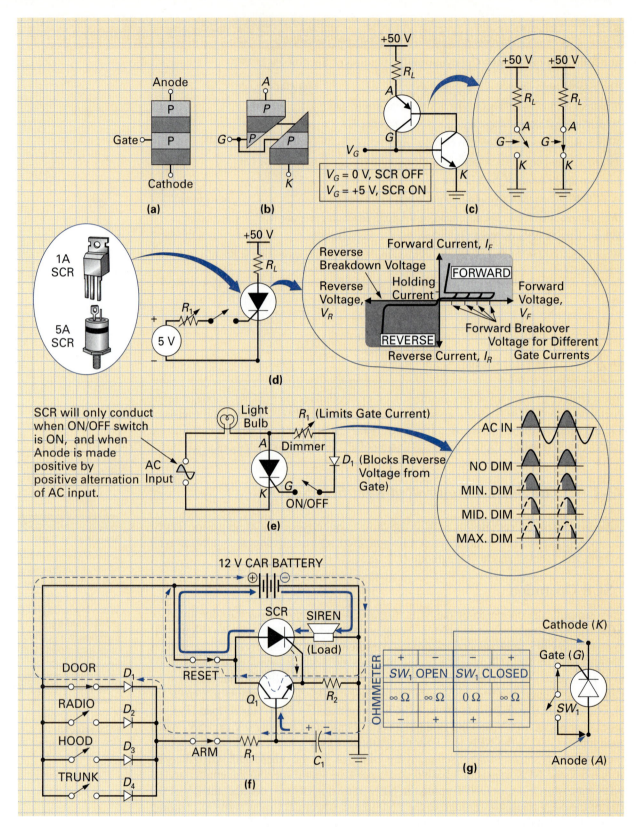

FIGURE 14-11 Silicon-Controlled Rectifier. (a) Construction. (b) Complementary Latch. (c) Closed, Open Latch Action. (d) Correctly Biased. (e) SCR Application—Basic Light Dimmer. (f) Application—Alarm System. (g) Testing SCRs.

SCR Characteristics

Figure 14-11(d) shows a correctly biased SCR; the left inset shows a typical low-power and high-power package, and the right inset shows a typical SCR characteristic curve. This graph plots the forward and reverse voltage applied across the SCR's anode-to-cathode against the current through the SCR between cathode and anode. Looking at the forward conduction quadrant, you can see that the voltage needed to turn ON an SCR is called the **forward breakover voltage.** An SCR's forward breakover voltage, or turn ON voltage, is inversely proportional to the value of gate current. For example, a larger gate current will cause the SCR to turn ON when only a small forward voltage is applied between the SCR's anode-to-cathode. If no gate current is applied, the forward voltage applied across the SCR's anode-to-cathode will have to be very large to make the SCR turn ON and conduct current between cathode and anode. Once the SCR is turned ON, a holding current latches the SCR's complementary latch ON, independent of the gate current. If the forward current between the SCR's cathode and anode falls below this minimum holding current value, the SCR will turn OFF because the NPN/PNP latch will not have enough current to keep each other ON. A gate trigger is therefore used to turn ON the SCR; however, once the SCR is ON, it can only be turned OFF by decreasing the anode-to-cathode voltage so that the cathode-to-anode current passing through the SCR drops below the holding current value.

In the reverse direction with the gate switch open, the SCR acts in almost the same way as a diode because a large reverse voltage is needed between anode and cathode to cause the SCR to break down.

SCR Applications

The SCR is used in a variety of electrical power applications, such as light dimmer circuits, motor-speed control circuits, battery charger circuits, temperature control systems, and power regulator circuits. From the characteristic curve in Figure 14-11(d), we discovered that the SCR will only conduct current in the forward direction, which is why it is classed as a **unidirectional device.** This means that if an ac signal is applied across the SCR, it will only respond to a gate trigger during the time that the ac alternation makes the anode positive with respect to the cathode. For example, Figure 14-11(e) shows how the SCR could be connected to form a simple light dimmer circuit. When the ON/OFF switch is open, the SCR will be OFF because its gate current is zero, and the light will be OFF. When the ON/OFF switch is closed, diode D_1 will connect a positive voltage to the gate of the SCR whenever the ac input is positive. The value of gate current applied to the SCR is controlled by the variable dimmer resistor (R_1), and therefore this resistor value will determine the SCR's forward breakover (turn ON) voltage. Referring to the waveforms in the inset in Figure 14-11(e), you can see that when the resistance of R_1 is zero (no dim), gate current will be maximum, the SCR will turn ON for the full positive half cycle of the ac input, and the average power delivered to the light bulb will be HIGH. As the resistance of R_1 is increased, the SCR gate current is decreased, causing the SCR to turn ON for less of the positive alternation and therefore the average power delivered to the light bulb to decrease. You may ask: Why don't we simply connect a variable resistor in series with the light bulb to vary the light bulb's current and therefore brightness? The problem with this arrangement is identical to the disadvantages of the previously discussed series dissipative regulator. The wattage rating, size, and cost of the variable resistor would be very large, and the circuit would be very inefficient because power is taken away from the light bulb by dissipating it away as heat from the resistor. Varying the SCR's ON/OFF time is a much more efficient system because we will only switch through to the light bulb the power that is desired and therefore vary the average power applied to the load in almost exactly the same way as the previously discussed switching regulator circuit.

Now that we have seen the SCR in an ac circuit application, let us now see how it could be used in a dc circuit application. Figure 14-11(f) shows a basic car alarm system. When both the ARM switch and RESET switch are closed, the alarm system is active and any of the four sensor switches (door, radio, hood, or trunk) will activate the alarm. For example, Figure 14-11(f) shows what will happen if the car door is opened when the alarm system is armed. Opening the door will cause capacitor C_1 to charge via diode D_1 and resistor R_1. After a short delay, the charge on C_1 is large enough to turn ON Q_1, which will switch the positive potential on its collector from the battery through to its emitter and to the gate of the SCR. This positive gate trigger will turn ON the SCR and activate the siren because the SCR is equivalent to a closed switch when ON and when triggered will connect the full 12-V battery across the siren. Once the SCR is turned ON, it will remain latched ON independent of the ARM switch and the sensor switches. Only by opening the RESET switch, which is hidden within the vehicle, can the siren be shut OFF. The values of R_1 and C_1 are chosen so that a small delay occurs before Q_1 and the SCR are triggered. This delay is included so that the vehicle owner has enough time to enter the car and disarm the alarm system by opening the ARM switch.

SCR Testing

Using the oscilloscope, you can monitor the gate trigger input and ON/OFF switching of an SCR while it is operating in circuit. If you suspect that the SCR is the cause of a circuit malfunction, you should remove the SCR and use the ohmmeter test circuit shown in Figure 14-11(g) to check for terminal-to-terminal opens and shorts. With this test, we will be using the ohmmeter's internal battery to apply different polarities to the different terminals of the SCR and SW_1 to either apply or disconnect gate current. To explain the ohmmeter response table in this illustration, you can see that if SW_1 is open (no gate current), the resistance between anode and cathode should be almost infinite ohms (actually about 250 kΩ), no matter what polarity is applied between anode and cathode. On the other hand, if SW_1 is closed and the anode is made positive with respect to the cathode, the gate will also be made positive due to SW_1, and so the SCR should turn ON and have a very low resistance between anode and cathode. If SW_1 is closed, and a reverse polarity is applied across the SCR (anode is made negative, cathode is made positive), the ohmmeter should once again read infinite ohms.

14-2-2   *The Triode AC Semiconductor Switch (TRIAC)*

The disadvantage of the SCR is that it is unidirectional, which means that it can only be activated when the applied anode-to-cathode voltage makes the anode positive with respect to the cathode and it will conduct current in one direction. As a result, the SCR can only control a dc supply voltage, or one-half cycle of the ac supply voltage. To gain control of the complete ac input cycle, we would need to connect two SCRs in parallel, facing in opposite directions, as shown in Figure 14-12(a). This is exactly what was done to construct the **triode ac semiconductor switch** or **TRIAC,** which has three terminals called main terminal 1 (MT_1), main terminal 2 (MT_2), and gate (G). The *p-n* doping for this **bidirectional device** is shown in Figure 14-12(b), and its schematic symbol is shown in Figure 14-12(c).

TRIAC Operation and Characteristics

Because the TRIAC is basically two SCRs connected in parallel, back-to-back, it comes as no surprise that its operation and characteristics are very similar to the SCR. The characteristic curve for a typical TRIAC is shown in Figure 14-12(d). Looking at the identical forward and reverse curves, you can see that the key difference with the TRIAC is that it can be triggered or activated by either a positive or negative input gate trigger. This means that the TRIAC can be used to control both the positive and negative alternation of an ac supply voltage. To explain this in more detail, Figure 14-12(e) shows how a TRIAC could be connected across an ac input. When the ON/OFF gate switch is open, the TRIAC will not receive a gate trigger and so it will remain OFF. When the ON/OFF gate switch is closed, the TRIAC will be triggered by the positive and negative cycles of the ac input, via R_1. By adjusting the resistance of R_1, we can control the TRIAC's value of gate current. By controlling gate current, we can control the TRIAC's turn-ON voltage (positive and negative breakover voltage), so that the ac input voltage can be chopped up to adjust the average value of voltage applied to the load.

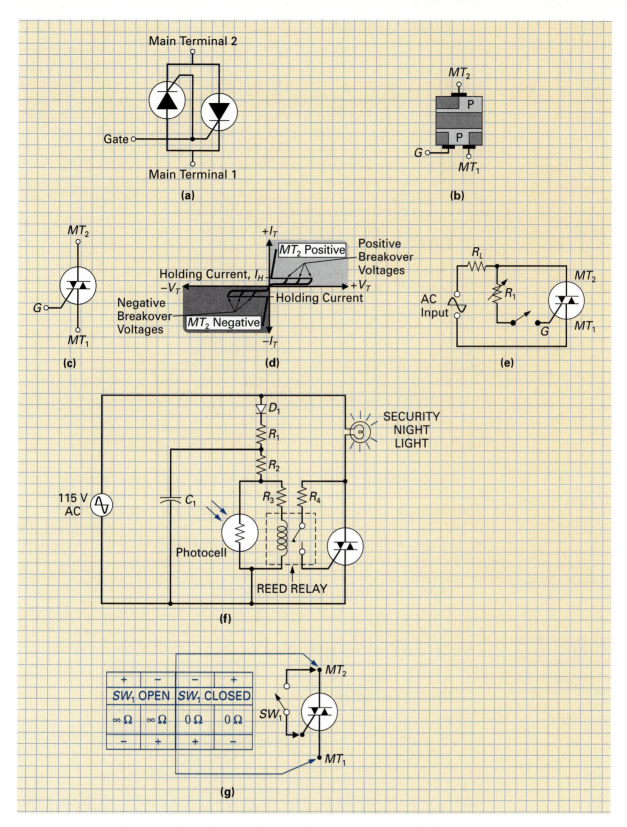

FIGURE 14-12 TRIAC. (a) TRIAC Equivalent Circuit. (b) Construction. (c) Schematic Symbol. (d) *V–I* Characteristics. (e) Application—AC TRIAC Switch. (f) Application—Automatic Night Light. (g) Testing TRIACs.

TRIAC Applications

Figure 14-12(f) shows how a TRIAC could be connected as an automatic night light for home or business security and safety. Later in this chapter, we will discuss the photocell in more detail. However, for now just think of it as a variable resistor that changes its resistance based on the amount of light present. During the day when the photocell is exposed to light, its resistance is less than a few ohms, because since it is in parallel with the energizing coil of a reed relay, most of the current will pass through the photocell, keeping the reed relay de-energized. As the sun goes down and the photocell is deprived of light, its resistance increases to a few megohms, and this high resistance will cause the current through the reed relay coil to increase and therefore the reed relay to energize. With the reed-relay switch closed, the TRIAC will be triggered by each half cycle of the ac supply voltage, causing it to turn ON and connect power to the night light.

TRIAC Testing

Using the oscilloscope, you can monitor the gate trigger input, and ON/OFF switching of a TRIAC, while it is operating in-circuit. If you suspect that the TRIAC is the cause of a circuit malfunction, you should remove the TRIAC and use the ohmmeter test circuit shown in Figure 14-12(g), to check for terminal-to-terminal opens and shorts. The ohmmeter response table in this illustration shows that if SW_1 is open (no gate current), the resistance between MT_2 and MT_1 should be almost infinite ohms (actually about 250 kΩ), no matter what polarity is applied. When switch 1 is closed, the gate of the TRIAC will receive a trigger. Because the TRIAC operates on either a positive or negative trigger, it should turn ON no matter what polarity is applied and therefore have a very low resistance between MT_2 and MT_1.

14-2-3   *The Diode AC Semiconductor Switch (DIAC)*

One disadvantage of the TRIAC is that its positive breakover voltage is usually slightly different from its negative breakover voltage. This nonsymmetrical trigger characteristic can be compensated for by using a **diode ac semiconductor switch** or **DIAC** to trigger a TRIAC. The DIAC's construction is shown in Figure 14-13(a), its schematic symbol in Figure 14-13(b), and its equivalent circuit in Figure 14-13(c). Equivalent to two back-to-back, series-connected junction diodes, the DIAC has two terminals.

DIAC Operation and Characteristics

Since the PNP regions of a DIAC are all equally doped, the DIAC will have the same forward and reverse characteristics, as shown in Figure 14-13(d). As a result, the DIAC is classed as a **symmetrical bidirectional switch,** which means that it will have the same value of breakover voltage in both the forward and reverse directions.

DIAC Applications

Figure 14-13(e) shows how a DIAC could be connected as a pulse-triggering device in a TRIAC ac power control circuit. The DIAC will turn ON when the capacitor has charged to either the positive or negative breakover voltage ($+V_{BO}$ or $-V_{BO}$). Once this voltage is reached, the DIAC turns ON and the capacitor discharges through the DIAC, triggering the TRIAC into conduction, which then connects the ac supply voltage across the load. The variable resistor R_1 is used to adjust the RC charge time constant, so the DIAC turn-ON time, and therefore TRIAC turn-ON time, can be changed.

DIAC Testing

Using the oscilloscope, you can monitor the ON/OFF switching of a DIAC while it is operating in-circuit. If you suspect that the DIAC is the cause of a circuit malfunction, you should remove the DIAC and check it with the ohmmeter, as seen in Figure 14-13(f).

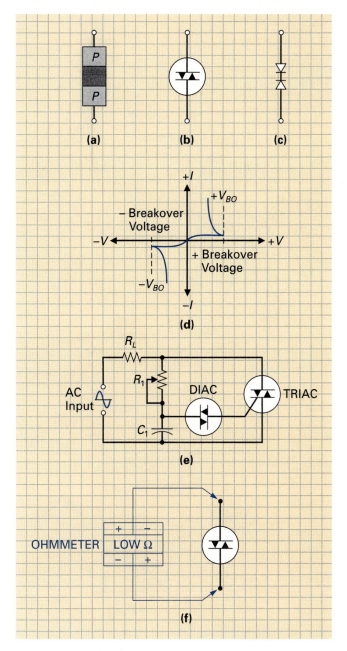

FIGURE 14-13 DIAC. (a) Construction. (b) Schematic Symbol. (c) Equivalent Circuit. (d) *V-I* Characteristics. (e) Application—TRIAC Control. (f) Testing DIACs.

Because the DIAC is basically two diodes connected back-to-back in series, the ohmmeter should show a low resistance reading between its terminals no matter what polarity is applied.

14-2-4 *The Unijunction Transistor (UJT)*

The **unijunction transistor** or **UJT** operates in a very different way to the SCR, TRIAC, and DIAC. Although it is given the name transistor, it is never used as an amplifying device like the BJT and FET; it is only ever used as a voltage-controlled switch. Figure 14-14(a) shows the construction of the UJT and illustrates how the uni (one) junction transistor derives its name from the fact that it has only one P-N junction. Looking at this illustration,

Unijunction Transistor or UJT

A P-N device that has an emitter connected to the P-N junction on one side of the bar and two bases at either end of the bar. Used primarily as a switching device.

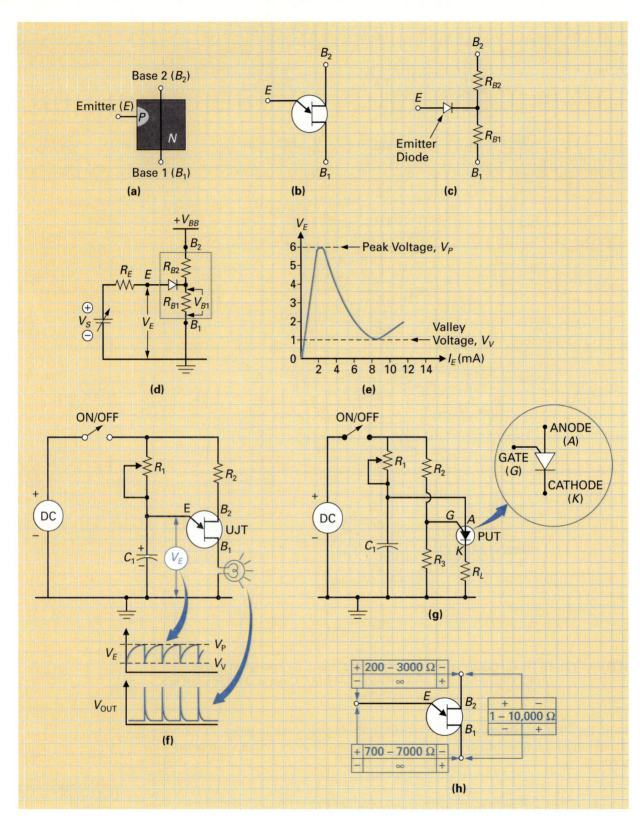

FIGURE 14-14 Unijunction Transistors. (a) Construction. (b) Schematic Symbol. (c) Equivalent Circuit. (d) Correctly Biased. (e) *V–I* Characteristic Curve. (f) Application—Relaxation Oscillator. (g) Programmable Unijunction Transistor (PUT). (h) Ohmmeter Resistances.

you can see that the UJT is a three-terminal device with an emitter lead (E) attached to a small p-type pellet that is fused into a bar of n-type silicon with contacts at either end [labeled base 1 (B_1) and base 2 (B_2)].

The schematic symbol for the UJT is shown in Figure 14-14(b). To remember whether the junction is a P-N or N-P type, think of the arrow as a junction diode: The emitter (diode–anode) is positive and the bar (diode–cathode) is negative.

UJT Operation and Characteristics

Figure 14-14(c) illustrates the UJT's equivalent circuit. The emitter-to-bar P-N junction is equivalent to a junction diode, and the bar is equivalent to a two-resistor voltage divider (R_{B1} and R_{B2}). Referring once again to the UJT construction in Figure 14-14(a), you can see that the emitter pellet is closer to terminal B_2 than B_1, and this is why the resistance of R_{B2} is smaller than the resistance of R_{B1}. Figure 14-14(d) shows how a UJT should be correctly biased. The voltage source V_{BB} is connected to make B_2 positive relative to B_1, and the input voltage V_S is connected to make the emitter positive with respect to B_1. The resistor R_E is used to limit emitter current. When V_S is zero, the UJT's emitter diode is OFF, and the resistance between B_2 and B_1 allows only a very small amount of current between ground and $+V_{BB}$. If the emitter supply voltage is increased so that V_E exceeds the voltage at B_1, the emitter diode will turn ON and inject holes into the p region. Flooding the lower half of the UJT with holes increases the amount of current flow through the UJT, dramatically reducing the resistance of R_{B1}. A lower R_{B1} resistance results in a lower V_{B1} voltage drop, and this further increases the E to B_1 P-N forward bias, permitting more holes to be injected into the n-type bar between E and B_1. Figure 14-14(e) graphically illustrates this action by plotting the UJT's emitter voltage (between E and B_1) against the UJT's emitter current. An increase in V_S, and therefore V_E, produces very little emitter current until the **peak voltage (V_P)** is reached. Beyond V_P, V_E has exceeded V_{B1} and the emitter diode is forward biased, causing I_E to increase and V_E to decrease due to the lower resistance of R_{B1}. This negative resistance region reaches a low point known as the **valley voltage (V_v),** which is a point at which V_E begins to increase and the UJT no longer exhibits a negative resistance. The point at which a UJT turns ON and increases the current between B_1 and B_2 can be controlled and used in switching applications.

UJT Applications

The UJT's negative resistance characteristic is useful in switching and timing applications. Figure 14-14(f) shows how a UJT could be connected to form a relaxation oscillator in an emergency flasher circuit. When the ON/OFF switch is closed, capacitor C_1 charges by resistor R_1. When the voltage across C_1 reaches the UJT's vp value, the UJT will turn ON and its resistance between E and B_1 will drop LOW. This low resistance will allow C_1 to discharge through the UJT's E-to-B_1 junction and into the flasher light bulb, causing it to momentarily flash. As C_1 discharges, its voltage decreases and this causes the UJT to turn OFF. The cycle then repeats because the off UJT will allow capacitor C_1 to begin charging toward V_p, at which time it will trigger the UJT and repeat the process. The circuit's repetition rate, or frequency, is determined by the UJT's V_p rating, the supply voltage, and the RC time constant. To change the flashing rate, the value of R_1 can be changed to vary the rate at which C_1 is charged and therefore how soon the UJT is triggered.

The Programmable UJT (PUT)

The **programmable unijunction transistor** or **PUT** is a variation on the basic UJT thyristor. This four-layer thyristor has three terminals, labeled cathode (K), anode (A), and gate (G). The key difference between the basic UJT and the PUT is that the PUT's peak voltage (V_p) can be controlled. Figure 14-14(g) shows how a PUT could also be connected to form a relaxation oscillator circuit. To differentiate the PUT's schematic symbol from the SCR, the gate input is connected into the anode side of the diode symbol instead of the cathode side. This circuit will produce exactly the same output waveform as the circuit shown in

Figure 14-14(f). The gate-to-cathode voltage is derived from R_3, which is connected with R_2 to form a voltage divider. This circuit will operate in exactly the same way as the previous relaxation oscillator, in that C_1 will charge via R_1 until the charge across C reaches the V_p value. In this circuit, however, the V_p trigger voltage is set by R_3. When the PUT's anode-to-cathode voltage exceeds the gate voltage by 0.7 V (single-diode voltage drop), the PUT will turn ON and C_1 will discharge through the PUT and develop an output pulse across R_L. To vary the frequency of this circuit, we can change the resistance of R_1 as before, or change the ratio of R_2 to R_3, which controls the V_p value of the PUT. For example, if R_3 is made larger than R_2, the gate voltage and therefore V_P voltage will be larger. A high V_P value will mean that C_1 will have to charge to a larger voltage before the PUT will turn ON. Increasing the time needed for C_1 to charge will decrease the triggering rate of the PUT and therefore decrease the circuit's frequency of operation.

UJT Testing

Using the oscilloscope, you can monitor the ON/OFF switching of a UJT while it is operating in-circuit. If you suspect that the UJT is the cause of a circuit malfunction, you should remove it and check it with the ohmmeter, as seen in Figure 14-14(h).

SELF-TEST EVALUATION POINT FOR SECTION 14-2

Use the following questions to test your understanding of Section 14-2.

1. The SCR's forward breakover voltage will _____ as gate current is increased.
2. The SCR can only be used to control dc power. (True/False)
3. The TRIAC is a _____ directional device that can be either positive or negative triggered.
4. Which device can be used to trigger the TRIAC to compensate for the TRIAC's nonsymmetrical triggering characteristic?
5. TRIACs should always be used instead of SCRs when we want to control the complete ac power cycle. (True/False)
6. The unijunction transistor has _____ P-N semiconductor junction(s).
7. The UJT can, like the bipolar and field effect transistors, function as either a switch or an amplifier. (True/False)
8. The PUT's peak voltage can be varied by changing the voltage between _____ and cathode.

14-3   TRANSDUCERS

Semiconductor Transducer

An electronic device that converts one form of energy to another.

A **semiconductor transducer** is an electronic device that converts one form of energy to another. Electronic transducers can be classified as either **input transducers** (such as the photodiode) that generate input control signals or **output transducers** (such as the LED) that convert output electrical signals to some other energy form. Input transducers are sensors that convert thermal, optical, mechanical, and magnetic energy variations into equivalent voltage and current variations. Output transducers, on the other hand, perform the exact opposite, converting voltage and current variations into optical or mechanical energy variations.

Input Transducer

A transducer that generates input control signals.

Output Transducers

Transducers that convert output electrical signals to some other energy form.

14-3-1   *Optoelectronic Transducers*

The LED and photodiode are the most frequently used optoelectronic devices; however, they are not the only semiconductor devices in the optoelectronic family. In this section, we will review the operation of the LED and photodiode, along with the operation of other types of semiconductor light transducers.

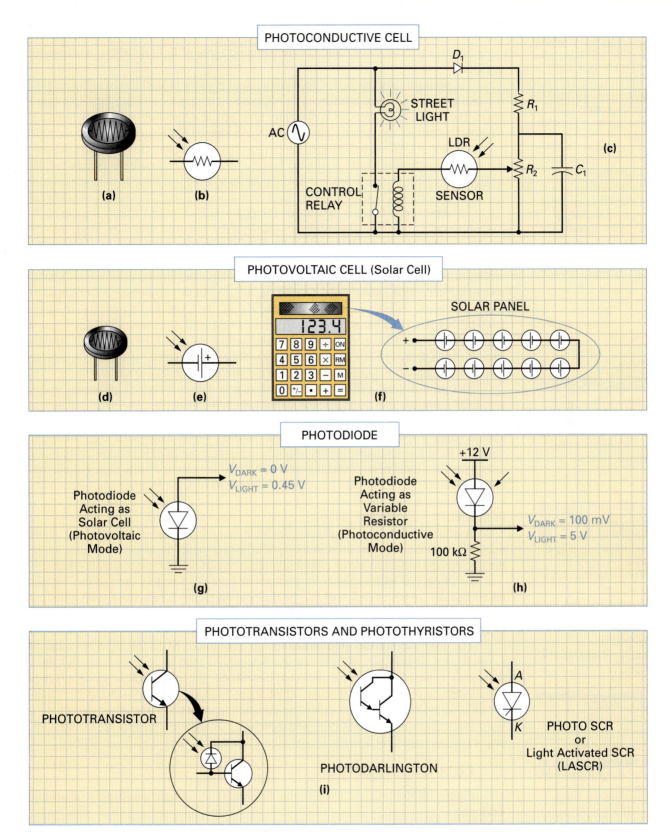

FIGURE 14-15 Light-Sensitive Devices.

Light-Sensitive Devices

Figure 14-15 illustrates some of the different types of **light-sensitive semiconductor devices.**

The **photoconductive cell** or **light-dependent resistor (LDR)** shown in Figure 14-15(a) is a two-terminal device that changes its resistance (conductance) when light (photo) is applied. The photoconductive cell is normally mounted in a metal or plastic case with a glass window that allows the sensed light to strike the S-shaped light-sensitive material (typically cadmium sulfide). When light strikes the photoconductive atoms, electrons are released into the conduction band and the resistance between the device's terminals is reduced. When light is not present, the electrons and holes recombine and the resistance is increased. A photoconductive cell will typically have a "dark resistance" of several hundred megohms and a "light resistance" of a few hundred ohms. The photoconductive cell's key advantage is that it can withstand a high operating voltage (typically a few hundred volts). Its disadvantages are that it responds slowly to changes in light level and that its power rating is generally low (typically a few hundred milliwatts). The schematic symbol for the photoconductive cell is shown in Figure 14-15(b). Figure 14-15(c) shows how the photoconductive cell could be connected to control a street light. During the day, the LDR's resistance is LOW due to the high light levels, and the current through D_1, R_1, R_2, the LDR, and the control relay coil is HIGH. This HIGH value of current will energize the coil and cause the relay's normally closed (*NC*) contacts to open and therefore the street light to be OFF. When dark, the resistance of the LDR will be HIGH, and the relay will be de-energized, its contacts will return to their normal condition, which is closed, and the street light will be ON.

The **photovoltaic cell** or **solar cell,** shown in Figure 14-15(d), generates a voltage across its terminals that will increase as the light level increases. The solar cell is usually made from silicon; its schematic symbol is shown in Figure 14-15(e). The solar cell is available as either a discrete device or as a solar panel in which many solar cells are interconnected to form a series-aiding power source, as shown in the calculator application in Figure 14-15(f). The output of a solar cell is normally rated in volts and milliamps. For example, a typical photovoltaic cell could generate 0.5 V and 40 mA. To increase the output voltage, simply connect solar cells in series; to increase the output current, simply connect solar cells in parallel.

A **photodiode** is a photodetecting or light-receiving device that contains a semiconductor P-N junction. When used in the "photovoltaic mode" as shown in Figure 14-15(g), the photodiode will generate an output voltage (voltaic) in response to a light (photo) input (i.e., it will operate like a solar cell). Photodiodes are most widely used in the "photoconductive mode," in which they will change their conductance (conductive) when light (photo) is applied (i.e., it will operate like an LDR). In this mode, the photodiode is reverse biased (*n*-type region is made positive, *p*-type region is made negative).

The **phototransistor, photodarlington,** and **photo-SCR** (or light-activated SCR, LASCR), are all examples of light-reactive devices. As seen in the phototransistor inset, all three devices basically include a photodiode to activate the device whenever light is present. The advantage of the phototransistor is that it can produce a higher output current than the photodiode; however, the photodiode has a faster response time. The advantage of the photodarlington is its high current gain, and therefore it is ideal in low-light applications. When larger current switching is needed (typically, a few amps), the LASCR can be used.

Light-Emitting Devices

Figure 14-16 illustrates some of the different types of **light-emitting semiconductor devices.**

The LED was covered in detail in Chapter 3; however, let us review its basic characteristics. Figure 14-16(a) shows the schematic symbols used to represent a **light-emitting diode** or **LED,** Figure 14-16(b) shows its typical physical appearance, and Figure 14-16(c) shows how the LED can be used as an ON/OFF indicator. The LED is basically a P-N junction diode, and like all semiconductor diodes it can be either forward biased or reverse biased. When forward biased, it will emit energy in response to a forward current. This emis-

Light-Sensitive Semiconductor Devices

Semiconductor devices that change their characteristics in response to light.

Photoconductive Cell or Light-Dependent Resistor (LDR)

A two-terminal device that changes its resistance when light is applied.

Photovoltaic Cell or Solar Cell

A device that generates a voltage across its terminal that will increase as the light level increases.

Photodiode

A photodetecting or light-receiving device that contains a semiconductor P-N junction.

PhotoTransistor, PhotoDarlington, and Photo-SCR

Examples of light-reactive devices.

Light-Emitting Semiconductor Devices

Semiconductor devices that will emit light when an electrical signal is applied.

Light-Emitting Diode or LED

A semiconductor diode that converts electric energy into electromagnetic radiation at visible and near-infrared frequencies when its P-N junction is forward biased.

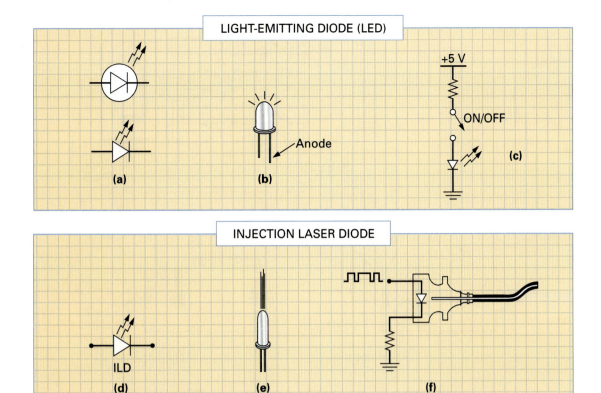

FIGURE 14-16   Light-Emitting Devices.

sion of energy may be in the form of heat energy, light energy, or both heat and light energy, depending on the type of semiconductor material used. The type of material also determines the color, and therefore frequency, of the light emitted. For example, different compounds are available that will cause the LED to emit red, yellow, green, blue, white, orange, or infrared light when it is forward biased.

Figure 14-16(d) shows the schematic symbol for the **injection laser diode** or **ILD,** Figure 14-16(e) shows its typical physical appearance, and Figure 14-16(f) shows how it is usually used in fiber-optic communication. The ILD differs from the LED in that it generates monochromatic (one-color, or one-frequency) light. Although it appears as though the LED is only generating one-color light, it is in fact emitting several wavelengths or different frequencies that combined make up a color. Generating only one frequency is ideal in applications such as fiber optics, where we want to keep the light beam tightly focused so that the beam can travel long distances down a very thin piece of glass or plastic fiber. To explain this point in more detail, let us compare the light from a light bulb to the light produced by an ILD. An ILD generates light of only one frequency (monochromatic), and because all of the small packets of light being generated are of the same frequency, they act like an organized army, marching together in the same direction as a tight concentrated beam. The light bulb, on the other hand, generates white light that is composed of every color or frequency (panchromatic), and because these frequencies are all different, they act completely disorganized and therefore radiate in all directions.

Optoelectronic Application—Character Displays

One of the biggest applications of LEDs is in the multisegment display discussed previously in Chapter 3.

The variety of character displays are also available as **liquid crystal displays (LCDs),** as shown in Figure 14-17. The two key differences between an LED and LCD display are that an LED display generates light while the LCD display controls light, and the LCD

Injection Laser Diode or ILD

A semiconductor P-N junction diode that uses a lasing action to increase and concentrate the light output.

Liquid Crystal Displays (LCDs)

A digital display having two sheets of glass separated by a sealed quantity of liquid crystal material. When a voltage is applied across the front and back electrodes, the liquid crystal's molecules become disorganized, causing the liquid to darken.

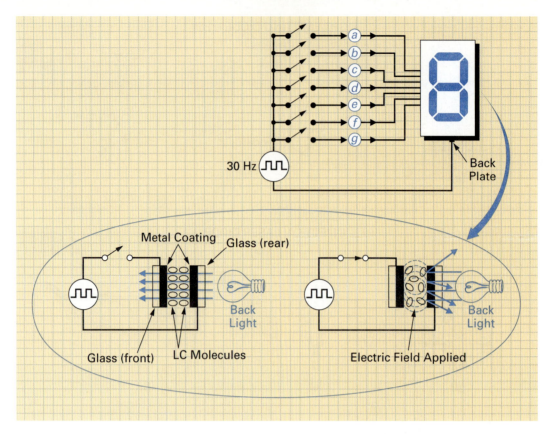

FIGURE 14-17   **Liquid Crystal Character Displays.**

display consumes a lot less power than an LED display. The low-power-consumption feature of the LCD display makes it ideal in portable battery-operated systems such as wristwatches, calculators, video games, portable test equipment, and so on. The LCD's only disadvantage is that the display is hard to see, but this can be compensated for by including backlighting to highlight the characters. The liquid crystal display contains two pieces of glass that act as a sandwich for a "nematic liquid" or liquid crystal material, as seen in the inset in Figure 14-17. The rear piece of glass is completely coated with a very thin layer of transparent metal; the front piece of glass is coated with the same transparent metal segments in the shape of the desired display. The operation of the liquid crystal display is explained in the two illustrations in the inset in Figure 14-17. When a segment switch is open, no electric field is generated between the two LCD metal plates, the nematic liquid molecules remain in their normal state, which is parallel to the plane of the glass, and so all of the backlighting passes through to the front display, making the segment invisible. When a segment switch is closed, an ac voltage is applied between the two metal plates, generating an electric field between the two LCD metal plates. This electric field will cause the nematic liquid molecules to turn by 90°, and so all of the backlighting for that segment is blocked, making the segment visible.

Optoelectronic Application—Optically Coupled Isolators

Up until this point, we have considered the optoelectronic emitter and detector as discrete or individual devices. There are, however, some devices available that include both a light-emitting and light-sensing device in one package. These devices are called **optically coupled isolators,** a name that describes their basic function: They are used to *optically couple two electrically isolated points.* To examine these devices in more detail, refer to the three basic types shown in Figure 14-18.

Figure 14-18(a) shows an **optically coupled isolator DIP module.** This optocoupler contains an infrared-emitting diode (IRED) and a silicon phototransistor and is generally

Optically Coupled Isolators

Devices that contain a light-emitting and light-sensing device in one package. They are used to optically couple two electrically isolated points.

Optically Coupled Isolator DIP Module

An optocoupler that contains an infrared-emitting diode and a silicon phototransistor and is generally used to transfer switching information between two electrically isolated points.

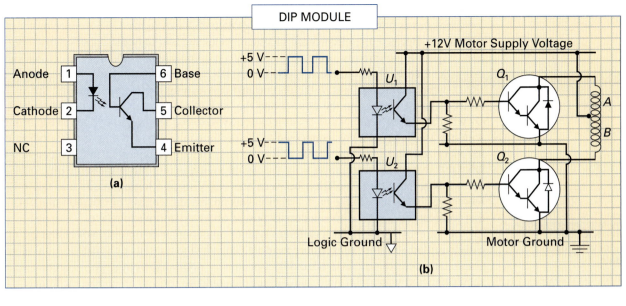

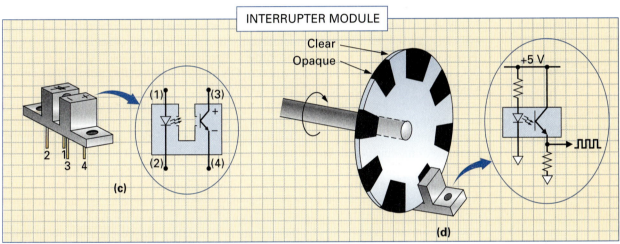

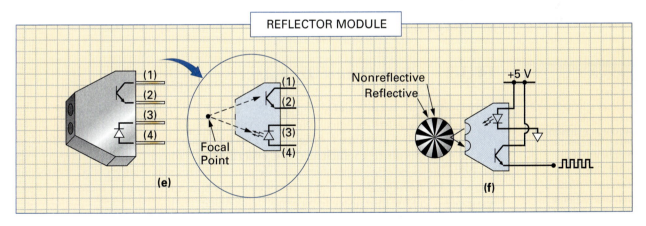

FIGURE 14-18   **Optoelectronic Application—Optically Coupled Isolators.**

used to transfer switching information between two electrically isolated points. Figure 14-18(b) shows how DIP optically coupled isolator ICs can be used in a stepper-motor control circuit. Because the circuits are identical, let us explain only the upper circuit's operation. Digital logic circuits generate the +5-V peak multiphase ON/OFF switching signal that is applied to IRED in the optocoupler. A HIGH input, for example, will turn ON the IRED, which will emit light to the phototransistor, turning it ON and switching the +12 V on its collector through to the emitter and the base of Q_1. This +12-V input to the base of Q_1 will cause it to turn ON and act as a closed switch, connecting ground to the top of the A winding. Because +12 V is connected to the center of the A/B winding, the A winding will be energized and the motor will move a step. The optocoupler, therefore, couples the ON/OFF switching information from the digital logic circuits and at the same time provides the necessary electrical isolation between the +5-V digital logic circuits and the +12-V motor supply circuits. In the past, isolation was provided by relays or isolation transformers that were larger in size, consumed more power, and were more expensive.

Figure 14-18(c) shows a different type of optocoupler or optoisolator called the **optically coupled isolator interrupter module.** This device consists of a matched and aligned emitter and detector and is used to detect opaque or nontransparent targets. Figure 14-18(d) shows how the interrupter optocoupler module could be used as an optical tachometer. The IRED is permanently ON, as seen in the inset in Figure 14-18(d) and emits a constant light beam toward the phototransistor. A transparent disc mounted to a shaft has opaque targets evenly spaced around the disc. As these opaque targets pass through the infrared beam, they will cause the phototransistor to momentarily turn OFF. On the other hand, a transparent section of the disc will not block the infrared light, and so the phototransistor will remain ON. As a result, the phototransistor will turn ON and OFF, generating pulses, the number of which is an indication of the shaft's speed of rotation.

Figure 14-18(e) shows another type of optocoupler called the **optically coupled isolator reflector module.** Like the interrupter module, this device consists of a matched and aligned emitter and detector, and it is also used to detect targets. Figure 14-18(f) shows how the reflector module could also be used as an optical tachometer. The disadvantage with the interrupter module is that the disc has to be exactly aligned between the emitter or detector sections. The reflector module does not have to be positioned so close to the target disc, and therefore the alignment is not so crucial. The target disc used for reflector modules is different in that it is composed of reflective and nonreflective target areas. As the disc rotates, almost no light will be reflected by the dark areas when they are present at the focal point of the IRED and phototransistor. However, when a reflective target is present at the focal point, light is reflected directly back to the phototransistor, turning it ON and therefore generating an output pulse.

14-3-2   *Temperature Transducers*

A **thermistor** is a semiconductor device that acts as a temperature-sensitive resistor. Figure 14-19(a) shows a few different types of thermistors. There are basically two different types: **positive temperature coefficient (PTC) thermistors** and the more frequently used **negative temperature coefficient (NTC) thermistors.** To explain the difference, Figure 14-19(b) shows how a temperature increase causes the resistance of an NTC thermistor to decrease. On the other hand, Figure 14-19(c) shows how a temperature increase causes the resistance of a PTC thermistor to increase.

Thermistors are used in a variety of applications. For example, an NTC could be used in a fire alarm circuit. When the ambient temperatures are LOW, the resistance of the NTC is HIGH, and therefore current cannot energize the alarm. If the ambient temperature increases to a HIGH level, however, the resistance of the thermistor drops LOW, and the alarm circuit is energized.

The PTC, on the other hand, could be used as a sort of circuit protection device (similar to a circuit breaker). With the PTC, the "switch temperature" (which is the temperature at which the resistance rapidly increases) can be varied by different construction techniques

---

Optically Coupled Isolator Interrupter Module

A device that consists of a matched and aligned emitter and detector and that is used to detect opaque or nontransparent targets.

Optically Coupled Isolator Reflector Module

A device that consists of a matched and aligned emitter and detector and that is used to detect targets.

Thermistor

A semiconductor device that acts as a temperature-sensitive resistor.

Positive Temperature Coefficient (PTC) Thermistors

A thermistor in which a temperature increase causes the resistance to increase.

Negative Temperature Coefficient (NTC) Thermistor

A thermistor in which a temperature increase causes the resistance to decrease.

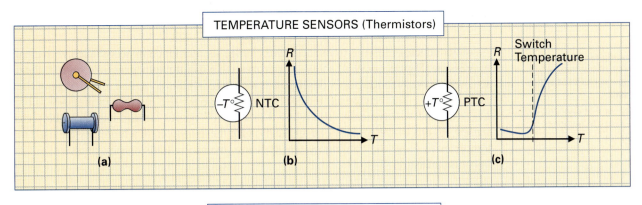

TEMPERATURE SENSORS (Thermistors)

(a)   (b)   (c)

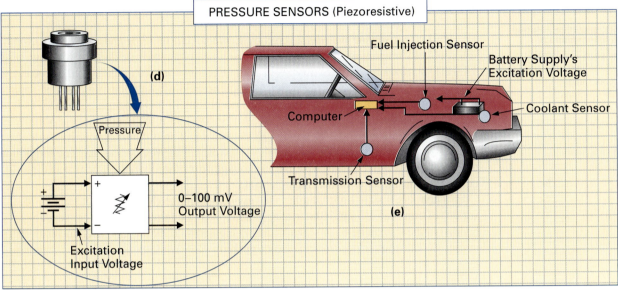

PRESSURE SENSORS (Piezoresistive)

(d)

Pressure

0–100 mV
Output Voltage

Excitation
Input Voltage

Fuel Injection Sensor

Battery Supply's
Excitation Voltage

Computer

Coolant Sensor

Transmission Sensor

(e)

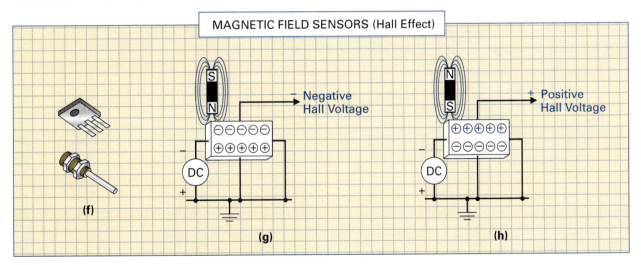

MAGNETIC FIELD SENSORS (Hall Effect)

Negative
Hall Voltage

Positive
Hall Voltage

DC

DC

(f)   (g)   (h)

FIGURE 14-19   **Semiconductor Transducers.**

from below 0°C to above 160°C. The PTC therefore could be used as a current-limiting circuit protection device because currents lower than a limiting value will not generate enough
heat to cause the PTC thermistor to switch to its high resistance state. On the other hand, circuit currents that go above the limiting value will generate enough heat to cause the PTC
thermistor to switch to its high-resistance state. Current surges therefore are limited to a safe
value until the surge is over.

14-3-3 *Pressure Transducers*

A semiconductor pressure transducer will change its resistance in accordance with changes in pressure. Figure 14-19(d) shows a **piezoresistive diaphragm pressure sensor.** Piezoresistance of a semiconductor is described as a change in resistance due to a change in the applied pressure. This device has a dc excitation voltage applied, as seen in the inset in Figure 14-19(d), and will typically generate a 0-to 100-mV output voltage based on the pressure sensed. Figure 14-19(e) shows an application for these devices in which they are used in automobiles to sense the cooling-system pressure, hydraulic-transmission pressure, and fuel-injection pressure.

14-3-4 *Magnetic Transducers*

The **Hall effect sensor** was discovered by Edward Hall in 1879 and is used in computers, automobiles, sewing machines, aircraft, machine tools, and medical equipment. Figure 14-19(f) shows a few different types of Hall effect sensors. To explain the Hall effect principle, Figure 14-19(g) shows that when a magnetic field whose polarity is north is applied to the sensor, it causes a separation of charges, generating in this example a negative Hall voltage output. On the other hand, Figure 14-19(h) shows that if the magnetic field polarity is reversed so that a south pole is applied to the sensor, it will cause a polarity separation of charges within the sensor that will result in a positive Hall voltage output. The amplitude of the generated positive or negative output voltage from the Hall effect sensor is directly dependent on the strength of the magnetic field.

Their small size, light weight, and ruggedness make the Hall effect sensors ideal in a variety of commercial and industrial applications. For example, Hall effect sensors are embedded in the human heart to serve as timing elements. They are also used to sense shaft rotation, camera shutter positioning, rotary position, flow rate, and so on.

SELF-TEST EVALUATION POINT FOR SECTION 14-3

Use the following questions to test your understanding of Section 14-3.

1. The photoconductive cell basically operates as a light-sensitive _____.
2. The photovoltaic cell converts light energy into _____ energy and is often referred to as a _____ cell.
3. What are the two modes of operation for the photodiode?
4. With the photoconductive cell, photodiode, phototransistor, photodarlington, photothyristor, and other light-sensitive devices, a light increase always causes a _____ in conductance.
5. LEDs convert _____ energy into _____ energy, whereas an ILD will convert _____ energy into _____ energy.
6. LED displays emit light and LCD displays control light. (True/False)
7. An optically coupled isolator DIP module will electrically connect two points, and isolating information from the two points. (True/False)
8. Would a PTC thermistor or NTC thermistor be ideal as a series current limiter?

REVIEW QUESTIONS

Multiple-Choice Questions

1. The ______________ multivibrator will produce a continuously alternating square wave or pulse-wave output.
   a. astable **b.** monostable **c.** bistable **d.** Schmitt
2. The ______________ multivibrator is also called a one-shot.
   a. astable **b.** monostable **c.** bistable **d.** Schmitt
3. Which multivibrator is also called a set-reset flip-flop?
   a. Astable **b.** Monostable **c.** Bistable **d.** Schmitt
4. The output pulse width of a 555 timer is determined by the externally connected ______________ and ______________.
   a. power supply, resistor
   b. load resistance, capacitor
   c. capacitor, load resistor
   d. input resistor, capacitor
5. The 555 timer consists of a ______________ resistor voltage divider, ______________ comparator(s), ______________ *R-S* flip-flop, an INVERTER, output stage and discharge transistor on a single IC.
   a. 2, 2, 2 **b.** 3, 2, 1 **c.** 1, 2, 3 **d.** 3, 1, 2
6. A monostable multivibrator will generally make use of a ______________ and ______________ circuit on the trigger input.
   a. integrator, clipper **c.** Schmitt, clipper
   b. differentiator, Schmitt **d.** differentiator, clipper
7. Which of the following thyristors would normally be used to control dc power?
   a. TRIAC **d.** DIAC
   b. SCR **e.** Both (b) and (c)
   c. UJT
8. Which of the following thyristors is a bidirectional device?
   a. TRIAC **c.** UJT
   b. SCR **d.** ENIAC
9. ______________ an SCR's gate current will ______________ its forward breakover voltage.
   a. Increasing, decrease **c.** Decreasing, increase
   b. Increasing, increase **d.** Decreasing, decrease
10. Which of the following thyristors has a symmetrical switching characteristic?
    a. TRIAC **d.** DIAC
    b. SCR **e.** Both (b) and (c)
    c. UJT
11. Which of the following light-sensitive devices will generate a voltage when light is applied?
    a. Photoconductive cell **d.** Phototransistor
    b. Photodiode **e.** Both (b) and (c)
    c. Photovoltaic cell
12. Which of the following is a monochromatic light source?
    a. LED **d.** LCD
    b. Light bulb **e.** Both (b) and (c)
    c. ILD
13. Which of the following devices would be best suited to detect whenever a coin has been inserted in a slot?
    a. Piezoresistive sensor **d.** Interrupter optoisolator
    b. Hall effect sensor **e.** Both (a) and (c)
    c. PTC thermistor
14. What advantage does a liquid crystal display have over an LED display?
    a. Consumes less power **d.** Both (a) and (c)
    b. Is easier to read **e.** Both (a) and (b)
    c. Is ideal in portable applications
15. What device could be used as a series-connected current limiter?
    a. Hall effect sensor **c.** PTC thermistor
    b. Piezoresistive sensor **d.** NTC thermistor
16. Which of the following sensors could be used to control the switching of current through a motor's stator windings by sensing the position of the motor's permanent magnet rotor?
    a. Hall effect sensor **c.** PTC thermistor
    b. Piezoresistive sensor **d.** NTC thermistor

Practice Problems

17. Identify the circuit shown in Figure 14-20.
18. Calculate the frequency of the output for the circuit shown in Figure 14-20.
19. Identify the circuit shown in Figure 14-21.
20. Calculate the width of the output pulse (t_2) from the circuit shown in Figure 14-21.
21. Comparing the input pulse to the output pulse in Figure 14-21, how much pulse stretching has actually occured?
22. What would be the approximate width of the trigger pulse at *TP*2 in Figure 14-21?
23. Identify the circuit shown in Figure 14-22.
24. Calculate the following in relation to the circuit shown in Figure 14-22:
    a. The positive half-cycle time
    b. The negative half-cycle time
    c. The cycle's period
    d. The cycle's frequency

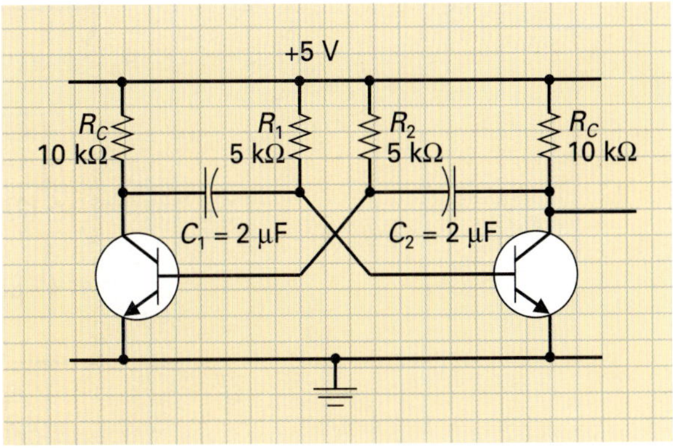

FIGURE 14-20 A Two-State Switching Circuit.

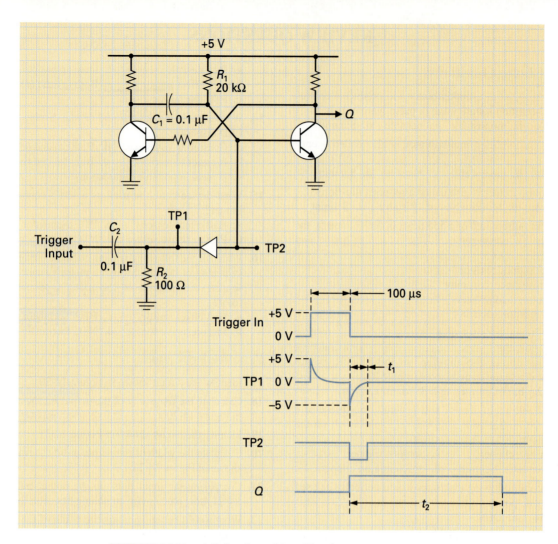

FIGURE 14-21   A Pulse-Stretching Circuit.

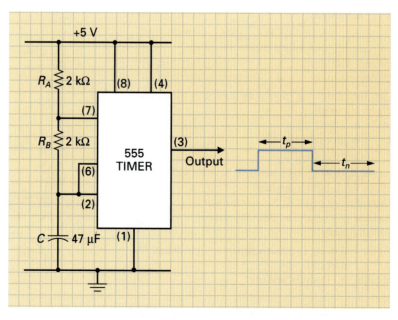

FIGURE 14-22   A 555 Timer Circuit.

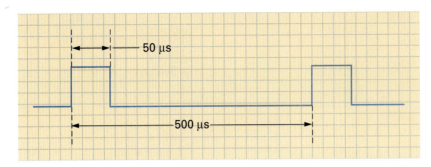

FIGURE 14-23 A Rectangular or Pulse Waveform.

25. What would be the duty cycle of the output waveform generated by the 555 timer circuit in Figure 14-22?

26. Referring to Figure 14-23, what is the waveform's
    a. pulse width
    c. frequency
    b. period
    d. duty cycle

27. Identify the schematic symbols shown in Figure 14-24.

To practice your circuit recognition and operation ability, refer to the circuit in Figure 14-25, and answer the following questions.

28. Describe the operation of the circuit when the input is normally HIGH.

29. Describe the operation of the circuit when the input is taken LOW.

30. How is the SCR turned ON, and how is the SCR turned OFF?

To practice your circuit recognition and operation ability, refer to the circuit in Figure 14-26, and answer the following questions.

31. Is the 555 timer connected to function as an astable, monostable, or bistable multivibrator?

32. Describe the basic operation of this circuit.

33. Why do you think a TRIAC is being used in this circuit instead of an SCR?

To practice your circuit recognition and operation ability, refer to the circuit in Figure 14-27(a) and (b), and answer the following questions.

34. Identify the light-emitting and light-sensitive devices used in these circuits.

35. Why was a different light-reactive device used for each of these circuits?

36. Why are these optically coupled isolators needed for computer output device control?

To practice your circuit recognition and operation ability, refer to the circuit in Figure 14-28, and answer the following questions.

37. Identify the light-sensitive device used in this circuit.

38. What is the function of this circuit?

39. Describe the operation of the circuit.

To practice your circuit recognition and operation ability, refer to the circuit in Figure 14-29 on p. 408, and answer the following questions.

40. Identify the light-sensitive device used in this circuit.

41. What is the function of this circuit?

42. Describe the operation of the circuit.

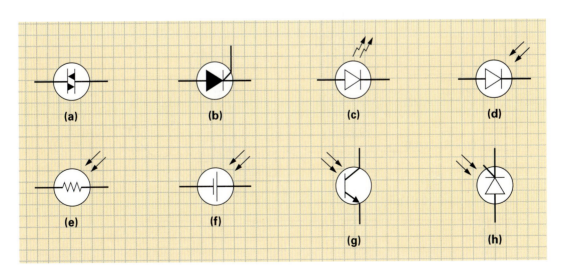

FIGURE 14-24 Schematic Symbols.

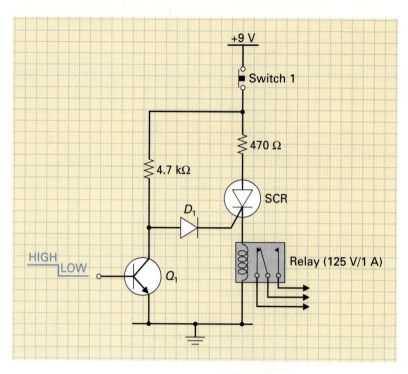

FIGURE 14-25 **Low-Power to High-Power ON/OFF Switching.**

To practice your circuit recognition and operation ability, refer to the circuit in Figure 14-30, and answer the following questions.

43. Identify the light-sensitive device used in this circuit.

44. What is the function of this circuit?

45. Describe the operation of the circuit.

To practice your circuit recognition and operation ability, refer to the circuit in Figure 14-31, and answer the following questions.

46. Identify the light-sensitive device used in this circuit.

47. What is the function of this circuit?

48. Briefly describe the function of the 741 op-amp in this circuit.

49. Why is a variable resistor connected to the 741 op-amp?

50. What is the normal voltage drop across an LED?

51. Describe how the bar graph display in this circuit operates.

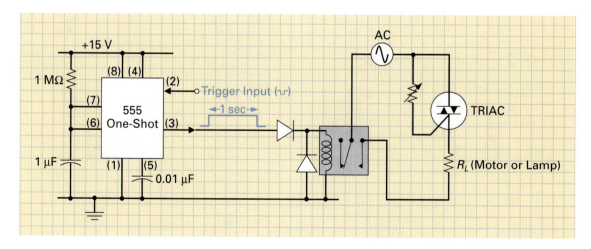

FIGURE 14-26 **One-Shot Timer Control of TRIAC.**

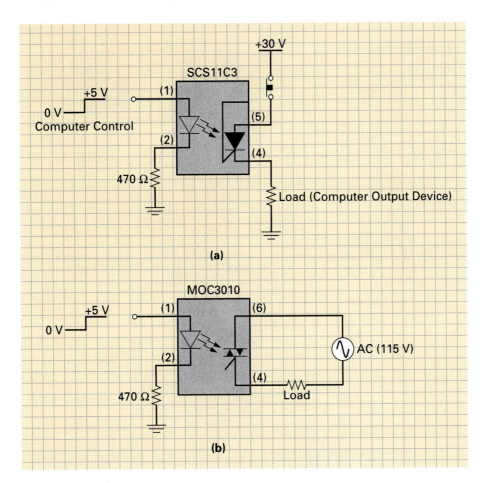

FIGURE 14-27   Computer Output Device Control.

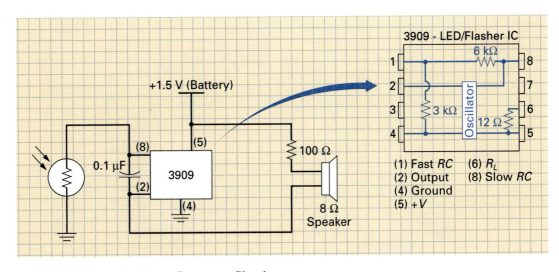

FIGURE 14-28   Light-Controlled Tone-Generator Circuit.

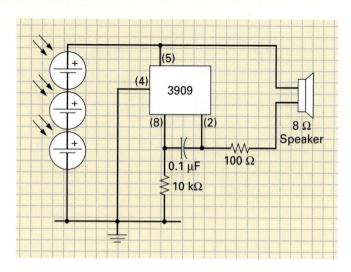

FIGURE 14-29 Solar-Powered Tone Generator.

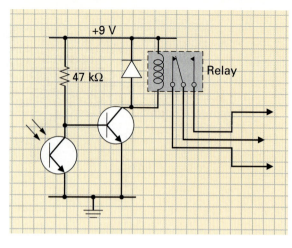

FIGURE 14-30 Optical ON/OFF Relay Control Circuit.

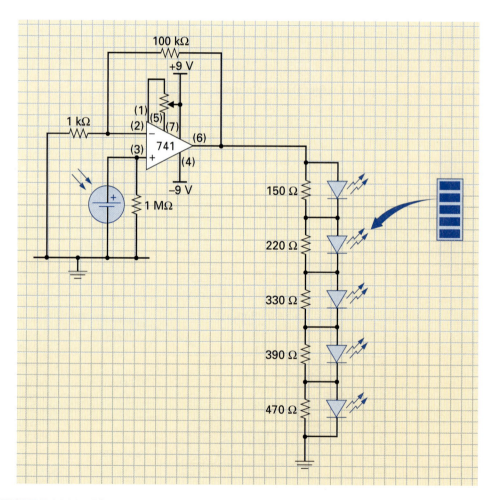

FIGURE 14-31 Lightmeter with Bar Graph Display.

Answers to Self-Test Evaluation Point Questions

STEP 1-1
1. Diodes, transistors and integrated circuits (ICs)
2. Current or voltage

STEP 1-2
1. They all have 4 valence electrons.
2. 1, 4
3. Covalent bond
4. Negative, decreases
5. Increases
6. Intrinsic, positive, negative

STEP 1-3
1. To increase their conductivity
2. Electrons, N
3. Holes, P
4. Electrons, holes

STEP 1-4
1. Depletion region
2. (a) 700 mV
3. True
4. Open, closed

STEP 2-1
1. Anode and cathode
2. Switch
3. Forward, reverse
4. OFF

STEP 2-2
1. 0.7 V
2. 0.7 V
3. One
4. The peak inverse voltage or maximum reverse voltage for a silicon diode is normally about 50 V.
5. Decreases

STEP 3-1
1. True
2. Reverse
3. Zener has a "Z" shaped bar instead of a straight bar.
4. No
5. Input voltage, load impedance

STEP 3-2
1. Electrons and holes
2. The semiconductor compound used
3. (b) 2 V
4. Series current limiting resistor
5. Forward current

STEP 4-1
1. Transformer, rectifier, filter, regulator
2. The rectifier
3. The transformer

STEP 4-2
1. Down

2. $V_S = \dfrac{N_S}{N_P} \times V_P$

$V_S = \dfrac{1}{5} \times 120 \text{ V rms} = 24 \text{ V rms}$

STEP 4-3
1. ac, pulsating dc
2. Half
3. 50 Hz
4. Bridge

STEP 4-4
1. (b) A low pass
2. Inversely proportional, $C\uparrow$, % ripple $\downarrow$
3. R in series, C_{in} shunt.
4. Parallel

STEP 4-5
1. To maintain a constant dc output voltage despite variations in the ac input voltage and the output load resistance
2. Small
3. Parallel
4. (a) $+12$ V   (b) -5 V   (c) $+15$ V   (d) -12 V

STEP 5-1
1. 325.2 V
2. The half-wave voltage doubler has two disadvantages. The first is that the ripple frequency of 60 Hz is difficult to filter, and the second is that an expensive output capacitor is required since it must have a voltage rating that is more than twice the peak of the ac input.
3. Voltage Tripler: $V_{dc} = 3 \times V_{Spk} = 3 \times 162.6 \text{ V} = 487.8 \text{ V}$
   Voltage Quadrupler: $V_{dc} = 4 \times V_{Spk} = 4 \times 162.6 \text{ V}$
   $= 650.4 \text{ V}$
4. Half-wave voltage doubler circuit = 60 Hz
   Full-wave voltage doubler circuit = 120 Hz

STEP 5-2
1. Series clippers and shunt clippers
2. Shunt
3. (d) Both the basic positive series and shunt clipper
4. Limiter

STEP 6-1

1. An LED converts an electrical input into a light output, whereas a photodiode changes its electrical characteristics in response to a light input.
2. Photovoltaic action is a process by which a device generates a voltage as a result of exposure to light radiation.
   Photoconduction is a process by which the conductance of a material is changed by electromagnetic radiation in the light spectrum.
3. Visible light band, orange/red
4. a. *p*-type, intrinsic layer, *n*-type (PIN) photodiode
   b. Integrated detector and preamplifier (IDP) receiver module
   c. Infrared emitting diode (IRED)
5. An increase in light will cause an increase in the conduction of a photodiode.

STEP 6-2

1. False
2. In applications requiring a voltage-controlled variable capacitor
3. True
4. When forward biased, a varactor diode will operate like an ordinary junction diode. The depletion region, and therefore the anode to cathode capacitance, will be zero.

STEP 6-3

1. True
2. True
3. Fast turn-on time
4. The metal oxide varistor (MOV)

STEP 7-1

1. NPN and PNP
2. Emitter, base, and collector
3. As a switch, and as a variable-resistor
4. The two-state switching action is used in digital circuits, while the variable-resistor action is used in analog circuits.

STEP 7-2

1. Current
2. Forward, reverse
3. (b)
4. Open switch
5. Closed switch
6. (a) Common-base
   (b) Common-collector
   (c) Common-emitter
7. Voltage-divider bias
8. Base-biasing

STEP 8-1

1. $\text{Gain}(A) = \dfrac{\text{Amplitude of Output Signal}}{\text{Amplitude of Input Signal}}$
2. $V_{\text{OUT}} = A_V \times V_{\text{IN}} = 75 \times 0.5 \text{ mV} = 37.5 \text{ mV}$
3. Voltage, Power, Current
4. The upper frequency limit at which the gain of the amplifier falls below half power.
5. A graph used to plot gain versus frequency.
6. Class A

STEP 8-2

1. True
2. Resistor R_E achieves circuit thermal stability through emitter feedback. With R_E included in circuit, V_{CC} is developed across R_C, Q_1's collector-emitter, and R_E. An increase in temperature will cause an increase in I_C, I_E, and therefore the voltage developed across R_E. An increase in V_{RE} will decrease the base-emitter bias voltage applied, causing Q_1's internal currents to decrease to their original values. A change in the output current (I_C) due to temperature will cause the emitter resistor to feed back a control voltage to the input.
3. The complementary dc amplifier arrangement enables us to cascade several stages to achieve high signal gain without distortion because the Q point, which is set by the previous stage, continually alternates between being more positive (slightly above the mid-point) or less positive (slightly below the mid-point).
4. Sum
5. Current
6. High, low

STEP 8-3

1. 20 Hz to 20,000 Hz
2. Voltage and power
3. Voltage, power
4. Capacitor C_E is connected in parallel with R_E to prevent signal variations at the emitter decreasing the gain of the amplifier stage.
5. AC signal coupling and dc bias isolation
6. Class B
7. Fewer circuit components are needed to construct a power amplifier because a phase splitter is not needed due to the complementary nature of the NPN and PNP transistors.

STEP 9-1

1. *LC*, Crystal, *RC*
2. Positive
3. (a) Transformer
   (b) Grounded tapped inductor
   (c) Two capacitors grounded at common connection

STEP 9-2

1. No frequency drift
2. High, low
3. Minimum, maximum
4. Series, parallel

STEP 9-3

1. 10 kHz
2. 180

STEP 10-1

1. (c)
2. (a) Radio frequency
   (b) Amplitude modulation
   (c) Federal Communications Commission
3. RF amplifier/multiplier circuit
4. $\text{BW} = 2 \times f_{AF(\text{HI})} = 2 \times 3.5 \text{ kHz} = 7 \text{ kHz}$
5. Good
6. Increased, decreased

STEP 10-2

1. (d)
2. RF amplifier/mixer
3. The IF amplifier is cheaper since it is fixed-tuned, and its narrow bandwidth gives it a higher gain.
4. Fixed-tuned

STEP 10-3

1. Televisions, monitors, radar and sonar displays, and so on.
2. (b)
3. To frequency compensate for a circuit's inherent stray-wire and transistor junction capacitance.
4. True

STEP 10-4

1. 1. First, connect a dc current meter between V_{CC} and the tuned circuit, as shown in Figure 10-11(c).
   2. Next, connect a frequency generator to the input of the amplifier, and set it to the desired frequency.
   3. Finally, adjust the variable inductor (L_1) until the meter shows a large dip or null in current.
2. When the resonant frequency of an amplifier's tuned circuit drifts off its designed setting

STEP 11-1

1. (a) Bipolar junction transistor
   (b) Junction field effect transistor
   (c) Metal oxide semiconductor field effect transistor
2. n-channel and p-channel
3. Current, voltage
4. Reverse
5. True
6. False
7. Transconductance
8. Very high, reverse

STEP 11-2

1. No need for $-V_{GG}$, temperature stability
2. R_S
3. Positive, 0 V, negative
4. Voltage divider

STEP 11-3

1. Common-source
2. Common-drain
3. Common-source
4. Common-drain

STEP 11-4

1. Its high input impedance
2. Because it responds well to the small signal voltages from the antenna, and because it is a low noise component

STEP 11-5

1. (a) 0.3 V dc (min), 3.0 dc (max)
   (b) 5 mA dc

STEP 12-1

1. Depletion, enhancement
2. True
3. ON
4. MOSFET
5. (d)
6. Dual-gate D-MOSFET

STEP 12-2

1. False
2. OFF
3. (c) Drain feedback biased
4. True

STEP 13-1

1. Integrated circuit
2. $-$, $+$, $+V$, $-V$, Output
3. TO5 Can, DIP package
4. The prefix indicates the manufacturer, the following 3 digit code indicates the op-amp type, the letter after the part number indicates the operating temperature range, and the final suffix code indicates the package type.

STEP 13-2

1. Yes
2. An open-loop op-amp circuit has no feedback, whereas a closed-loop op-amp circuit does have a feedback path.
3. High gain, very high input impedance, and very low output impedance
4. Differential amplifier, darlington-pair voltage amplifier, emitter-follower output amplifier
5. (a) Comparator
6. Negative or degenerative feedback will lower the op-amp's gain to
   a. prevent output waveform distortion,
   b. prevent the amplifier from going into oscillation, and
   c. reduce the gain of the op-amp to a consistent value.

STEP 13-3

1. Voltage follower
2. Summing amplifier
3. The differentiator circuit has an input capacitor and a feedback resistor, while the integrator circuit has an input resistor and a feedback capacitor.
4. Differential or difference amplifier
5. See Figure 13-11(b)
6. An active filter circuit will filter and amplify the signal input, while a passive filter will only filter the signal input.

STEP 14-1

1. Free-running
2. Mono-stable
3. Bistable
4. Because it can be latched in the set or reset condition
5. The three 5kΩ voltage divider
6. Astable, monostable

STEP 14-2

1. Decrease
2. False
3. Bidirectional
4. DIAC
5. True
6. One
7. False
8. Gate

STEP 14-3

1. Resistor
2. Electrical, solar
3. Photoconductive and photovoltaic
4. Decrease
5. Electrical to light, electrical to light
6. True
7. False
8. PTC

Chapter 1

1. a   **9.** b
3. d   **11.** b
5. d   **13.** b
7. d   **15.** c

17. (a) P-N junction is forward biased. $I = V_S - V_{P-N}/R =$
5 V − 0.7 Ω = 13 mA
    (b) P-N junction is forward biased. $I = V_S - V_{P-N}/R =$
15 V −0.3 V/15 KΩ = 980μA
    (c) P-N junction is reverse biased. $I = 0$ A

19. (a) $V_{RI} = V_S - V_{P-N} = 5$ V − 0.7 V = 4.3 V
    (b) $V_{RI} = V_S - V_{P-N} = 15$ V − 0.3 V = 14.7 V
    (c) Since the P-N junction is open, or reverse biased, all of
the applied voltage will appear across this series open.
Therefore, the voltage across $R_1 = 0$ V.

Chapter 2

1. c
3. d
5. c
7. d
9. b

11. (a) reverse biased
    (b) reverse biased
    (c) reverse biased
    (d) forward biased
    (e) forward biased
    (f) reverse biased

13. (a) $V_{Diode} = 10$ V
    (b) $V_{Diode} = 0.7$ V

15.

SWITCH POSITION	A	B	C
1.	Logic 1	Logic 0	Logic 1
2.	Logic 0	Logic 1	Logic 0
3.	Logic 0	Logic 0	Logic 0

Logic 0 = 0 V, Logic 1 = +5 V
    (a) With D_1 open, the code generated in position 1 will
change since the output line C cannot be pulled low.
Therefore, the code generated when the switch is in
position 1 with D_1 permanently open will be:

SWITCH POSITION	A	B	C
1.	Logic 1	Logic 0	Logic 1
2.	Logic 0	Logic 1	Logic 0
3.	Logic 0	Logic 0	Logic 0

Output line C will give a high in switch position 1.
    (b) Yes, since the maximum operation current will be 25.8
mA when the switch is in position 3 and D_5, D_6 and D_7
are on. Each parallel branch in this case will have the
following value of current:
$I = V_S - V_{Diode}/R$
$I = 5$ V − 0.7 V/500Ω
$I = 8.6$ mA
Since there are three branches, the total current will be
3 × 8.6 mA = 25.8 mA.
A 200 mA fuse will therefore blow whenever the switch
is put in position 3, because a current of 25.8 mA will be
drawn.

Chapter 3

1. c
3. a
5. d
7. c
9. d

11. (a) Polarity correct (+ → cathode, − → anode)
    (b) Polarity incorrect (+ → anode, − → cathode)
    (c) Polarity incorrect
    (d) Polarity correct
    (e) Polarity correct
    (f) Polarity for both D_1 and D_2 are correct

13. (a) $I_S = V_{in} - V_Z/R_S = 10$ V − 6.8 V/200Ω = 16 mA
    (b) Since the zener diode is forward biased, we will assume
a 0.7 V forward voltage drop. Therefore
$I_S = V_{in} - V_Z/R_S = 20$ V − 0.7 V/570 Ω = 33.9 mA
    (c) Since the zener diode is forward biased, we will assume
a 0.7 V forward voltage drop. Therefore
$I_S = V_{in} - V_Z/R_S = 5$ V − 0.7 V/400 Ω = 10.75 mA
    (d) Since the input voltage is not large enough to send the
zener into its reverse zener breakdown region, the circuit
current will be equal to that of the reverse leakage
current, which is almost zero.
    (e) $I_S = V_{in} - V_Z/R_S = 6$ V − 4.7 V/200 Ω = 6.5 mA
    (f) For zener D_1, $I_S = V_{in} - V_Z/R_S = 10$ V
− 6.8 V/220 Ω = 14.5 mA
For zener D_2, since the input voltage is not large enough
to send the zener into its reverse zener breakdown region,
the circuit current will be equal to that of the reverse
leakage current, which is almost zero.

15. $P_D = I_{ZM} \times V_Z = 6.5$ mA × 4.7 V = 30.55 mW
A 50 mW zener diode would be adequate in this
application.

17. By reversing the input voltage polarity, and the zener
diode's orientation. For example, if the polarity of the 10 V
supply voltage was reversed so that the negative terminal
connects to R_S, and the zener's connection was also reversed
so that its anode was connected to point X, the output
voltage will be 5.6 V. Similarly, if the polarity of the 20 V
supply voltage was reversed so that the positive terminal
connects to R_S, and the zener's connection was also reversed
so that its cathode was connected to point Y, the output
voltage will be +12 V.

19. (a) Forward biased (+ → anode, − → cathode)
    (b) Forward biased
    (c) D_1 is reverse biased, D_2 is forward biased

21. Yes

23. Input = +10 V, Output = Green; Input = −10 V,
Output = Red

25. Input = +10 V
$I_{LED-GREEN} = V_S - V_{LED}/R_1 + R_2 = 10$ V
− 2.5 V/100 Ω + 200 Ω = 15 mA
Input = −10 V
$I_{LED-RED} = V_{Source} - V_{D1} - V_{LED}/R_2 = 10$ V
− 0.7 V − 2.5 V/200 Ω = 34 mA

Chapter 4

1. a
3. a
5. c
7. c
9. d
11. $V_P = V_{rms} \times 1.414 = 240\ V \times 1.414 = 339.4\ V_{peak}$
$V_S = N_S/N_P \times V_P = 1/19 \times 339.4\ V = 17.86\ V_{peak}$

13. DC ripple frequency out = AC ripple frequency in = 60 Hz
15. Full wave output ripple frequency = $2 \times$ Input Freq. = $2 \times 60\ Hz = 120\ Hz$

Chapter 5

1. b
3. d
5. c
7. b
9. d
11. (a) Positive half-wave voltage doubler circuit
$V_{primary\ pk} = 115\ V \times 1.414 = 162.6\ V$
$V_{S\ pk} = \dfrac{N_S}{N_P} \times V_P = \dfrac{1}{3} \times 162.6\ V = 54.2\ V$
$V_{out} = 2 \times V_{S\ pk} = 2 \times 54.2\ V = 108.4\ V$
(b) Negative half-wave voltage doubler circuit
$V_{primary\ pk} = 115\ V \times 1.414 = 162.6\ V$
$V_{S\ pk} = \dfrac{N_S}{V_P} \times V_P = \dfrac{1}{2} \times 162.6\ V = 81.3\ V$
$V_{out} = 2 \times V_{S\ pk} = 2 \times 81.3\ V = 162.6\ V$
13. A *dual output power supply circuit* is a circuit that provides both a positive and negative dc output voltage. In this example, the full-wave voltage doubler circuit will generate a positive voltage at TP1 (+162.6 V) with respect to ground, and a negative voltage at TP2 (−162.6 V) with respect to ground. The voltage between TP1 and TP2 will therefore be the difference between these two opposite voltages +162.6 V − (−162.6 V) = 325.2 V
15. $+V_{pk} = +V_{in\ pk} - 0.7\ V = +12\ V - 0.7\ V = +11.3\ V$
$-V_{pk} = 0\ V$
17. A biased negative series clipper circuit
$+V_{pk} = +V_{in\ pk} - 0.7\ V = +12\ V - 0.7\ V = +11.3\ V$
$-V_{pk} = +V_{DC} = +5\ V$
19. A shunt connected clipper diode is included to protect each driver circuit from the counter emf transient that will be generated by each coil of the stepper motor. In this application, when a drive circuit's output is switched to 0 V, the 12 V supply is connected across an 18 Ω winding generating a circuit current of 666 mA (12 V/18 Ω). In this condition, the associated diode is reverse biased. The counter emf is generally equal but opposite to the applied voltage, and therefore when the diode is forward biased by the counter emf, the forward current through the diode will be about 666 mA. Therefore, the diode's current rating must be larger than this value.

Chapter 6

1. b
3. a
5. b
7. d
9. d
11. (a) $\lambda = c/f = 3 \times 10^{17}/6 \times 10^{17} = 500\ nm$
(b) $\lambda = c/f = 3 \times 10^{18}/1 \times 10^{15} = 3000\ Angstroms$
(c) $\lambda = c/f = 3 \times 10^{14}/1 \times 10^{14} = 3\ microns$

13. Ratio = $I_L/I_D = 20\ \mu A/20\ nA = 1000$

15. Dark current = 20 nA:
$V_R = I_D \times R = 20\ nA \times 500\ k\Omega = 10\ mV$
Light current = 20 μA:
$V_R = I_L \times R = 20\ \mu A \times 500\ k\Omega = 10\ V$
17. Correctly (reverse) biased
19. Circuit 6-21(a) would use the bi-directional transient suppressor diode in Figure 6-21(e).
Circuit 6-21(b) would use the uni-directional transient suppressor diode in Figure 6-21(c).

Chapter 7

1. d   **11.** d
3. a   **13.** d
5. b   **15.** a
7. b   **17.** c
9. a   **19.** d
21. a. NPN, x = base, y = collector, z = emitter.
b. PNP, x = base, y = collector, z = emitter.
23. a. $I_E = 25\ mA$, $I_C = 24.6\ mA$, $I_B = I_E - I_C = 0.4\ mA$ or 400 μA
b. $I_B = 600\ \mu A$, $I_C = 14\ mA$, $I_E = I_B + I_C = 14.6\ mA$
c. $I_E = 4.1\ mA$, $I_B = 56.7\ \mu A$, $I_C = I_E - I_B = 4.04\ mA$
25. a. Common-collector circuit, since the input is applied to the base and the output is taken from the emitter (collector is common to both input and output).
b. Common-emitter circuit, since the input is applied to the base and the output is taken from the collector (emitter is common to both input and output).
c. Common-emitter circuit, since the input is applied to the base and the output is taken from the collector (emitter is common to both input and output).
d. Common-emitter circuit, since the input is applied to the base and the output is taken from the collector (emitter is common to both input and output).
27. a. $V_RB = 20\ V - 0.7\ V = 19.3\ V$
$I_B = V_RB/R_B = 19.3\ V/2.5\ M\Omega = 7.72\ \mu A$
b. $I_C = I_B \times B_{DC} = 7.72\ \mu A \times 150 = 1.16\ mA$
c. $V_{R_C} = I_C \times R_C = 1.16\ mA \times 10\ k\Omega = 11.6\ V$
$V_{CE} = V_{CC} - V_{R_C} = 20\ V - 11.6\ V = 8.4\ V$
29. Saturation Point,
$$I_{C(Sat)} = \frac{V_{CC}}{R_C + R_E}$$
$$= \frac{20\ V}{4.7\ k\Omega + 1.1\ k\Omega}$$
$$= 3.45\ mA$$
Q Point,
$I_C = 1.7\ mA$, $V_{CE} = 10.14\ V$
Cutoff Point,
$V_{CE(cutoff)} = V_{CC} = 20\ V$

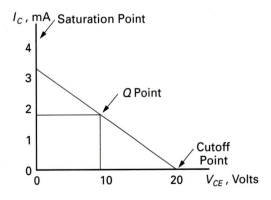

Chapter 8

1. a   11. d
3. d   13. b
5. d   15. b
7. a   17. d
9. d   19. b

21. **a.** $A_V = \dfrac{V_{OUT}}{V_{IN}} = \dfrac{7.5\ \text{mV}}{150\ \mu\text{V}} = 50$

    b. $A_V = \dfrac{\Delta V_{OUT}}{\Delta V_{IN}} = \dfrac{9\ \text{V} - 1\ \text{V}}{6\ \text{mV} - 2\ \text{mV}} = \dfrac{8\ \text{V}}{4\ \text{mV}} = 2000$

    c. $V_{OUT} = A_V \times V_{IN} = 15 \times 2\ \text{mV} = 30\ \text{mV}$

    d. $V_{OUT(pk)} = A_V \times V_{IN(pk)} = 50 \times 6\ \text{mV} = 300\ \text{mV (pk)}$

23. **a.** $A_P = \dfrac{P_{OUT}}{P_{IN}} = \dfrac{15\ \text{W}}{25\ \text{mW}} = 600$

    b. $P_{IN} = V_{IN} \times I_{IN} = 25\ \text{mV} \times 150\ \mu\text{A} = 3.7\ \mu\text{W}$
    $P_{OUT} = A_P \times P_{IN} = 50 \times 3.7\ \mu\text{W} = 185\ \mu\text{W}$

    c. $A_P = A_V \times A_I = 30 \times 15 = 450$

    d. $P_{OUT(max)} = A_P \times P_{IN(max)} = 75 \times 10\ \text{mW} = 750\ \text{mW}$
    $P_{OUT(min)} = A_P \times P_{IN(min)} = 75 \times 4\ \text{mW} = 300\ \text{mW}$

25. **a.** $R_C = \dfrac{R_3 \times R_L}{R_3 + R_L} = \dfrac{3.7\ \text{k}\Omega \times 10\ \text{k}\Omega}{3.7\ \text{k}\Omega + 10\ \text{k}\Omega} = \dfrac{\text{k}\Omega}{6.96\ \text{k}\Omega} = 2.7\ \text{k}\Omega$

    $A_V = \dfrac{R_C}{R_E} = \dfrac{2.7\ \text{k}\Omega}{1.2\ \text{k}\Omega} = 2.25$

    The output signal voltage will be 2.25 times greater than the input signal voltage.

    b. $V_{OUT} = A_V \times V_{IN} = 2.25 \times 3.5\ \text{V} = 7.9\ \text{V}$

    c. $V_{R2} = \dfrac{R_2}{R_1 + R_2} \times V_{CC} = 1.8\ \text{V}$

    $I_{R2} = \dfrac{V_{R2}}{R_2} = \dfrac{1.8\ \text{V}}{2.2\ \text{k}\Omega} = 818\ \mu\text{A}$

    $I_{R2} \cong I_{R1}$ (I_B can be ignored), therefore $I_{R1} = 818\ \mu\text{A}$
    $V_E = V_B - 0.7\ \text{V} = 1.8\ \text{V} - 0.7\ \text{V} = 1.1\ \text{V}$

    $I_E = \dfrac{V_E}{R_E} = \dfrac{1.1\ \text{V}}{1.2\ \text{k}\Omega} = 0.92\ \text{mA}$

    $I_E \cong I_C$, therefore $I_C = 0.92\ \text{mA}$

    $P_L = \dfrac{V_L^2}{R_L} = \dfrac{7.9\ \text{V}^2}{10\ \text{k}\Omega} = 6.25\ \text{mW}$

    $I_{CC} = I_{R1} + I_C = 818\ \mu\text{A} + 0.92\ \text{mA} = 1.74\ \text{mA}$
    $P_S = V_{CC} \times I_{CC} = 10\ \text{V} \times 1.74\ \text{mA} = 17.4\ \text{mW}$

    $\eta = \dfrac{P_L}{P_S} \times 100\% = \dfrac{6.24\ \text{mW}}{17.4\ \text{mW}} \times 100\% = 35.86\%$

27. $\text{CMRR} = V_{VD}/A_{CM} = 6000/0.33 = 18{,}181.92$

29. $P_L = \dfrac{V_L^2}{R_L} = \dfrac{4.2\ \text{V}^2}{11\ \text{k}\Omega} = 17.64\ \text{mW}$

    $P_S = V_{CC} \times I_{CC} = 5\ \text{V} \times 20\ \text{mA} = 100\ \text{mW}$

    $\eta = \dfrac{P_L}{P_S} \times 100\% = \dfrac{17.64\ \text{mW}}{100\ \text{mW}} \times 100\% = 17.64\%$

Chapter 9

1. b
3. c
5. b
7. b
9. a
11. **(a)** Armstrong oscillator

    $f_R = 1/2\pi\sqrt{L_1 \times C_1} = 1/2\pi\sqrt{5\ \mu\text{H} \times 1\ \mu\text{F}} = 71.2\ \text{kHz}$
    (b) Hartley oscillator
    $L_T = L_1 + L_2 = 0.47\ \text{mH} + 0.2\ \text{mH} = 0.67\ \text{H}$
    $f_R = 1/2\pi\sqrt{L_T \times C_1} =$
    $1/2\pi\sqrt{0.67\ \text{mH} \times 0.03\ \mu\text{F}} = 35.5\ \text{kHz}$

13. $f_{R(Low)} = 1/2\pi\sqrt{L_1 \times C_3} =$
    $1/2\pi\sqrt{0.47\ \text{mH} \times 15\ \text{nF}} = 60\ \text{kHz}$
    $f_{R(High)} = 1/2\pi\sqrt{L_1 \times C_3} =$
    $1/2\pi\sqrt{0.47\ \text{mH} \times 0.1\ \text{nF}} = 734\ \text{kHz}$

15. **(a)** An *RC* phase-shift oscillator
    $f_R = 1/2 \times \pi \times \sqrt{6 \times R \times C} =$
    $1/2\pi \times 2.45 \times 5\ \text{k}\Omega \times 0.2\ \mu\text{F} = 65\ \text{Hz}$
    (b) A Wein-Bridge oscillator
    $f_R = 1/2\pi\sqrt{C_1 \times R_1 \times C_2 \times R_2} =$
    $1/6.28 \times \sqrt{0.1\ \mu\text{F} \times 1\ \text{k}\Omega \times 0.47\ \mu\text{F} \times 5\ \text{k}\Omega}$
    $= 328.3\ \text{Hz}$

Chapter 10

1. c   9. b
3. d   11. d
5. c   13. a
7. a   15. c

17. **(a)** $f_R = 410.9\ \text{kHz}$
    $X_L = 2\pi f L = 2\pi \times 410.9\ \text{kHz} \times 1.5\ \text{mH} = 3.87\ \text{k}\Omega$
    $Q = X_L/R_W = 3.87\ \text{k}\Omega/18\ \Omega = 215$
    (b) $f_R = 838.8\ \text{KHz}$
    $X_L = 2\pi f L = 2\pi \times 838.8\ \text{kHz} \times 0.6\ \text{mH} = 3.16\ \text{k}\Omega$
    $Q = X_L/R_W = 3.16\ \text{k}\Omega/25\ \Omega = 126.4$

19. $f_{H(CUTOFF)} = 1/2\pi R_C C_T = 1/2 \times \pi \times 1.3\ \text{k}\Omega \times 160\ \text{pF}$
    $= 0.159/(1.3 \times 10^3) \times (160 \times 10^{-12}) = 765.2\ \text{kHz}$

Chapter 11

1. d   9. a
3. b   11. a
5. b   13. b
7. a   15. c

17. **(a)** $\delta_m = \Delta I_D/\Delta V_{GS} = 8\ \text{mA} - 2\ \text{mA}/(-3.5\ \text{V}) -$
    $(-0.5\ \text{V}) = 6\ \text{mA}/3\ \text{V} = 2\ \text{millisiemens}$
    (b) $A_v = \delta_m \times R_D = 2\ \text{mS} \times 10\ \text{k}\Omega = 20$

19. **(a)** $V_S = I_S \times R_S = 2\ \text{mA} \times 390\ \Omega = 0.78\ \text{V}$
    (b) $V_{GS} = V_G - V_S = 0\ \text{V} - 0.78\ \text{V} = 0.78\ \text{V}$
    $V_{DS} = V_{DD} - (V_{RD} + V_{RS}) = 10\ \text{V} - [(I_D R_D) + 2\ \text{V})]$
    (c) $= 10\ \text{V} - [(2\ \text{mA} \times 2\ \text{k}\Omega) + 0.78\ \text{V}) = 10\ \text{V}$
    $- (4 + 0.78) = 15\ \text{V} - 4.78\ \text{V} = 10.22\ \text{V}$

21. **(a)** Common-drain
    (b) Slightly less than 1 (output = 0.98 × input)
    (c) Impedence matching

23. Opens and shorts between terminals, JFET transconductance/gain, and I_{DSS} and leakage current values.

25. **(a)** *n*-channel JFET is ok
    (b) Gate-source check ok, gate-drain open. Since P-N (gate-channel) junction seems ok, drain terminal may be disconnected internally.
    (c) JFET bad; gate-channel short

Chapter 12

1. b
3. d
5. c
7. b
9. c

11. **(a)** $I_{D(sat)} = \dfrac{V_{DD}}{R_D} = \dfrac{6\ \text{V}}{1\ \text{k}\Omega} = 6\ \text{mA}$

    (b) $V_{DS(OFF)} = V_{DD} = 6\ \text{V}$

13. Figure 12-21 uses zero biasing
    Figure 12-22 uses drain feedback biasing

15. (a) Q_1 is always ON due to the $+10$ V drain feedback bias, and therefore it will always have a drain-to-source ON resistance $R_{DS(ON)}$ of 1 kΩ.

(b) When $V_{IN} = +5$ V, Q_2 is ON and has a drain-to-source ON resistance of 300 Ω. Therefore, V_{OUT} equals

$$V_{OUT} = \frac{R_{Q_2}}{R_{Q_1} + R_{Q_2}} \times V_{DD} =$$

$$\frac{300\ \Omega}{1000\ \Omega + 300\ \Omega} \times 10\ V = 0.23 \times 10\ V = 2.3\ V$$

(c) When $V_{IN} = 0$ V, Q_2 is OFF and therefore
$V_{OUT} = V_{DD} = 10$ V

17. The foam is conductive to ensure all pins are at the same potential, and therefore the resistance should be 0 Ω.

19. When $V_{GS} = 0$ V, Q_1 will be ON and I_D should be about half of its maximum since a V_{GS} of zero volts puts the Q point at a mid-point on the load line. Q_1 should therefore have some resistance, and therefore voltage drop, between drain and source. Since no voltage is reaching the drain and $V_{DD} = 20$ V, we would have to suspect that R_D is open.

Chapter 13

1. b
3. d
5. a
7. c
9. e
11. a. LM = National Semiconductor
318 = Op-Amp Type
C = Commercial 0 to 70° C
N = Plastic DIP with longer leads
b. NE = Signetics
101 = Op-Amp Type
C = Commercial 0 to 70°C
D = Plastic DIP
13. A single polarity supply ($+5$ V) inverting op-amp. The inverting amplifier would have to have a negative input because the output can only be between 0 V ($-V$ supply voltage) and $+5$ V ($+V$ supply voltage).
15. a. Differentiator (see waveforms in Figure 13-13)
b. Integrator (see waveforms in Figure 13-14)
17. a. Summing Amplifier
$V_{OUT} = -(V_{IN1} + V_{IN2}) = -(3\ V + 6\ V) = -9\ V$
b. Difference Amplifier
$V_{OUT} = V_{IN2} - V_{IN1} = 10\ V - 6.5\ V = 3.5\ V$
19. $f_C = 1/2\pi RC = 1/2\pi \times 10\ k\Omega \times 0.1\ \mu F = 159.2$ Hz

Chapter 14

1. a
3. c
5. b
7. e

9. a
11. e
13. d
15. c
17. A square-wave astable multivibrator circuit.
19. A monostable multivibrator circuit
21. Ratio = Output/Input = 1.4 ms or 1400 μS/100 μs = 14
The output is 14 times longer than the input.
23. A 555 timer IC connected as an astable multivibrator circuit
25. Duty cycle = $P_W t \times 100\% =$
131.6 ms./197.4 ms $\times 100\% = 66.67\%$
27. a. DIAC
b. SCR
c. LED or ILD
d. Photodiode
e. Photoresistor
f. Solar cell
g. Phototransistor
h. LASCR
29. A low input will turn OFF Q_1, causing its collector output to go HIGH, which will trigger the SCR ON, and connect power to the relay's energizing coil
31. Monostable
33. Because the TRIAC can be used to control the full ac cycle
35. The SCR was used to connect dc power to a load, while the TRIAC was used to connect ac power to a load.
37. Photoresistor
39. As the light level changes, the resistance of the photoresistor changes, changing the oscillator's RC frequency-determining value, and therefore changing the frequency at the output of the 3909 being applied to the speaker.
41. To power a constant-frequency tone generator circuit using solar cells.
43. Phototransistor
45. When light is present, phototransistor is ON. NPN transistor is OFF, relay is de-energized, normally closed contacts are closed, normally open contacts are open. When light is not present, phototransistor is OFF, NPN transistor is ON, relay is energized, normally closed contacts open, normally open contacts closed.
47. To have a bargraph display represent the level of ambient light present.
49. To control the dc output offset of the op-amp
51. When a voltage of 2 V is dropped across a resistor, its associated LED will turn ON. The larger value resistors in the bargraph voltage divider will develop a larger voltage drop. This means that the resistor at the bottom of the rung will turn ON first, when only a small voltage is applied from the 741. As the light level increases, the voltage out of the 741 increases, the voltage drop across all of the resistors increases proportionally, and the LEDs turn ON in order from the bottom up.

Index

A

AC charge and discharge, 253
AC meter, 218
Air core, 317
Alternating current, 193
American wire gauge, 42
Ammeter, 15
Amp, 12
Ampere, 12
Ampère, Andre, 12
Ampere-hours, 86
Amplitude, 201
Atom, 3
Atomic number, 3
Atomic weight, 5
Attenuate, 277
Average value, 203

B

Babbage, Charles, 230
Ballast resistor, 71
Bandpass filter, 394
Bands, 6
Band-stop filter, 395
Bandwidth, 384
Battery, 18
B-H curve, 291
Bleeder current, 176
Boot, Henry, 209
Branch current, 137
Breadboard, 95
Breakdown voltage, 27, 237

C

Cable, 45
Calibration, 222
Capacitance meter, 275
Capacitive reactance, 259
Capacitive time constant, 248

Capacitor, 230

Capacitor, 230
Capacitor coding, 247
Capacitor phase relationship, 255
Capacitors in parallel, 239
Capacitors in series, 240
Center tapped transformer, 361
Christie, S., 160
Circuit breaker, 90
Closed circuit, 28
Coercive force, 294
Coil, 289
Cold resistance, 71
Communication, 197
Compound, 9
Conductance, 25
Conductor, 24
Conventional flow, 14
Coulomb, 11
Coulomb, Charles, 10
Counter EMF, 311
Cray, Seymour, 31
Current, 9
Current divider, 139
Current in parallel, 137
Current in series circuit, 103
Current in series-parallel, 169
Current ratio, 353
Cutoff frequency, 384

D

Dielectric strength, 27
Differentiator, 282
DIP, 59
Direct current, 193
Dissipation, 55
Dot convention, 357
Dry cell, 84
Du Fay, Charles, 25
Dual in-line package, 59
Duty cycle, 212

Static, 12
Steinmetz, Charles, 2
Step-down transformer, 352
Step-up transformer, 351
Subatomic, 3
Substrate, 57
Surface mount technology, 61
Switch, 93

T

Tank circuit, 388
Tapered, 62
Testing capacitors, 273
Testing inductors, 338
Testing resistors, 73
Testing transformers, 366
Thermocouple, 79
Thick film resistor, 59
Time constant, 249, 320
Tolerance, 56
Transducer, 197
Transformer, 195, 345
Transformer loading, 348
Transformer ratings, 365
Transformer types, 358
Triangular wave, 217
Triggering, 222
Troubleshooting parallel circuits, 153
Troubleshooting series circuits, 127
Troubleshooting series-parallel, 181
True power, 268
Tuned circuit, 391
Turns ratio, 350

V

Variable resistor, 61
Variable value inductors, 319
Variable voltage divider, 118
Varian, Russell and Sigurd, 209
Vector, 201
Volta, Alessandro, 14
Voltage in parallel, 135
Voltage in series circuit, 108
Voltage in series-parallel, 167
Voltage ratio, 351
Voltage, 15
Voltaic cell, 83
Voltmeter, 20

W

Watt, 46
Watt, James, 46
Wattage rating, 55
Wavelength, 208
Weber, 302
Weber, Eduard, 302
Wheatstone bridge, 177
Wheatstone, Charles, 160
Wireless communication, 198
Wires, 45
Wirewound resistor, 58
Work, 45
Wozniak, Stephen, 54

Z

Zworykin, Vladymir, 221